Development and Application of Polymer Scaffolds

Development and Application of Polymer Scaffolds

Guest Editors

Wang Guo
Hui You
Pei Feng
Yu Long

MDPI

Basel • Beijing • Wuhan • Barcelona • Belgrade • Novi Sad • Cluj • Manchester

Guest Editors

Wang Guo
School of Mechanical
Engineering
Guangxi University
Nanning
China

Hui You
School of Mechanical
Engineering
Guangxi University
Nanning
China

Pei Feng
College of Mechanical and
Electrical Engineering
Central South University
Changsha
China

Yu Long
School of Mechanical
Engineering
Guangxi University
Nanning
China

Editorial Office
MDPI AG
Grosspeteranlage 5
4052 Basel, Switzerland

This is a reprint of the Special Issue, published open access by the journal *Polymers* (ISSN 2073-4360), freely accessible at: https://www.mdpi.com/journal/polymers/special_issues/TS7AA7MSI8.

For citation purposes, cite each article independently as indicated on the article page online and as indicated below:

Lastname, A.A.; Lastname, B.B. Article Title. *Journal Name* **Year**, *Volume Number*, Page Range.

ISBN 978-3-7258-5259-8 (Hbk)
ISBN 978-3-7258-5260-4 (PDF)
https://doi.org/10.3390/books978-3-7258-5260-4

Contents

polymers

MDPI

Editorial

Development and Application of Polymer Scaffolds

Wang Guo [1,2]

[1] State Key Laboratory of Featured Metal Materials and Life-Cycle Safety for Composite Structures, School of Mechanical Engineering, Guangxi University, Nanning 530004, China; guowang@gxu.edu.cn or drwangguo@outlook.com
[2] Institute of Laser Intelligent Manufacturing and Precision Processing, Guangxi Key Laboratory of Manufacturing System and Advanced Manufacturing Technology, School of Mechanical Engineering, Guangxi University, Nanning 530004, China

1. The Evolving Landscape of Polymer Scaffolds in Regenerative Medicine

It is with great pleasure that we introduce this Special Issue of *Polymers*, entitled "Development and Application of Polymer Scaffolds" (https://www.mdpi.com/journal/polymers/special_issues/TS7AA7MSI8 accessed on 18 August 2025). This collection was conceived to capture the dynamic and rapidly advancing research in the design, fabrication, functionalization, and application of polymer-based scaffolds, which stand at the forefront of tissue engineering and regenerative medicine.

Polymer scaffolds have become indispensable tools in the quest to repair and regenerate damaged tissues. They serve as temporary, three-dimensional templates that mimic the native extracellular matrix (ECM), providing the necessary structural support and guidance for cells to adhere, proliferate, differentiate, and ultimately form new, functional tissue. The widespread adoption of polymers for this purpose is rooted in their remarkable versatility; their properties, including biocompatibility, biodegradability, mechanical strength, and processability, can be precisely tailored to meet the demanding requirements of diverse biological environments.

The field has undergone a significant evolution. Early research primarily focused on creating scaffolds that were biocompatible, possessed an interconnected porous architecture to facilitate nutrient and waste transport, and degraded at a rate commensurate with new tissue formation. While these foundational principles remain critical, the contemporary paradigm has shifted towards the development of "bio-instructive" or "functional" scaffolds. These advanced constructs are no longer passive supports but are engineered to be active participants in the regenerative cascade. They can present specific biological cues to direct cell fate, release therapeutic agents in a controlled manner, or respond dynamically to physiological stimuli, thereby orchestrating a more efficient and targeted healing response.

This Special Issue brings together a collection of 14 peer-reviewed articles that exemplify this paradigm shift. The compilation features 11 original research articles, two comprehensive reviews, and one systematic review, reflecting a healthy balance between novel discovery and the critical synthesis of existing knowledge. The composition of this collection itself speaks to the maturation of the polymer scaffold field. While novel, highly specialized research continues to push the frontiers of what is possible, the inclusion of systematic reviews on topics such as 3D printing in prosthodontics and hydrogels for gingival regeneration indicates that certain sub-domains have amassed a sufficient body of literature to identify established trends and define clear future directions—a hallmark of a maturing scientific discipline. This collection, contributed by research groups from around the globe, offers a comprehensive snapshot of the current State of the Art and is intended to serve as a valuable resource for the scientific community. The global and collaborative nature of this research is highlighted by the diverse origins of the contributions, as summarized in Table 1.

Table 1. Statistical overview of the contributions to the Special Issue.

Statistic	Count
Total Authors	98
Contributing Institutions (First Level)	24
Contributing Countries	17

2. Highlights of the Special Issue: A Thematic Overview of Contributions

The 14 articles published in this Special Issue cover a broad spectrum of research, from fundamental materials science to application-driven engineering. For ease of navigation, the contributions are summarized in Table 2 before being discussed in thematic detail.

Table 2. Summary of the articles published in the Special Issue "Development and Application of Polymer Scaffolds" (Papers ordered by publication date, most recent first).

Ref.	First Author	Corresponding Author	First Affiliation (Institution, Country)	Title	Article Type	Key Focus/Contribution
[1]	Gutiérrez, B.	Dias, F.J.	Universidad de La Frontera, Chile	Comprehensive Development of a Cellulose Acetate and Soy Protein-Based Scaffold for Nerve Regeneration	Article	Fabrication and biocompatibility of a natural polymer-based conduit for peripheral nerve repair.
[2]	Song, X.	Cameron, R.E.	University of Cambridge, UK	Controlling the Architecture of Freeze-Dried Collagen Scaffolds with Ultrasound-Induced Nucleation	Article	A novel processing technique to control pore architecture and improve reproducibility in collagen scaffolds.
[3]	Guo, W.	Guo, W.; You, H.; Long, Y.	Guangxi University, China	Magnesium Hydroxide as a Versatile Nanofiller for 3D-Printed PLA Bone Scaffolds	Article	Multifunctional nanofiller to simultaneously improve mechanical, degradation, and osteogenic properties of PLA.
[4]	Alassaf, M.	Aouak, T.	Qassim University, Saudi Arabia	Mevacor/Poly(vinyl acetate/2-hydroxyethyl methacrylate) as Solid Solution: Preparation, Solubility Enhancement and Drug Delivery	Article	A novel copolymer system for enhancing the solubility and controlled delivery of a hydrophobic drug.
[5]	Deshpande, M.V.	Deshpande, M.V.; King, M.W.	North Carolina State University, USA	Degradation of Poly(ε-caprolactone) Resorbable Multifilament Yarn under Physiological Conditions	Article	In-depth analysis of PCL yarn degradation, providing crucial data for scaffold design.
[6]	Menotti, F.	Palmero, P.; Allizond, V.	Politecnico di Torino, Italy	Tuning of Silver Content on the Antibacterial and Biological Properties of Poly(ε-caprolactone)/Biphasic Calcium Phosphate 3D-Scaffolds for Bone Tissue Engineering	Article	Functionalization of bone scaffolds with silver to impart antibacterial properties without compromising cytocompatibility.

Table 2. *Cont.*

Ref.	First Author	Corresponding Author	First Affiliation (Institution, Country)	Title	Article Type	Key Focus/Contribution
[7]	To, J.	Tam, J.P.	Nanyang Technological University, Singapore	Design of Potent and Salt-Insensitive Antimicrobial Branched Peptides	Article	A molecular design approach to creating highly effective antimicrobial peptides for biomaterial functionalization.
[8]	Botvin, V.	Botvin, V.; Surmenev, R.	Tomsk Polytechnic University, Russia	Effect of Fe_3O_4 Nanoparticles Modified by Citric and Oleic Acids on the Physicochemical and Magnetic Properties of Hybrid Electrospun P(VDF-TrFE) Scaffolds	Article	Fabrication of magnetoactive scaffolds for potential applications in stimuli-responsive tissue engineering.
[9]	Zdraveva, E.	Zdraveva, E.; Mijovic, B.	University of Zagreb, Croatia	The Reliability of PCL/Anti-VEGF Electrospun Scaffolds to Support Limbal Stem Cells for Corneal Repair	Article	A drug-eluting scaffold designed to prevent neovascularization and support stem cells for corneal regeneration.
[10]	Sangkatip, R.	Sriseubsai, W.	King Mongkut's University of Technology North Bangkok, Thailand	Gelatin/$Na_2Ti_3O_7$ Nanocomposite Scaffolds: Mechanical Properties and Characterization for Tissue Engineering Applications	Article	Development of a nanocomposite hydrogel with optimized mechanical and swelling properties.
[11]	Pereira, P.	Coelho, J.F.J.; Fonseca, A.C.	University of Coimbra, Portugal	In Vitro Evaluation of Biphasic Calcium Phosphate Scaffolds Derived from Cuttlefish Bone Coated with Poly(ester urea): for Bone Tissue Regeneration	Article	Use of a marine-derived biomaterial coated with a novel polymer to enhance osteogenic differentiation of MSCs.
[12]	Dimitrova, M.	Dimitrova, M.	Medical University of Plovdiv, Bulgaria	Recent Advances in 3D Printing of Polymers for Application in Prosthodontics	Review	A comprehensive review of the State of the Art in 3D printing for dental prosthetic applications.
[13]	Kováč, J.	Žiaran, S.	Comenius University, Slovakia	Bioadhesive and Injectable Hydrogels and Their Correlation with Mesenchymal Stem Cells Differentiation for Cartilage Repair: A Mini-Review	Review	A focused review on the critical role of bioadhesion in injectable hydrogels for cartilage regeneration.
[14]	Hutomo, D.I.	Amir, L.	Universitas Indonesia, Indonesia	Hydrogel-Based Biomaterial as a Scaffold for Gingival Regeneration: A Systematic Review of In Vitro Studies	Systematic Review	A systematic analysis of the literature on hydrogel scaffolds for periodontal soft tissue engineering.

2.1. Innovations in Scaffold Fabrication and Structural Control

Control over a scaffold's internal architecture is paramount, as it directly influences cell infiltration, nutrient transport, and mechanical performance. The contributions in this Special Issue highlight two parallel streams of innovation in scaffold manufacturing. On one hand, established techniques are being refined for greater precision and reliability, which is essential for translating laboratory successes into consistent products. On the other hand, advanced digital fabrication methods are being leveraged to create structures of unprecedented complexity and personalization.

Representing the first stream, Song et al. [2] address a fundamental challenge in a classic fabrication method: the stochastic nature of ice nucleation during freeze-drying. By introducing low-frequency ultrasound, they demonstrate the ability to trigger nucleation on demand in collagen slurries. This novel process enables precise control over the resulting pore size and distribution, significantly improving the structural reproducibility of scaffolds—a critical step towards reliable, large-scale manufacturing.

In the second stream, the review by Dimitrova et al. [12] provides a comprehensive overview of how additive manufacturing (AM), or 3D printing, is revolutionizing the field of prosthodontics. This work underscores the transformative power of AM to fabricate patient-specific implants and devices with complex geometries that are unattainable with conventional methods. This theme resonates strongly within the broader tissue engineering field, where 3D printing is increasingly used to create anatomically accurate constructs. Together, these papers show that progress is advancing on two crucial fronts: making established, cost-effective techniques more precise and making complex, digital techniques more accessible and clinically relevant.

2.2. Functionalization of Scaffolds for Enhanced Biological Performance

A central theme of this Special Issue is the strategic functionalization of scaffolds to imbue them with bioactive properties that actively direct the healing process. The works collected here demonstrate a sophisticated design philosophy, where additives are selected not just to add a single feature, but to solve a constellation of interrelated material and biological challenges.

A compelling example of this integrated approach is presented by Guo et al. [3]. Polylactic acid (PLA) is a widely used biomaterial, but it suffers from several interconnected drawbacks: its slow degradation produces acidic byproducts that can cause inflammation, and its mechanical properties are often insufficient for load-bearing applications. By incorporating magnesium hydroxide ($Mg(OH)_2$) nanoparticles into 3D-printed PLA scaffolds, the authors address all these issues simultaneously. The alkaline nature of $Mg(OH)_2$ neutralizes the acidic degradation products, which in turn accelerates the hydrolysis of PLA for a more favorable degradation profile. Furthermore, the released magnesium ions (Mg^{2+}) are known to be osteoinductive, and the nanoparticles act as a reinforcing agent, improving the scaffold's mechanical strength. This work exemplifies a shift from single-feature enhancement to integrated system optimization.

Post-implantation infection is a major cause of failure for biomedical devices. Two articles in this Special Issue tackle this challenge from different angles. Menotti et al. [6] carefully tune the concentration of silver in poly(ε-caprolactone)/biphasic calcium phosphate (PCL/BCP) scaffolds. They identify a therapeutic window that provides potent antibacterial activity against common pathogens while remaining non-toxic to bone cells, highlighting the delicate balance required for cytocompatible functionalization. Approaching the problem from a molecular level, To et al. [7] report the design of potent and salt-insensitive antimicrobial branched peptides. This work offers a highly specific and sophisticated biological alternative to traditional antimicrobial agents like metallic ions.

Finally, pushing towards the development of "smart" scaffolds, Botvin et al. [8] fabricate magnetoactive scaffolds by incorporating iron oxide (Fe_3O_4) nanoparticles into electrospun polymer fibers. Such materials have the potential for remotely triggered drug release, mechanical or electrical stimulation of cells, or localized hyperthermia therapy, representing a significant step towards dynamic, responsive regenerative devices.

2.3. Tailored Scaffolds for Specific Tissue Engineering Applications

The diversity of tissues in the human body necessitates scaffold designs that are meticulously tailored to specific biological and mechanical requirements. The articles in this collection showcase this application-driven approach, reflecting an evolution in the concept of bio-inspiration. The field is moving beyond simply mimicking the static physical structure of the ECM to designing scaffolds that actively support and guide the unique biological processes essential for the regeneration of a specific tissue.

For nerve regeneration, Gutiérrez et al. [1] report the development of a nerve guide conduit fabricated from natural polymers—cellulose acetate and soy protein. Their comprehensive characterization focuses on properties critical for peripheral nerve repair, including suturability, flexibility, porosity conducive to Schwann cell migration, and excellent biocompatibility.

In the realm of corneal repair, Zdraveva et al. [9] address a key failure mode: abnormal blood vessel growth (neovascularization) into the transparent cornea. Their work presents a dual-function electrospun PCL scaffold. It not only provides a supportive substrate for limbal stem cells, but also delivers an anti-vascular endothelial growth factor (anti-VEGF) drug in a sustained manner. This design actively intervenes in a detrimental biological process, showcasing a functional approach to bio-inspiration that goes beyond simple structural mimicry.

For bone and cartilage regeneration, several innovative strategies are presented. Pereira et al. [11] demonstrate remarkable resourcefulness by using cuttlefish bone as a natural, sustainable source for BCP scaffolds. These marine-derived scaffolds are then coated with a novel poly(ester urea), which is shown to significantly enhance the osteogenic differentiation of human mesenchymal stromal cells. The state of the field for challenging tissues is synthesized in two key reviews. Kováč et al. [13] provide a mini-review on injectable and bioadhesive hydrogels for cartilage repair, highlighting the critical role of adhesion in ensuring scaffold integration and function. Concurrently, Hutomo et al. [14] present a systematic review of the in vitro evidence for hydrogel-based biomaterials in gingival regeneration, consolidating the current knowledge base for this important application in periodontics.

2.4. Advancing Drug Delivery and Material Degradation Studies

The final theme focuses on the dynamic nature of scaffolds, either as controlled release platforms or through their own degradation and resorption over time. The inclusion of fundamental materials science studies alongside application-focused work highlights a mature understanding within the field: successful clinical translation is built upon a rigorous, quantitative understanding of underlying material behaviors.

Alassaf et al. [4] develop a novel copolymer system specifically as a drug carrier. By creating a solid solution of the hydrophobic drug Mevacor within a poly(vinyl acetate-co-2-hydroxyethyl methacrylate) matrix, they significantly enhance the drug's solubility and achieve controlled release profiles. This work underscores the potential of polymer scaffolds to address longstanding pharmaceutical challenges.

In a crucial contribution to the foundational knowledge, Deshpande et al. [5] conduct a meticulous long-term study on the in vitro degradation of PCL multifilament yarn under physiological conditions. While not a cell-based study, this work provides indispensable data for the design of any load-bearing or long-term implant. Accurately predicting the rate at which a scaffold loses its mechanical integrity and is resorbed by the body is essential for ensuring its function and preventing clinical complications. Such foundational studies are the bedrock upon which successful regenerative therapies are built.

Similarly, the work by Sangkatip et al. [10] on gelatin/$Na_2Ti_3O_7$ nanocomposite scaffolds focuses on the systematic optimization of the material's composition to achieve desirable mechanical properties and swelling behavior. This detailed characterization is a vital prerequisite for more advanced applications, such as controlled drug delivery, where the degree of swelling directly dictates the rate of drug diffusion and release.

3. Concluding Remarks and Future Perspectives

The 14 articles gathered in this Special Issue collectively paint a vibrant picture of a field characterized by rapid innovation and increasing sophistication. The overarching trend is the clear evolution of polymer scaffolds from passive, structural templates into active, multifunctional, and bio-instructive systems designed to orchestrate complex biological responses.

Building upon the excellent work presented here, several exciting frontiers are emerging. The multi-functionalization demonstrated by Guo et al. [3] and the stimuli-responsive capabilities shown by Botvin et al. [8] point towards a future of "smart" scaffolds with integrated diagnostic and therapeutic functions. One can envision future implants that sense the local biochemical environment (e.g., pH, inflammatory markers) and respond by delivering appropriate therapeutic agents on demand. The natural progression from 3D printing, as reviewed by Dimitrova et al., [12] leads to 4D bioprinting, where printed constructs can change their shape, properties, or function over time in response to stimuli. This could enable the fabrication of self-assembling devices or scaffolds that dynamically soften as new tissue forms, better matching the mechanical environment of regeneration. Furthermore, the development of advanced materials, such as the novel poly(ester urea) used by Pereira et al., [11] will continue to be a driving force. The future likely lies in hybrid bio-inks that combine the bioactivity of natural polymers with the processability and mechanical robustness of synthetic polymers, enabling the direct printing of cell-laden, living tissue constructs.

Despite this immense promise, significant hurdles remain on the path to widespread clinical translation. Key challenges include developing robust and scalable manufacturing processes, navigating complex and costly regulatory pathways, and conducting the large-scale, long-term clinical trials required to definitively prove safety and efficacy.

In closing, we extend our sincere gratitude to all the authors for their outstanding contributions, to the many expert reviewers for their diligent and constructive feedback, and to the editorial team at *Polymers* for their invaluable support in bringing this Special Issue to fruition. We hope that this collection will serve as both a valuable resource and a source of inspiration for researchers, clinicians, and engineers as we continue to advance the development and application of polymer scaffolds for the benefit of human health.

Acknowledgments: The Guest Editors would like to thank all the authors for their high-quality submissions and valuable contributions to this Special Issue. We are also deeply grateful to the numerous expert reviewers whose voluntary efforts and insightful comments ensured the scientific rigor of the published works. Finally, we express our sincere appreciation to the editorial staff of *Polymers* for their professional and efficient support throughout the entire process.

Conflicts of Interest: The author declares no conflicts of interest.

References

1. Gutiérrez, B.; González-Quijón, M.E.; Martínez-Rodríguez, P.; Alarcón-Apablaza, J.; Godoy, K.; Cury, D.P.; Lezcano, M.F.; Vargas-Chávez, D.; Dias, F.J. Comprehensive Development of a Cellulose Acetate and Soy Protein-Based Scaffold for Nerve Regeneration. *Polymers* **2024**, *16*, 216. [CrossRef] [PubMed]
2. Song, X.; Philpott, M.A.; Best, S.M.; Cameron, R.E. Controlling the Architecture of Freeze-Dried Collagen Scaffolds with Ultrasound-Induced Nucleation. *Polymers* **2024**, *16*, 213. [CrossRef] [PubMed]

3. Guo, W.; Bu, W.; Mao, Y.; Wang, E.; Yang, Y.; Liu, C.; Guo, F.; Mai, H.; You, H.; Long, Y. Magnesium Hydroxide as a Versatile Nanofiller for 3D-Printed PLA Bone Scaffolds. *Polymers* **2024**, *16*, 198. [CrossRef] [PubMed]
4. Alassaf, M.; Alqahtani, S.M.; Al Khulaifi, R.S.; Saeed, W.S.; Alsubaie, F.S.; Semlali, A.; Aouak, T. Mevacor/Poly(Vinyl Acetate/2-Hydroxyethyl Methacrylate) as Solid Solution: Preparation, Solubility Enhancement and Drug Delivery. *Polymers* **2023**, *15*, 3927. [CrossRef] [PubMed]
5. Deshpande, M.V.; Girase, A.; King, M.W. Degradation of Poly(ε-Caprolactone) Resorbable Multifilament Yarn under Physiological Conditions. *Polymers* **2023**, *15*, 3819. [CrossRef] [PubMed]
6. Menotti, F.; Scutera, S.; Coppola, B.; Longo, F.; Mandras, N.; Cavallo, L.; Comini, S.; Sparti, R.; Fiume, E.; Cuffini, A.M.; et al. Tuning of Silver Content on the Antibacterial and Biological Properties of Poly(ε-Caprolactone)/Biphasic Calcium Phosphate 3D-Scaffolds for Bone Tissue Engineering. *Polymers* **2023**, *15*, 3618. [CrossRef] [PubMed]
7. To, J.; Zhang, X.; Tam, J.P. Design of Potent and Salt-Insensitive Antimicrobial Branched Peptides. *Polymers* **2023**, *15*, 3594. [CrossRef] [PubMed]
8. Botvin, V.; Fetisova, A.; Mukhortova, Y.; Wagner, D.; Kazantsev, S.; Surmeneva, M.; Kholkin, A.; Surmenev, R. Effect of Fe_3O_4 Nanoparticles Modified by Citric and Oleic Acids on the Physicochemical and Magnetic Properties of Hybrid Electrospun P(VDF-TrFE) Scaffolds. *Polymers* **2023**, *15*, 3135. [CrossRef] [PubMed]
9. Zdraveva, E.; Dolenec, T.; Tominac Trcin, M.; Govorčin Bajsić, E.; Holjevac Grgurić, T.; Tomljenović, A.; Dekaris, I.; Jelić, J.; Mijovic, B. The Reliability of PCL/Anti-VEGF Electrospun Scaffolds to Support Limbal Stem Cells for Corneal Repair. *Polymers* **2023**, *15*, 2663. [CrossRef] [PubMed]
10. Sangkatip, R.; Jongwuttanaruk, K.; Sriseubsai, W. Gelatin/$Na_2Ti_3O_7$ Nanocomposite Scaffolds: Mechanical Properties and Characterization for Tissue Engineering Applications. *Polymers* **2023**, *15*, 2322. [CrossRef] [PubMed]
11. Pereira, P.; Neto, A.S.; Rodrigues, A.S.; Barros, I.; Miranda, C.; Ramalho-Santos, J.; Pereira De Almeida, L.; Ferreira, J.M.F.; Coelho, J.F.J.; Fonseca, A.C. In Vitro Evaluation of Biphasic Calcium Phosphate Scaffolds Derived from Cuttlefish Bone Coated with Poly(Ester Urea) for Bone Tissue Regeneration. *Polymers* **2023**, *15*, 2256. [CrossRef] [PubMed]
12. Dimitrova, M.; Vlahova, A.; Kalachev, Y.; Zlatev, S.; Kazakova, R.; Capodiferro, S. Recent Advances in 3D Printing of Polymers for Application in Prosthodontics. *Polymers* **2023**, *15*, 4525. [CrossRef] [PubMed]
13. Kováč, J.; Priščáková, P.; Gbelcová, H.; Heydari, A.; Žiaran, S. Bioadhesive and Injectable Hydrogels and Their Correlation with Mesenchymal Stem Cells Differentiation for Cartilage Repair: A Mini-Review. *Polymers* **2023**, *15*, 4228. [CrossRef] [PubMed]
14. Hutomo, D.I.; Amir, L.; Suniarti, D.F.; Bachtiar, E.W.; Soeroso, Y. Hydrogel-Based Biomaterial as a Scaffold for Gingival Regeneration: A Systematic Review of In Vitro Studies. *Polymers* **2023**, *15*, 2591. [CrossRef] [PubMed]

polymers

MDPI

Article

Magnesium Hydroxide as a Versatile Nanofiller for 3D-Printed PLA Bone Scaffolds

Wang Guo [1,2,*], Wenlang Bu [1,2], Yufeng Mao [1,2], Enyu Wang [1,2], Yanjuan Yang [1,2], Chao Liu [1,2], Feng Guo [3,4], Huaming Mai [3,4], Hui You [1,2,*] and Yu Long [1,2,*]

1 State Key Laboratory of Featured Metal Materials and Life-Cycle Safety for Composite Structures, Guangxi University, Nanning 530004, China; 2111301002@st.gxu.edu.cn (W.B.); 2211393019@st.gxu.edu.cn (Y.M.); 2211301064@st.gxu.edu.cn (E.W.); 2111301074@st.gxu.edu.cn (Y.Y.); 2111301033@st.gxu.edu.cn (C.L.)
2 Guangxi Key Laboratory of Manufacturing System and Advanced Manufacturing Technology, School of Mechanical Engineering, Guangxi University, Nanning 530004, China
3 Guangxi Key Laboratory of Oral and Maxillofacial Rehabilitation and Reconstruction, Guangxi Medical University, Nanning 530021, China; ndgf@163.com (F.G.); huamingmai@163.com (H.M.)
4 Department of Oral and Maxillofacial Surgery, College of Stomatology, Guangxi Medical University, Nanning 530021, China
* Correspondence: guowang@gxu.edu.cn or drwangguo@outlook.com (W.G.); hyou@gxu.edu.cn or usmlhy@iim.ac.cn (H.Y.); longyu@gxu.edu.cn (Y.L.)

Abstract: Polylactic acid (PLA) has attracted much attention in bone tissue engineering due to its good biocompatibility and processability, but it still faces problems such as a slow degradation rate, acidic degradation product, weak biomineralization ability, and poor cell response, which limits its wider application in developing bone scaffolds. In this study, $Mg(OH)_2$ nanoparticles were employed as a versatile nanofiller for developing PLA/$Mg(OH)_2$ composite bone scaffolds using fused deposition modeling (FDM) 3D printing technology, and its mechanical, degradation, and biological properties were evaluated. The mechanical tests revealed that a 5 wt% addition of $Mg(OH)_2$ improved the tensile and compressive strengths of the PLA scaffold by 20.50% and 63.97%, respectively. The soaking experiment in phosphate buffered solution (PBS) revealed that the alkaline degradation products of $Mg(OH)_2$ neutralized the acidic degradation products of PLA, thus accelerating the degradation of PLA. The weight loss rate of the PLA/20$Mg(OH)_2$ scaffold (15.40%) was significantly higher than that of PLA (0.15%) on day 28. Meanwhile, the composite scaffolds showed long-term Mg^{2+} release for more than 28 days. The simulated body fluid (SBF) immersion experiment indicated that $Mg(OH)_2$ promoted the deposition of apatite and improved the biomineralization of PLA scaffolds. The cell culture of bone marrow mesenchymal stem cells (BMSCs) indicated that adding 5 wt% $Mg(OH)_2$ effectively improved cell responses, including adhesion, proliferation, and osteogenic differentiation, due to the release of Mg^{2+}. This study suggests that $Mg(OH)_2$ can simultaneously address various issues related to polymer scaffolds, including degradation, mechanical properties, and cell interaction, having promising applications in tissue engineering.

Keywords: fused deposition modeling (FDM); magnesium hydroxide ($Mg(OH)_2$); polylactic acid (PLA); bone scaffold; degradation properties; biological properties; mechanical properties

1. Introduction

Bone scaffolds act as a temporary extracellular matrix and can provide a suitable physical, chemical, and biological microenvironment for cell growth and new bone regeneration. Therefore, they need to have a good degradation ability, biological activity, cell response, and suitable mechanical properties [1]. At present, there are many different types of materials applied in bone scaffolds including polymers, ceramics, and metals; they each have their own advantages and disadvantages [2–4]. Synthetic polymers are currently the most widely studied scaffold materials, especially biodegradable polymer materials [5]. PLA, a type of aliphatic polyester, has good biocompatibility, biodegradability, and processability. It can be completely degraded into carbon dioxide and water and then excreted through the body's metabolism [6], thus being approved by the U.S. Food and Drug Administration (FDA) for clinical use [7]. Currently, PLA has been widely used in tissue engineering and drug delivery systems [8].

However, there were still some problems that hindered the application of PLA in bone tissue engineering. Firstly, the degradation rate of bone scaffolds should match the rate of new bone growth [9], but the degradation rate of PLA is much slower; its in vivo degradation lasts for more than four years [10], which is much slower than the regeneration rate of bone defects; it takes about several months to years [11]. In a word, a much slower degradation of PLA is unbeneficial to provide the adequate space for new bone formation. Secondly, the accumulation of acidic products produced by PLA degradation may lead to aseptic inflammatory reactions [12]. In terms of biological performance, PLA, similar to aliphatic polymers, faces the problems of a weak biological activity and cell response [13]. In addition, the mechanical properties of PLA need to be improved to provide a more stable mechanical support during bone regeneration and reconstruction [14].

To better apply PLA in developing bone scaffolds, much effort has been devoted to improving PLA. Polymer blending, which involves adding a polymer with a faster degradation rate, such as PGA [15] or PEG [16], is a common method to improve the degradation rate of PLA-based scaffolds. However, the accelerated degradation rate of the composite scaffold is primarily caused by the rapid degradation of the added polymer and does not fundamentally alter the degradation rate of PLA itself. Adding hydroxyapatite, the main inorganic component of bone, is one common method to improve the biomineralization ability of PLA bone scaffolds [17]. However, the biomineralization ability primarily depends on the speed of ion exchange. Since hydroxyapatite degrades slowly and has a slow ion exchange rate, its ability to enhance the biomineralization of PLA is limited. Loading bone-related growth factors, such as BMP [18] and VEGF [19], onto bone scaffolds is one common method to improve the cell response and promote new bone growth. However, as biological molecules, these growth factors have unstable physicochemical properties and are susceptible to inactivation and loss of effectiveness during loading and post-processing procedures [20].

Metal Mg, with a high reactivity, can degrade fast in an aqueous solution, producing magnesium hydroxide ($MgOH_2$) and releasing Mg^{2+}. $Mg(OH)_2$ is alkaline and can be used to neutralize the acidic degradation products of polyester. The Mg element, an important macroelement widely existing in bone tissue, plays a significant role in bone metabolism and growth by affecting the cell activity of osteoblasts and osteoclasts through activating some pathways [21]. Thus, metal Mg has the potential to improve the degradation properties and cell responses of polyesters [22]. However, the degradation of pure Mg will release hydrogen gas, resulting in the formation of hydrogen gas pockets ($Mg + 2H_2O = Mg^{2+} + 2OH^- + H_2\uparrow$) [23,24], which are harmful to bone growth. Additionally, the chemical activity and degradation rate of metal Mg are much higher; after being soaked in an aqueous solution for several days, it can completely degrade in only one to two weeks, which can lead to a significant pH change of the solution along with the generation of large quantities of hydrogen gas.

$Mg(OH)_2$ is a basic hydroxide of magnesium that, unlike metallic Mg, does not undergo a hydrogen evolution reaction upon degradation. It can degrade and release Mg^{2+} and OH^- in a mild manner. If $Mg(OH)_2$ was incorporated into polyesters, a sustained release of Mg^{2+} may be more beneficial for cell proliferation and osteogenic differentiation. On the other hand, the release of OH^- may accelerate the degradation rates of polyesters and neutralize the acid products. In addition, as a rigid inorganic material, $Mg(OH)_2$ can act as a reinforcing filler to improve the mechanical properties of polymers [25]. Therefore, if we incorporate $Mg(OH)_2$ nanoparticles into PLA, this is expected to simultaneously improve the degradation, biomineralization, cell response, and mechanical properties of the PLA.

The preparation process is another important factor in determining the performance of bone scaffolds besides the material design. Traditional preparation methods for bone scaffolds mainly include solvent casting [26], gas foaming [27], freeze-drying [28], phase separation [29], etc. It is difficult for these methods to build a complex, personalized bone scaffold and precisely control the porous structure. Three-dimensional printing technologies, based on the principle of discrete accumulation, could theoretically realize the precise forming of complex structures, which could accurately meet the controllable building of porous structures and personalized shapes for bone scaffolds [30,31]. Due to the wide range of processable materials, low material and use cost, environmental protection, and nontoxicity, fused deposition modeling (FDM), as the most common 3D printing technology, offers the potential for design and manufacturing in the combination of polymers and the biomedical field [32,33]. Furthermore, FDM has shown obvious competitiveness in the preparation of biodegradable polymer bone scaffolds including PLA [34–36]. However, to the best of our knowledge, there are few studies currently available on PLA/$Mg(OH)_2$ porous bone scaffolds, let alone those using FDM 3D printing technology.

In this study, we incorporated $Mg(OH)_2$ nanoparticles into PLA, used FDM 3D printing technology to prepare PLA/$Mg(OH)_2$ composite bone scaffolds with various contents, and then investigated their comprehensive performances. The phase composition of PLA/$Mg(OH)_2$ was detected using XRD and FTIR. The mechanical properties of PLA/$Mg(OH)_2$ composite scaffolds were tested using tensile and compressive tests. The degradation and Mg^{2+} release behaviors of the composite scaffolds were evaluated using a PBS immersion test, and the degradation mechanism was discussed. The biomineralization of the PLA/$Mg(OH)_2$ composite scaffolds was evaluated using the SBF immersion test. The cell response behaviors of BMSCs on the composite scaffolds were evaluated in the aspects of cell adhesion, proliferation, and differentiation.

2. Materials and Methods

2.1. Materials

The materials used in the experiments are mainly included: PLA (REVODE, Zhejiang Hisun Biomaterials Co., Ltd., Taizhou, China); $Mg(OH)_2$ nanopowders (~200 nm, as shown in Figure S1) (Beijing Deke Science and Technology Co., Ltd., Beijing, China); dichloromethane (CH_2Cl_2, $\geq$99.5%, Sichuan Xilong Science Co., Ltd., Chengdu, China); simulated body fluid (SBF) (R24167, 1.5$\times$, Shanghai Yuanye Bio-Technology Co., Ltd., Shanghai, China); rabbit bone marrow mesenchymal stem cells (BMSCs) (RBXMX-01001) and BMSC complete culture medium (RBXMX-80011) (Cyagen Biosciences, Inc., Guangzhou, China); DAPI staining solution (C1005), actin-tracker red-rhodamine (C2207S), BCIP/NBT alkaline phosphatase color development kit (C3206) (Shanghai Beyotime Biotechnology Co., Ltd., Shanghai, China); and phosphate buffered solution (Yida Technology Co., Ltd., Quanzhou, China).

2.2. 3D Printing of the PLA/$Mg(OH)_2$ Scaffolds

The schematic diagram of the preparation process for the PLA/$Mg(OH)_2$ composite filaments and scaffold is shown in Figure S2. Firstly, the PLA/$Mg(OH)_2$ composites were prepared using a solvent mixing method in a fume hood: (1) PLA granules were dissolved in dichloromethane (DCM) (w/v = 1 g/10 mL) with magnetic stirring for 1 h at 40 °C; (2) $Mg(OH)_2$ nanopowders were dispersed in DCM (w/v = 1 g/10 mL) using an ultrasonic dispersing treatment at room temperature for 15 min; (3) the $Mg(OH)_2$ suspension was poured into the PLA solution, and then underwent magnetic stirring until most of the solvent DCM was volatilized; (4) the PLA/$Mg(OH)_2$ mixed suspension was poured into a stainless steel plate and air-dried overnight for further solvent volatilizing and film formation; (5) the films were cut into fragments, and then dried in a drying oven at 55 °C for 24 h to reach constant weight; (6) finally, the fragments were used for subsequent filament preparation. The weight fractions of $Mg(OH)_2$ in the composites were as follows: 0 wt%,

2.5 wt%, 5 wt%, 7.5 wt%, 10 wt%, and 20 wt%, and the corresponding composites were labeled as PLA, PLA/2.5$Mg(OH)_2$, PLA/5$Mg(OH)_2$, PLA/7.5$Mg(OH)_2$, PLA/10$Mg(OH)_2$, and PLA/20$Mg(OH)_2$, respectively.

The PLA/$Mg(OH)_2$ filaments were prepared using a single-screw extruder (Wellzoom C, Shenzhen Misida Technology Co., Ltd., Shenzhen, China). The temperatures of the extruding and plasticizing zones of the extruder were 165 °C and 170 °C, respectively. The filaments with a diameter of 1.75 ± 0.1 mm were obtained by adjusting the extrusion speed and pulling speed.

The PLA/$Mg(OH)_2$ scaffold samples were prepared using an FDM 3D printer (Ender-3 S1 Pro, Shenzhen Creality 3D Technology Co., Ltd., Shenzhen, China). The 3D model of the PLA/$Mg(OH)_2$ porous scaffold was designed using SolidWorks 2022, as shown in Figure S3. Considering that our research focuses on the effect of $Mg(OH)_2$ content on the performance of composite scaffolds, we have kept the designed porosity constant at 50%. The main process parameters are set as follows: nozzle diameter = 0.4 mm (this size has a balance of printing accuracy, production rate, and printing stability according to pre-experiment), layer height = 0.2 mm, printing speed = 30 mm/s, nozzle temperature = 190 °C, hot bed temperature = 55 °C, and filling density = 100%. Other parameters are shown in Table S1.

2.3. Physical and Chemical Properties

The three-dimensional macroscopic appearance of the scaffolds was imaged with a digital camera; the top and side macroscopic appearances of the scaffolds were observed with an industrial camera (TD2KHU, Shenzhen Sanqiang Taida Optical Instrument Co., Ltd., Shenzhen, China); the microscopic appearance of the struts of the scaffolds was observed using SEM (SU8020, HITACHI, Tokyo, Japan).

The porosity of the PLA/$Mg(OH)_2$ scaffold was determined according to the liquid displacement method [37]: (1) First, calculate the volume (V) based on the actual length, width, and height of the scaffold. (2) Place the scaffold in a 5 mL graduated cylinder, inject deionized water into the graduated cylinder until it reaches 4 mL using a syringe, and record the amount of water consumed (ΔV) in the syringe at this point. (3) At this point, the volume of the solid part of the scaffold can be calculated as $(4 - \Delta V)$. Therefore, the actual porosity of the scaffold can be calculated using the following formula:

$$Porosity = \left(1 - \frac{4 - \triangle V}{V}\right) \times 100\%$$

The functional groups of the raw $Mg(OH)_2$ nanopowders and composite samples were measured using the Fourier transform infrared spectrometer (FTIR, iCAV9, Tianjin Energy Spectrum Technology Co., Ltd., Tianjin, China). The FTIR spectra were recorded using 32 scans at a resolution of 8 cm^{-1}. The phase composition and crystal structure of PLA/$Mg(OH)_2$ scaffolds were analyzed using XRD (D8, Bruker Co., Karlsruhe, Germany) with a scanning range of 2θ = 5–80° and a speed of 8°/min.

To evaluate the hydrophilicity of PLA/$Mg(OH)_2$, the water contact angle (WCA) was measured using a contact angle measuring instrument (SDC-200S, Dongguan SINDIN Precision Instrument Co., Ltd., Dongguan, China). Firstly, 5 μL of distilled water was dropped on the composite samples which were prepared using an FDM printer (50 × 10 × 1 mm^3), then the shape of the water drop was captured using the camera equipped with the measuring instrument, and finally, the water contact angle was measured and recorded.

2.4. Mechanical Properties

The mechanical properties of the PLA/$Mg(OH)_2$ composites were tested using a universal material testing machine (AGS-X, Shimadzu, Kyoto, Japan), in terms of tensile and compressive properties. The tensile sample was 1BB type (referred to as GB-T 1040.2-2006), as shown in Figure S4; the loading speed was set at 2.5 mm/min. The compression test used 10 × 10 × 10 mm^3 porous scaffolds with a pore and strut size of 800 μm and theoretical porosity of 50% (as shown in Figure S3), and the loading speed was set at 1 mm/min. The cross-sectional morphology of the PLA/$Mg(OH)_2$ samples treated with liquid nitrogen fracture was observed using SEM.

2.5. Degradation Properties

The degradation properties and Mg^{2+} release behavior of the PLA/$Mg(OH)_2$ scaffolds were evaluated using a PBS immersion test, and then the water absorption, mass loss, pH, and Mg^{2+} concentration were determined. The main experiment steps are as follows: (1) Each scaffold (10 × 10 × 3.2 mm^3) was dried in a drying oven to a constant weight, and then the dry weight was recorded with an analytical balance (accuracy: 1.0 mg) as m_0; (2) the dried samples were separately added into centrifugal tubes contained PBS solution, and incubated in the drying oven at 37.5 °C for 7, 14, and 28 days; (3) after the scheduled time, the scaffold samples and soaking solutions were separately collected; the surface water of the scaffold samples was absorbed with filter paper, and then their wet weight was recorded using an electronic balance (accuracy: 1.0 mg) as M_t; the pH and Mg^{2+} concentration of the soaking solutions were measured using a digital pH meter and an inductively coupled plasma atomic emission spectrometer (ICP-AES) (ICP700T, Suzhou Booway Instrument Technology Co., Ltd., Suzhou, China), respectively; (4) the wet scaffold samples were dried in a drying oven to a constant weight, and then the dry weight was recorded with an analytical balance as m_t. The mass loss and water absorption were calculated according to the following equations, respectively:

$$\text{Mass loss } (\%) = \frac{m_0 - m_t}{m_0} \times 100\%$$

$$\text{Water absorption } (\%) = \frac{M_t - m_t}{m_t} \times 100\%$$

where m_0 and m_t are the dry weights of the scaffold samples before and after immersion, respectively; M_t is the wet weight of scaffold samples after immersion.

Additionally, the surface morphology of the scaffold samples after immersion was observed using SEM, and the phase composition was analyzed with XRD using mapping mode.

2.6. Biomineralization

The biomineralization of the scaffolds (the ability to deposit hydroxyapatite) was evaluated by immersing the PLA/$Mg(OH)_2$ scaffolds in SBF. First, the SBF powders were dissolved in 900 mL of deionized water at 37 °C, and the pH of the solution was titrated to 7.4 with dilute hydrochloric acid (2 mol/L). The scaffolds were weighed and put into a centrifuge tube, and the SBF was dripped into the centrifuge tube through a pipette gun (the ratio of the sample mass to the solution volume was w/v = 1 g/30 mL), and then we put the centrifuge tube into a vacuum oven at 37.5 °C for incubation. The scaffolds were taken out on days 7, 14, and 28 and then dried to a constant weight. The surface morphology of PLA and PLA/$Mg(OH)_2$ scaffolds after the immersion test was observed using SEM. The deposition of apatite on the surface of the scaffolds was analyzed using XRD.

2.7. Cell Responses

The cell responses of PLA/$Mg(OH)_2$ scaffolds were evaluated through cell culture of BMSCs in terms of cell adhesion, proliferation and osteogenic differentiation using fluorescence staining and ALP staining. The scaffold samples (10 × 10 × 1.6 mm^3) were sterilized by ethylene oxide, and then immersed in PBS overnight for wetting. BMSCs were cultured in a complete medium and seeded on the scaffold samples in a 24-well plate, and then incubated at 37 °C for different times.

For the fluorescence staining, the scaffold samples were treated as follows after a scheduled time (1, 3, and 7 days): (1) gently washed with PBS 3 times for 5 min and then fixed by 4% formaldehyde solution in PBS; (2) added to 0.1% Triton X-100 in PBS to increase permeability; (3) stained with phalloidin solution; (4) stained with DAPI solution; and (5) finally, the scaffolds were observed and imaged with fluorescence microscopy to evaluate the cell proliferation.

For ALP staining, the scaffold samples were treated as follows: (1) after the cells adhered, the culture medium was replaced with an osteogenic induction medium containing vitamin C (0.2 mM), dexamethasone (10^{-7} M), and β-glycerophosphate (10 mM); (2) after the osteogenic induction time reached 7 and 14 days, the cells were stained with BCIP/NBT ALP staining solution; after incubating for 30 min, the staining solution was sucked off, and the color reaction was terminated by washing with triple distilled water; (3) finally, the stained cells were observed using an inverted phase-contrast microscope (Eclipse Ts2, Nikon, Tokyo, Japan).

2.8. Statistical Analysis

Quantitative data were expressed as the mean ± standard deviation. The Student's *t*-test was performed to determine the statistical significance between groups, and it was considered to be significantly different when $p < 0.05$.

3. Results and Discussion

3.1. Morphology and Porosity of PLA/$Mg(OH)_2$ Scaffolds

The macroscopic and microscopic morphology of porous PLA/$Mg(OH)_2$ scaffolds are shown in Figure 1. From the macroscopic morphology (Figure 1(a1–a6)), all the scaffolds showed a well-ordered and interconnected three-dimensional porous structure; the color of the scaffolds turned milky white after adding $Mg(OH)_2$. From the top view (Figure 1(b1–b6)), the pore size was approximately 800 μm, and the strut size was also approximately 800 μm. From the side view (Figure 1(c1–c6)), the interlayer bonding between the extruding fibers was tight. The SEM images (Figure 1(d1–d6)) show that the strut is dense without an apparent defect, indicating a good printing quality; additionally, the surface of the struts became rough after adding $Mg(OH)_2$, which might positively affect the hydrophilicity and cell growth of the scaffolds.

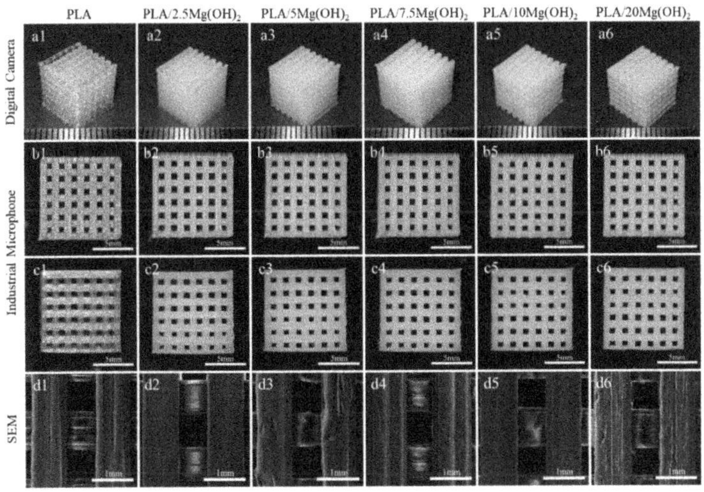

Figure 1. Morphology characterization of FDM 3D-printed PLA/$Mg(OH)_2$ scaffolds: three-dimensional macroscopic appearance with digital camera (**a1–a6**); top view (**b1–b6**) and side view (**c1–c6**) with digital camera; the strut appearance with SEM (**d1–d6**). Notes: the prepared filaments have a good FDM printing quality, and the FDM 3D printed scaffolds show a well-ordered and interconnected three-dimensional porous structure.

The actual porosities of all scaffolds are slightly lower than the designed theoretical porosity (~50%), as shown in Table S2. The reason is due to the expansion swelling effect, which is a common phenomenon in the melt extrusion of polymer materials [38]. In the FDM printing process, when the polymer material is extruded from the hot nozzle, the melt exhibits an expanded state, leading to an increase in the size of the extruded filament. The factors influencing the expansion swelling effect include the process parameters such as printing speed, extrusion temperature, nozzle diameter, as well as the rheological properties of the material [39]. It was known that the inorganic particles influenced the rheological properties of the polymer composites. An increase in the content of inorganic particles leads to an increase in the viscosity of composite materials [40]. In this study, when $Mg(OH)_2$ particles were added to the PLA matrix, the viscosity of the melt increased. Ultimately, this leads to an increased expansion swelling of the composite filaments, resulting in an increase in the strut size and a decrease in the pore size and actual porosity of the scaffold.

3.2. Physical and Chemical Characterizations

The phase composition of the PLA/$Mg(OH)_2$ scaffolds was analyzed using XRD (Figure 2a). The diffraction peaks of $Mg(OH)_2$ were at 2θ = 18.6° (001), 32.8° (100), 38.0° (101), 50.8° (102), 58.7° (110), 62.2° (111), 68.2° (103), and 72.1° (201). PLA showed diffraction peaks at 16.9°. There appeared to be $Mg(OH)_2$ peaks in the PLA/$Mg(OH)_2$ scaffolds, and the peak intensity increased with the content. In addition, there were no obvious peaks of impurity. The XRD result indicated that $Mg(OH)_2$ was successfully composited with the PLA, without the formation of impurities during the melt extrusion process [41].

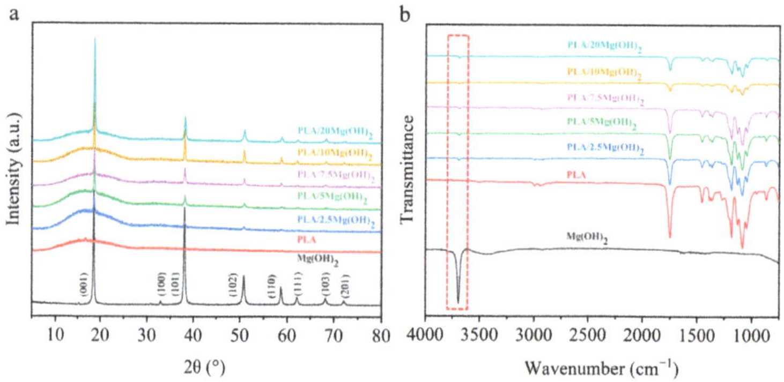

Figure 2. XRD patterns (**a**) and FTIR spectra (**b**) of PLA/$Mg(OH)_2$. Notes: the results indicate $Mg(OH)_2$ was composited with PLA without impurity formation after screw extrusion and FDM 3D printing.

The functional group structures of PLA/$Mg(OH)_2$ composites were analyzed using FTIR, as shown in Figure 2b. $Mg(OH)_2$ powders showed an obvious peak at 3691 cm^{-1} (the red dashed box labeled in Figure 2b) caused by an OH^- stretching vibration. Compared with the PLA, the peaks of $Mg(OH)_2$ were observed in the PLA/$Mg(OH)_2$ composites, and its peak intensity increased with the content. The presence of OH^- may enable $Mg(OH)_2$ to form hydrogen bond interactions with PLA [42], which may positively affect the mechanical properties of PLA/$Mg(OH)_2$ composites.

Hydrophilicity is an essential property of biomaterials, which affects cell attachment, spread, and proliferation [43]. The hydrophilicity of PLA/$Mg(OH)_2$ composites was evaluated using a water contact angle (WCA) test (Figure 3). PLA showed a WCA of 86.90°, located between hydrophilic and hydrophobic. The water contact angle gradually decreased with the content of $Mg(OH)_2$. For PLA/20$Mg(OH)_2$, the WCA decreased to 66.99°, which was 19.91° lower than PLA, indicating a significantly improved hydrophilicity.

The results indicated that the hydrophilicity of PLA could be effectively improved by $Mg(OH)_2$ and was closely related to the content.

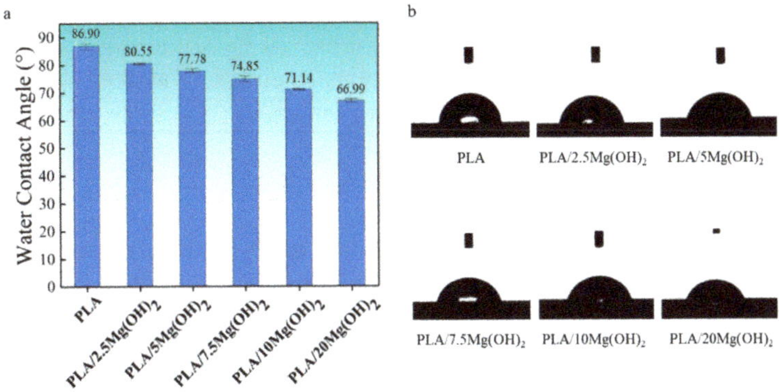

Figure 3. Water contact angle data (**a**) and images (**b**) of PLA/$Mg(OH)_2$ composites. Notes: the hydrophilicity of PLA was effectively improved by $Mg(OH)_2$.

3.3. Mechanical Properties

Mechanical properties are significant for bone scaffolds as they provide mechanical support for bone defects after implantation [44]. The effect of $Mg(OH)_2$ on the mechanical properties of PLA was evaluated using tensile and compression tests, as shown in Figure 4. The tensile strength of PLA/$Mg(OH)_2$ composites increased gradually with the $Mg(OH)_2$ content, reaching a peak value of 71.94 MPa at a 5 wt% content, which was significantly ($p < 0.05$) increased by 20.50% compared with PLA (59.72 MPa). However, when the content further increased to a 7.5–20 wt%, the tensile strength decreased instead. Nevertheless, the tensile strength of PLA/20$Mg(OH)_2$ was still higher than that of PLA. The tensile modulus of PLA/$Mg(OH)_2$ increased continuously with the content of $Mg(OH)_2$ increasing, achieving a maximum value of 1602.71 MPa at a 20 wt%. The elongation at break increased first and then decreased with the content of $Mg(OH)_2$ (Figure 4a), reaching a peak value at 5 wt%.

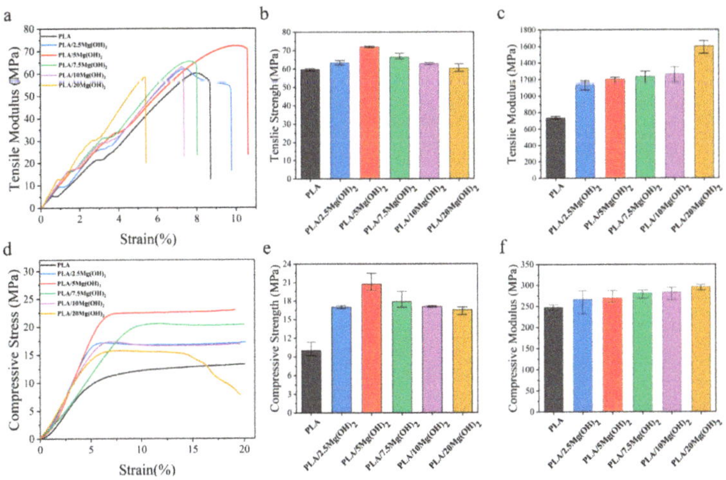

Figure 4. Mechanical properties of PLA/$Mg(OH)_2$: typical tensile stress-strain curves (**a**), tensile strength (**b**), tensile modulus (**c**); typical compressive stress-strain curves (**d**), compressive strength (**e**), compressive modulus (**f**). Notes: as the content of $Mg(OH)_2$ increased, the strength and compressive modulus of PLA/$Mg(OH)_2$ increased first and then decreased, with an optimal content at 5 wt%.

The compressive properties of PLA/$Mg(OH)_2$ scaffolds are shown in Figure 4d–f. The compressive stress-strain curves could be divided into elastic and plastic stages (Figure 4d). In the elastic stage, the curves were approximately linear, with stress approximately proportional to the strain. During this stage, the scaffolds underwent an elastic deformation under compressive stress. As the compressive stress further increased, the curves entered the plastic stage, where the pore structure of the scaffold gradually compressed, and the scaffold gradually densified.

The compressive strength and modulus of the scaffolds were determined from the stress-strain curves, as shown in Figure 4e,f. The compressive strength of PLA/$Mg(OH)_2$ scaffolds increased gradually with the content of $Mg(OH)_2$ (Figure 4e), reaching a peak value of 19.66 MPa at a 5 wt%, which was significantly increased by 63.97% ($p < 0.05$) compared with PLA (11.99 MPa). However, with the content further increasing to a 7.5–20 wt%, the compressive strength decreased, but all the composite scaffolds were still stronger than PLA. The trend of the compressive modulus (Figure 4f) was similar to that of the compressive strength: as the content of $Mg(OH)_2$ increased, the compressive modulus of PLA/$Mg(OH)_2$ scaffolds increased to a peak value of 296.73 MPa at a 20 wt%, which was 19.87% higher than PLA (247.55 MPa). The mechanical test results demonstrated that $Mg(OH)_2$ could significantly strengthen the PLA, and the strengthening effect was associated with the content. Natural bone can be classified into two types: cortical bone and trabecular (spongy) bone. Their mechanical strengths are approximately 100–200 MPa and 2–20 MPa, respectively [45]. In our results, the compressive strength of porous scaffolds ranged from 11.99 MPa to 19.66 MPa, meeting the strength requirements for spongy bone but showing a gap for cortical bone.

The liquid nitrogen fracture morphology of the scaffolds was observed using SEM (Figure 5). The fracture surface of pure PLA was smooth without obvious deformation (Figure 5a), indicating a brittle fracture behavior. After adding $Mg(OH)_2$ (Figure 5b,c), the fracture morphology of the PLA matrix became rough. This result indicated that $Mg(OH)_2$ could strengthen the PLA matrix. When the content of $Mg(OH)_2$ was added at relatively low levels (2.5 wt% and 5 wt%, Figure 5b,c), the dispersion of $Mg(OH)_2$ nanoparticles in the PLA matrix was relatively uniform, and their interfacial adhesion with the PLA matrix was relatively good, which played a crucial role in the enhancement of the mechanical properties. However, when the $Mg(OH)_2$ was added at excessive levels (7.5–20 wt% in this study, Figure 5d–f), the dispersion of the nanoparticles began to deteriorate, with large numbers of nanoparticles agglomerating together, leading to large amounts of defects and stress concentration [46]. As a result, the mechanical enhancing effect of $Mg(OH)_2$ nanoparticles was weakened.

The content and distribution of nanofillers have significant effects on the mechanical properties of polymer nanocomposites [47]. At an appropriate content (no more than a 5 wt% in this study), $Mg(OH)_2$ nanoparticles can be relatively dispersed in the PLA matrix and act as rigid phases; meanwhile, $Mg(OH)_2$ may form good interfacial adhesion with PLA due to the hydrogen bond interaction [48], thus effectively absorbing stress from the relatively soft PLA matrix. As a result, the resistance to deformation and fracture of the matrix was strengthened, contributing to significant increases in the mechanical properties of PLA/$Mg(OH)_2$. However, when at a relatively high content, $Mg(OH)_2$ nanoparticles tend to form much agglomeration, thus weakening the mechanical strengthening effects.

Considering that the mechanical performance is optimal at a $Mg(OH)_2$ content of 5 wt%, we have chosen to proceed with subsequent degradation, biomineralization, and cell culture experiments using this content. Additionally, to explore the influence of $Mg(OH)_2$ content on PLA/$Mg(OH)_2$, we have also included a 20 wt% content as a control in subsequent experiments.

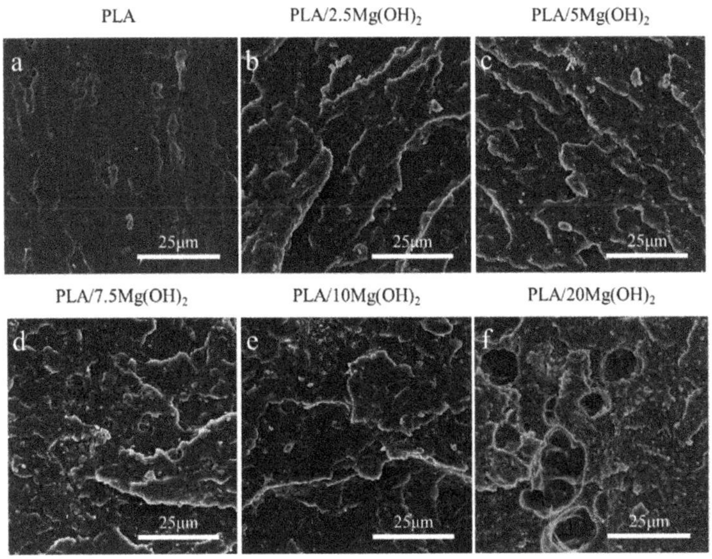

Figure 5. SEM images of liquid nitrogen fracture morphology for PLA/$Mg(OH)_2$ composites: PLA (**a**), PLA/2.5$Mg(OH)_2$ (**b**), PLA/5$Mg(OH)_2$ (**c**), PLA/7.5$Mg(OH)_2$ (**d**), PLA/10$Mg(OH)_2$ (**e**), PLA/20$Mg(OH)_2$ (**f**). Notes: at a low content (2.5 and 5 wt%), $Mg(OH)_2$ particles can disperse relatively uniformly, thus exerting well mechanical strengthening effects; while at a high content (7.5–20 wt%), $Mg(OH)_2$ particles tend to form obvious aggregates with poor bonding with PLA matrix.

3.4. Degradation Properties

The degradation properties and Mg^{2+} release behavior of PLA/$Mg(OH)_2$ scaffolds were tested with PBS immersion for 7, 14, and 28 days. The results of the water absorption rate, mass loss rate, pH value, and Mg^{2+} concentration are shown in Figure 6. The water absorption rate gradually increased with immersion time for all the scaffolds (Figure 6a). More importantly, the water absorption rate of PLA/$Mg(OH)_2$ scaffolds was significantly higher than that of PLA at each time point, and it increased with the $Mg(OH)_2$ content. On day 28, the water absorption rate of the PLA/20$Mg(OH)_2$ scaffold reached 94.19%, which was significantly higher than that of the PLA scaffold (8.06%) ($p < 0.05$). For bone scaffolds, a high water absorption capacity is usually beneficial for nutrient delivery and cell affinity, thus helping to maintain metabolism and promoting cell proliferation [49]. Additionally, a high water absorption capacity could promote the degradation of PLA, which is more conducive to the growth of bone tissues.

The mass loss of PLA/$Mg(OH)_2$ scaffolds is shown in Figure 6b. During the 28-day PBS immersion, the mass of the PLA scaffolds hardly changed. By contrast, PLA/$Mg(OH)_2$ scaffolds showed an obvious mass loss, which was positively associated with the immersion time and $Mg(OH)_2$ content. After 28 days of degradation, the mass loss rates of PLA, PLA/5$Mg(OH)_2$, and PLA/20$Mg(OH)_2$ scaffolds were 0.15%, 3.12%, and 15.40%, respectively.

The pH value change of the degradation medium PBS is shown in Figure 6c. During the 28-day PBS immersion, the pH value of the PLA scaffolds fluctuated little and remained almost unchanged at a pH = 7.4, indicating that almost no signs of degradation of the PLA occurred. The pH value of the PLA/5$Mg(OH)_2$ scaffold showed a trend of increasing slightly at first and then decreasing, but remained nearly neutral at a pH = 7.4. However, the pH value of the PLA/20$Mg(OH)_2$ scaffold continually decreased gradually with the immersion time, reaching 6.49 on day 28. The results of mass loss and pH indicated that $Mg(OH)_2$ could effectively promote the degradation of PLA.

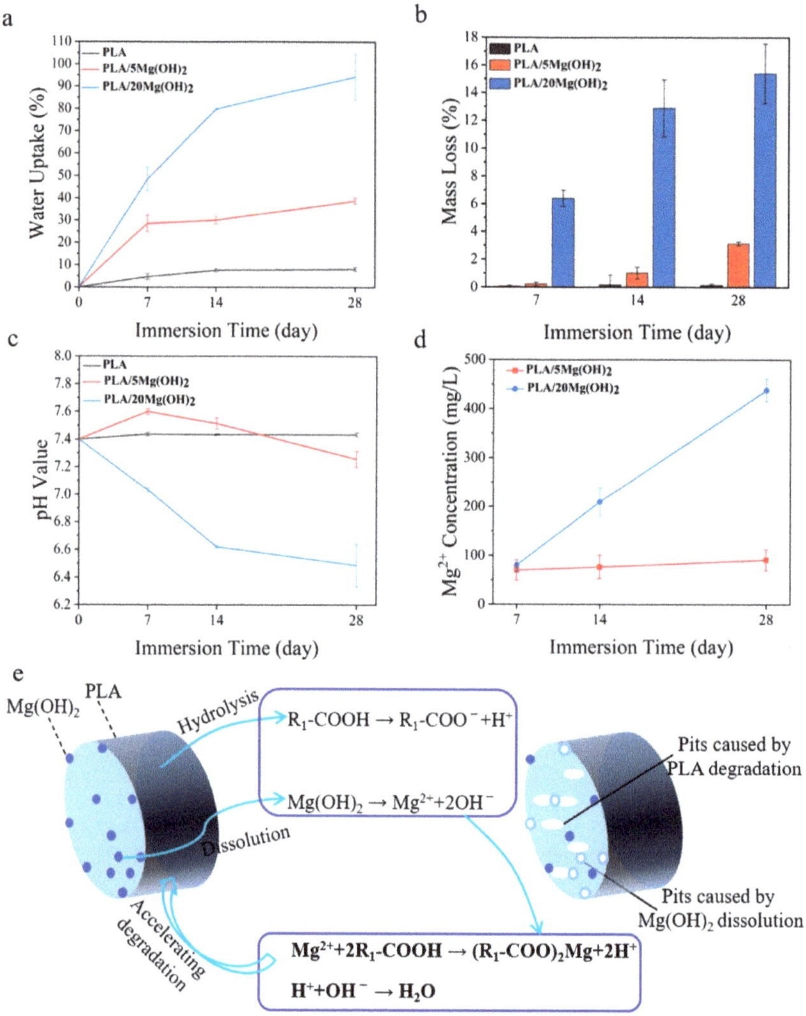

Figure 6. The degradation properties of PLA/Mg(OH)$_2$: water absorption (**a**), mass loss (**b**), pH value (**c**), Mg^{2+} release concentration from scaffolds (**d**), the schematic diagram illustrating the mechanism by which Mg(OH)$_2$ accelerated the degradation of PLA scaffolds (**e**). Notes: Mg(OH)$_2$ accelerated the degradation of PLA by adjusting the acidity and alkalinity in the aqueous environment.

The Mg^{2+} release behavior of PLA/Mg(OH)$_2$ scaffolds is shown in Figure 6d. As the degradation time prolonged, the Mg^{2+} concentration released from the scaffolds gradually increased. The PLA/20Mg(OH)$_2$ scaffold released a higher amount of Mg^{2+} than the PLA/5Mg(OH)$_2$ at each period.

The mechanism by which Mg(OH)$_2$ promoted the degradation of PLA scaffolds is schematically illustrated in Figure 6e. In the PBS immersion, under the action of water molecules, the ester bonds of polylactic acid are cleaved to produce hydrolysis products with carboxyl and hydroxyl end groups (Equation (1)):

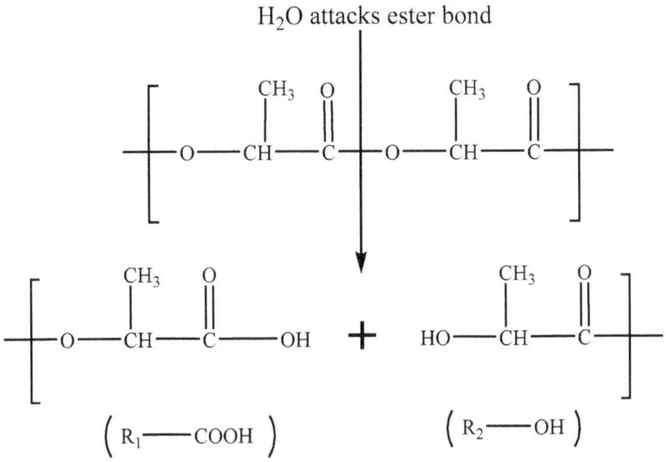

(1)

Equation (2) refers to the further dissociation of carboxyl groups that produce H^+ ions, which lower the pH value in the environment. This has a "self-catalytic" effect on the degradation of PLA. Meanwhile, it promotes the dissolution of alkaline $Mg(OH)_2$ in the solution, releasing Mg^{2+} and OH^- ions, as described in Equation (3).

$$R_1\text{-}COOH \rightarrow R_1\text{-}COO^- + H^+ \quad (2)$$

$$Mg(OH)_2 = Mg^{2+} + 2OH^- \quad (3)$$

In return, the two types of products can consume the hydrolysis products of PLA (Equation (4)) and neutralize the acidic products of PLA (Equation (5)). This acid-base neutralization reaction continuously promotes the degradation of the PLA matrix. Why was the pH of the PLA/5$Mg(OH)_2$ scaffold > 7.4 before 7 days, while the pH of the PLA/20$Mg(OH)_2$ scaffold was < 7.4? The possible reasons can be explained as follows: before 7 days, the PLA/5$Mg(OH)_2$ scaffold, due to its small degradation amount, had little catalytic degradation effect on the PLA, resulting in PLA with almost no degradation, which meant the acidic effect caused by the PLA degradation was still less than the alkaline effect caused by the degradation of $Mg(OH)_2$ itself. Therefore, the pH remained alkaline before 7 days. In contrast, in the PLA/20$Mg(OH)_2$ scaffold, there was a significant degradation of $Mg(OH)_2$, leading to a substantial catalytic effect on the PLA degradation, even resulting in the acidic effect caused by the PLA degradation exceeding the alkaline effect caused by the degradation of $Mg(OH)_2$ itself. As a result, the pH became acidic before 7 days. After 7 days, as the degradation of $Mg(OH)_2$ continued, the catalytic effect of both the 5 wt% and 20 wt% scaffolds on PLA increased significantly, accelerating the degradation of the PLA. Thereby, the acidic effect produced by the PLA degradation surpasses the alkaline effect of $Mg(OH)_2$ degradation, resulting in a pH < 7.4 after 7 days. The pH change trends indicated that $Mg(OH)_2$ particles could accelerate the degradation of PLA by adjusting the acidity and alkalinity in the aqueous environment.

$$Mg^{2+} + 2R_1\text{-}COOH \rightarrow (R_1\text{-}COO)_2Mg + 2H^+ \quad (4)$$

$$H^+ + OH^- \rightarrow H_2O \quad (5)$$

The surface morphology of PLA/$Mg(OH)_2$ scaffolds after 7, 14, and 28-day PBS immersion was observed using SEM, as shown in Figure 7a–c. For the PLA scaffold, during the whole 28-day immersion period, its surface remained smooth and flat, with almost no cracks and pits. For PLA/$Mg(OH)_2$ scaffolds, during the initial 14 days, there were some white particles exposed on the surface of the PLA/5$Mg(OH)_2$ scaffold (Figure 7(b1–b4)). According to the XRD analysis (Figure 7(d1)), the main components were $Mg(OH)_2$ and $Mg_3(PO_4)_2{\cdot}5H_2O$. The detachment of $Mg(OH)_2$ and magnesium salts from the PLA matrix during degradation may be responsible for the formation of cracks and pits on the surface. For the PLA/20$Mg(OH)_2$ scaffold (Figure 7(c1–c6)), there appeared to be whisker-like mixtures after only 7 days. According to the XRD analysis (Figure 7(d2)), the mixtures mainly consisted of magnesium salts ($Mg_3(PO_4)_2{\cdot}5H_2O$ and $Mg(H_2PO_4)_2$) and $Mg(OH)_2$. The deposits may be generated by the reaction between the Mg^{2+} released from $Mg(OH)_2$ and the $PO_4{}^{3-}$ and $HPO_4{}^{2-}$ in the PBS solution. The results showed that $Mg(OH)_2$ significantly improved the degradation performance of PLA.

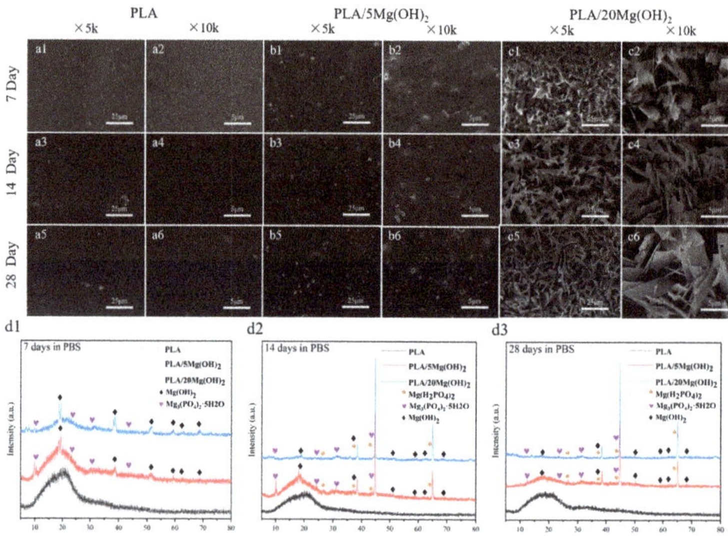

Figure 7. Degradation morphology and product analysis of PLA/$Mg(OH)_2$ scaffolds after immersing in PBS for 7, 14, and 28 days. SEM morphology of PLA (**a1–a6**); PLA/5$Mg(OH)_2$ (**b1–b6**); PLA/20$Mg(OH)_2$ (**c1–c6**); XRD patterns after degradation (**d1–d3**). Notes: With the increase in $Mg(OH)_2$ content and experimental time, the mixed crystalline whisker-like substance of magnesium dihydrogen phosphate and magnesium hydroxide appeared on the surface of the scaffolds.

3.5. Biomineralization

The biomineralization ability of PLA/$Mg(OH)_2$ scaffolds was investigated using an SBF immersion for 7, 14, and 28 days, and the surface morphology was observed using SEM (Figure 8). There was only a small amount of white, spherical mineral deposits on the surface of the PLA scaffold during the immersion (Figure 8(a1–a6)); for the PLA/5$Mg(OH)_2$ scaffold (Figure 8b), there were more spherical mineral deposits which increased in number and volume with the immersion time. On day 28, they were particularly tightly combined, with a finely textured cauliflower-shaped appearance that was observed using high-magnification SEM (Figure 8(b6)). On the PLA/20$Mg(OH)_2$ scaffold (Figure 8c), lots of spherical mineral deposits could be observed after only 7 days (Figure 8(c1,c2)). As the immersion time was prolonged to 14 days, the number and density of the deposits increased, and the volume became larger with small white minerals constantly nucleating and growing on the large deposits, showing a phenomenon of multi-layer deposition (Figure 8(c3,c4)). On day 28, there were white deposits of different sizes growing together and overlapping layer by layer, becoming thick and dense (Figure 8(c5,c6)).

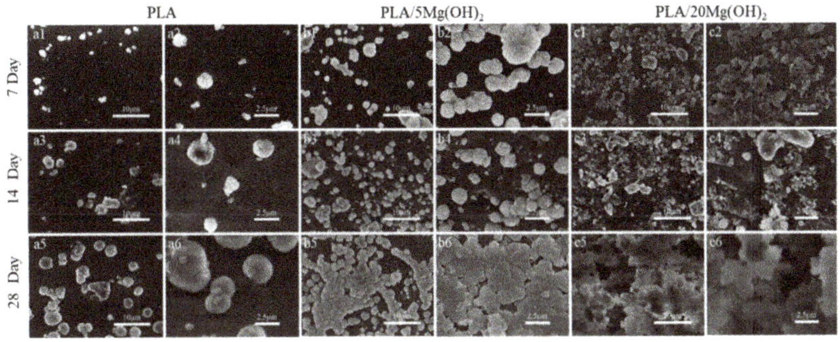

Figure 8. SEM surface morphology of PLA/$Mg(OH)_2$ after immersion in SBF for 7, 14, and 28 days. SEM morphology of PLA (**a1–a6**); PLA/5$Mg(OH)_2$ (**b1–b6**); PLA/20$Mg(OH)_2$ (**c1–c6**). Notes: With the increase in $Mg(OH)_2$ content and immersion time, the quantity and volume of white spherical mineral deposits on the surface of the scaffold gradually increased.

The elemental analysis of the white spherical deposits was performed using EDS (Figure 9(a1–a3)). The EDS spectra showed that the inorganic elements were mainly Ca and P, indicating that the deposits were mainly minerals containing Ca and P; a small amount of Mg was also detected due to the dissolution of $Mg(OH)_2$. The XRD analysis (Figure 9(b1–b3)) further demonstrated that the deposits were mainly composed of calcium magnesium phosphate ($Ca_3Mg_3(PO_4)_4$, Mg-substituted apatite). With the increase in the immersion time, the peaks at $2\theta = 10.50°$ and $31.70°$ of Mg-substituted apatite ($Ca_3Mg_3(PO_4)_4$) indicated further deposition and growth. Additionally, the XRD patterns indicated that the amount of Mg-substituted apatite increased with the content of $Mg(OH)_2$. In addition, the EDS spectra also showed that the peak intensity of Ca and P elements were positively related to the $Mg(OH)_2$ content (Figure 9(a3)). This may be attributed to $Mg(OH)_2$ and Mg^{2+} as sites for the nucleation of apatite. The apatite plays significant roles in stimulating cell metabolism, bone regeneration, and chemical bonding between the biomaterial and bone tissues [50,51]. In a word, the results demonstrated that adding $Mg(OH)_2$ could effectively improve the apatite-forming ability and biomineralization of PLA scaffolds. Based on the above experimental results, the PLA/5$Mg(OH)_2$ scaffold with the best mechanical properties and the PLA/20$Mg(OH)_2$ with the best degradation performance and biomineralization were selected for subsequent cell culture experiments.

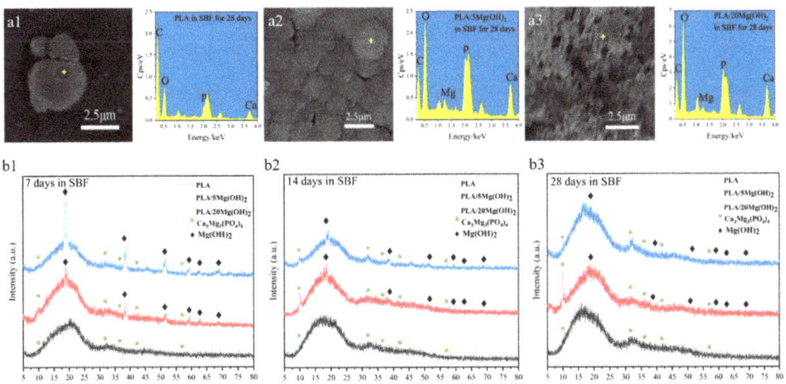

Figure 9. Element and phase analysis of mineral deposits on PLA/$Mg(OH)_2$ after immersion in SBF for 7, 14, and 28 days: EDS spectra (**a1–a3**) and XRD patterns (**b1–b3**). Notes: EDS indicated the mineral deposits consisted mainly of Ca, P, and Mg elements, and XRD further indicated the mineral deposits consisted mainly of $Ca_3Mg_3(PO_4)_4$ (Mg-substituted apatite).

3.6. Cell Responses

BMSCs were used to evaluate the cell response behaviors on PLA/$Mg(OH)_2$ scaffolds. The fluorescence staining images after 1, 3, and 7 days of cell culture are shown in Figure 10. It was obvious that the number of cells on the PLA/5$Mg(OH)_2$ scaffold (Figure 10(a4–a6)) was more than that on the PLA at each time point, with a better cell spreading morphology (Figure 10(b4–b6)). On day 7, the cells covered most of the strut (Figure 10(c6)). There was a "strange" phenomenon where the phalloidin staining (cell actin) was abnormal for the PLA/20$Mg(OH)_2$ scaffold (Figure 10(b7–b9)), which may have resulted from the damage of excessive $Mg(OH)_2$ and alkalinity on the staining reaction and cells. The results showed that an appropriate content of $Mg(OH)_2$ was beneficial to enhance cell adhesion and proliferation.

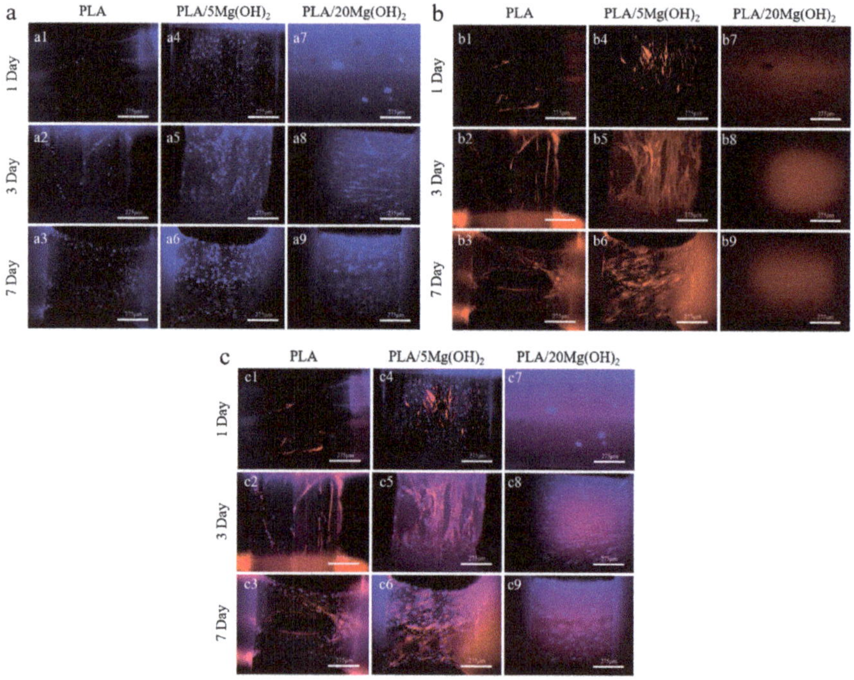

Figure 10. Fluorescence staining images of PLA and PLA/$Mg(OH)_2$ scaffolds: DAPI (**a1–a9**), phalloidin (**b1–b9**), merged (**c1–c9**). Notes: Observations on the struts of the composite scaffold reveal a greater number of cell nuclei. The strut with PLA/5$Mg(OH)_2$ shows the highest density of spread-out actin, while no actin is observed in PLA/20$Mg(OH)_2$, indicating that the appropriate addition of $Mg(OH)_2$ promotes cell proliferation and growth.

Alkaline phosphatase (ALP) is an enzyme responsible for the hydrolysis of phosphate esters into organophosphates, which is an early marker of osteoblast differentiation and bone matrix formation. BMSCs were used to evaluate the osteogenic differentiation using ALP staining (Figure 11). Compared with the PLA group (Figure 11a,d), the PLA/5$Mg(OH)_2$ group (Figure 11b,e) showed a more positive ALP staining, and the stained cells increased with the culture time. This result indicated that $Mg(OH)_2$ could improve the osteogenic differentiation of BMSCs. However, in the PLA/20$Mg(OH)_2$ group (Figure 11c,f), there were no signs of cells, with a large amount of whisker-like substances appearing, which may have resulted from the precipitation of magnesium salts produced by the dissolution of excessive amounts of $Mg(OH)_2$. As a result, the pH of the culture medium significantly changed, and the cell growth was hindered.

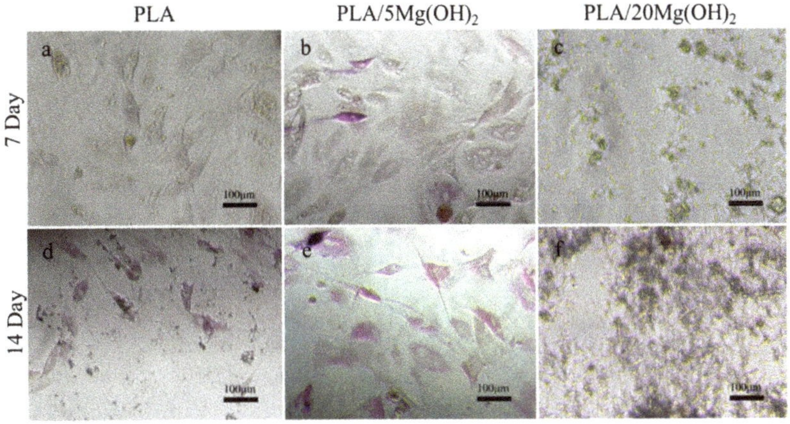

Figure 11. ALP staining images of BMSCs after culturing for 7 and 14 days: PLA (**a**,**d**), PLA/5$Mg(OH)_2$ (**b**,**e**), PLA/20$Mg(OH)_2$ (**c**,**f**). Notes: More cells with a higher degree of staining are observed in PLA/5$Mg(OH)_2$, but only a large amount of whisker-like substances are seen in PLA/20$Mg(OH)_2$, indicating that the appropriate addition of $Mg(OH)_2$ is beneficial for enhancing the osteogenic differentiation ability of cells.

A neutral or weakly alkaline microenvironment is more suitable for cell growth, while neither acid nor alkaline conditions are favorable [52]. When $Mg(OH)_2$ is added at an appropriate content (5 wt%), its dissolution can create a weakly alkaline microenvironment and release a moderate amount of Mg^{2+}, thus promoting the cell response. However, when the addition of $Mg(OH)_2$ was excessive (20 wt% in this study), the dissolution of large amounts of $Mg(OH)_2$ led to a relatively higher alkalinity. The abnormal phenomenon of phalloidin staining in PLA/20$Mg(OH)_2$ demonstrated the alkalinity was excessively high and was harmful to the cells.

It was reported that Mg^{2+} could regulate the expression of genes related to cell proliferation and osteogenic differentiation, including ALP, Col-1, Runx2, OCN, OPN, etc. [53,54]. In this study, PLA/5$Mg(OH)_2$ can release Mg^{2+} into the culture medium without remarkably changing the pH, thus positively involving the cell metabolism and stimulating better cell adhesion, proliferation, and osteogenic differentiation. However, in PLA/20$Mg(OH)_2$, the dissolution of excessive amounts of $Mg(OH)_2$ resulted in a significant pH change in the culture medium, which negatively affected the cell responses and even prevailed over the positive stimulation of Mg^{2+} release. In summary, an appropriate content of $Mg(OH)_2$ (~5 wt% in this study) can effectively improve the cell responses of PLA scaffolds.

Based on the experimental results above, it was indicated that $Mg(OH)_2$ can simultaneously address various issues related to the PLA scaffold, including degradation, mechanical, and biological properties. The PLA/$Mg(OH)_2$ scaffold with good comprehensive properties may have the potential to fill and repair bone defects in clinic. While promoting bone tissue regeneration, they gradually degrade, making space available for new bone regeneration. This can reduce the need for autografts and allografts, avoiding the disadvantages of multiple surgeries, limited donor availability, complications at the donor site, and immune rejection.

However, the current study also has some limitations. The following issues are worth exploring in future research: (1) Investigate the effects of different $Mg(OH)_2$ particle sizes and shapes on the properties of the PLA/$Mg(OH)_2$ scaffolds; (2) Explore the effects of more different contents of $Mg(OH)_2$ on the biocompatibility and biodegradation of the PLA/$Mg(OH)_2$ scaffolds; (3) Evaluate the long-term degradation and stability of the PLA/$Mg(OH)_2$ scaffolds in vivo.

4. Conclusions

In this study, we prepared PLA/$Mg(OH)_2$ composite bone scaffolds using FDM technology and investigated their comprehensive performances. The results showed that adding $Mg(OH)_2$ improved the mechanical, degradation, and biological properties of PLA: (1) The addition of 5 wt% $Mg(OH)_2$ resulted in optimal mechanical properties, with a 63.97% increase in the compressive strength and a 20.50% increase in the tensile strength compared to PLA. (2) $Mg(OH)_2$ remarkably accelerated the degradation rate of PLA through an acid–base neutralization reaction; the mass loss of the scaffold after 28 days of immersion was increased from 0.15% to 15.40% by the addition of 20 wt% $Mg(OH)_2$. (3) The composite scaffolds showed a long-lasting Mg^{2+} release of more than 28 days due to the encapsulation effects of the PLA matrix on the $Mg(OH)_2$ nanoparticles. (4) $Mg(OH)_2$ improved the nucleation and growth of apatite and enhanced the biomineralization ability of PLA scaffolds. (5) PLA/5$Mg(OH)_2$ exhibited better cell adhesion, proliferation, and osteogenic differentiation of BMSCs due to Mg^{2+} release and a suitable pH microenvironment. The above research results showed that the $Mg(OH)_2$ could be used as a multifunctional filler to enhance the mechanical, degradation, and biological properties of polymer bone scaffolds.

Supplementary Materials: The following supporting information can be downloaded at: https://www.mdpi.com/article/10.3390/polym16020198/s1, Table S1: FDM 3D printing parameters; Table S2: Actual porosity data of the PLA/$Mg(OH)_2$ scaffolds (determined according to the liquid displacement method); Figure S1: Representative SEM morphology of $Mg(OH)_2$; Figure S2: The schematic diagram of the preparation process for PLA/$Mg(OH)_2$ composite, filament and scaffold; Figure S3: Schematic diagram of sample for the compression test; Figure S4: Schematic diagram of 1BB-type sample for the tensile test.

Author Contributions: W.G.: conceptualization, investigation, methodology, formal analysis, visualization, data curation, supervision, writing—original draft, writing—reviewing and editing, funding acquisition, and project administration. W.B.: conceptualization, investigation, methodology, formal analysis, visualization, data curation, writing—original draft, and writing—reviewing and editing. Y.M.: investigation, methodology, and visualization. E.W.: investigation, methodology, and visualization. Y.Y.: methodology and visualization. C.L.: methodology and visualization. F.G.: investigation, methodology, visualization, data curation, and validation. H.M.: methodology. H.Y.: methodology and supervision. Y.L.: methodology and supervision. All authors have read and agreed to the published version of the manuscript.

Funding: This work was supported by the following: (1) Natural Science Foundation of China (52205342); (2) Base and Talent Special Project of Guangxi Science and Technology Plan Project (Gui Ke AD22035188); (3) Guangxi Natural Science Foundation (2023GXNSFBA026338); (4) National Key R&D Program of China (2021YFE0203500); (5) Special Fund for Local Scientific and Technological Development Guided by the Central Government (GKZY21195029); (6) Guangxi Bagui Scholars Project (2019A02); (7) Guangxi University Innovation and Development Multiplication Plan Project (202201343, 202201369); (8) National Natural Science Foundation of China (62304057, 82071098); (9) Guangxi High-Level Medical Talent Training Plan ('139' plan) (G201901005).

Data Availability Statement: Data will be made available on request.

Acknowledgments: We appreciate the valuable suggestions provided by several anonymous reviewers, which helped us improve the manuscript.

Conflicts of Interest: The authors declare no conflict of interest.

References

1. Collins, M.N.; Ren, G.; Young, K.; Pina, S.; Reis, R.L.; Oliveira, J.M. Scaffold Fabrication Technologies and Structure/Function Properties in Bone Tissue Engineering. *Adv. Funct. Mater.* **2021**, *31*, 2010609. [CrossRef]
2. Shao, H.; Zhang, Q.; Sun, M.; Wu, M.; Sun, X.; Wang, Q.; Tong, S. Effects of Hydroxyapatite-Coated Porous Titanium Scaffolds Functionalized by Exosomes on the Regeneration and Repair of Irregular Bone. *Front. Bioeng. Biotechnol.* **2023**, *11*, 1283811. [CrossRef]
3. Luo, Y.; Peng, X.; Cheng, C.; Deng, Y.; Lei, N.; Feng, S.; Yu, X. 3D Molybdenum Disulfide Nanospheres Loaded with Metformin to Enhance SCPP Scaffolds for Bone Regeneration. *ACS Appl. Mater. Interfaces* **2023**. [CrossRef]

4. Yang, J.; Wang, H.; Huang, W.; Peng, K.; Shi, R.; Tian, W.; Lin, L.; Yuan, J.; Yao, W.; Ma, X.; et al. A Natural Polymer-Based Hydrogel with Shape Controllability and High Toughness and Its Application to Efficient Osteochondral Regeneration. *Mater. Horiz.* **2023**, *10*, 3797–3806. [CrossRef] [PubMed]
5. Krobot, S.; Melcova, V.; Mencik, P.; Kontarova, S.; Rampichova, M.; Hedvicakova, V.; Mojzisova, E.; Baco, A.; Prikryl, R. Poly(3-Hydroxybutyrate) (PHB) and Polycaprolactone (PCL) Based Blends for Tissue Engineering and Bone Medical Applications Processed by FDM 3D Printing. *Polymers* **2023**, *15*, 2404. [CrossRef] [PubMed]
6. Meng, Y.; Zhu, Y.; Zhou, L.; Meng, X.; Yang, Y.; Zhao, R.; Xia, J.; Yang, B.; Lu, Y.; Wu, H.; et al. Artificial Nacre with High Toughness Amplification Factor: Residual Stress-Engineering Sparks Enhanced Extrinsic Toughening Mechanisms. *Adv. Mater.* **2022**, *34*, 2108267. [CrossRef] [PubMed]
7. Feng, Z.; Wang, D.; Zheng, Y.; Zhao, L.; Xu, T.; Guo, Z.; Hussain, M.I.; Zeng, J.; Lou, L.; Sun, Y.; et al. A Novel Waterborne Polyurethane with Biodegradability and High Flexibility for 3D Printing. *Biofabrication* **2020**, *12*, 035015. [CrossRef] [PubMed]
8. Asadi, M.; Salehi, Z.; Akrami, M.; Hosseinpour, M.; Jockenhövel, S.; Ghazanfari, S. 3D Printed pH-Responsive Tablets Containing N-Acetylglucosamine-Loaded Methylcellulose Hydrogel for Colon Drug Delivery Applications. *Int. J. Pharm.* **2023**, *645*, 123366. [CrossRef] [PubMed]
9. Shuai, C.; Li, S.; Peng, S.; Feng, P.; Lai, Y.; Gao, C. Biodegradable Metallic Bone Implants. *Mater. Chem. Front.* **2019**, *3*, 544–562. [CrossRef]
10. Aprile, P.; Letourneur, D.; Simon-Yarza, T. Membranes for Guided Bone Regeneration: A Road from Bench to Bedside. *Adv. Healthc. Mater.* **2020**, *9*, 2000707. [CrossRef]
11. Sculean, A.; Stavropoulos, A.; Bosshardt, D.D. Self-Regenerative Capacity of Intra-Oral Bone Defects. *J. Clin. Periodontol.* **2019**, *46*, 70–81. [CrossRef] [PubMed]
12. Li, J.; Wang, C.; Gao, G.; Yin, X.; Pu, X.; Shi, B.; Liu, Y.; Huang, Z.; Wang, J.; Li, J.; et al. MBG/PGA-PCL Composite Scaffolds Provide Highly Tunable Degradation and Osteogenic Features. *Bioact. Mater.* **2022**, *15*, 53–67. [CrossRef]
13. Ma, Y.-F.; Yan, X.-Z. Periodontal Guided Tissue Regeneration Membranes: Limitations and Possible Solutions for the Bottleneck Analysis. *Tissue Eng. Part B Rev.* **2023**, *29*, 532–544. [CrossRef] [PubMed]
14. Guo, W.; Yang, Y.; Liu, C.; Bu, W.; Guo, F.; Li, J.; Wang, E.; Peng, Z.; Mai, H.; You, H.; et al. 3D Printed TPMS Structural PLA/GO Scaffold: Process Parameter Optimization, Porous Structure, Mechanical and Biological Properties. *J. Mech. Behav. Biomed. Mater.* **2023**, *142*, 105848. [CrossRef] [PubMed]
15. Shuai, C.; Yang, W.; Feng, P.; Peng, S.; Pan, H. Accelerated Degradation of HAP/PLLA Bone Scaffold by PGA Blending Facilitates Bioactivity and Osteoconductivity. *Bioact. Mater.* **2021**, *6*, 490–502. [CrossRef] [PubMed]
16. Yao, M.; Liu, L.; Ma, C.; Zhang, H.; Zhang, Y.; Song, R.; Fang, Z.; Song, P. A Lysine-Derived Flame Retardant for Improved Flame Retardancy, Crystallinity, and Aqueous-Phase Degradation of Polylactide. *Chem. Eng. J.* **2023**, *462*, 142189. [CrossRef]
17. Custodio, C.L.; Broñola, P.J.M.; Cayabyab, S.R.; Lagura, V.U.; Celorico, J.R.; Basilia, B.A. Powder Loading Effects on the Physicochemical and Mechanical Properties of 3D Printed Poly Lactic Acid/Hydroxyapatite Biocomposites. *Int. J. Bioprint.* **2021**, *7*, 326. [CrossRef]
18. Xue, Y.; Li, Y.; Zhang, D.; Xu, W.; Ning, C.; Han, D. Calcium Phosphate Silicate Microspheres with Soybean Lecithin as a Sustained-Release Bone Morphogenetic Protein-Delivery System for Bone Tissue Regeneration. *ACS Biomater. Sci. Eng.* **2023**, *9*, 2596–2607. [CrossRef]
19. Salehi, S.; Tavakoli, M.; Mirhaj, M.; Varshosaz, J.; Labbaf, S.; Karbasi, S.; Jafarpour, F.; Kazemi, N.; Salehi, S.; Mehrjoo, M.; et al. A 3D Printed Polylactic Acid-Baghdadite Nanocomposite Scaffold Coated with Microporous Chitosan-VEGF for Bone Regeneration Applications. *Carbohydr. Polym.* **2023**, *312*, 120787. [CrossRef]
20. Li, D.; Yang, Z.; Zhao, X.; Luo, Y.; Ou, Y.; Kang, P.; Tian, M. A Bone Regeneration Strategy via Dual Delivery of Demineralized Bone Matrix Powder and Hypoxia-Pretreated Bone Marrow Stromal Cells Using an Injectable Self-Healing Hydrogel. *J. Mater. Chem. B* **2021**, *9*, 479–493. [CrossRef]
21. Li, D.; Dai, D.; Xiong, G.; Lan, S.; Zhang, C. Composite Nanocoatings of Biomedical Magnesium Alloy Implants: Advantages, Mechanisms, and Design Strategies. *Adv. Sci.* **2023**, *10*, 2300658. [CrossRef] [PubMed]
22. Xia, X.; Huang, J.; Wei, J.; Jin, S.; Zou, Q.; Zuo, Y.; Li, J.; Li, Y. Magnesium Oxide Regulates the Degradation Behaviors and Improves the Osteogenesis of Poly(Lactide-Co-Glycolide) Composite Scaffolds. *Compos. Sci. Technol.* **2022**, *222*, 109368. [CrossRef]
23. Parande, G.; Manakari, V.; Sharma Kopparthy, S.D.; Gupta, M. A Study on the Effect of Low-Cost Eggshell Reinforcement on the Immersion, Damping and Mechanical Properties of Magnesium–Zinc Alloy. *Compos. Part B Eng.* **2020**, *182*, 107650. [CrossRef]
24. Xu, B.; Song, Y.; Yang, K.; Li, Y.; Chen, B.; Liao, X.; Jia, Q. Magnesium Metal and Its Corrosion Products: Promising Materials for Tumor Interventional Therapy. *J. Magnes. Alloys* **2023**, *11*, 763–775. [CrossRef]
25. Yao, W.; Xu, X.; Zhou, J.; Chen, M.; Meng, Q.; Wang, P.; Zhang, B.; Zhou, Y.; Ding, Y. Mechanically Robust and Flame-Retarded EPDM Composites with High Loading of $Mg(OH)_2$ Based on Reversible Crosslinking Network from Diels-Alder Reactions. *Polym. Degrad. Stab.* **2022**, *202*, 110029. [CrossRef]
26. Ferrández-Montero, A.; Lieblich, M.; González-Carrasco, J.L.; Benavente, R.; Lorenzo, V.; Detsch, R.; Boccaccini, A.R.; Ferrari, B. Development of Biocompatible and Fully Bioabsorbable PLA/Mg Films for Tissue Regeneration Applications. *Acta Biomater.* **2019**, *98*, 114–124. [CrossRef] [PubMed]

27. Choi, W.J.; Hwang, K.S.; Kwon, H.J.; Lee, C.; Kim, C.H.; Kim, T.H.; Heo, S.W.; Kim, J.-H.; Lee, J.-Y. Rapid Development of Dual Porous Poly(Lactic Acid) Foam Using Fused Deposition Modeling (FDM) 3D Printing for Medical Scaffold Application. *Mater. Sci. Eng. C* **2020**, *110*, 110693. [CrossRef] [PubMed]
28. Wu, M.; Chen, F.; Wu, P.; Yang, Z.; Zhang, S.; Xiao, L.; Deng, Z.; Zhang, C.; Chen, Y.; Cai, L. Nanoclay Mineral-Reinforced Macroporous Nanocomposite Scaffolds for in Situ Bone Regeneration: In Vitro and in Vivo Studies. *Mater. Des.* **2021**, *205*, 109734. [CrossRef]
29. Gandolfi, M.G.; Zamparini, F.; Degli Esposti, M.; Chiellini, F.; Fava, F.; Fabbri, P.; Taddei, P.; Prati, C. Highly Porous Polycaprolactone Scaffolds Doped with Calcium Silicate and Dicalcium Phosphate Dihydrate Designed for Bone Regeneration. *Mater. Sci. Eng. C Mater. Biol. Appl.* **2019**, *102*, 341–361. [CrossRef]
30. Thurzo, A.; Gálfiová, P.; Nováková, Z.V.; Polák, Š.; Varga, I.; Strunga, M.; Urban, R.; Surovková, J.; Leško, L'.; Hajdúchová, Z.; et al. Fabrication and In Vitro Characterization of Novel Hydroxyapatite Scaffolds 3D Printed Using Polyvinyl Alcohol as a Thermoplastic Binder. *Int. J. Mol. Sci.* **2022**, *23*, 14870. [CrossRef]
31. Perez-Davila, S.; Garrido-Gulias, N.; Gonzalez-Rodriguez, L.; Lopez-Alvarez, M.; Serra, J.; Lopez-Periago, J.E.; Gonzalez, P. Physicochemical Properties of 3D-Printed Polylactic Acid/Hydroxyapatite Scaffolds. *Polymers* **2023**, *15*, 2849. [CrossRef] [PubMed]
32. Rastpeiman, S.; Panahi, Z.; Akrami, M.; Haririan, I.; Asadi, M. Facile Fabrication of an Extended-Release Tablet of Ticagrelor Using Three Dimensional Printing Technology. *J. Biomed. Mater. Res. Part A* **2024**, *112*, 20–30. [CrossRef] [PubMed]
33. Abdollahi, A.; Ansari, Z.; Akrami, M.; Haririan, I.; Dashti-Khavidaki, S.; Irani, M.; Kamankesh, M.; Ghobadi, E. Additive Manufacturing of an Extended-Release Tablet of Tacrolimus. *Materials* **2023**, *16*, 4927. [CrossRef] [PubMed]
34. Mohammed, A.H.; Kovacev, N.; Elshaer, A.; Melaibari, A.A.; Iqbal, J.; Hassanin, H.; Essa, K.; Memic, A. Preparation of Polylactic Acid/Calcium Peroxide Composite Filaments for Fused Deposition Modelling. *Polymers* **2023**, *15*, 2229. [CrossRef] [PubMed]
35. Buj-Corral, I.; Sanz-Fraile, H.; Ulldemolins, A.; Tejo-Otero, A.; Dominguez-Fernandez, A.; Almendros, I.; Otero, J. Characterization of 3D Printed Metal-PLA Composite Scaffolds for Biomedical Applications. *Polymers* **2022**, *14*, 2754. [CrossRef] [PubMed]
36. Kontogianni, G.-I.; Bonatti, A.F.; De Maria, C.; Naseem, R.; Melo, P.; Coelho, C.; Vozzi, G.; Dalgarno, K.; Quadros, P.; Vitale-Brovarone, C.; et al. Promotion of In Vitro Osteogenic Activity by Melt Extrusion-Based PLLA/PCL/PHBV Scaffolds Enriched with Nano-Hydroxyapatite and Strontium Substituted Nano-Hydroxyapatite. *Polymers* **2023**, *15*, 1052. [CrossRef] [PubMed]
37. Guo, W.; Liu, C.; Bu, W.; Yang, Y.; Guo, F.; Li, J.; Wang, E.; Mao, Y.; Mai, H.; You, H.; et al. 3D Printing of Polylactic Acid/Boron Nitride Bone Scaffolds: Mechanical Properties, Biomineralization Ability and Cell Responses. *Ceram. Int.* **2023**, *49*, 25886–25898. [CrossRef]
38. Colon, A.R.; Kazmer, D.O.; Peterson, A.M.; Seppala, J.E. Characterization of Die-Swell in Thermoplastic Material Extrusion. *Addit. Manuf.* **2023**, *73*, 103700. [CrossRef]
39. Schuller, T.; Fanzio, P.; Galindo-Rosales, F.J. Analysis of the Importance of Shear-Induced Elastic Stresses in Material Extrusion. *Addit. Manuf.* **2022**, *57*, 102952. [CrossRef]
40. Ferrandez-Montero, A.; Lieblich, M.; Benavente, R.; González-Carrasco, J.L.; Ferrari, B. New Approach to Improve Polymer-Mg Interface in Biodegradable PLA/Mg Composites through Particle Surface Modification. *Surf. Coat. Technol.* **2020**, *383*, 125285. [CrossRef]
41. Borgohain, X.; Rashid, H. Rapid and Enhanced Adsorptive Mitigation of Groundwater Fluoride by Mg(OH)(2) Nanoflakes. *Environ. Sci. Pollut. Res. Int.* **2022**, *29*, 70056–70069. [CrossRef] [PubMed]
42. Chen, S.; Li, Z.; Wu, Y.; Mahmood, N.; Lortie, F.; Bernard, J.; Binder, W.H.; Zhu, J. Hydrogen-Bonded Supramolecular Polymer Adhesives: Straightforward Synthesis and Strong Substrate Interaction. *Angew. Chem. Int. Edit.* **2022**, *61*, e202203876. [CrossRef] [PubMed]
43. Chen, W.; Nichols, L.; Brinkley, F.; Bohna, K.; Tian, W.; Priddy, M.W.; Priddy, L.B. Alkali Treatment Facilitates Functional Nano-Hydroxyapatite Coating of 3D Printed Polylactic Acid Scaffolds. *Mater. Sci. Eng. C* **2021**, *120*, 111686. [CrossRef] [PubMed]
44. Lin, C.-Y.; Kang, J.-H. Mechanical Properties of Compact Bone Defined by the Stress-Strain Curve Measured Using Uniaxial Tensile Test: A Concise Review and Practical Guide. *Materials* **2021**, *14*, 4224. [CrossRef] [PubMed]
45. Ge, M.; Xie, D.; Yang, Y.; Liang, H.; Jiao, C.; Ye, Y.; Wu, J.; Yu, H.; Tian, Z. Additive Manufacturing a Novel Bioceramic Bone Scaffold with $MgO/Ca_3(PO_4)_2$: Microstructure, Mechanical Property and Controllable Degradation Behavior. *J. Mater. Res. Technol.* **2023**, *27*, 2249–2263. [CrossRef]
46. Han, L.; Wang, J.; Chen, Y.; Huang, Y.; Liu, Y.; Wang, Z. Fabrication and Mechanical Properties of WC Nanoparticle Dispersion-Strengthened Copper. *Mater. Sci. Eng. A* **2021**, *817*, 141274. [CrossRef]
47. Matsumoto, T.; Mori, S.; Ohashi, T.; He, C.; Nishino, T. Stress-Transfer Analyses in Cellulose Nanofiber/Montmorillonite Nanocomposites with X-ray Diffraction and Chemical Interaction between Cellulose Nanofiber and Montmorillonite. *Cellulose* **2022**, *29*, 2949–2960. [CrossRef]
48. Yu, L.; Liu, Y.; Feng, P.; Shuai, C.; Peng, S.; Min, A. Organically Modified Montmorillonite Improves Interfacial Compatibility between PLLA and PGA in Bone Scaffold. *Polym. Degrad. Stabil.* **2020**, *182*, 109394. [CrossRef]
49. Sayed, M.; Mahmoud, E.M.; Saber, S.M.; Raafat, S.N.; Gomaa, S.M.; Naga, S.M. Effect of the Injectable Alginate/Nano-Hydroxyapatite and the Silica/Nano-Hydroxyapatite Composites on the Stem Cells: A Comparative Study. *J. Non-Cryst. Solids* **2023**, *610*, 122327. [CrossRef]

50. Xiao, X.; Liu, Z.; Shu, R.; Wang, J.; Zhu, X.; Bai, D.; Lin, H. Periodontal Bone Regeneration with a Degradable Thermoplastic HA/PLCL Bone Graft. *J. Mater. Chem. B* **2023**, *11*, 772–786. [CrossRef]
51. Guo, F.; Wang, E.; Yang, Y.; Mao, Y.; Liu, C.; Bu, W.; Li, P.; Zhao, L.; Jin, Q.; Liu, B.; et al. A Natural Biomineral for Enhancing the Biomineralization and Cell Response of 3D Printed Polylactic Acid Bone Scaffolds. *Int. J. Biol. Macromol.* **2023**, *242*, 124728. [CrossRef] [PubMed]
52. Mao, L.; Yin, Y.; Zhang, L.; Chen, X.; Wang, X.; Chen, F.; Liu, C. Regulation of Inflammatory Response and Osteogenesis to Citrate-Based Biomaterials through Incorporation of Alkaline Fragments. *Adv. Healthc. Mater.* **2022**, *11*, 2101590. [CrossRef] [PubMed]
53. Dong, Q.; Zhang, M.; Zhou, X.; Shao, Y.; Li, J.; Wang, L.; Chu, C.; Xue, F.; Yao, Q.; Bai, J. 3D-Printed Mg-Incorporated PCL-Based Scaffolds: A Promising Approach for Bone Healing. *Mater. Sci. Eng. C* **2021**, *129*, 112372. [CrossRef] [PubMed]
54. Zhao, S.; Xie, K.; Guo, Y.; Tan, J.; Wu, J.; Yang, Y.; Fu, P.; Wang, L.; Jiang, W.; Hao, Y. Fabrication and Biological Activity of 3D-Printed Polycaprolactone/Magnesium Porous Scaffolds for Critical Size Bone Defect Repair. *ACS Biomater. Sci. Eng.* **2020**, *6*, 5120–5131. [CrossRef]

polymers

MDPI

Article

Controlling the Architecture of Freeze-Dried Collagen Scaffolds with Ultrasound-Induced Nucleation

Xinyuan Song, Matthew A. Philpott, Serena M. Best and Ruth E. Cameron *

Department of Materials Science and Metallurgy, University of Cambridge, Cambridge CB3 0FS, UK; xs284@cam.ac.uk (X.S.); smb51@cam.ac.uk (S.M.B.)
* Correspondence: rec11@cam.ac.uk

Abstract: Collagen is a naturally occurring polymer that can be freeze-dried to create 3D porous scaffold architectures for potential application in tissue engineering. The process comprises the freezing of water in an aqueous slurry followed by sublimation of the ice via a pre-determined temperature–pressure regime and these parameters determine the arrangement, shape and size of the ice crystals. However, ice nucleation is a stochastic process, and this has significant and inherent limitations on the ability to control scaffold structures both within and between the fabrication batches. In this paper, we demonstrate that it is possible to overcome the disadvantages of the stochastic process via the use of low-frequency ultrasound (40 kHz) to trigger nucleation, on-demand, in type I insoluble bovine collagen slurries. The application of ultrasound was found to define the nucleation temperature of collagen slurries, precisely tailoring the pore architecture and providing important new structural and mechanistic insights. The parameter space includes reduction in average pore size and narrowing of pore size distributions while maintaining the percolation diameter. A set of core principles are identified that highlight the huge potential of ultrasound to finely tune the scaffold architecture and revolutionise the reproducibility of the scaffold fabrication protocol.

Keywords: freeze-drying; collagen scaffolds; ice nucleation; ultrasound; porous architecture; tissue engineering

1. Introduction

Collagen is a natural polymer that can be processed to form three-dimensional porous scaffolds. These structures are used widely as substrates for tissue engineering, providing mechanical support for cell attachment and stimulating desirable cellular interactions [1]. Freeze-drying, also known as lyophilisation, is a popular process for producing these types of scaffolds. An aqueous slurry of collagen is frozen, creating an interlinked network of ice crystals with the collagen excluded to the edges of the crystals [2]. Then, the pressure is reduced, and the temperature is raised to induce sublimation, creating a porous scaffold with pores replacing the ice crystals. Therefore, the initial freezing process should be controlled carefully to form desirable ice crystal morphology and final scaffold architecture [3,4].

The solidification of ice is a two-stage process which involves ice nucleation and crystal growth. Nucleation proceeds with the initial formation of a crystal when the temperature is lower than the equilibrium freezing temperature (T_E). Below T_E, the system appears in a metastable state where many clusters of different radii may form, but without a long-range order [5,6]. Under the Classical Nucleation Theory, each nucleus encounters an energetic barrier to formation due to factors including loss of entropy, increased strain energy and increased interfacial energy at the solid–liquid interface [7,8]. Only crystals larger than the critical radius may form a nucleus and grow spontaneously. Primary nucleation can be either homogeneous or heterogeneous. Heterogenous nucleation takes place in the presence of solutes, impurities, grain boundaries or dislocations, reducing the energy barrier for nucleation [9]. Afterwards, secondary nucleation can proceed when new crystals are produced from pre-existing crystals [10].

The temperature and location at which nucleation occurs have significant impact on the architecture of the ice-templated scaffold. Scaffolds of 0.5 wt% collagen have significantly larger pore sizes than those produced from 1 wt% collagen [11]. It is believed that higher collagen content provides additional surface area and acts as a nucleation site for heterogeneous nucleation, forming many small ice crystals and replicating scaffolds with smaller pores [4,12,13]. The thermal gradient within the sample at nucleation influences the isotropy of scaffold architecture. If there is a large thermal gradient inside the slurry and regions of the slurry stay above the equilibrium freezing temperature at nucleation, an anisotropic structure can be produced [4,11]. In practical terms, uneven slurry cooling can be achieved by increasing the filling height of the slurry and hence distance from the cooling shelf or changing the mould material to have a significantly different thermal conductivity from the slurry [4,11].

After nucleation, the crystal growth proceeds, and the growth rate can be affected by the effectiveness of latent heat removal [14,15]. Pawelec et al. introduced a thermal parameter, "time at equilibrium", defined as the time that the slurry spends around the equilibrium freezing temperature [16]. The cut-off point for the equilibrium temperature was set as -1.5 °C when large-scale molecular movement during the ice–water transition process is completed. The pore size of an ice-templated collagen scaffold is closely correlated with the time at equilibrium [16]. There are a number of practical ways to affect the efficiency of latent heat release in a freezing slurry and therefore lengthen the time at equilibrium: a higher cooling shelf temperature is set, a lower collagen concentration is used and the mould contact area with the heat sink is reduced [4,11]. Moreover, with ineffective latent heat removal, crystallisation proceeds slowly, allowing annealing before the end of solidification [17]. In annealing, also known as Ostwald ripening, small crystals are dissolved and incorporated into larger ones, which leads to the formation of larger crystals.

Tuning the processing parameters during freeze-drying can tailor the scaffold architecture for different applications. However, scaffolds fabricated in the same batch or under identical external thermal conditions may still exhibit differing architectures as the result of the stochastic nature of ice nucleation [4,18–20]. Figure 1 summarises the key arguments explaining how the stochastic ice nucleation leads to large scaffold structural variations. Controlling the temperature of nucleation would allow greater control of scaffold reproducibility. The addition of nucleation agents can narrow the range of temperature at which nucleation occurs [19]. Also, multiple methods to stimulate nucleation have been explored, including seeding, mechanical methods (shaking, taping and agitation), and the applications of ultrasound, laser, electric and magnetic fields [21–25]. Ultrasound has emerged as particularly promising due to its low cost, ease of operation, and avoidance of ionising radiation, high temperature and direct contact with the sample.

Ultrasound is defined as an acoustic wave with a frequency greater than 20 kHz [26]. When ultrasound is applied to a liquid, zones of low and high pressure are generated, and microscopic voids are formed in the liquid in a process known as cavitation [26–28]. Bubbles that experience periodic growth and compression over many cycles are defined as stable (or repetitive) cavitation bubbles [29,30]. In contrast, bubbles which grow and collapse violently within a small number of cycles are known as unstable (transient) cavitation bubbles [31].

Ultrasound vibration has been shown to facilitate the phase transition from supercooled water to ice [32]. According to Hickling's theory, vigorous collapsing of cavitation bubbles can create local zones of high pressure exceeding 5 GPa over a period of nanoseconds resulting in high degrees of supercooling, which acts as the driving force for nucleation [27]. Furthermore, ultrasound can break down pre-existing ice dendrites into smaller fragments, initiating secondary nucleation [32,33]. Zhang et al. suggested the flow streams caused by the cavitation bubbles may promote nucleation [34]. Dodds et al. suggested that molecular aggregates form due to the pressure gradient around microbubbles, causing nucleation [35]. The exact mechanism is still under debate, and it is likely that multiple mechanisms may be involved in the process of ultrasound-induced nucleation.

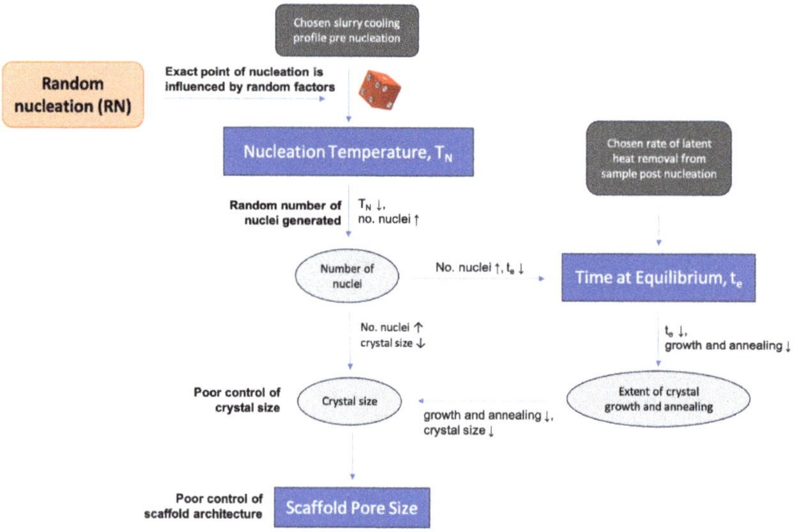

Figure 1. A flowchart showing how the random nucleation of ice in the freezing collagen slurry leads to poor control over scaffold architecture.

Ultrasound technologies have been applied in the fields of biomaterial fabrication and drug development, improving precision and reproducibility [30,36–38]. In 2022, Xiao et al. reported the use of an ultrasound-assisted freeze-drying technique to fabricate polyimide aerogels, where the ultrasound was applied to the gel during the entire freezing phase. Aerogel prepared using this new technique had smaller pores and a more consistent structure compared with that prepared using the traditional method. The pore structure can be further controlled by tuning ultrasonic power [39]. Kiani et al. explored the effect of ultrasound on ice crystal formation in frozen agar gels. Their finding suggests that the temperature to apply ultrasound to the supercooled slurry was the key to tailoring the ice crystal size, while no major difference was observed when tuning the ultrasound power, exposure time and duty cycle [40].

In this study, ultrasound was incorporated into the standard freeze-drying protocol to fabricate collagen scaffolds for tissue engineering applications. The hypothesis leading to this work was that using ultrasound to trigger nucleation at a particular temperature can improve the control over scaffold architecture and the reproducibility of the scaffold fabrication process. This work aimed to first explore the stochastic nature of ice nucleation, measuring the nucleation temperature of collagen slurries of 0.5, 1 and 1.5 wt%. Then, the ability of ultrasound to trigger ice nucleation in 1 wt% collagen slurry across a range of temperatures was explored. Finally, the impact of the ultrasound-induced nucleation method on the thermal profiles of the freezing slurry and resultant scaffold architecture was investigated. The knowledge gained from this study can serve as a fundamental guideline for fabricating ice-templated scaffolds with ultrasound to form desirable structures.

2. Materials and Methods

2.1. Collagen Slurry Preparation and Thermocouples Set-Up

Collagen suspensions of 0.5, 1 and 1.5 wt% were prepared by hydrating collagen powder, type I collagen derived from Dervo bovine dermal collagen (Collagen Solutions, Glasgow, UK), in 0.05 M acetic acid (Sigma-Aldrich, Dorset, UK). The suspension was left in a 4 °C fridge for at least 72 h for hydration. Then, the suspension was blended at 18,000 rpm for 2 min, left to rest for 1 min and blended again at 22,000 rpm for 1 min. Afterwards, the slurry was centrifuged at 2500 rpm for 5 min to remove air bubbles.

CytoOne 12-well cell culture plates were used as moulds. For each well, a type K thermocouple (Pico Technology, Saint Neots, UK) was fixed at 10 mm height of the well wall. Thermocouples were then plugged into a portable temperature data logger (Omega, Manchester, UK). Next, 4 mL of collagen slurry was pipetted into each well of the plate and left in the fridge until the temperature stabilised.

2.2. Random Nucleation of Ice

The stochastic nucleation temperatures of collagen slurries with different concentrations were measured. The cell culture plate containing collagen slurry was placed into a −20 °C spark-free laboratory-grade chest freezer (WolfLabs, York, UK), and the slurry was left to freeze naturally. The temperature profile was measured using a type K thermocouple and recorded by a temperature data logger. The nucleation of ice was identified by a rapid increase in temperature to 0 °C on the release of latent heat. Nucleation temperature was defined as the last temperature recorded before a characteristic jump in temperature reading, as shown in Figure 2.

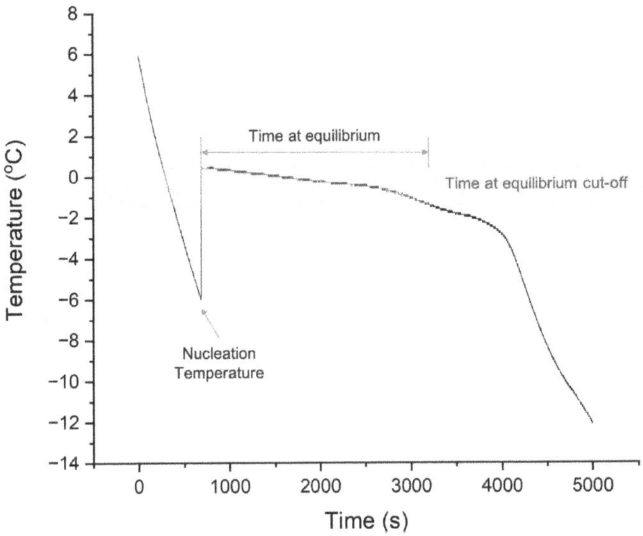

Figure 2. A typical thermal profile of slurry during freezing, with the nucleation temperature and time at equilibrium defined.

2.3. Scaffold Fabrication

In this study, 1 wt% collagen slurry was used for scaffold fabrication. As shown in Figure 3, collagen scaffolds were produced using a two-step freeze-drying process: freezing in a precooled commercial ultrasonic bath inside a −20 °C freezer, followed by drying in a VirTis AdVantage freeze dryer (Biopharma Process Systems, Winchester, UK). During the freezing process, the temperature profile of the slurry was recorded, identifying the nucleation temperature and time at equilibrium, as illustrated in Figure 2. The time at equilibrium, defined by Pawelec et al., measures the time the slurry spent above the equilibrium cut-off temperature (−1.5 °C) [16].

Two different ice nucleation methods were applied in the initial freezing step. (1) Ultrasound-induced nucleation: An ultrasonic cleaning bath (40 kHz frequency and 0.2 Wcm^{-2} intensity) was placed inside the −20 °C freezer. To reduce the attenuation effect of ultrasound, the bath was filled with 200 mL ethanol (50% v/v). Once the temperature inside the ultrasonic bath stabilised to around −16 °C, the cell culture plate with collagen slurry was transferred from the 4 °C fridge into the ultrasonic bath. The temperature at the top of the slurry was carefully monitored. Ultrasound was switched on for 3 s to induce ice nucleation at the chosen test temperature while minimising potential thermal effects. Samples were then left to freeze for more than 60 min. (2) Random nucleation: A similar experimental setup was used, but no ultrasound was applied, and the slurry was left to freeze naturally, with the temperature profile recorded.

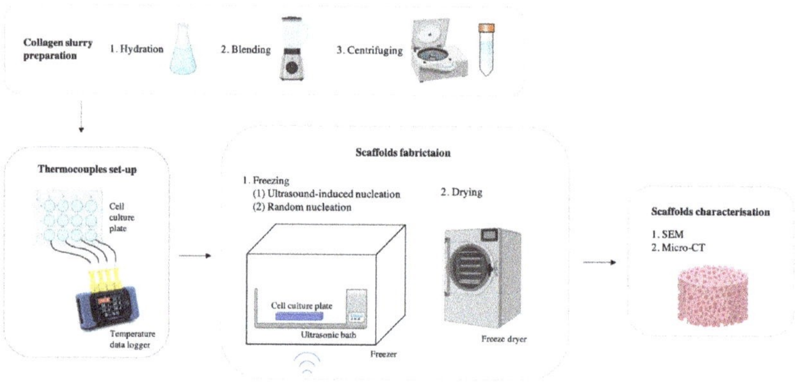

Figure 3. Schematic showing the experimental procedure.

The freeze dryer shelf was manually cooled to −20 °C, and the frozen collagen slurry was transferred to a freeze dryer for the drying step. The pressure was reduced to 80 mTorr over 30 min, the temperature increased to 0 °C over 30 min, and these conditions were held for 1200 min for ice sublimation. The scaffold fabrication process was repeated for both ultrasound-induced and random nucleation groups to obtain samples nucleating across a wide range of temperatures.

2.4. Scanning Electron Microscopy (SEM)

SEM was used to visualise the morphology of the scaffold in 2D. Horizontal and vertical cross-sections were cut from the centre of the scaffold using a scalpel. Then, samples were mounted on the sample holder using carbon tape and sputter-coated with a layer of platinum for 2 min at 40 mA. Scaffolds were imaged using the FEI Nova Nano SEM system (FEI Company, Hillsboro, OR, USA) under a secondary electron mode, with an aperture size of 30 μm, an operating voltage of 5 kV and a spot size of 3 nm.

2.5. Micro-Computed Tomography (Micro-CT)

Three-dimensional quantitative structural characterisation of scaffolds was carried out using micro-CT. Offeddu et al. suggested little difference in the pore architecture between dry and hydrated scaffolds, yet higher background noise was observed during the scanning of hydrated scaffolds [41]. Therefore, in this study, dry scaffolds were directly characterised without hydration. Cylindrical samples were cut from scaffolds using an 8 mm biopsy punch. Then, they were imaged under a micro-CT (Skyscan 1272, Bruker, Kontich, Belgium) with an operating voltage of 25 kV, a current of 93 μA and a voxel size of 3 μm^3 (5 μm^3 for scaffolds with large pores). The scan was performed under 180° rotation, with a rotation step size of 0.2 and an exposure time of 1830 ms. Three-dimensional image reconstruction was carried out using NRecon (Version 1.6.9.8, Bruker, Kontich, Belgium) with a threshold range from 0 to 0.06. The scaffold architecture was visualised in 3D using the program DataViewer (Version 1.5.0, Bruker, Kontich, Belgium). Afterwards, the dataset of each scaffold was analysed using CTAnalyser (Version 1.20.3.0, Bruker, Kontich, Belgium). Cubic volumes of interest (VOIs) of 1 mm^3 were selected from each sample. The images were thresholded under the Otsu method and de-speckled using the sweep function. Three-dimensional structural analysis was performed to measure the scaffold pore size and degree of anisotropy. Pore size was calculated using a sphere fitting algorithm, which is defined as the largest diameter of a sphere which can be bounded by a pore [42]. The degree of anisotropy is defined as the ratio between the maximum and the minimum radius of the mean intercept length (MIL) [43]. MIL distribution was calculated by overlaying parallel lines in different directions on the 3D reconstructed data.

The percolation diameter is a scale-invariant property describing the transport pathway through the scaffold [44]. It is defined as the size of the largest sphere able to travel through an infinitely large scaffold without obstructions. The shrink wrap feature in CTAn was used to identify the volume accessible to a virtual sphere of different sizes. From Equation (1), for objects with different diameters (d) invading the scaffold, their linear penetration depths in the z-direction (L) can be measured. Then, by extrapolating the line to an infinite sample size, the percolation diameter (d_c) was obtained. In this study, the bitmaps saved from shrink wrap operation were directly analysed using the "segmented percolation method" developed by Nair et al. which excludes the anomalous features within the dataset and calculates percolation diameter using a bespoke Python script [45].

$$L = L_0(d - d_c)^{-0.88}. \quad (1)$$

3. Results

3.1. Nucleation Temperature Distribution in Collagen Slurry

The random nucleation temperatures of 0.5, 1.0 and 1.5 wt% collagen slurries were measured and organised into histograms, as shown in Figure 4. The nucleation of ice in collagen slurries was found to occur across a broad range of temperatures.

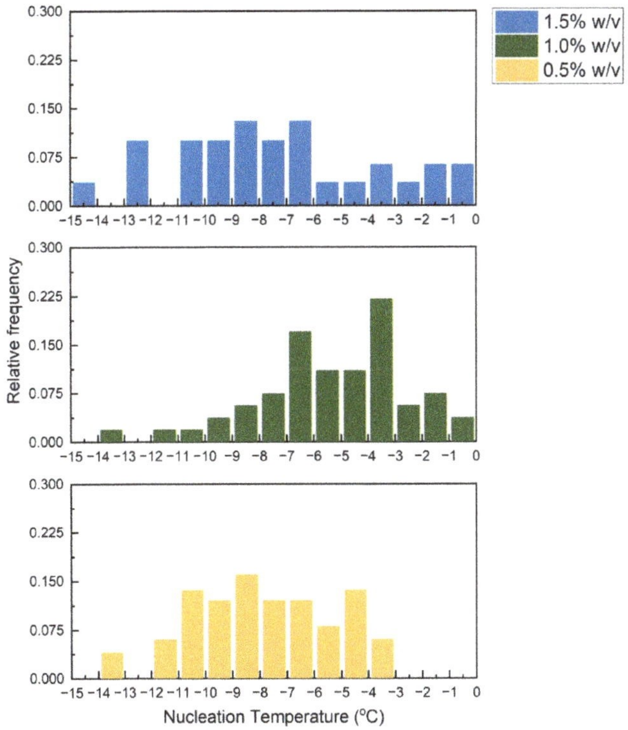

Figure 4. Histograms showing the distribution of nucleation temperatures of 1.5%, 1.0% and 0.5% *w/v* collagen slurry. Sample count = 39, 54 and 41, respectively. Average nucleation temperature = −7.4, −5.2 and −8.0, respectively. Range of nucleation temperature = 13.1, 13.7 and 10.1, respectively.

3.2. Thermal Profile of Freezing Collagen Slurry

For the following study, 1 wt% collagen slurry was chosen due to its wide applicability in scaffold fabrication. Ultrasound was applied to the supercooled collagen slurry to stimulate ice nucleation at different temperatures. Figure 5 shows the thermal profiles of four representative samples with ultrasound-triggered ice nucleation at −1.7, −3.8, −5.8 and −10.1 °C, respectively. The onset of nucleation was characterised by an instantaneous jump in temperature to around the equilibrium freezing temperature.

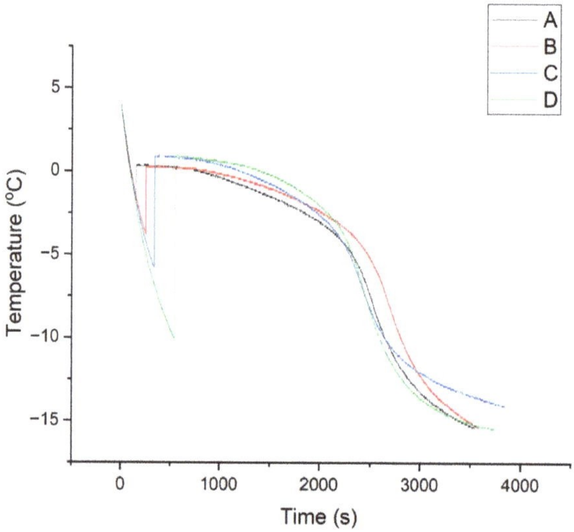

Figure 5. Thermal profiles of ultrasound-triggered ice nucleation and growth in 1 wt% collagen slurry. Ultrasound was applied to stimulate nucleation at (**A**) −1.7 °C, (**B**) −3.8 °C, (**C**) −5.8 °C and (**D**) −10.1 °C.

3.3. Scaffold Architecture Visualisation

Scaffolds templated by ice that nucleated at different temperatures were visualised using micro-CT and SEM in a dry state. As shown in Figure 6, with a more negative nucleation temperature, the scaffolds displayed smaller, more isotropic pores with a narrow distribution of sizes. If the ice nucleated at a temperature above about −4 °C, a coarse, anisotropic, plate-like structure was observed. Scaffolds with ultrasound-induced nucleation had smaller pores than those nucleated randomly. A smooth pore wall was observed in scaffolds produced under both nucleation methods.

3.4. Relating Scaffold Architecture to Thermal Parameters

Figure 7A shows the time at equilibrium plotted against nucleation temperature. For random nucleation (RN), the slurry tended to spend less time at equilibrium if it nucleated at a more negative nucleation temperature. In comparison, for ultrasound-induced nucleation (UIN), the time at equilibrium stayed broadly constant with different nucleation temperatures. For nucleation temperature above about −6 °C, collagen slurries with RN spent a significantly longer time at equilibrium than those with UIN, as illustrated in Figure 7B,C.

The average scaffold pore size was plotted against nucleation temperature and time at equilibrium for the two nucleation methods. As shown in Figure 8A, there was clearly a positive correlation between the average pore size of the scaffold and nucleation temperature. With UIN, the scaffolds generally had a smaller pore size across the entire nucleation temperature range. The difference in scaffold pore size between the two nucleation methods was found to be more significant with small undercooling. The average scaffold pore size was also plotted against time at equilibrium in Figure 8B. For the RN group, scaffolds generally had a larger pore size if the slurry spent longer time at equilibrium during freezing. For the UIN group, the data points centred around smaller time at equilibrium and pore size.

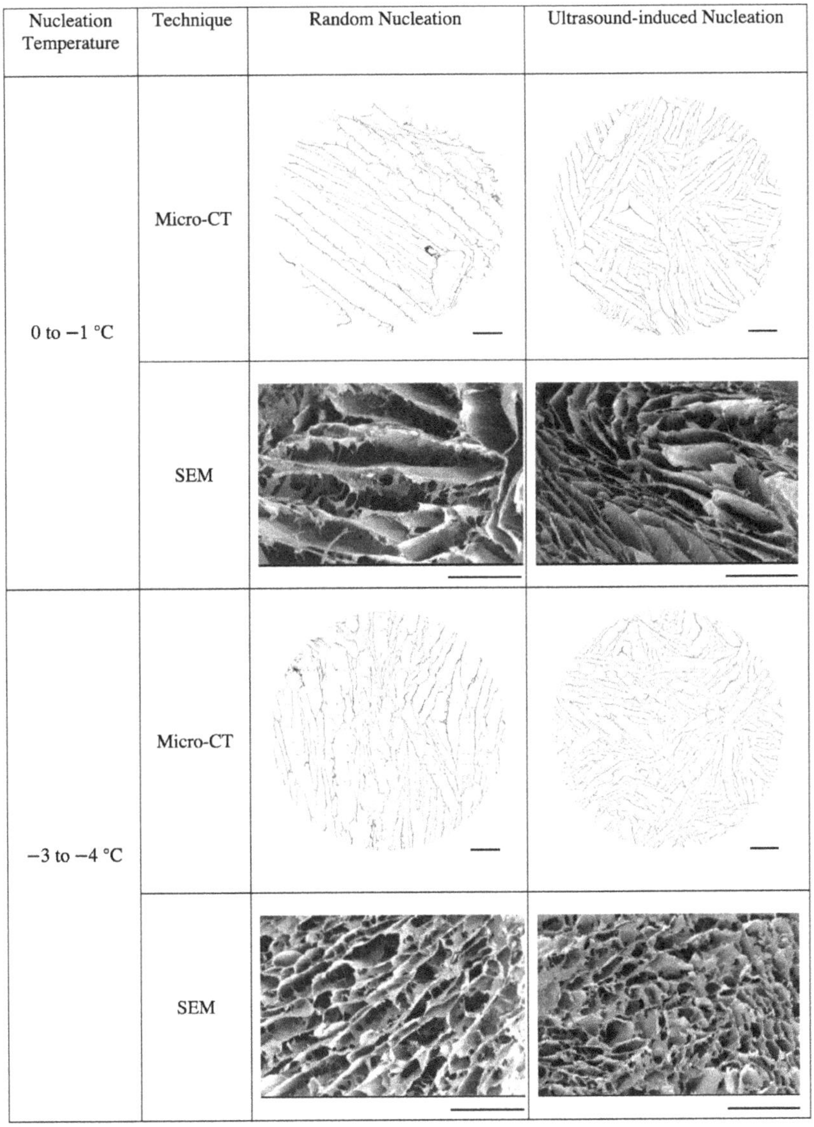

Figure 6. *Cont.*

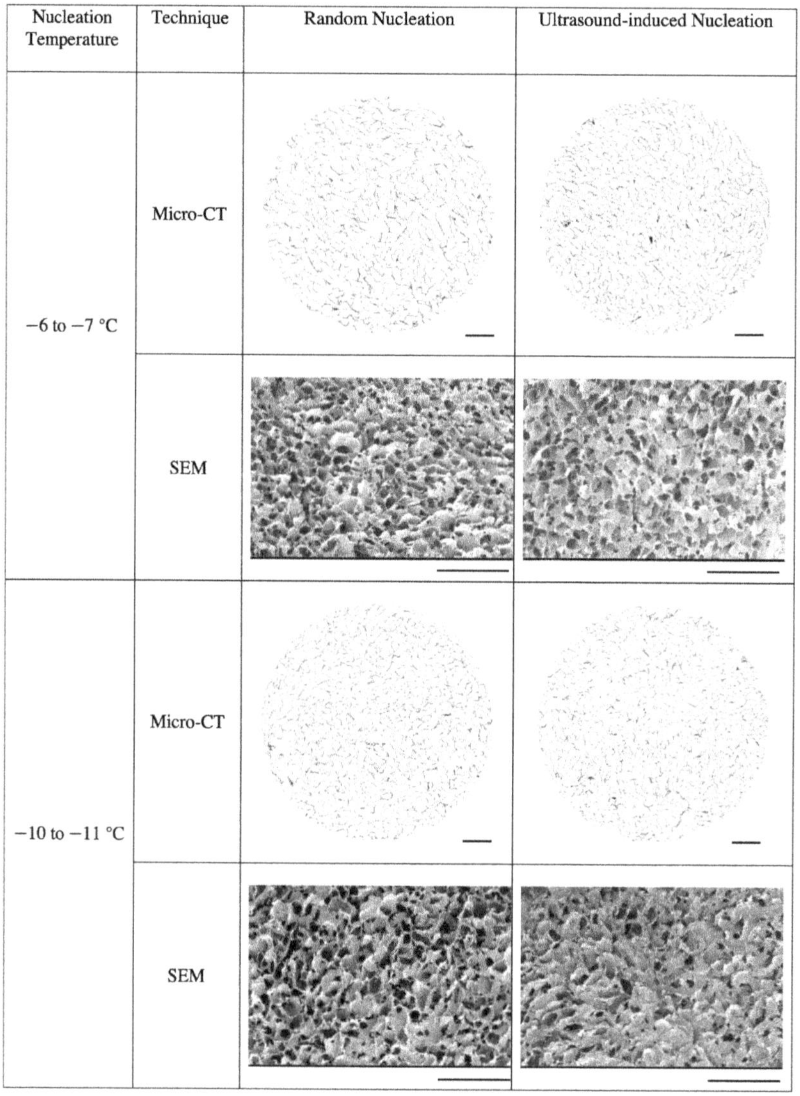

Figure 6. Reconstructed micro-computed tomography scan and scanning electron microscopy images of collagen scaffolds produced under random and ultrasound-induced nucleation at 0 to −1 °C, −3 to −4 °C, −6 to −7 °C and −10 to −11 °C. Images were taken at the horizontal cross-sections at the centre of the scaffold. Scale bar = 1 mm.

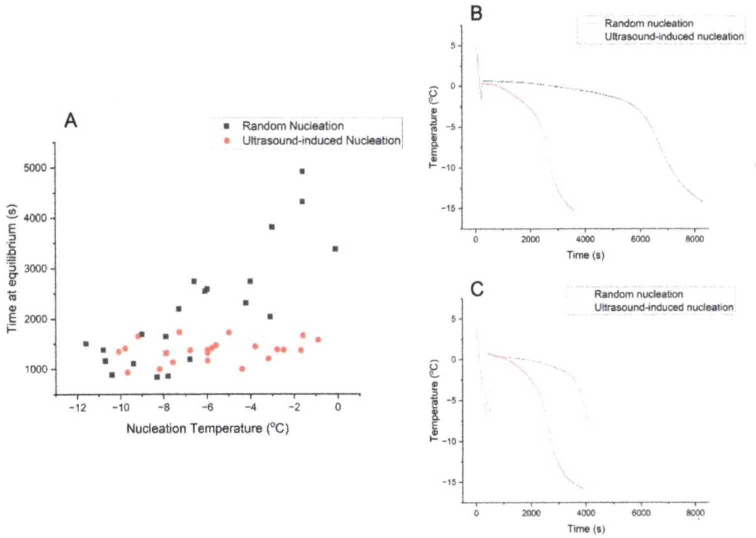

Figure 7. (**A**) Time at equilibrium against nucleation temperature. (**B**) Thermal profile of collagen slurries with ice nucleating at around −2 °C. (**C**) Thermal profile of collagen slurries with ice nucleating at around −6 °C.

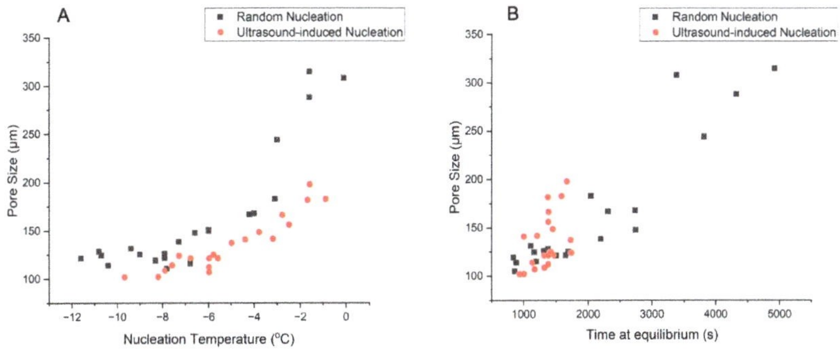

Figure 8. (**A**) Average scaffold pore size against nucleation temperature. (**B**) Average scaffold pore size against time at equilibrium.

The pore size distribution of collagen scaffolds produced with ice nucleating at different temperatures was plotted in Figure 9. For samples with a lower ice nucleation temperature, the width of the pore size distribution was smaller. With UIN, the scaffolds had a narrower distribution of pore sizes, implying the formation of a more uniform, consistent pore architecture.

Figure 10 depicts the average scaffold degree of anisotropy with respect to ice nucleation temperature. Scaffolds usually had a smaller degree of anisotropy if ice nucleated at a lower temperature, implying the formation of a more isotropic architecture. No significant difference was observed between RN and UIN groups.

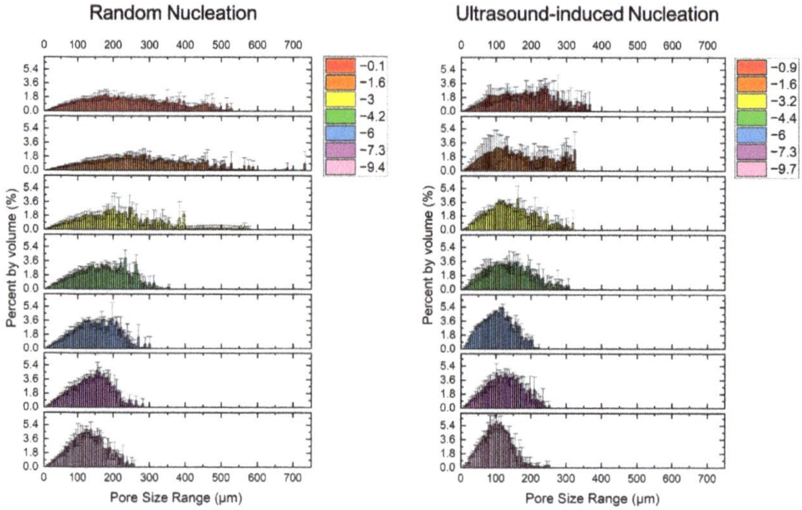

Figure 9. Pore size distribution of scaffolds underwent random and ultrasound-induced nucleation at different nucleation temperatures (°C). Columns and error bars represent the average percentage of pores with a pore size in any given 6 μm interval and standard deviation measured from three volumes of interest (VOIs) along the height of the scaffold.

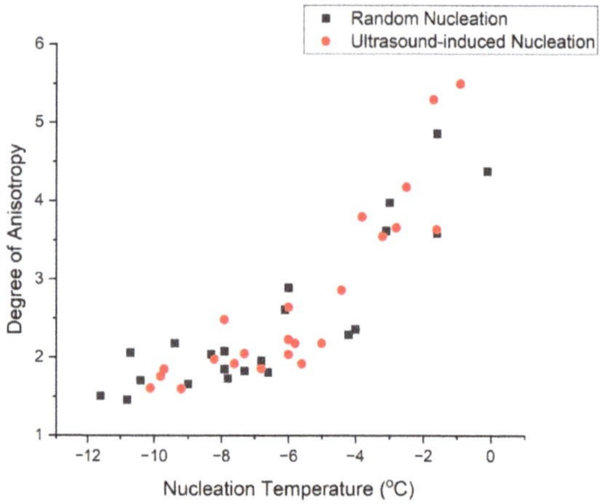

Figure 10. Scaffold degree of anisotropy against nucleation temperature.

The results of scaffold percolation diameter versus nucleation temperature are shown in Figure 11. The general trend suggests that at a lower nucleation temperature, scaffolds tended to have smaller percolation diameters. However, there was a relatively large amount of scatter in the data and no significant difference was observed between RN and UIN groups.

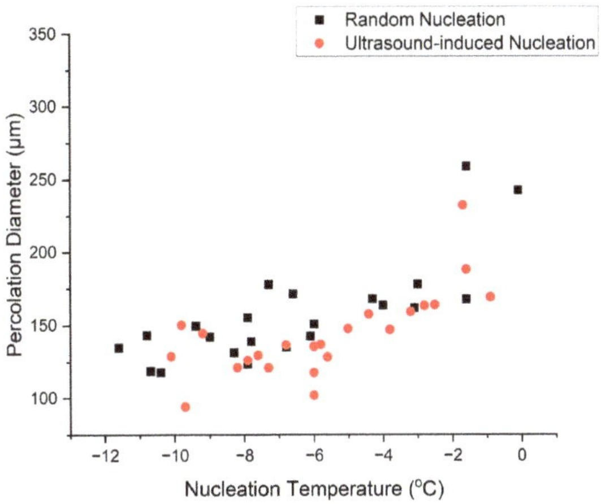

Figure 11. Scaffold percolation diameter against nucleation temperature.

4. Discussion

4.1. Stochastic Nature of Ice Nucleation

Figure 4 reveals that the nucleation temperatures in collagen slurries of different concentrations were spread over a wide range of values, which confirmed the large variability in ice nucleation temperature under identical conditions. The random nucleation event then led to large variations in scaffold architecture, as depicted in Figure 6. The distribution of nucleation temperatures in 1 wt% collagen slurry was centred around a higher temperature than samples based on 0.5 wt% slurry. The higher concentration of solids provided more nucleation sites, reducing the amount of undercooling required for nucleation. However, in 1.5 wt% collagen slurries, the increased viscosity was likely to have suppressed the formation of nuclei [46]. This effect might have outweighed the effects of the presence of increased nucleation sites, shifting the nucleation events to higher degrees of undercooling. This finding highlights the need to control the ice nucleation event for reproducibility in scaffold fabrication.

4.2. Ultrasound-Triggered Nucleation

The application of ultrasound to stimulate ice nucleation in a supercooled 1% collagen slurry is illustrated in Figure 5, resulting in an instantaneous increase in slurry temperature to the equilibrium freezing temperature. Therefore, the effectiveness of ultrasound-induced nucleation in collagen slurry was confirmed. This could be explained by Hickling's theory, as depicted in Figure 12: the application of ultrasound leads to the formation of microbubbles [27]. Then, the fluctuating ultrasound wave causes the collapse of cavitating bubbles. This generates local zones of high pressure which raises the equilibrium freezing temperature of the water, providing the driving force of ice nucleation. Also, the flow streams formed due to the movement of stable cavitation bubbles can promote nucleation [34]. This finding was later confirmed by Chow et al. with microscopy studies [33]. In addition, Dodds et al. believed that the huge pressure gradient generated by an oscillating bubble helps the diffusion of particles, forming large molecular aggregates. Thus, the cavitation bubbles acted as a cluster attachment reactor, facilitating the formation of a nucleus [35,47]. It is likely that multiple mechanisms play a role in the process of ultrasound-induced nucleation.

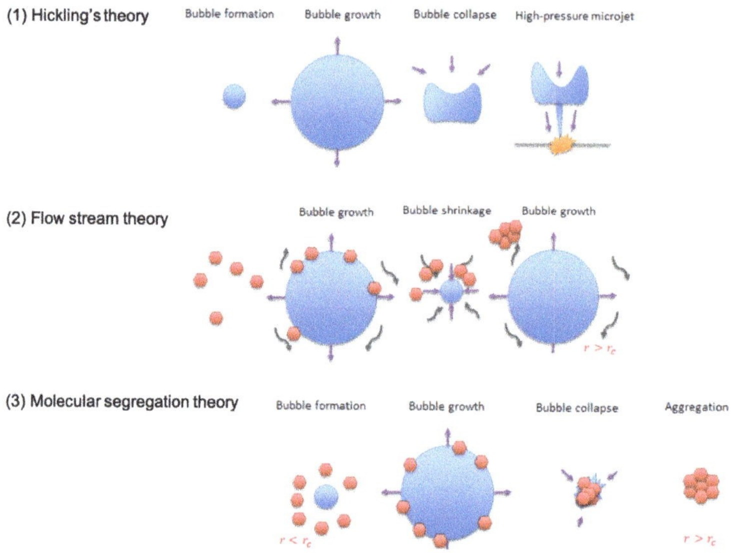

Figure 12. Schematic representation of gas bubble behaviour when ultrasound was applied by three different theories. (1) Hickling's theory: graph adapted from "Nucleation of freezing by cavity collapse and its relation to cavitation damage" [27]. (2) Flow stream theory: graph adapted from "Ultrasonic-induced nucleation of ice in water containing air bubbles" [34]. (3) Molecular segregation theory: graph adapted from "The effect of ultrasound on crystallisation-precipitation processes: some examples and a new segregation model" [35]. Red hexagons represent molecules, and blue circles represent bubbles. r = radius, r_c = critical radius for nucleation.

4.3. *Ultrasound-Induced Nucleation and Random Nucleation: Correlating Thermal Parameters to Scaffold Architecture*

4.3.1. Pore Morphology

For nucleation temperatures ranging from 0 to −4 °C, large anisotropic plate-like structures were formed. At lower nucleation temperatures, scaffolds tended to have smaller and equiaxed pores. There was no significant difference in pore morphology between the two nucleation methods. These findings were consistent with those of Nakagawa et al. reporting that in 10% frozen mannitol solutions, large, dendritic ice crystals were found in samples nucleated at higher temperatures, whereas smaller and more uniform ice crystals were presented in samples nucleated at lower temperatures. They believed that the degree of undercooling affects the number of nuclei formed in the slurry, determining final crystal morphology [20].

4.3.2. Time at Equilibrium against Nucleation Temperature

Time at equilibrium and nucleation temperature are two key thermal parameters during the slurry freezing process. The nucleation temperature measures the amount of undercooling before crystal nucleation. Time at equilibrium reflects how long the slurry spends in the crystal growth phase [16]. As shown in Figure 7, for the random nucleation group, the slurry which nucleated at a lower nucleation temperature tended to have a shorter time at equilibrium. With a larger undercooling, there is a larger driving force for nucleation. Therefore, as more nuclei are present, the crystal growth phase is shortened due to the limited space available [16]. However, with ultrasound, the time at equilibrium remained largely constant across a wide range of nucleation temperatures. With a small undercooling, the ultrasound effectively promotes the formation of nuclei, reducing the time required for subsequent crystal growth. With a larger undercooling, the number of nuclei did not increase further with ultrasound. As a result, the time at equilibrium did not reduce further upon the application of ultrasound.

4.3.3. Pore Size against Nucleation Temperature and Time at Equilibrium

As shown in Figure 8A, scaffolds had a smaller average pore size if ice nucleated at a lower temperature. At a greater undercooling, there is a larger driving force for nucleation and a smaller driving force for growth [48]. Therefore, the ice crystals are smaller, replicating a scaffold with smaller pores. Compared with random nucleation, slurries that underwent ultrasound-induced nucleation generally exhibited smaller average pore sizes. The difference between the two nucleation methods was more obvious with a smaller undercooling. The application of ultrasound could introduce more ice nuclei simultaneously in the supercooled sample, which limited the space for crystal growth [26,49]. Therefore, many small ice crystals formed, leading to scaffolds with smaller pore sizes.

As shown in Figure 8B, for random nucleation, the scaffolds had smaller pore sizes if the slurry spent a shorter time at equilibrium. This was in good agreement with the results of Pawelec et al., who suggested a universal relationship between the scaffold pore size and time at equilibrium [16]. With ultrasound, more nucleation sites were formed, shortening the time required for subsequent crystal growth [50]. Therefore, a shorter time at equilibrium and smaller average pore sizes were observed for samples with ultrasound-induced nucleation.

4.3.4. Pore Size Distribution

Scaffolds produced by freeze-drying naturally contain a distribution of pores with different sizes. In Figure 9, it can be seen that as nucleation temperature reduces, the scaffold tends to have a smaller distribution or pore sizes, suggesting a more homogenous structure. Scaffolds formed with random and ultrasound-induced nucleation of ice at a similar nucleation temperature range were compared. The pore size distribution was centred around a smaller value and was narrower for samples formed with ultrasound-induced nucleation. Ultrasound simultaneously induces ice nucleation across the entire supercooled slurry, producing ice crystals of similar sizes. This proves a finer control of the pore size and greater reproducibility of scaffolds.

4.3.5. Degree of Anisotropy against Nucleation Temperature

As shown in Figure 10, the average degree of anisotropy of the collagen scaffold decreases if the ice nucleates at a more negative nucleation temperature. This implies the formation of a more isotropic scaffold arrangement at a lower nucleation temperature. A similar trend was observed in ultrasound-induced nucleation groups. This indicates that while ultrasound can affect the nucleation and hence the pore size; there is little impact on pore anisotropy. This is consistent with Nakagawa's findings [20].

4.3.6. Percolation Diameter against Nucleation Temperature

Percolation diameter measures the size of the largest sphere penetrating the scaffold without obstruction, a valid predictor of the extent of cell transport through the scaffold. With lower nucleation temperatures, the scaffold tended to have smaller percolation diameters, as illustrated in Figure 11. The percolation diameter depends on the size, shape and number of pathways of connected pores. Scaffolds with smaller pores might tend to have narrower pathways of connected pores, impeding the movement of large objects. The application of ultrasound does not appear to impact the scaffold percolation diameter significantly. Ultrasound-assisted nucleation might allow a degree of independent control of scaffold pore size and interconnectivity.

4.4. Impact of Ultrasound-Trigged Nucleation on Ice Solidification and Final Scaffold Structure

In freeze drying, the ice crystal structure is a negative template of the resulting scaffold architecture. As summarised in Figure 1, the formation of ice crystals proceeds in two stages:

nucleation and growth. Nucleation is stochastic and can happen across a wide range of temperatures under the same external condition. After nucleation, the crystal growth and annealing proceed, which can be quantified by the time at equilibrium [16]. Overall, the ice crystal size is influenced by both nucleation temperature and time at equilibrium. These two thermal parameters are related: with a larger driving force (lower nucleation temperature), more nuclei are formed, limiting crystal growth (shorter time at equilibrium).

In this study, we demonstrate the ability of ultrasound to trigger ice nucleation instantaneously at certain temperatures due to the cavitation effect. As shown in Figure 13, with ultrasound, more nuclei are formed, and the crystal growth stage is shortened, forming smaller and homogeneous ice crystals. Therefore, ultrasound-induced nucleation can produce scaffolds with smaller average pore size and narrower pore size distribution. In summary, the incorporation of ultrasound into the conventional freeze-drying protocol can improve the control and consistency of scaffold architecture. We can apply ultrasound to induce nucleation at different temperatures, tuning the scaffold architecture for specific tissue engineering applications.

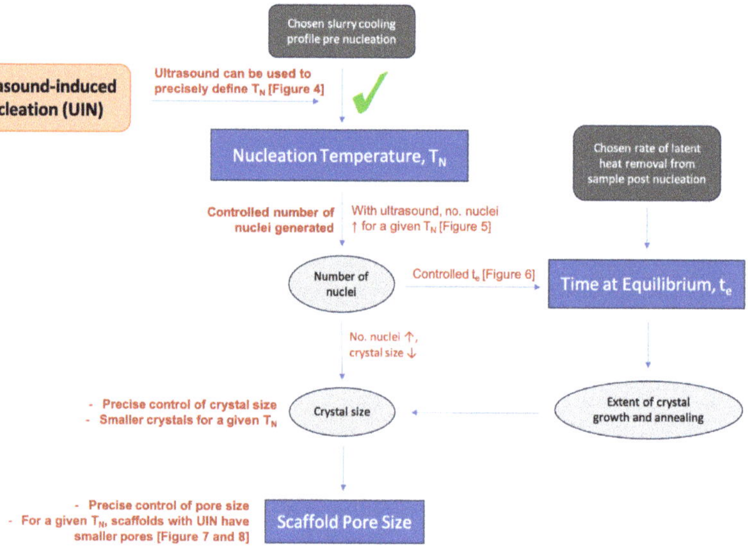

Figure 13. A flowchart illustrating how application of ultrasound affects the solidification process, final ice crystal size and scaffold pore structure.

5. Conclusions

In standard freeze-drying protocols to fabricate 3D porous collagen scaffolds, ice nucleation occurs across a wide range of temperatures. The large variability in nucleation temperature of ice can lead to poor reproducibility. At a small undercooling, large, anisotropic plate-like structures are formed, while with larger undercooling, smaller and homogenous pores are formed. The stochastic nature of nucleation means that scaffolds show large structural variations even under the same fabrication protocol. Nowadays, ultrasound may be used as a method to trigger nucleation at a particular temperature. Ultrasound-induced nucleation is characterised by a sharp increase in temperature to the equilibrium temperature. Scaffolds with ultrasound-induced nucleation have a smaller average pore size and a narrower distribution of pore sizes compared with those formed with stochastic nucleation. However, the overall scaffold anisotropy and interconnectivity remained unchanged. The incorporation of ultrasound-trigged nucleation into the standard freeze-drying protocol therefore provides a route for finer control over scaffold architecture.

Author Contributions: Conceptualization, X.S., M.A.P., S.M.B. and R.E.C.; Data curation, X.S. and M.A.P.; Methodology, X.S. and M.A.P.; Supervision, S.M.B. and R.E.C.; Writing—original draft, X.S.; Writing—review and editing, S.M.B. and R.E.C. All authors have read and agreed to the published version of the manuscript.

Funding: X.S. was funded by Benefactors' scholarship, St John's College, Cambridge. R.E.C. and S.M.B. were funded by EPSRC fellowship (EP/N019938/1). M.A.P. was supported by a studentship funded by the EPSRC.

Institutional Review Board Statement: Not applicable.

Data Availability Statement: Our research data are openly available in Apollo at https://doi.org/10.17863/CAM.99991 (accessed on 29 November 2023), reference number 99991.

Acknowledgments: Graphical abstract and Figure 3 were created with BioRender.com (accessed on 8 August 2023).

Conflicts of Interest: The authors declare no conflict of interest.

References

1. Chan, B.P.; Leong, K.W. Scaffolding in tissue engineering: General approaches and tissue-specific considerations. *Eur. Spine J.* **2008**, *17*, 467–479. [CrossRef] [PubMed]
2. Wegst, U.G.K.; Schecter, M.; Donius, A.E.; Hunger, P.M. Biomaterials by freeze casting. *Philos. Trans. R. Soc. A Math. Phys. Eng. Sci.* **2010**, *368*, 2099–2121. [CrossRef] [PubMed]
3. Schoof, H.; Apel, J.; Heschel, I.; Rau, G. Control of pore structure and size in freeze-dried collagen sponges. *J. Biomed. Mater. Res.* **2001**, *58*, 352–357. [CrossRef] [PubMed]
4. Pawelec, K.M.; Husmann, A.; Best, S.M.; Cameron, R.E. Ice-templated structures for biomedical tissue repair: From physics to final scaffolds. *Appl. Phys. Rev.* **2014**, *1*, 021301. [CrossRef]
5. Elliott, R. Solidification of cast irons. In *Cast Iron Technology*; Elsevier: Amsterdam, The Netherlands, 1988; pp. 91–125. [CrossRef]
6. Matsumoto, M.; Saito, S.; Ohmine, I. Molecular dynamics simulation of the ice nucleation and growth process leading to water freezing. *Nature* **2002**, *416*, 409–413. [CrossRef] [PubMed]
7. Gautam, A.; Komal, P.; Gautam, P.; Sharma, A.; Kumar, N.; Jung, J.P. Recent Trends in Noble Metal Nanoparticles for Colorimetric Chemical Sensing and Micro-Electronic Packaging Applications. *Metals* **2021**, *11*, 329. [CrossRef]
8. Akyurt, M.; Zaki, G.; Habeebullah, B. Freezing phenomena in ice–water systems. *Energy Convers. Manag.* **2002**, *43*, 1773–1789. [CrossRef]
9. Heneghan, A.F.; Wilson, P.W.; Haymet, A.D.J. Heterogeneous nucleation of supercooled water, and the effect of an added catalyst. *Proc. Natl. Acad. Sci. USA* **2002**, *99*, 9631–9634. [CrossRef]
10. Salas, G.; Costo, R.; del Puerto Morales, M. Synthesis of Inorganic Nanoparticles. In *Frontiers of Nanoscience*; Elsevier: Amsterdam, The Netherlands, 2012; pp. 35–79. [CrossRef]
11. Pawelec, K.; Husmann, A.; Best, S.; Cameron, R. Understanding anisotropy and architecture in ice-templated biopolymer scaffolds. *Mater. Sci. Eng. C* **2014**, *37*, 141–147. [CrossRef]
12. Deville, S.; Maire, E.; Lasalle, A.; Bogner, A.; Gauthier, C.; Leloup, J.; Guizard, C. Influence of Particle Size on Ice Nucleation and Growth during the Ice-Templating Process: Rapid Communications of the American Ceramic Society. *J. Am. Ceram. Soc.* **2010**, *93*, 2507–2510. [CrossRef]
13. Lasalle, A.; Guizard, C.; Leloup, J.; Deville, S.; Maire, E.; Bogner, A.; Gauthier, C.; Adrien, J.; Courtois, L. Ice-Templating of Alumina Suspensions: Effect of Supercooling and Crystal Growth During the Initial Freezing Regime. *J. Am. Ceram. Soc.* **2011**, *95*, 799–804. [CrossRef]
14. Hindmarsh, J.; Russell, A.; Chen, X. Measuring dendritic growth in undercooled sucrose solution droplets. *J. Cryst. Growth* **2005**, *285*, 236–248. [CrossRef]
15. Hallett, J. Experimental Studies of the Crystallization of Supercooled Water. *J. Atmos. Sci.* **1964**, *21*, 671–682. [CrossRef]
16. Pawelec, K.M.; Husmann, A.; Best, S.M.; Cameron, R.E. A design protocol for tailoring ice-templated scaffold structure. *J. R. Soc. Interface* **2014**, *11*, 20130958. [CrossRef]
17. Searles, J.A.; Carpenter, J.F.; Randolph, T.W. The ice nucleation temperature determines the primary drying rate of lyophilization for samples frozen on a temperature-controlled shelf. *J. Pharm. Sci.* **2001**, *90*, 860–871. [CrossRef] [PubMed]
18. Kiani, H.; Zhang, Z.; Delgado, A.; Sun, D.-W. Ultrasound assisted nucleation of some liquid and solid model foods during freezing. *Food Res. Int.* **2011**, *44*, 2915–2921. [CrossRef]
19. Wilson, P.; Heneghan, A.; Haymet, A. Ice nucleation in nature: Supercooling point (SCP) measurements and the role of heterogeneous nucleation. *Cryobiology* **2003**, *46*, 88–98. [CrossRef]
20. Nakagawa, K.; Hottot, A.; Vessot, S.; Andrieu, J. Influence of controlled nucleation by ultrasounds on ice morphology of frozen formulations for pharmaceutical proteins freeze-drying. *Chem. Eng. Process.-Process Intensif.* **2006**, *45*, 783–791. [CrossRef]

21. Candoni, N.; Hammadi, Z.; Grossier, R.; Ildefonso, M.; Zhang, S.; Morin, R.; Veesler, S. Addressing the Stochasticity of Nucleation: Practical Approaches. In *Advances in Organic Crystal Chemistry*; Tamura, R., Miyata, M., Eds.; Springer: Tokyo, Japan, 2015; pp. 95–113. [CrossRef]
22. Barber, E.R.; Ward, M.R.; Ward, A.D.; Alexander, A.J. Laser-induced nucleation promotes crystal growth of anhydrous sodium bromide. *CrystEngComm* **2021**, *23*, 8451–8461. [CrossRef]
23. Allahyarov, E.; Sandomirski, K.; Egelhaaf, S.; Löwen, H. Crystallization seeds favour crystallization only during initial growth. *Nat. Commun.* **2015**, *6*, 7110. [CrossRef]
24. Saclier, M.; Peczalski, R.; Andrieu, J. Effect of ultrasonically induced nucleation on ice crystals' size and shape during freezing in vials. *Chem. Eng. Sci.* **2010**, *65*, 3064–3071. [CrossRef]
25. Dalvi-Isfahan, M.; Hamdami, N.; Xanthakis, E.; Le-Bail, A. Review on the control of ice nucleation by ultrasound waves, electric and magnetic fields. *J. Food Eng.* **2017**, *195*, 222–234. [CrossRef]
26. Leong, T.; Ashokkumar, M.; Kentish, S. The fundamentals of power ultrasound—A review. *Acoust. Aust.* **2011**, *39*, 54–63.
27. Hickling, R. Nucleation of Freezing by Cavity Collapse and its Relation to Cavitation Damage. *Nature* **1965**, *206*, 915–917. [CrossRef]
28. Feng, H.; Barbosa-Canovas, G.; Weiss, J. (Eds.) Ultrasound Technologies for Food and Bioprocessing. In *Food Engineering Series*; Springer: New York, NY, USA, 2011. [CrossRef]
29. Morey, M.D.; Deshpande, N.S.; Barigou, M. Foam Destabilization by Mechanical and Ultrasonic Vibrations. *J. Colloid Interface Sci.* **1999**, *219*, 90–98. [CrossRef]
30. Ahmed, S.E.; Martins, A.M.; Husseini, G.A. The use of ultrasound to release chemotherapeutic drugs from micelles and liposomes. *J. Drug Target.* **2014**, *23*, 16–42. [CrossRef]
31. Yasui, K. Influence of ultrasonic frequency on multibubble sonoluminescence. *J. Acoust. Soc. Am.* **2002**, *112*, 1405–1413. [CrossRef]
32. Inada, T.; Zhang, X.; Yabe, A.; Kozawa, Y. Active control of phase change from supercooled water to ice by ultrasonic vibration 1. Control of freezing temperature. *Int. J. Heat Mass Transf.* **2001**, *44*, 4523–4531. [CrossRef]
33. Chow, R.; Blindt, R.; Chivers, R.; Povey, M. The sonocrystallisation of ice in sucrose solutions: Primary and secondary nucleation. *Ultrasonics* **2003**, *41*, 595–604. [CrossRef]
34. Zhang, X.; Inada, T.; Tezuka, A. Ultrasonic-induced nucleation of ice in water containing air bubbles. *Ultrason. Sonochem.* **2002**, *10*, 71–76. [CrossRef]
35. Dodds, J.; Espitalier, F.; Louisnard, O.; Grossier, R.; David, R.; Hassoun, M.; Baillon, F.; Gatumel, C.; Lyczko, N. The Effect of Ultrasound on Crystallisation-Precipitation Processes: Some Examples and a New Segregation Model. *Part. Part. Syst. Charact.* **2007**, *24*, 18–28. [CrossRef]
36. Dalecki, D.; Hocking, D.C. Ultrasound Technologies for Biomaterials Fabrication and Imaging. *Ann. Biomed. Eng.* **2014**, *43*, 747–761. [CrossRef] [PubMed]
37. Watson, N.; Johal, R.; Glover, Z.; Reinwald, Y.; White, L.; Ghaemmaghami, A.; Morgan, S.; Rose, F.; Povey, M.; Parker, N. Post-processing of polymer foam tissue scaffolds with high power ultrasound: A route to increased pore interconnectivity, pore size and fluid transport. *Mater. Sci. Eng. C* **2013**, *33*, 4825–4832. [CrossRef] [PubMed]
38. Sirsi, S.R.; Borden, M.A. State-of-the-art materials for ultrasound-triggered drug delivery. *Adv. Drug Deliv. Rev.* **2014**, *72*, 3–14. [CrossRef] [PubMed]
39. Xiao, H.; Lv, J.-B.; Tan, W.; He, X.; Chen, M.-H.; Zeng, K.; Hu, J.-H.; Yang, G. Ultrasound-assisted freeze-drying process for polyimide aerogels. *Chem. Eng. J.* **2022**, *450*, 138344. [CrossRef]
40. Kiani, H.; Zhang, Z.; Sun, D.-W. Effect of ultrasound irradiation on ice crystal size distribution in frozen agar gel samples. *Innov. Food Sci. Emerg. Technol.* **2013**, *18*, 126–131. [CrossRef]
41. Offeddu, G.; Ashworth, J.; Cameron, R.; Oyen, M. Structural determinants of hydration, mechanics and fluid flow in freeze-dried collagen scaffolds. *Acta Biomater.* **2016**, *41*, 193–203. [CrossRef]
42. Meek, M.C.; Best, S.; Cameron, R. The effects of despeckling filters on pore size measurements in collagen scaffold micro-CT data. *J. Microsc.* **2021**, *284*, 142–156. [CrossRef]
43. Jiang, Y.; Zhao, J.; White, D.L.; Genant, H.K. Micro CT and Micro MR imaging of 3D architecture of animal skeleton. *J. Musculoskelet. Neuronal Interact.* **2000**, *1*, 45–51.
44. Ashworth, J.C.; Mehr, M.; Buxton, P.G.; Best, S.M.; Cameron, R.E. Cell Invasion in Collagen Scaffold Architectures Characterized by Percolation Theory. *Adv. Healthc. Mater.* **2015**, *4*, 1317–1321. [CrossRef]
45. Nair, M.; Shepherd, J.H.; Best, S.M.; Cameron, R.E. MicroCT analysis of connectivity in porous structures: Optimizing data acquisition and analytical methods in the context of tissue engineering. *J. R. Soc. Interface* **2020**, *17*, 20190833. [CrossRef] [PubMed]
46. Nishiwaki, M.; Toramaru, A. Inclusion of Viscosity Into Classical Homogeneous Nucleation Theory for Water Bubbles in Silicate Melts: Reexamination of Bubble Number Density in Ascending Magmas. *J. Geophys. Res. Solid Earth* **2019**, *124*, 8250–8266. [CrossRef]
47. Grossier, R.; Louisnard, O.; Vargas, Y. Mixture segregation by an inertial cavitation bubble. *Ultrason. Sonochem.* **2007**, *14*, 431–437. [CrossRef]
48. Shu, Q.; Li, Q.; Medeiros, S.L.S.; Klug, J.L. Development of Non-reactive F-Free Mold Fluxes for High Aluminum Steels: Non-isothermal Crystallization Kinetics for Devitrification. *Met. Mater. Trans. B* **2020**, *51*, 1169–1180. [CrossRef]

49. Fang, L.; Gao, Z.; Wu, S.; Jia, S.; Wang, J.; Rohani, S.; Gong, J. Ultrasound-assisted solution crystallization of fotagliptin benzoate: Process intensification and crystal product optimization. *Ultrason. Sonochem.* **2021**, *76*, 105634. [CrossRef]
50. De Castro, M.L.D.; Priego-Capote, F. Ultrasound-assisted crystallization (sonocrystallization). *Ultrason. Sonochem.* **2007**, *14*, 717–724. [CrossRef]

polymers

MDPI

Review

Recent Advances in 3D Printing of Polymers for Application in Prosthodontics

Mariya Dimitrova [1,*], Angelina Vlahova [1,2], Yavor Kalachev [1], Stefan Zlatev [1,2], Rada Kazakova [1,2] and Saverio Capodiferro [3]

[1] Department of Prosthetic Dentistry, Faculty of Dental Medicine, Medical University of Plovdiv, 4000 Plovdiv, Bulgaria; angelina.vlahova@mu-plovdiv.bg (A.V.); yavor.kalachev@mu-plovdiv.bg (Y.K.); stefan.zlatev@mu-plovdiv.bg (S.Z.); rada.kazakova@mu-plovdiv.bg (R.K.)

[2] CAD/CAM Center of Dental Medicine, Research Institute, Medical University of Plovdiv, 4000 Plovdiv, Bulgaria

[3] Department of Interdisciplinary Medicine, Aldo Moro, University of Bari, 70100 Bari, Italy; saverio.capodiferro@uniba.it

* Correspondence: maria.dimitrova@mu-plovdiv.bg

Abstract: Contemporary mass media frequently depict 3D printing as a technology with widespread utilization in the creation of dental prosthetics. This paper endeavors to provide an evidence-based assessment of the current scope of 3D printing's integration within dental laboratories and practices. Its primary objective is to offer a systematic evaluation of the existing applications of 3D-printing technology within the realm of dental prosthetic restorations. Furthermore, this article delves into potential prospects, while also critically examining the sustained relevance of conventional dental laboratory services and manufacturing procedures. The central focus of this article is to expound upon the extent to which 3D printing is presently harnessed for crafting dental prosthetic appliances. By presenting verifiable data and factual insights, this article aspires to elucidate the actual implementation of 3D printing in prosthetic dentistry and its seamless integration into dental practices. The aim of this narrative review is twofold: firstly, to provide an informed and unbiased evaluation of the role that 3D printing currently plays within dental laboratories and practices; and secondly, to instigate contemplation on the transformative potential of this technology, both in terms of its contemporary impact and its future implications, while maintaining a balanced consideration of traditional dental approaches.

Keywords: 3D printing; prosthodontics; additive manufacturing; polymers; digital; prosthetic restorations

1. Introduction

The advent of industrial-scale additive manufacturing, more commonly known as 3D printing, saw its initial appearance on the market during the early 1980s [1]. Visionaries in the realm of 3D printing include notable figures like Charles W. Hull (the visionary behind 3D Systems), S. Scott Crump (the mind behind Stratasys), and the pioneering duo of Hans J. Langer and Hans Steinbichler (the founders of EOS). It was Charles W. Hull who obtained the patent for the first 3D printer in 1986 [2]. In their nascent phase, these 3D printers primarily found their utility in the domain of rapid prototyping [3,4].

Nevertheless, the evolution of this technology surged forward with remarkable speed in the subsequent years [5]. Notably, the expiry of the patent for the fused deposition modeling (FDM) process in 2009 propelled 3D printers into the realm of consumers, catalyzing a significant penetration into this sector [6,7]. This transformative momentum extended its influence into the realm of dentistry as well [8]. The dimensions of printing units shrank while costs diminished, resulting in an altered landscape of applicability [9,10]. Simultaneously, the gamut of printable materials broadened to encompass a diverse array, ranging from plastics and metals to ceramics and even biological tissues [11]. These rapid prototyping methods can be classified based on the specific types of materials employed, such as plastics, metals, or powders [12,13].

Additive manufacturing (AM) procedures involve the step-by-step creation of items according to three-dimensional models [3]. The phrase commonly employed interchangeably with all additive techniques is 3D printing. As per the EN ISO/ASTM 52900:2021 standard for terminology, an AM process is defined as the "procedure of combining materials to construct objects using 3D model data, typically layer by layer, in contrast to subtractive manufacturing techniques" [9].

The speed of progress in the evolution of digital dental manufacturing has become truly remarkable [14]. Subtractive methods have achieved remarkable levels of both efficiency and precision in achieving accurate fits, while additive techniques such as 3D printing are gaining prominence at an escalating rate [15]. The merging of various manufacturing approaches, such as the pairing of laser sintering with CNC machining, or the integration of digital design and 3D printing alongside traditional analog pressing, vividly showcases the immense possibilities that lie ahead [16,17].

Societal shifts are exerting transformative influences on the field of dental technology, much like they do in other sectors [18]. One of these profound changes pertains to a scarcity of a proficient workforce. Notably, there is an ongoing decline in the number of individuals pursuing training in dental technology [19], even as the demand for dental prostheses continues to surge due to shifting demographics [8]. Furthermore, patients are increasingly constrained by time limitations prompted by heightened expectations in their professional spheres, curtailing their capacity to undergo dental procedures [20].

In the face of these challenges, the digital revolution emerges as a potential solution, as digital procedures often stand out for their efficiency [21]. Within the dental laboratory, the integration of digital processes offers advantages like heightened precision and reproducibility, alongside enhanced material characteristics and user convenience [22]. This technological transition holds the potential to address these challenges by harnessing the inherent efficiencies of digital methodologies [23,24].

The intriguing fusion of a digitalized work environment and a hands-on artisanal craft renders dental technology a compelling choice for young individuals seeking a diverse and multifaceted professional journey [25]. Numerous dental laboratories have adeptly embraced the delicate equilibrium between preserving age-old craftsmanship and embracing the digital realm, finding harmony between tradition and disruption, and reconciling established values with the need for necessary changes [26,27]. Within this unfolding narrative, the role of 3D printing as a digital manufacturing process takes center stage, illuminating a pivotal facet of this evolution [28]. To put it in simplified terms, the workflow involves a dental technician generating a digital dataset on a computer and creating a three-dimensional entity through computer-aided design (CAD) [29]. This digital blueprint is then transmitted to a 3D printer, where it metamorphoses into a tangible object [30].

An inherent advantage of additive manufacturing techniques lies in their capacity to visualize and actualize three-dimensional concepts on a screen, enabling the realization of a virtually boundless array of forms and intricacies [31]. An intriguing facet often overlooked is that the mechanical and aesthetic attributes of the fabricated item can be subtly tailored during the 3D construction process [32]. This distinct capability is absent in subtractive manufacturing, wherein the material characteristics are predetermined by the supplier of the prefabricated material [33,34]. The confluence of customization opportunities and the expeditious availability of digitally conceived items, often at reduced costs, positions additive manufacturing as a pivotal cornerstone within the realm of digital dentistry [35,36].

Polymeric materials play a significant role in the realm of dentistry, offering a wide array of applications stemming from their distinct surface qualities, mechanical and biological attributes, simplified processing, and cost-effectiveness [4]. Among the frequently employed polymers in dental applications are polymethyl methacrylate (PMMA), polyurethane (PU), polyethylene (PE), polycarbonate (PC), polyetheretherketone (PEEK), polyethylene glycol (PEG), polydimethylsiloxane (PDMS), polylactic acid (PLA), poly(e-caprolactone) (PCL), acrylonitrile butadiene styrene (ABS), and polypropylene (PP) [2].

While their mechanical traits are linked to the inherent properties of the bulk material, their interaction with oral tissues heavily relies on surface characteristics. This justifies the utilization of polymer coatings to enhance their biocompatibility [7]. The applications of these polymers span nearly all sectors of dentistry, encompassing direct restorative procedures, prosthodontics, orthodontics, and even implantology. Notably, synthetic PEEK has emerged as a potential implant material [15]. Leveraging 3D printing, intricately detailed custom facial prostheses made from polymers can be readily produced [26]. Furthermore, polymers have been instrumental in crafting scaffolds that contribute to bone structure regeneration, along with the development of tissues resembling dentin and pulp. They also find utility in producing membranes for guided tissue regeneration and as carriers for drug delivery in the treatment of various oral and periodontal conditions [21].

The objective of this narrative review is to provide an overview of the current advanced status of additive manufacturing within prosthodontics, with a specific focus on polymeric materials, and to outline potential future perspectives for this modern technology. To achieve this, a comprehensive search was conducted across prominent databases including PubMed, Web of Science, and EMBASE (Figure 1).

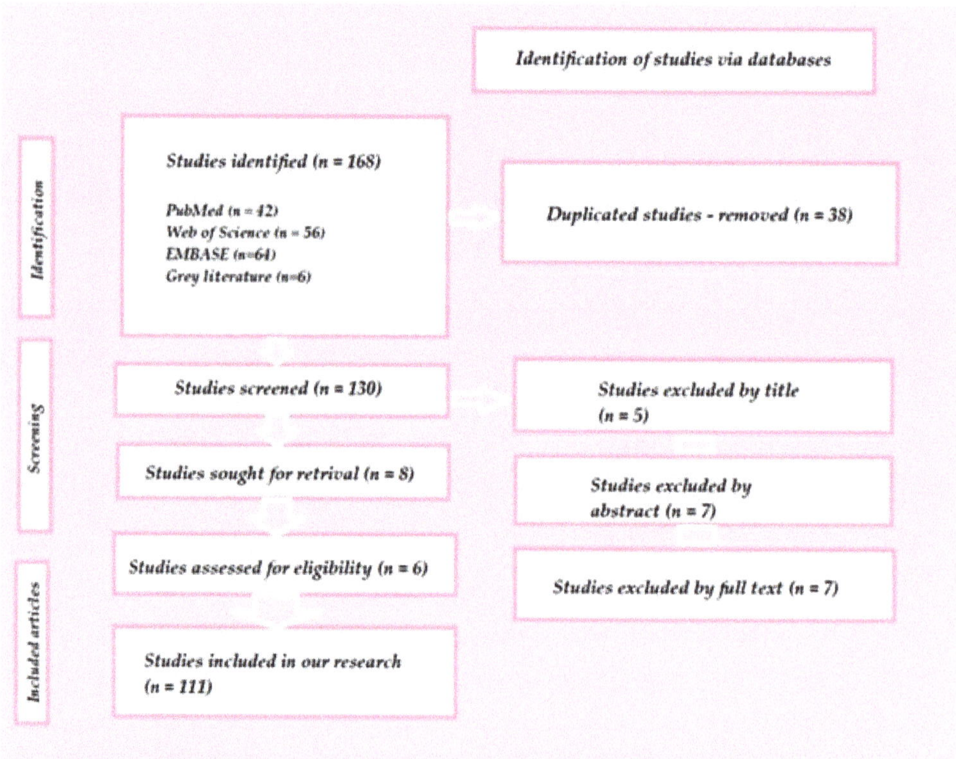

Figure 1. Design of the conducted study.

The scope of the search encompassed articles that documented the utilization of 3D-printed polymers in prosthetic dentistry. This was accomplished by employing a combination of diverse keywords, such as "dentistry" OR "digital dentistry", AND "polymers", AND "3D printing" OR "rapid prototyping" OR "additive manufacturing" OR "digital prosthodontics". The inclusion criteria ensured that only full-text articles written in English were considered for analysis. In addition to the electronic search, a supplementary manual exploration of relevant citations and references was carried out to further augment the comprehensiveness of the review.

2. General Techniques for 3D Printing in Prosthetic Dentistry

Among the various dental specialties, prosthodontics appears to have reaped substantial benefits from the advancements in 3D-printing technology. This is particularly evident in both fixed and removable denture fabrication, which can now be seamlessly executed through a fully digital workflow, resulting in precise and well-fitting prostheses. Polymers have proven to be well-suited for additive manufacturing processes involved in creating temporary crowns, denture bases, and artificial teeth. There have even been efforts to explore the 3D printing of polymeric permanent crowns and bridges. The range of applications of 3D-printed polymers in prosthodontics extends to crafting custom trays, patterns, try-ins, dental bite registrations, and various types of models. Several frequently employed polymers in the field of prosthetic dentistry include polymethylmethacrylate (PMMA), polylactic acid (PLA), polyetheretherketone (PEEK), and acrylonitrile butadiene styrene (ABS). Their molecular structures and general applications in the prosthodontic field are presented in Figures 2 and 3.

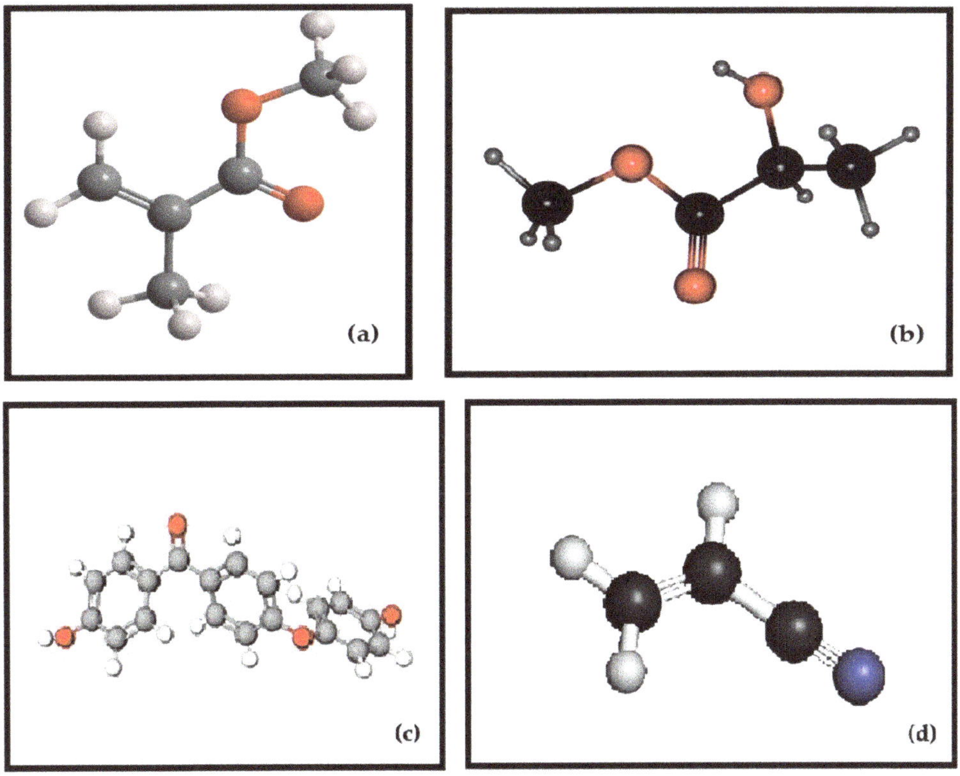

Figure 2. Molecular structures of the polymers for 3D Printing in Prosthodontics—(**a**) PMMA; (**b**) PLA; (**c**) PEEK; (**d**) ABS.

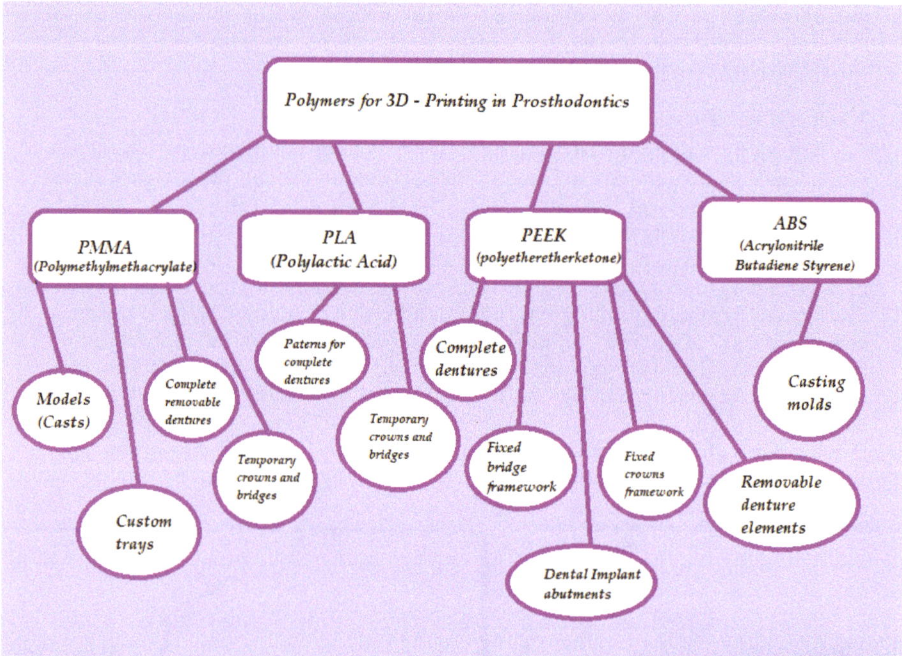

Figure 3. Three-dimensional-printed polymers used in prosthetic dentistry and their application.

Multiple 3D printing methods are available for the additive fabrication of prosthetic restorations (Figures 3–5).

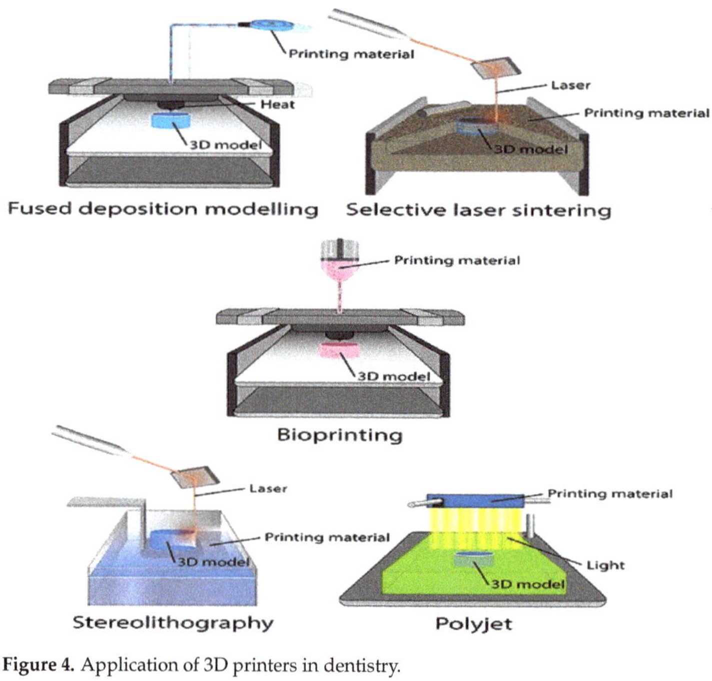

Figure 4. Application of 3D printers in dentistry.

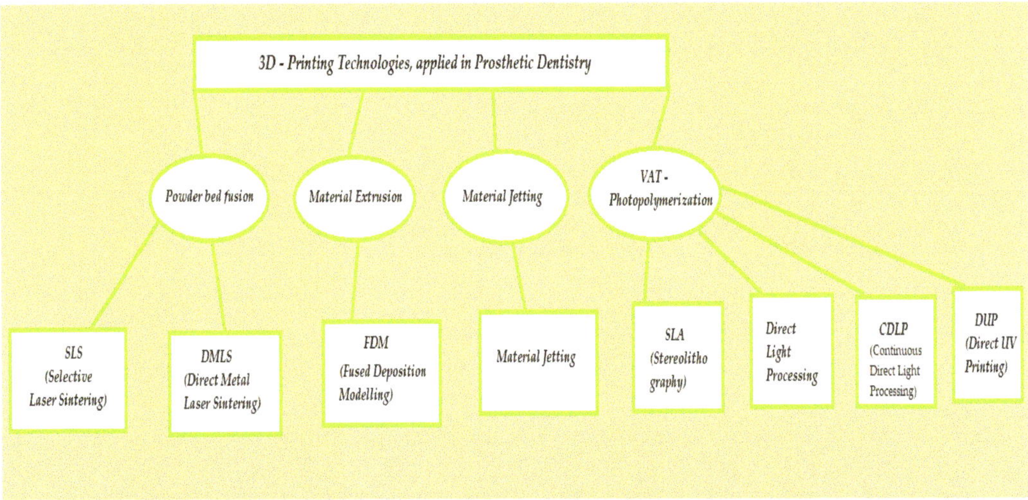

Figure 5. Three-dimensional-printing technologies, applied in prosthetic dentistry.

When these methods are juxtaposed, they reveal distinct attributes, encompassing speed, precision, scale, and procedural stability, contingent upon the underlying technology employed [37,38]. At present, within the domain of prosthetic dentistry, stereolithographic techniques emerge as the prevailing choice [39]. These encompass conventional stereolithography, laser-based solidification (stereolithography, SLA), and mask exposure methods (digital light processing, DLP). In both these methodologies, the object is solidified within a reservoir of photopolymer through the influence of light [40,41].

Starting in 2018, an array of 3D printers has entered the market that utilizes economical liquid crystal displays (LCD) [42]. This technology is termed direct ultraviolet printing (DUP) and exploits the LCD screens for precise pixel-wise exposure of the construction platform [43]. The typical choice for illumination is UV LEDs within a wavelength range of 395 to 405 nm. Moreover, direct 3D printing processes, also known as material jetting (MJT), have found application within dental contexts. Notably, there is a noteworthy technique known as multi-material 3D printing by Stratasys, which enables the simultaneous processing of diverse colors and materials with varying properties in a single build [5,44].

However, within the dental field, material extrusion (MEX) processes, including techniques such as fused-filament fabrication (FFF) or fused deposition modeling (FDM), currently face drawbacks as indicated by recent findings [45]. This is attributed to the extended printing times and challenges in attaining higher resolutions associated with these processes [46]. Among the diverse technologies mentioned within the realm of plastics, it is apparent that SLA, DLP, and MJT stand out as particularly compelling from both a technical and economic perspective [47,48]. Table 1 presents the most commonly used polymer materials in prosthodontics, organized according to their processing and their properties (Table 1).

The pioneering 3D printing systems to enter the market were stereolithographic setups that employ laser beams to solidify liquid substances [49]. Charles Hull, as early as the 1980s, filed a patent application for the initial stereolithography printer [50]. In their initial iterations, these devices were notably extensive and came with a high price tag [51]. In contrast, the latest wave of stereolithographic printers has become significantly more cost-effective. For the past five years, Formlabs (Sommerville, MA, USA) has provided a 3D printer tailored for dental applications [52]. This budget-friendly system serves as an excellent point of entry into 3D-printing technology, although it should be noted that the printing process might take longer compared to DLP printers [53,54].

Table 1. Three-dimensional-printed polymers, systemized by processing method and characteristics.

Type of Polymer	Processing Method	Characteristics
Methacrylic Acid (Formlabs: Dental SG)	Stereolithography	Elastic Modulus: 1670 MPa; Orientations of 0° to 90°.
Polylactic acid (PLA)	Fused deposition modeling	Tensile Strength: 28–56 Mpa; Elastic Modulus: 2000 Mpa; Orientations of 0° to 90°
Polyether ether ketone (PEEK)	Fused deposition modeling	Tensile Strength: 58–85 Mpa Elastic Modulus: 3000–4100 Mpa;
Acrylonitrile butadiene styrene (ABS)	Fused deposition modeling	Tensile Strength: 27–31 Mpa; Layer height: 0.05–0.14 mm; Processed at 210–240 °C. data

In conjunction with stereolithography, digital light processing (DLP) stands out as one of the most widely adopted additive manufacturing methods within the current dental domain [55]. The configuration of a DLP printer closely resembles that of an SLA printer, with the primary distinction being the source of illumination utilized [56,57]. Unlike the SLA printer that employs a laser beam for curing the photopolymer, DLP printers employ projection technology developed by Texas Instruments [58]. In this method, short-wavelength light (currently operating at wavelengths of 380 nm and 405 nm) is directed through a central digital micromirror device (DMD), which constitutes the core of DLP technology [59,60]. This setup employs micro mirrors with an approximate edge length of 16 μm that can be individually tilted under the influence of electrostatic forces, enabling the light to be optically directed onto the build platform [61]. This platform is situated within a translucent reservoir of photopolymer (also known as a photopolymer bath), or onto an absorptive surface [62,63]. The light, channeled through the DMD, projects the exposure mask onto the build platform through an optical lens. This prompts the photopolymer to solidify in the exposed regions [64]. After each mask projection, the build platform ascends along the z-axis, allowing fresh material to flow into the area beneath the object, thereby preparing it for exposure with the next mask [65,66]. In the realm of DLP technology, the time required for fabrication is predominantly determined by the object's dimension along the z-axis, rendering it relatively independent of the object's overall complexity or geometry [67].

In DLP (digital light processing) printers, the process involves using tiny mirrors to create individual image points or pixels [68]. However, the number of these micromirrors on a DMD (digital micromirror device) is limited. This limitation becomes evident when enlarging the build platform, which consequently increases the lengths of the edges along the x and y axes [69,70]. This expansion results in reduced precision. Despite this challenge, there are three methods currently employed to achieve larger build platforms. Inexpensive DLP printers utilize DMD chips with lower resolutions, leading to a smaller physical size. By using high-resolution DMD chips (e.g., HD 1920 × 1080 pixels), greater object accuracy can be attained without increasing the printer's footprint [71]. Even more impressive, the use of 4K DMD chips (3840 × 2160 pixels) enables the combination of high resolutions with a sizable build area, as seen in the Rapid Shape D70+ printer. However, the cost of 4K DMD chips remains a significant obstacle.

The approach of running two DLP projectors with HD resolution in parallel results in a seam or "joint" on the build platform due to the utilization of two light sources. This joint prevents the printing of objects that span the projection field. For instance, the Rapid Shape D40 II printer employs this method [72]. To enhance the capabilities of DLP printers, innovations are applied in detaching objects from the material vat during the build process (W2P Engineering, Vienna, Austria). This detachment occurs after each cycle of exposure when the build platform is raised along the z-axis [73]. Four distinct techniques are employed to achieve these fixed intervals. The build platform follows a predefined path within a specific timeframe after each exposure cycle. This approach maintains a consistent path-to-time ratio throughout the build process, even if fewer support structures could allow for earlier detachment. While straightforward, this method does not alter the overall duration of the building processes [74].

Force sensors are utilized to measure the detachment force required (Rapid Shape, Heimsheim, Germany). Smart control technology calculates an optimal path-to-time ratio, thus accelerating the building process. An important benefit is the precise and controlled separation process. The patented Force Feedback technology is exemplified in printers like the Rapid Shape D30 [75].

Vat Deflection Feedback System (VDFS; W2P, Vienna, Austria) employs an additional sensor to expedite the building process. By allowing the deformation of the material tray (FlexVat), the detachment force is minimized. This leads to heightened printing speed and quality [76].

Carbon3D (CDLP; Carbon3D, Redwood City, CA, USA) introduced the patented continuous liquid interface production (CLIP) technology, categorized as a CDLP process. Unlike the step-by-step object buildup in conventional DLP printers, CLIP employs a continuous build process that does not necessitate the usual detachment steps. This is achievable due to an oxygen-rich zone ("dead zone") immediately above the build platform, where the photopolymer does not cure. Oxygen is introduced into this zone through an oxygen-permeable window [77]. As adhesion between the object and the build platform is absent, a continuous build process becomes feasible. The outcome is rapid build speeds, precise object formation, and uninterrupted object geometries along the z-axis. Examples of dental applications encompass digitally manufactured Lucitone Digital Print denture bases from Dentsply Sirona (York, PA, USA), as well as bite splints produced using Carbon3D printers like KeyPrint and KeySplint Soft Clear [78].

DLP projectors that move underneath the material vat can expose a larger area for printing. A key advantage of this approach is the absence of a joint line on the printed object. As a result, the entire build platform can be used at full resolution, leading to higher precision, increased printing accuracy, and optimal utilization of the printer's capacity [79].

MovingLight (Prodways Group, Paris, France) is a technology based on the DLP process but stands out due to its dynamic projector movement. Unlike its competitors, the projector is not fixed in a single position within the printer; instead, it moves across the entire working area in multiple steps [80]. This innovative movement results in high resolutions (42 μm) and remarkable accuracy, even with a substantial build platform. Examples of printers utilizing this technology include Prodways' ProMaker LD10 Dental Plus, LD10 Dental Models, LD20 Dental Plus, and LD20 Dental Models. The latter two models incorporate two movable projector heads, reducing printing times by an additional 40%. For instance, these printers can produce 55 dental arches in approximately 1 h [81,82].

Another approach is the material jetting technique, in which the material is deposited directly onto the build platform using a print head, resembling the process of 2D printing. Following this deposition, the material undergoes curing during an intermediate exposure step, gradually constructing the object layer by layer. The most prominent example of this approach is the Polyjet method, developed by Stratasys in Eden Prairie, MN, USA [83]. This method is characterized by its rapid building process and remarkable precision. Notably, it supports multi-material 3D printing, allowing for the creation of objects using up to five distinct materials in a palette of over 500,000 colors. Stratasys offers products like the J720 Dental and J750 Digital Anatomy printers, both capable of operating in multi-material and multicolor modes [84].

At present, polymers stand as the predominant choice for 3D printing within the dental field. They are employed to create both permanent and removable dentures, along with a diverse array of dental devices, dental implants, and tissue formations.

3. Applications of 3D-Printed Polymers in Fixed Prosthodontics

3.1. Fabricating of 3D Models Using Intraoral Scan

Thanks to the impressive efficiency and precision of DLP printers, they excel in fabricating master models and segmented models from intraoral scan data. This capability finds significant utility in various applications, with a notable one being in oral implantology [85]. In this context, producing models for oral implantology using DLP printers holds promise. A crucial aspect of this process involves accurately placing laboratory analogs within the printed model, as this directly impacts the fit of restorations, both proximally and occlusally (Figure 6) [86].

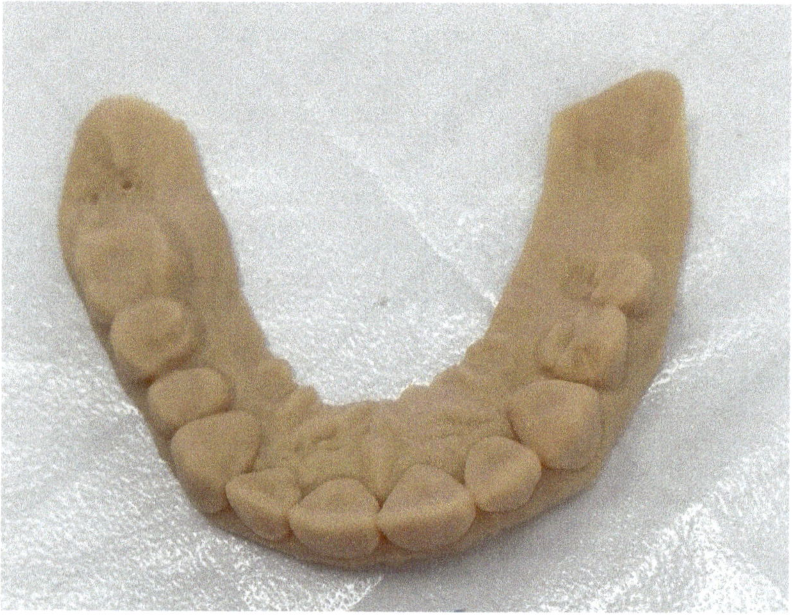

Figure 6. Three-dimensional–printed model of an upper jaw (origin of the Figure: author's own clinical case, no copyright issue).

3.2. Fabricating of Temporary Crowns and Fixed Partial Dentures

Temporary crowns and bridges play a crucial role in the process of treating fixed partial dentures (FPDs). Their primary purposes include safeguarding the pulp, ensuring alignment with the bite, preserving tooth placement, preventing fractures, and enduring functional pressures [87]. On the aesthetic front, they must exhibit a lasting color and the appropriate level of translucency. Provisional crowns need to exhibit exceptional functionality along with strong aesthetic and preventive qualities. Moreover, they serve as prototypes for future fixed partial dentures (FPDs) [88].

Currently, a diverse range of materials and technologies are employed in the fabrication of temporary FPDs to ensure simple production, pleasing aesthetics, and relatively elevated levels of hardness and strength [89]. The materials used for crafting provisional crowns and bridges can be divided into two primary categories: (1) methacrylate polymers and (2) composites. Polymethyl methacrylate (PMMA), polymethyl methacrylate, polyvinyl ethyl methacrylate, bisphenol A glycidyl methacrylate, urethane dimethacrylate, and others are the key materials in use [90]. The methods employed for producing temporary crowns and bridges encompass traditional techniques, involving heat-cured PMMA, as well as the relatively modern CAD-CAM milling of PMMA blanks [91]. Heat-cured acrylic polymers exhibit greater strength and resistance to wear compared to self-cured alternatives. Additionally, they display consistent color and are easily treatable on the surface. Temporary structures created from these materials can effectively fulfill their function over extended periods (Figure 7).

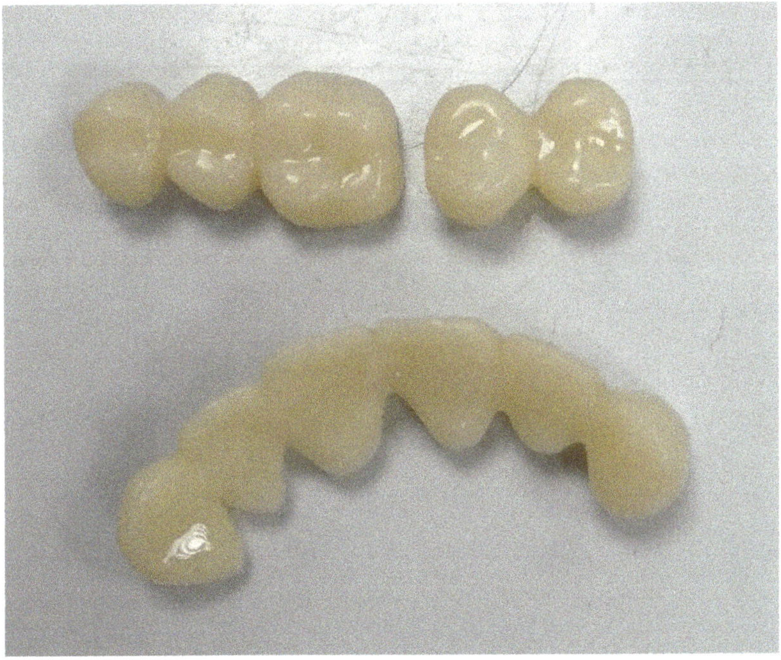

Figure 7. Three-dimensional-printed provisional bridges (origin of the Figure: author's own clinical case, no copyright issue), (Temporary CB Resin, Formlabs Dental, Somerville, MA, USA).

Utilizing CAD-CAM systems to mill provisional prostheses enables the utilization of high-density polymers, ensuring relatively robust mechanical characteristics and biocompatibility [92]. This results in the rapid creation of temporary restorations with notable precision concerning anatomical form, adherence to the teeth, and occlusal interactions [93]. These clear advantages offered by CAD-CAM systems have led to their extensive adoption within dental clinics and laboratories in recent times. Incorporating 3D-printing technology into the CAM module, along with the partial or complete digitization of processes right from the initial impression, can significantly reduce production time and deliver temporary prosthetic restorations of requisite accuracy and quality [94].

In the research by Digholkar S. et al. [95], an investigation was conducted into the microhardness and flexural strength of substances used in the production of conventional temporary fixed partial dentures (FPDs). These materials were derived from heat-cured polymer, PMMA produced through CAD-CAM milling, and a microhybrid photo-cured composite fabricated using 3D-printing technology. Their findings revealed that among the three materials, CAD-CAM-milled polymer exhibited the highest bending strength at 104.20 Mpa [96]. Conversely, in terms of microhardness, the 3D-printed composite displayed the greatest hardness at 32.77 HKN, attributed to the presence of filler. Tahayeri A. and collaborators generated samples using the commercially available NextDent C & B Vertex Dental polymer intended for temporary crowns and bridges [97]. They utilized a cost-effective Form 1+ printer from FormLabs, based on laser stereolithography (SLA) principles. Despite the limited precision of the printing system, they established that SLA-fabricated materials for temporary restorations possessed the essential mechanical characteristics for intraoral application [98].

Mai HN et al. [99] indicated that crowns produced through CAD-CAM systems (both milling and polymer-jet 3D printing) exhibited greater fitting accuracy compared to those created using matrices. This heightened accuracy, particularly in the occlusal region, was attributed to the significant enhancement in precision achieved through polymer-jet 3D printing. Additionally, Kim DY et al. [100] pointed out that the precision of dental crown fittings is impacted by the quantity of samples manufactured through microstereolithography. The most precise details are obtained when three pieces are printed on a single platform.

3.3. Fabricating Permanent Crowns, Inlays, Onlays, and Veneers

A ceramic-infused resin (Permanent Crown Resin, Formlabs Dental, USA) with a natural tooth color has been designed for 3D printing of durable single crowns, inlays, onlays, and veneers (Figure 8) [101]. The Permanent Crown Resin creates resilient and accurate long-term dental restorations, offered in four VITA Classical shades. The minimal water absorption and smooth surface guarantee that restorations crafted from this resin resist aging, discoloration, and plaque buildup [102].

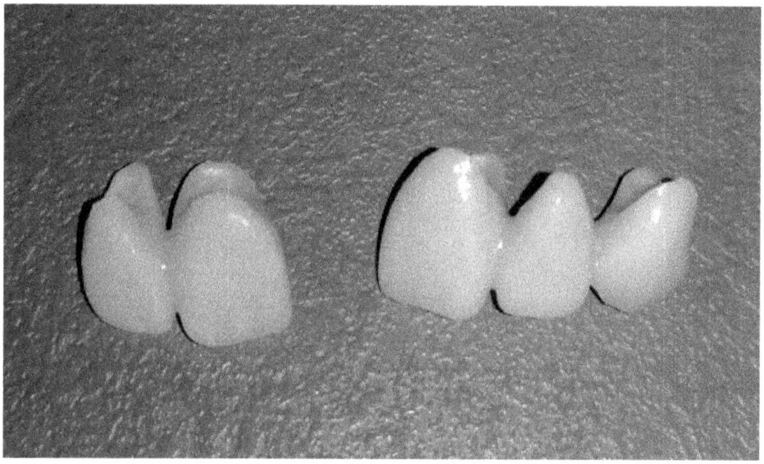

Figure 8. Permanent resin block crowns (origin of the Figure: author's own clinical case, no copyright issue), (Permanent Crown Resin, Formlabs Dental, USA).

3.4. Fabricating of Permanent Post and Core Restorations

When dealing with significant damage to the crown part of a tooth, necessitating added retention, insertion of a post into the root becomes essential to secure both the core and the restoration (Figure 9) [5,103]. The decision between utilizing a post that is tailor-made or one that is pre-manufactured depends on several considerations, including the shape of the canal, the amount of tooth structure remaining, and the chosen method of restoration [27].

Custom-designed posts are predominantly crafted from materials such as metal, zirconia, and composite materials reinforced with fibers.

As per the existing literature, the ideal choice of material for crafting post and core components should possess an elastic modulus that aligns with the natural flexural dynamics of the root [4,6,10]. Employing materials with biomechanical properties resembling dentin could also serve to mitigate the risks of root fractures or detachment [21]. One such category of materials demonstrating these attributes is advanced polymers like polyetheretherketone (PEEK) [104].

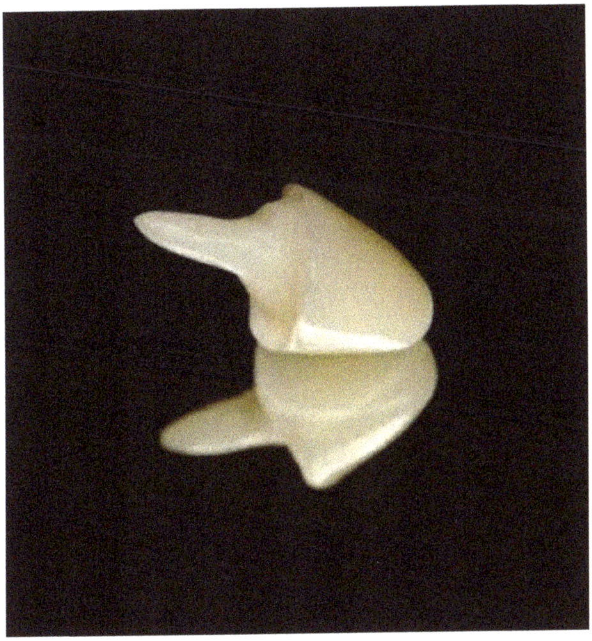

Figure 9. Post and core restoration, fabricated by PEEK (origin of the Figure: author's own clinical case, no copyright issue).

Past research has indicated that polyetheretherketone (PEEK) stands out as a promising biocompatible substance characterized by effective shock absorption, mechanical robustness, and notable resistance to both heat and chemicals [17]. Polyetheretherketone (PEEK) stands as a high-performance semi-crystalline polymer renowned for its impressive biocompatibility and strong processability [2]. PEEK exhibits substantial promise as a material for oral prosthetics due to its lightweight nature and relatively lower modulus (3–4 Gpa), rendering it a viable substitute for traditional Co-Cr alloy (230 Gpa) and Ti (104 Gpa). Limited studies have explored the application of PEEK in post and core assemblies [15,22]. In theory, utilizing PEEK and similar materials for post and core structures could potentially reduce the occurrence of root fractures [105]. To substantiate this hypothesis, additional laboratory and clinical investigations are imperative.

Beyond the appropriate biomaterial, the integration of a digital workflow through CAD/CAM methodologies also assumes significance in refining the creation of custom-designed post and core elements [9,13]. This approach permits the digital planning of anatomical structures, followed by milling or 3D printing [4]. The utilization of CAD/CAM technology not only streamlines the post and core fabrication process, saving time, but also enhances reliability [11]. Moreover, CAD/CAM-enabled post and core systems enable meticulous control over design aspects and the thickness of the cement layer [4].

Three-dimensional printing demonstrates a remarkable capability in handling thermoplastics, including PEEK, showcasing heightened production efficiency and minimal material wastage in contrast to traditional subtractive techniques [106]. Recent investigations have introduced the concept of dual-nozzle printing technology for processing diverse materials, thereby potentially supporting the realization of dual-color 3D printing for oral prosthetic materials that emulate tooth and gingiva shades [13]. Although 3D-printed PEEK has been successfully employed in various cases involving implants, artificial ribs, and frameworks for removable prostheses, its use remains relatively limited, and the application of dual-color PEEK printing in dental contexts remains unexplored [12,16]. A constraint of 3D-printed PEEK is its inherent brown color, negatively impacting aesthetics and hindering broader adoption [19].

According to the study of Chen et al. [107], this limitation can be overcome by fabricating a speech aid prosthesis framework using titanium dioxide (TiO2)/PEEK composite, achieving augmented mechanical strength and enhanced aesthetics. Removable dental prostheses necessitate both tooth-colored and gingival-colored components to mimic the textures of hard and soft oral tissues [108]. However, a monochromatic PEEK material falls short of capturing both tooth and gingiva colors simultaneously [14]. Apart from compromised aesthetics, PEEK frameworks require additional post-printing procedures like casting and molding to create the final prosthesis. Conventional manufacturing processes are not only time-intensive but also technologically demanding. Moreover, the interface between distinct parts cannot be eliminated as effectively as in one-piece printing with a dual-nozzle setup.

3.5. *Fabricating of Drilling Stents for Guided Implantology*

Recent software advancements have facilitated the integration of volumetric data sets derived from radiology (DICOM) with surface data sets (STL) obtained from laboratory or intraoral scanners. This integration enables the enhancement of implant placement, considering anatomical, surgical, and prosthetic considerations. Subsequently, these planned positions are executed using a surgical template that is placed within the patient's oral cavity [8]. DLP printing technology stands out in this context due to its ability to rapidly produce these templates with cost-effectiveness. Unlike subtractive techniques, DLP printing allows for the creation of intricate three-dimensional geometries without limitations on design possibilities [13].

4. Applications of 3D-Printed Polymers in Removable Prosthodontics

4.1. *Manufacturing of Custom Trays*

DLP printing technology presents a compelling option for producing personalized impression trays, primarily due to its swift processing. CAD software (DentalCAD 3.1 Rijeka, Exocad GmbH, Darmstadt, Germany) solutions available in the market enable the design of custom impression trays with optimal fit parameters in a streamlined manner, leading to significant time savings [11]. This is especially true when virtual undercut blocking and precise dimensional adjustments are factored in. Ensuring the impression's removal without irreversible deformation is critical [63]. Despite the technical benefits, it is worth noting that materials currently designated for crafting functional impression trays come at a high cost [86]. As a result, their practicality is primarily seen in the realm of implant impression trays. To make the most of this technology, it is advisable to integrate it with digital implant planning. In this scenario, digital models are already available, and the planned implant positions can serve as a foundation for constructing the trays.

4.2. *Manufacturing of Denture Bases for Removable Dentures*

Removable dentures are possible to be created using stereolithography (SLA) with 3D printers, which exhibit enhanced precision when the printing direction is inclined at a 45° angle [84]. When it comes to complete prosthetics, the concept of the impression technique holds appeal, but there is still a need for advancements in both materials and methodologies [6,85]. The complete digital production of removable partial dentures (RPDs) is presently limited to cases falling within the Kennedy III/IV classifications. For partially edentulous scenarios classified as Kennedy class I/II, the digital impression technique struggles to accurately capture the base edges and the displacement of the mucosa when the prosthesis applies pressure [86].

The process of printing through sintering or laser fusion (SLS) is quicker compared to alternative methods, albeit with higher costs. Within dental prosthetics, 3D printing serves multiple purposes: generating a model (from wax or plastic) that shall be transformed into the final prosthesis, or directly crafting definitive components from metal, resin, or ceramic materials (Figure 10) [87]. As of now, the extrusion technique, well-suited for thermoplastic polymers, finds primary use with PEEK [19]. Through an in vitro assessment of flexural strength (FS) values for six different prosthesis base resins, the following hierarchy was established: Machined resins (AvaDent and Polident) demonstrated superior results, trailed by a traditional heat-cured molded resin (Vertex) and a 3D-printed resin (NextDent, 3D Systems, Soesterberg, The Netherlands). In comparison, polyamide and another 3D-printed resin (Harz) exhibited notably lower values of flexural strength in contrast to standard resins [88].

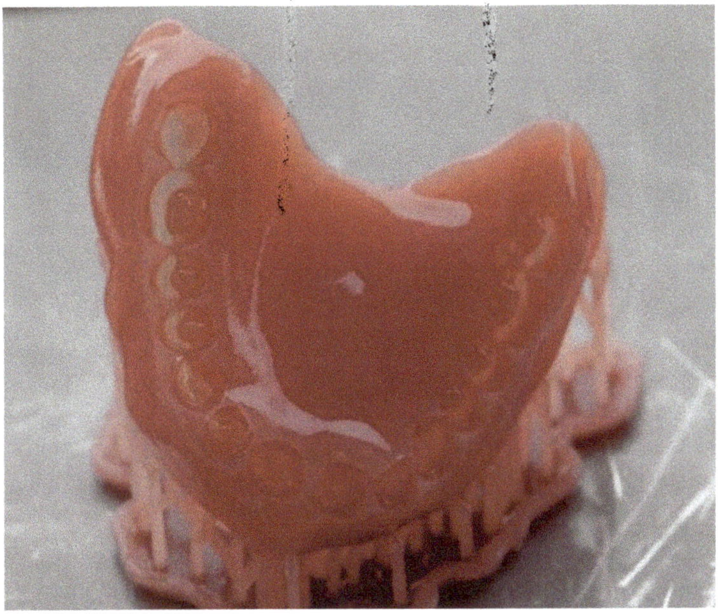

Figure 10. Three-dimensional-printed denture base for an upper complete removable denture (origin of the Figure: author's own clinical case, no copyright issue), (NextDent, 3D Systems, Soesterberg, The Netherlands).

4.3. Manufacturing of Artificial Teeth for Removable Dentures

The STL file format is a recognized standard for conveying three-dimensional object geometry through triangular representations, commonly used for transmitting information to 3D printers [16]. NextDent Denture 3D + is a biocompatible dental resin utilized for various components of removable dentures, encompassing both the denture base and teeth [19]. The physical and mechanical attributes of this material closely resemble those of traditional acrylic resin applied in removable dentures [20]. The production of three-dimensional denture teeth involves a methacrylate-based photopolymerized resin, which is processed through 3D printing techniques [21]. The denture teeth and denture base are fabricated separately via 3D printing, followed by joining the printed teeth to the printed denture base using a light-cured bonding agent and undergoing a final additional polymerization step (Figure 11) [22].

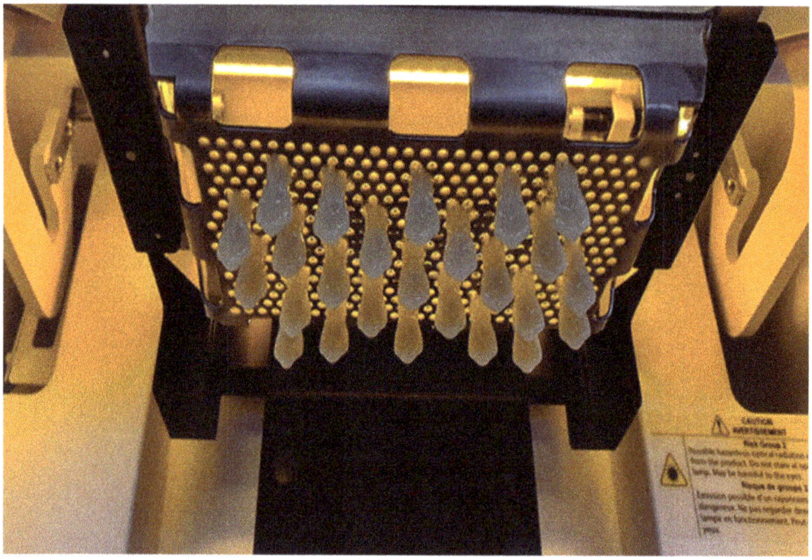

Figure 11. Three-dimensional printing of denture artificial teeth in the 3D printer NextDent 5100 (origin of the Figure: author's own clinical case, no copyright issue), (NextDent, 3D Systems, Sosterberg, The Netherlands).

4.4. Manufacturing of Occlusal Splints

In addition to the traditional scatter-and-press approach and subtractive milling technique, another avenue for crafting precisely fitting occlusal splints is through 3D printing (Figure 12) [109]. However, the accuracy of the overall production, as well as the material quality, long-term stability, and biocompatibility, play pivotal roles. Despite this, there is currently no substantial clinical experience with occlusal splints created through additive manufacturing. It is crucial to delve into the leaching behavior of additively manufactured occlusal splints both in laboratory settings and within the oral environment [64]. To determine the most effective long-term results, it would be valuable to compare this method with existing procedures [110].

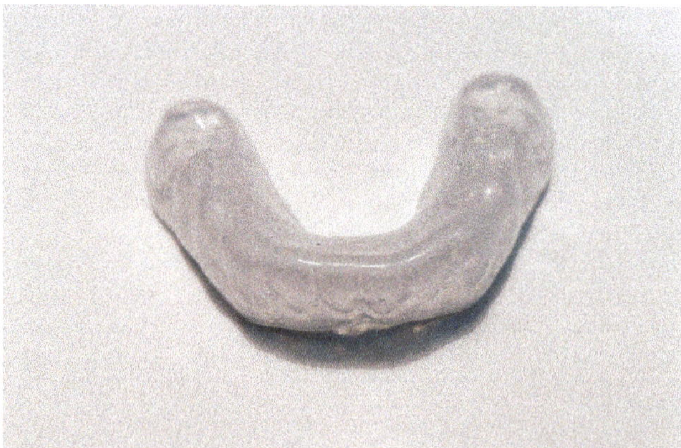

Figure 12. Three-dimensional–printed occlusal splint for the upper jaw (origin of the Figure: author's own clinical case, no copyright issue), (NextDent Ortho Flex, 3D Systems, Soesterberg, The Netherlands).

The standards are notably high for uniformity and biocompatibility, standards often achieved by high-performance polymers produced through subtractive CAD/CAM methods, such as milled splints [34]. Elements like the arrangement and alignment of objects and their impact on precision, stability, and durability must also be thoroughly studied. The orientation of the objects on the build platform, and therefore the layering direction, appears to be particularly significant in this context. Initial research indicates that 3D-printed occlusal splints exhibit accuracy levels similar to CAD/CAM-milled splints, yet they demonstrate increased material wear and less favorable material properties [65,67].

5. Drawbacks of the Applications of Additive Manufacturing in Prosthetic Dentistry

Emerging technologies, particularly those related to component detachment, can effectively address the speed issue and result in exceptionally rapid construction speeds. However, constraints also exist in terms of the range of materials compatible with 3D printing [6]. Within the realm of polymers, dental technology predominantly relies on printers using photopolymer-based materials, especially those associated with VAT polymerization techniques (SLA, DLP, DUP). This narrows down the spectrum of resins that can be employed, creating notable drawbacks when compared to conventional manufacturing processes like CNC technologies and analog methods [10].

A potential solution to this predicament is the "drop-on-demand" technology, wherein medical-grade thermoplastics are melted from granulates and applied in a molten state, drop by drop, onto the build platform. However, the resulting surface quality achievable with this technique significantly diverges from the outcomes seen in filament printers [55]. Furthermore, there is a dearth of data concerning the behavior of 3D-printed devices or restorations within the oral environment. Limited information exists regarding plaque formation, material elution patterns, and the general biocompatibility of 3D-printed polymer materials [64,77]. Consequently, more extensive, and specific data on these aspects are urgently required.

Acrylic resins are the outcome of a polymerization process, wherein monomers undergo a heat, light, or chemically activated addition reaction to form stable polymers [111,112]. Despite the process, complete conversion from monomers to polymers is unattainable, leading to the presence of residual monomers and potentially harmful chemical by-products, such as methyl methacrylate, formaldehyde, methacrylic acid, benzoic acid, or dibutyl phthalate within the material [113–116]. These substances leach into the surrounding saliva through diffusion and subsequently interact with the host mucosa, potentially causing cytotoxic effects [117,118].

The degree of residual monomers is influenced by the polymerization method employed in the resin material's manufacturing process. Dental materials produced by milling highly polymerized resin blanks exhibit residual monomer levels comparable to conventional materials [119]. In contrast, 3D-printed dental devices, cured through a light-based stepwise process, have been found to contain elevated levels of residual monomers [120]. Consequently, it seems plausible that 3D-printed oral splint materials may exert greater cytotoxicity than milled ones; however, surprisingly, scientific evidence supporting this notion is limited [121].

In essence, a distinction can be drawn between 3D printers designed for hobbyist use and those intended for professional applications. Practical experience has revealed that budget-friendly printers for hobbyists often yield unsatisfactory printing outcomes, particularly evident in the layering effect in FFF (fused-filament fabrication) printers due to filament fibers [29]. Consequently, the devices tailored for dental purposes tend to be more expensive yet produce satisfactory results. Nonetheless, even professional-grade printers exhibit some degree of layering effect in the Z-direction. This effect is largely influenced by the thickness of each individual layer. Thinner layers result in a reduced layering effect but also lead to extended processing times. Additionally, there are limitations concerning the maximum attainable print speed and the dimensions of the print area [101].

6. Conclusions

Additive processes offer a notable advantage by allowing the customization of an object's properties during the construction phase. This influence extends to both mechanical and aesthetic attributes. In contrast, subtractive processes determine these qualities based on the characteristics of the milling blank used in manufacturing. Consequently, 3D printing empowers users with a wide array of choices even during the initial design stages. On the other hand, the precision and efficiency of subtractive machining are exceptionally high, which makes a combination of both manufacturing techniques a logical consideration. 3D-printing technologies have already been embraced in dental laboratories and practices for producing removable partial dentures (RPDs) using CoCr. Furthermore, an increasing number of publications are emerging on the additive manufacturing of complete dentures. The initial outcomes related to mechanical strength, fit, and surface quality are encouraging. Given that denture bases have extensive contact with the oral mucosa, meticulous evaluation of biocompatibility is necessary. Specifically, elution behavior and cytotoxicity must be thoroughly examined before reaching a conclusive assessment.

Author Contributions: Conceptualization, M.D.; methodology, M.D.; software, A.V. and Y.K; validation, A.V., R.K. and S.C.; formal analysis, R.K., Y.K. and S.Z.; investigation, M.D.; resources, Y.K., S.Z. and S.C.; data curation, M.D.; writing—original draft preparation, M.D.; writing—review and editing, A.V.; visualization, R.K.; supervision, A.V., S.Z. and S.C.; project administration, Y.K. and S.C.; funding acquisition, S.Z. and R.K. All authors have read and agreed to the published version of the manuscript.

Funding: This research was funded by Medical University of Plovdiv, Bulgaria.

Institutional Review Board Statement: Not applicable.

Data Availability Statement: Data are contained within the article.

Acknowledgments: We are grateful for the support of the CAD/CAM Center of Dental Medicine, Research Institute, Medical University of Plovdiv, Bulgaria.

Conflicts of Interest: The authors declare no conflict of interest.

Abbreviations

ABS	acrylonitrile butadiene styrene
AM	additive manufacturing
CAD/CAM	computer-aided design/computer-aided manufacturing
CD	complete dentures
ISO	International Organization of Standardization
PC	polycarbonate
PCL	poly(e-caprolactone)
PDMS	polydimethylsiloxane
PE	polyethylene
PEEK	polyetheretherketone
PEG	polyethylene glycol
PLA	polylactic acid
PMMA	polymethyl methacrylate
PP	polypropylene
PU	polyurethane
STL	stereolithography, standard triangle language, standard tessellation language
3D	three-dimensional

References

1. Thomé, T.; Erhardt, M.C.G.; Leme, A.A.; Al Bakri, I.; Bedran-Russo, A.K.; Bertassoni, L.E. Emerging Polymers in Dentistry. In *Advanced Polymers in Medicine*; Puoci, F., Ed.; Springer: Cham, Switzerland, 2015; pp. 265–296.
2. Rokaya, D.; Srimaneepong, V.; Sapkota, J.; Qin, J.; Siraleartmukul, K.; Siriwongrungson, V. Polymeric materials and films in dentistry: An overview. *J. Adv. Res.* **2018**, *14*, 25–34. [CrossRef]

3. Smith, J.R.; Lamprou, D.A. Polymer coatings for biomedical applications: A review. *Trans. IMF* **2014**, *92*, 9–19. [CrossRef]
4. Nathanael, A.J.; Oh, T.H. Biopolymer Coatings for Biomedical Applications. *Polymers* **2020**, *12*, 3061. [CrossRef] [PubMed]
5. Najeeb, S.; Zafar, M.S.; Khurshid, Z.; Siddiqui, F. Applications of polyetheretherketone (PEEK) in oral implantology and prosthodontics. *J. Prosthodont. Res.* **2016**, *60*, 12–19. [CrossRef] [PubMed]
6. Pillai, S.; Upadhyay, A.; Khayambashi, P.; Farooq, I.; Sabri, H.; Tarar, M.; Lee, K.T.; Harb, I.; Zhou, S.; Wang, Y.; et al. Dental 3D-Printing: Transferring Art from the Laboratories to the Clinics. *Polymers* **2021**, *13*, 157. [CrossRef]
7. Abduo, J.; Lyons, K.; Bennamoun, M. Trends in Computer-Aided Manufacturing in Prosthodontics: A Review of the Available Streams. *Int. J. Dent.* **2014**, *2014*, 783948. [CrossRef]
8. Clark Ligon, S.; Liska, R.; Stampfl, J.; Gurr, M.; Mulhaupt, R. Polymers for 3D Printing and Customized Additive Manufacturing. *Chem. Rev.* **2017**, *117*, 10212–10290. [CrossRef]
9. Dimitrova, M.; Chuchulska, B.; Zlatev, S.; Kazakova, R. Colour Stability of 3D-Printed and Prefabricated Denture Teeth after Immersion in Different Colouring Agents—An In Vitro Study. *Polymers* **2022**, *14*, 3125. [CrossRef]
10. Tomova, Z.; Zhekov, Y.; Alexandrov, G.; Vlahova, A.; Vasileva, E. Application of CAD/CAM technologies and Materials for Prosthetic Restoration of Severely Damaged Teeth-Clinical Cases. *Aust. Dent. J.* **2023**, *ahead of print*. [CrossRef]
11. Ujfalusi, Z.; Pentek, A.; Told, R.; Schiffer, A.; Nyitrai, M.; Maroti, P. Detailed Thermal Characterization of Acrylonitrile Butadiene Styrene and Polylactic Acid Based Carbon Composites Used in Additive Manufacturing. *Polymers* **2020**, *12*, 2960. [CrossRef]
12. Heo, H.; Jin, Y.; Yang, D.; Wier, C.; Minard, A.; Dahotre, N.B.; Neogi, A. Manufacturing and Characterization of Hybrid Bulk Voxelated Biomaterials Printed by Digital Anatomy 3D Printing. *Polymers* **2021**, *13*, 123. [CrossRef]
13. Fully Guided Surgical Stent. Highly Accurate Guides for All Patient Cases and Implant Procedures. Available online: https://www.hiradentallab.com/fully-guided-surgical-stent/ (accessed on 11 November 2023).
14. Torabi, K.; Farjood, E.; Hamedani, E. Rapid prototyping technologies and their applications in prosthodontics, a review of literature. *J. Dent.* **2015**, *16*, 1–9.
15. Stansbury, J.W.; Idacavage, M.J. 3D printing with polymers: Challenges among expanding options and opportunities. *Dent. Mater.* **2016**, *32*, 54–64. [CrossRef]
16. Theus, A.S.; Ning, L.; Hwang, B.; Gil, C.; Chen, S.; Wombwell, A.; Mehta, R.; Serpooshan, V. Bioprintability: Physiomechanical and Biological Requirements of Materials for 3D Bioprinting Processes. *Polymers* **2020**, *12*, 2262. [CrossRef]
17. Georgiev, Z.; Alexandrov, S.; Tomova, Z. Experimental setting for testing the bond strength between metal alloy and PEEK. *J. IMAB Annu. Proc. Sci. Pap.* **2021**, *27*, 3881–3884. [CrossRef]
18. Nold, J.; Wesemann, C.; Rieg, L.; Binder, L.; Witkowski, S.; Spies, B.C.; Kohal, R.J. Does Printing Orientation Matter? In-Vitro Fracture Strength of Temporary Fixed Dental Prostheses after a 1-Year Simulation in the Artificial Mouth. *Materials* **2021**, *14*, 259. [CrossRef] [PubMed]
19. Groth, C.; Kravitz, N.D.; Jones, P.E.; Graham, J.W.; Redmond, W.R. Three-dimensional printing technology. *J. Clin. Orthod.* **2014**, *48*, 475–485. [PubMed]
20. Etemad-Shahidi, Y.; Qallandar, O.B.; Evenden, J.; Alifui-Segbaya, F.; Ahmed, K.E. Accuracy of 3-Dimensionally Printed Full-Arch Dental Models: A Systematic Review. *J. Clin. Med.* **2020**, *9*, 3357. [CrossRef] [PubMed]
21. Schweiger, J.; Edelhoff, D.; Güth, J.-F. 3D Printing in Digital Prosthetic Dentistry: An Overview of Recent Developments in Additive Manufacturing. *J. Clin. Med.* **2021**, *10*, 2010. [CrossRef] [PubMed]
22. Herschdorfer, L.; Negreiros, W.; Gallucci, G.; Hamilton, A. Comparison of the accuracy of implants placed with CAD-CAM surgical templates manufactured with various 3D printers: An in vitro study. *J. Prosthet. Dent.* **2020**, *125*, 905–910. [CrossRef]
23. Piedra Cascon, W.; Parra Nunez, A.; Charlen Diez, I.; Revilla León, M. Laboratory workflow to obtain long-term injected resin composite interim restorations from an additive manufactured esthetic diagnostic template. *J. Esthet. Restor. Dent.* **2019**, *31*, 13–19. [CrossRef] [PubMed]
24. Al Mortadi, N.; Jones, Q.; Eggbeer, D.; Lewis, J.; Williams, R.J. Fabrication of a resin appliance with alloy components using digital technology without an analog impression. *Am. J. Orthod. Dentofac. Orthop.* **2015**, *148*, 862–867. [CrossRef] [PubMed]
25. Elomaa, L.; Teixeira, S.; Hakala, R.; Korhonen, H.; Grijpma, D.W.; Seppala, J.V. Preparation of poly(epsilon-caprolactone)-based tissue engineering scaffolds by stereolithography. *Acta Biomater.* **2011**, *7*, 3850–3856. [CrossRef] [PubMed]
26. Kim, K.; Yeatts, A.; Dean, D.; Fisher, J.P. Stereolithographic bone scaffold design parameters: Osteogenic differentiation and signal expression. *Tissue Eng. Part B Rev.* **2010**, *16*, 523–539. [CrossRef]
27. Skoog, S.A.; Goering, P.L.; Narayan, R.J. Stereolithography in tissue engineering. *J. Mater. Sci. Mater. Med.* **2014**, *25*, 845–856. [CrossRef]
28. Arifin, N.; Sudin, I.; Ngadiman, N.H.A.; Ishak, M.S.A. A Comprehensive Review of Biopolymer Fabrication in Additive Manufacturing Processing for 3D-Tissue-Engineering Scaffolds. *Polymers* **2022**, *14*, 2119. [CrossRef] [PubMed]
29. Osman, R.B.; Alharbi, N.; Wismeijer, D. Build angle: Does it influence the accuracy of 3D-printed dental restorations using digital light-processing technology? *Int. J. Prosthodont.* **2017**, *30*, 182–188. [CrossRef]
30. Ausiello, P.; Gloria, A.; Maietta, S.; Watts, D.C.; Martorelli, M. Stress Distributions for Hybrid Composite Endodontic Post Designs with and without a Ferrule: FEA Study. *Polymers* **2020**, *12*, 1836. [CrossRef]
31. de Andrade, G.-S.; Tribst, J.-P.-M.; Dal Piva, A.-M.O.; Bottino, M.-A.; Borges, A.-L.-S.; Valandro, L.-F.; Özcan, M. A Study on Stress Distribution to Cement Layer and Root Dentin for Post and Cores Made of CAD/CAM Materials with Different Elasticity Modulus in the Absence of Ferrule. *J. Clin. Exp. Dent.* **2019**, *11*, e1–e8. [CrossRef]

32. Ruschel, G.H.; Gomes, É.A.; Silva-Sousa, Y.T.; Pinelli, R.G.P.; Sousa-Neto, M.D.; Pereira, G.K.R.; Spazzin, A.O. Mechanical Properties and Superficial Characterization of a Milled CAD-CAM Glass Fiber Post. *J. Mech. Behav. Biomed. Mater.* **2018**, *82*, 187–192. [CrossRef]
33. Bilgin, M.S.; Erdem, A.; Dilber, E.; Ersoy, İ. Comparison of Fracture Resistance between Cast, CAD/CAM Milling, and Direct Metal Laser Sintering Metal Post Systems. *J. Prosthodont. Res.* **2016**, *60*, 23–28. [CrossRef]
34. Lalama, M.; Rocha, M.G.; O'Neill, E.; Zoidis, P. Polyetheretherketone (PEEK) Post and Core Restorations: A 3D Accuracy Analysis between Heat-Pressed and CAD-CAM Fabrication Methods. *J. Prosthodont.* **2022**, *31*, 537–542. [CrossRef]
35. Teixeira, K.N.; Duque, T.M.; Maia, H.P.; Gonçalves, T. Fracture Resistance and Failure Mode of Custom-Made Post-and-Cores of Polyetheretherketone and Nano-Ceramic Composite. *Oper. Dent.* **2020**, *45*, 506–515. [CrossRef]
36. Uctasli, S.; Boz, Y.; Sungur, S.; Vallittu, P.K.; Garoushi, S.; Lassila, L. Influence of Post-Core and Crown Type on the Fracture Resistance of Incisors Submitted to Quasistatic Loading. *Polymers* **2021**, *13*, 1130. [CrossRef]
37. Iemsaengchairat, R.; Aksornmuang, J. Fracture Resistance of Thin Wall Endodontically Treated Teeth without Ferrules Restored with Various Techniques. *J. Esthet. Restor. Dent.* **2022**, *34*, 670–679. [CrossRef] [PubMed]
38. Sadowsky, S.J. Comments Regarding: Batista VES, Bitencourt SB, Bastos NA, Pellizzer EP, Goiato MC, Dos Santos DM. Influence of the Ferrule Effect on the Failure of Fiber-Reinforced Composite Post-and-Core Restorations: A Systematic Review and Meta-Analysis. J Prosthet Dent 2020;123:239-45. *J. Prosthet. Dent.* **2021**, *125*, 554.
39. Garcia, P.P.; Wambier, L.M.; de Geus, J.L.; da Cunha, L.F.; Correr, G.M.; Gonzaga, C.C. Do Anterior and Posterior Teeth Treated with Post-and-Core Restorations Have Similar Failure Rates? A Systematic Review and Meta-Analysis. *J. Prosthet. Dent.* **2019**, *121*, 887–894.e4. [CrossRef] [PubMed]
40. Mosharraf, R.; Fathi, A.; Botshekan, S.S. Comparative Evaluation of the Effect of Core Type and Antirotational Post on Stress Distribution in an Endodontically Treated Maxillary First Molar: FEA. *Int. J. Dent.* **2022**, *2022*, 4336980. [CrossRef] [PubMed]
41. Al-Qarni, F.D. Customized Post and Cores Fabricated with CAD/CAM Technology: A Literature Review. *Int. J. Gen. Med.* **2022**, *15*, 4771–4779. [CrossRef] [PubMed]
42. Lee, K.-S.; Shin, J.-H.; Kim, J.-E.; Kim, J.-H.; Lee, W.-C.; Shin, S.-W.; Lee, J.-Y. Biomechanical Evaluation of a Tooth Restored with High Performance Polymer PEKK Post-Core System: A 3D Finite Element Analysis. *Biomed Res. Int.* **2017**, *2017*, 1373127. [CrossRef] [PubMed]
43. Dimitrova, M.; Vlahova, A.; Hristov, I.; Kazakova, R.; Chuchulska, B.; Kazakov, S.; Forte, M.; Granberg, V.; Barile, G.; Capodiferro, S.; et al. Evaluation of Water Sorption and Solubility of 3D-Printed, CAD/CAM Milled, and PMMA Denture Base Materials Subjected to Artificial Aging. *J. Compos. Sci.* **2023**, *7*, 339. [CrossRef]
44. Heboyan, A.; Zafar, M.S.; Karobari, M.I.; Tribst, J.P.M. Insights into Polymeric Materials for Prosthodontics and Dental Implantology. *Materials* **2022**, *15*, 5383. [CrossRef] [PubMed]
45. Hadjieva, H.; Dimova, M.; Hadjieva, E.; Todorov, S. Changes in the vertical dimension of occlusion during the different periods of complete denture wear—A comparative study. *J. IMAB Annu. Proc. Sci. Pap.* **2014**, *20*, 546–549. [CrossRef]
46. Falcão Spina, D.R.; da Costa, R.G.; Correr, G.M.; Rached, R.N. Scanning of Root Canal Impression for the Fabrication of a Resin CAD-CAM-Customized Post-and-Core. *J. Prosthet. Dent.* **2018**, *120*, 242–245. [CrossRef] [PubMed]
47. Zheng, Z.; Sun, J.; Jiang, L.; Wu, Y.; He, J.; Ruan, W.; Yan, W. Influence of Margin Design and Restorative Material on the Stress Distribution of Endocrowns: A 3D Finite Element Analysis. *BMC Oral Health* **2022**, *22*, 30. [CrossRef]
48. Lee, H.; Son, K.B.D.; Lee, D.-H.; Kim, S.-Y.; Lee, K.-B. Comparison of wear of interim crowns in accordance with the build angle of digital light processing 3-dimensional printing: A pilot in vivo study. *Res. Sq.* **2022**, *9*, 417. [CrossRef]
49. Williams, R.J.; Bibb, R.; Rafik, T. A technique for fabricating patterns for removable partial denture frameworks using digitized casts and electronic surveying. *J. Prosthet. Dent.* **2004**, *91*, 85–88. [CrossRef] [PubMed]
50. Lin, W.-S.; Harris, B.T.; Pellerito, J.; Morton, D. Fabrication of an interim complete removable dental prosthesis with an in-office digital light processing three-dimensional printer: A proof-of-concept technique. *J. Prosthet. Dent.* **2018**, *120*, 331–334. [CrossRef]
51. Bilgin, M.S.; Erdem, A.; Aglarci, O.S.; Dilber, E. Fabricating complete dentures with CAD/CAM and RP technologies. *J. Prosthodont.* **2015**, *24*, 576–579. [CrossRef]
52. Revilla-León, M.; Besné-Torre, A.; Sánchez-Rubio, J.L.; Fábrega, J.J.; Özcan, M. Digital tools and 3D printing technologies integrated into the workflow of restorative treatment: A clinical report. *J. Prosthet. Dent.* **2019**, *121*, 3–8. [CrossRef]
53. Revilla-León, M.; Sánchez-Rubio, J.L.; Oteo-Calatayud, J.; Özcan, M. Impression technique for a complete-arch prosthesis with multiple implants using additive manufacturing technologies. *J. Prosthet. Dent.* **2017**, *117*, 714–720. [CrossRef]
54. Bagheri Saed, A.; Behravesh, A.H.; Hasannia, S.; Alavinasab Ardebili, S.A.; Akhoundi, B.; Pourghayoumi, M. Functionalized poly l-lactic acid synthesis and optimization of process parameters for 3D printing of porous scaffolds via digital light processing (DLP) method. *J. Manuf. Process.* **2020**, *56*, 550–561. [CrossRef]
55. Prasad, A.; Kandasubramanian, B. Fused Deposition Processing Polycaprolactone of Composites for Biomedical Applications. *Polym. Plast. Technol. Mater.* **2019**, *58*, 1365–1398. [CrossRef]
56. Rocha, C.R.; Perez, A.R.T.; Roberson, D.A.; Shemelya, C.M.; MacDonald, E.; Wicker, R.B. Novel ABS-based binary and ternary polymer blends for material extrusion 3D printing. *J. Mater. Res.* **2014**, *29*, 1859–1866. [CrossRef]
57. Rebong, R.E.; Stewart, K.T.; Utreja, A.; Ghoneima, A.A. Accuracy of three-dimensional dental resin models created by fused deposition modeling, stereolithography, and Polyjet prototype technologies: A comparative study. *Angle Orthod.* **2018**, *88*, 363–369. [CrossRef]

58. Pan, Y.; Zhou, C.; Chen, Y. A Fast Mask Projection Stereolithography Process for Fabricating Digital Models in Minutes. *J. Manuf. Sci. Eng.* **2012**, *134*, 051011. [CrossRef]
59. Hazeveld, A.; Huddleston Slater, J.J.R.; Ren, Y. Accuracy and reproducibility of dental replica models reconstructed by different rapid prototyping techniques. *Am. J. Orthod. Dentofac. Orthop.* **2014**, *145*, 108–115. [CrossRef] [PubMed]
60. Greco, G.B.; Popi, D.; Di Stefano, D.A. Accuracy of 3-dimensional printing of dental casts: A proposal for quality standardization. *J. Prosthet. Dent.* **2022**, *127*, 899–910. [CrossRef]
61. Alharbi, N.; Osman, R.; Wismeijer, D. Effects of build direction on the mechanical properties of 3D-printed complete coverage interim dental restorations. *J. Prosthet. Dent.* **2016**, *115*, 760–767. [CrossRef]
62. Li, X.; Xie, B.; Jin, J.; Chai, Y.; Chen, Y. 3D Printing Temporary Crown and Bridge by Temperature Controlled Mask Image Projection Stereolithography. *Procedia Manuf.* **2018**, *26*, 1023–1033. [CrossRef]
63. Dikova, T. Production of high-quality temporary crowns and bridges by stereolithography. *Scr. Sci. Med. Dent.* **2019**, *5*, 33–38. [CrossRef]
64. Dimitrova, M.; Corsalini, M.; Kazakova, R.; Vlahova, A.; Chuchulska, B.; Barile, G.; Capodiferro, S.; Kazakov, S. Comparison between Conventional PMMA and 3D Printed Resins for Denture Bases: A Narrative Review. *J. Compos. Sci.* **2022**, *6*, 87. [CrossRef]
65. Hada, T.; Kanazawa, M.; Iwaki, M.; Arakida, T.; Soeda, Y.; Katheng, A.; Otake, R.; Minakuchi, S. Effect of printing direction on the accuracy of 3D-printed dentures using stereolithography technology. *Materials* **2020**, *13*, 3405. [CrossRef] [PubMed]
66. Piedra Cascon, W.; Revilla-León, M. Digital workflow for the design and additively manufacture of a splinted framework and custom tray for the impression of multiple implants: A dental technique. *J. Prosthet. Dent.* **2018**, *120*, 805–811. [CrossRef] [PubMed]
67. Makvandi, P.; Esposito Corcione, C.; Paladini, F.; Gallo, A.L.; Montagna, F.; Jamaledin, R.; Pollini, M.; Maffezzoli, A. Antimicrobial modified hydroxyapatite composite dental bite by stereolithography. *Polym. Adv. Technol.* **2018**, *29*, 364–371. [CrossRef]
68. Yu, J.H.; Wang, Y.T.; Lin, C.L. Customized surgical template fabrication under biomechanical consideration by integrating CBCT image, CAD system and finite element analysis. *Dent. Mater. J.* **2018**, *37*, 6–14. [CrossRef] [PubMed]
69. Sun, Y.; Luebbers, H.; Agbaje, J.O.; Schepers, S.; Politis, C.; Van Slycke, S.; Vrielinck, L. Accuracy of dental implant placement using CBCT-derived mucosa-supported stereolithographic template. *Clin. Implant Dent. Relat. Res.* **2015**, *17*, 862–870. [CrossRef] [PubMed]
70. Ersoy, A.E.; Turkyilmaz, I.; Ozan, O.; McGlumphy, E.A. Reliability of implant placement with stereolithographic surgical guides generated from computed tomography: Clinical data from 94 implants. *J. Periodontol.* **2008**, *79*, 1339–1345. [CrossRef]
71. Cassetta, M.; Di Carlo, S.; Pranno, N.; Sorrentino, V.; Di Giorgio, G.; Pompa, G. The use of stereolithographic surgical templates in oral implantology. *Ann. Ital. Chir.* **2013**, *84*, 589–593.
72. Ozan, O.; Turkyilmaz, I.; Ersoy, A.E.; McGlumphy, E.A.; Rosenstiel, S.F. Clinical accuracy of 3 different types of computed tomography-derived stereolithographic surgical guides in implant placement. *J. Oral Maxillofac. Surg.* **2009**, *67*, 394–401. [CrossRef]
73. D'haese, J.; Van De Velde, T.; Komiyama, A.; Hultin, M.; De Bruyn, H. Accuracy and complications using computer-designed stereolithographic surgical guides for oral rehabilitation by means of dental implants: A review of the literature. *Clin. Implant Dent. Relat. Res.* **2012**, *14*, 321–335. [CrossRef]
74. Werz, S.M.; Zeichner, S.J.; Berg, B.I.; Zeilhofer, H.F.; Thieringer, F. 3D printed surgical simulation models as educational tool by maxillofacial surgeons. *Eur. J. Dent. Educ.* **2018**, *22*, e500–e505. [CrossRef] [PubMed]
75. Park, M.E.; Shin, S.Y. Three-dimensional comparative study on the accuracy and reproducibility of dental casts fabricated by 3D printers. *J. Prosthet. Dent.* **2018**, *119*, 861.e1–861.e7. [CrossRef] [PubMed]
76. Mai, H.-N.; Lee, K.-B.; Lee, D.-H. Fit of interim crowns fabricated using photopolymer-jetting 3D printing. *J. Prosthet. Dent.* **2017**, *118*, 208–215. [CrossRef]
77. Bibb, R.; Eggbeer, D.; Williams, R. Rapid manufacture of removable partial denture frameworks. *Rapid Prototyp. J.* **2006**, *12*, 95–99. [CrossRef]
78. Bibb, R.J.; Eggbeer, D.; Williams, R.; Woodward, A. Trial fitting of a removable partial denture framework made using computer-aided design and rapid prototyping techniques. *Proc. Inst. Mech. Eng. H J. Eng. Med.* **2006**, *220*, 793–797. [CrossRef] [PubMed]
79. Inokoshi, M.; Kanazawa, M.; Minakuchi, S. Evaluation of a complete denture trial method applying rapid prototyping. *Dent. Mater. J.* **2012**, *31*, 40–46. [CrossRef] [PubMed]
80. Zhang, Q.; Bei, H.-P.; Zhao, M.; Dong, Z.; Zhao, X. Shedding light on 3D printing: Printing photo-crosslinkable constructs for tissue engineering. *Biomaterials* **2022**, *286*, 121566. [CrossRef]
81. Roi, A.; Ardelean, L.C.; Roi, C.I.; Boia, E.-R.; Boia, S.; Rusu, L.-C. Oral Bone Tissue Engineering: Advanced Biomaterials for Cell Adhesion, Proliferation and Differentiation. *Materials* **2019**, *12*, 2296. [CrossRef]
82. Papaioannou, T.G.; Manolesou, D.; Dimakakos, E.; Tsoucalas, G.; Vavuranakis, M.; Tousoulis, D. 3D Bioprinting Methods and Techniques: Applications on Artificial Blood Vessel Fabrication. *Acta Cardiol. Sin.* **2019**, *35*, 284–289.
83. Ozbolat, I.T.; Hospodiuk, M. Current advances and future perspectives in extrusion-based bioprinting. *Biomaterials* **2016**, *76*, 321–343. [CrossRef]
84. Kyle, S.; Jessop, Z.M.; Al-Sabah, A.; Whitaker, I.S. 'Printability' of Candidate Biomaterials for Extrusion Based 3D Printing: State-of-the-Art. *Adv. Healthc. Mater.* **2017**, *6*, 1700264. [CrossRef]

85. He, Y.; Yang, F.; Zhao, H.; Gao, Q.; Xia, B.; Fu, J. Research on the printability of hydrogels in 3D bioprinting. *Sci. Rep.* **2016**, *6*, 29977. [CrossRef] [PubMed]
86. Hölzl, K.; Lin, S.; Tytgat, L.; van Vlierberghe, S.; Gu, L.; Ovsianikov, A. Bioink properties before, during and after 3D bioprinting. *Biofabrication* **2016**, *8*, 032002. [CrossRef] [PubMed]
87. New Shade NextDent Denture 3D+ Classic Pink is Now Available! Available online: https://nextdent.com/new-shade-nextdent-denture-3d-classic-pink-is-now-available/ (accessed on 7 November 2023).
88. Ning, L.; Chen, X. A brief review of extrusion-based tissue scaffold bio-printing. *Biotechnol. J.* **2017**, *12*, 1600671. [CrossRef] [PubMed]
89. Costa, P.F.; Vaquette, C.; Zhang, Q.; Reis, R.L.; Ivanovski, S.; Hutmacher, D.W. Advanced tissue engineering scaffold design for regeneration of the complex hierarchical periodontal structure. *J. Clin. Periodontol.* **2014**, *41*, 283–294. [CrossRef]
90. Sun, W.; Starly, B.; Daly, A.C.; Burdick, J.A.; Groll, J.; Skeldon, G.; Shu, W.; Sakai, Y.; Shinohara, M.; Nishikawa, M.; et al. The bioprinting roadmap. *Biofabrication* **2020**, *12*, 022002. [CrossRef]
91. Raman, R.; Bashir, R. Stereolithographic 3D Bioprinting for Biomedical Applications. In *Essentials of 3D Biofabrication and Translation*; Atala, A., Yoo, J.J., Eds.; Academic Press: Boston, MA, USA, 2015; pp. 89–121.
92. Kim, S.H.; Yeon, Y.K.; Lee, J.M.; Chao, J.R.; Lee, Y.J.; Seo, Y.B.; Sultan, M.T.; Lee, O.J.; Lee, J.S.; Yoon, S.I.; et al. Precisely printable and biocompatible silk fibroin bioink for digital light processing 3D printing. *Nat. Commun.* **2018**, *9*, 1620. [CrossRef] [PubMed]
93. Rider, P.; Kacarevic, Z.P.; Alkildani, S.; Retnasingh, S.; Barbeck, M. Bioprinting of tissue engineering scaffolds. *J. Tissue Eng.* **2018**, *9*, 2041731418802090. [CrossRef]
94. Dimova-Gabrovska, M.; Dimitrova, D.; Mitronin, V. Removable Prosthetic Treatment in children—Literature review. *J. IMAB Annu. Proc. Sci. Pap.* **2018**, *24*, 2172–2176. [CrossRef]
95. Saunders, R.E.; Derby, B. Inkjet printing biomaterials for tissue engineering: Bioprinting. *Int. Mater. Rev.* **2014**, *59*, 430–448. [CrossRef]
96. Guillotin, B.; Souquet, A.; Catros, S.; Duocastella, M.; Pippenger, B.; Bellance, S.; Bareille, R.; Remy, M.; Bordenave, L.; Amedee, J.; et al. Laser-assisted bioprinting of engineered tissue with high cell density and microscale organization. *Biomaterials* **2010**, *31*, 7250–7256. [CrossRef]
97. Guillotin, B.; Ali, M.; Ducom, A.; Catros, S.; Keriquel, V.; Souquet, A.; Remy, M.; Fricain, J.-C.; Guillemot, F. Laser-Assiste Bioprinting for Tissue Engineering. In *Biofabrication Micro- and Nano-Fabrication, Printing, Patterning, and Assemblies*; Forgacs, G., Sun, W., Eds.; William Andrew Publishing: Boston, MA, USA, 2013; pp. 95–118.
98. An, J.; Teoh, J.E.M.; Suntornnond, R.; Chua, C.K. Design and 3D Printing of Scaffolds and Tissues. *Engineering* **2015**, *1*, 261–268. [CrossRef]
99. Tzeng, J.-J.; Yang, T.-S.; Lee, W.-F.; Chen, H.; Chang, H.-M. Mechanical Properties and Biocompatibility of Urethane Acrylate-Based 3D-Printed Denture Base Resin. *Polymers* **2021**, *13*, 822. [CrossRef] [PubMed]
100. Deng, Y.; Li, J.; He, Z.; Hong, J.; Bao, J. Urethane acrylate-based photosensitive resin for three-dimensional printing of stereolitho graphic elastomer. *J. Appl. Polym. Sci.* **2020**, *137*, 49294. [CrossRef]
101. Permanent Crown Resin 1 kg (0.7 L). Available online: https://formlabs.com/store/materials/permanent-crown-resin-us/ (accessed on 8 November 2023).
102. Khorsandi, D.; Fahimipour, A.; Abasian, P.; Saber, S.S.; Seyedi, M.; Ghanavati, S.; Ahmad, A.; De Stephanis, A.A.; Taghavinezhad-dilami, F.; Leonova, A.; et al. 3D and 4D printing in dentistry and maxillofacial surgery: Printing techniques, materials, and applications. *Acta Biomater.* **2021**, *122*, 26–49. [CrossRef] [PubMed]
103. Réalisation D'un Inlay-Core en PEEK: Une Nouvelle Indication Pour ce Matériau. Available online: https://www.information-dentaire.fr/formations/realisation-d-un-inlay-core-en-peekune-nouvelle-indication-pour-ce-materiau/ (accessed on 10 November 2023).
104. Rusu, L.-C.; Ardelean, L.C.; Jitariu, A.-A.; Miu, C.A.; Streian, C.G. An Insight into the Structural Diversity and Clinical Applicability of Polyurethanes in Biomedicine. *Polymers* **2020**, *12*, 1197. [CrossRef]
105. Chen, H.; Lee, S.-Y.; Lin, Y.-M. Synthesis and formulation of PCL-based urethane acrylates for DLP 3D printers. *Polymers* **2020**, *12*, 1500. [CrossRef]
106. Ngo, T.; Kashani, A.; Imbalzano, G.; Nguyen, K.; Hui, D. Additive manufacturing (3D printing): A review of materials, methods, applications and challenges. *Compos. Part B Eng.* **2018**, *143*, 172–196. [CrossRef]
107. Molinero-Mourelle, P.; Canals, S.; Gómez-Polo, M.; Fernanda Solá-Ruiz, M.; del Río Highsmith, J.; Celemín Viñuela, A.; Solá-Ruiz, M.F.; Viñuela, A.C. Polylactic acid as a material for three-dimensional printing of provisional restorations. *Int. J. Prosthodont.* **2018**, *31*, 349–350. [CrossRef]
108. Benli, M.; Eker-Gümü¸s, B.; Kahraman, Y.; Huck, O.; Özcan, M. Can polylactic acid be a CAD/CAM material for provisional crown restorations in terms of fit and fracture strength? *Dent. Mater. J.* **2021**, *40*, 772–780. [CrossRef]
109. 3D Printed Splints vs. Thermoformed. Available online: https://www.designerdentallab.com/single-post/2018/01/09/3d-printed-splints-vs-thermoformed (accessed on 6 November 2023).
110. Ilyas, R.A.; Zuhri, M.Y.M.; Aisyah, H.A.; Asyraf, M.R.M.; Hassan, S.A.; Zainudin, E.S.; Sapuan, S.M.; Sharma, S.; Bangar, S.P.; Jumaidin, R.; et al. Natural Fiber-Reinforced Polylactic Acid, Polylactic Acid Blends and Their Composites for Advanced Applications. *Polymers* **2022**, *14*, 202. [CrossRef] [PubMed]

111. Bayraktar, G.; Guvener, B.; Bural, C.; Uresin, Y. Influence of polymerization method, curing process, and length of time of storage in water on the residual methyl methacrylate content in dental acrylic resins. *J. Biomed. Mater. Res. B Appl. Biomater.* **2006**, *76*, 340–345. [CrossRef] [PubMed]
112. Braden, M. Some Aspects of the Chemistry and Physics of Dental Resins. *Adv. Dent. Res.* **1988**, *2*, 93–97. [CrossRef] [PubMed]
113. Gautam, R.; Singh, R.D.; Sharma, V.P.; Siddhartha, R.; Chand, P.; Kumar, R. Biocompatibility of polymethylmethacrylate resins used in dentistry. *J. Biomed. Mater. Res. Part B Appl. Biomater.* **2012**, *100*, 1444–1450. [CrossRef] [PubMed]
114. Goiato, M.C.; Freitas, E.; dos Santos, D.; de Medeiros, R.; Sonego, M. Acrylic resin cytotoxicity for denture base—Literature review. *Adv. Clin. Exp. Med.* **2015**, *24*, 679–686. [CrossRef] [PubMed]
115. Singh, R.D.; Gautam, R.; Siddhartha, R.; Singh, B.P.; Chand, P.; Sharma, V.P.; Jurel, S.K. High performance liquid chromatographic determination of residual monomer released from heat-cured acrylic resin. An in vivo study. *J. Prosthodont.* **2013**, *22*, 358–361. [CrossRef]
116. Wedekind, L.; Güth, J.-F.; Schweiger, J.; Kollmuss, M.; Reichl, F.-X.; Edelhoff, D.; Högg, C. Elution behavior of a 3D-printed, milled and conventional resin-based occlusal splint material. *Dent. Mater.* **2021**, *37*, 701–710. [CrossRef]
117. Kurt, A.; Altintas, S.H.; Kiziltas, M.V.; Tekkeli, S.E.; Guler, E.M.; Kocyigit, A.; Usumez, A. Evaluation of residual monomer release and toxicity of self-adhesive resin cements. *Dent. Mater. J.* **2018**, *37*, 40–48. [CrossRef]
118. Urban, V.M.; Machado, A.L.; Vergani, C.E.; Giampaolo, E.T.; Pavarina, A.C.; de Almeida, F.G.; Cass, Q.B. Effect of water-bath post-polymerization on the mechanical properties, degree of conversion, and leaching of residual compounds of hard chairside reline resins. *Dent. Mater.* **2009**, *25*, 662–671. [CrossRef]
119. Steinmassl, P.-A.; Wiedemair, V.; Huck, C.; Klaunzer, F.; Steinmassl, O.; Grunert, I.; Dumfahrt, H. Do CAD/CAM dentures really release less monomer than conventional dentures? *Clin. Oral Investig.* **2017**, *21*, 1697–1705. [CrossRef]
120. Zhang, C.-Y.; Cheng, Y.-L.; Tong, X.-W.; Yu, H.; Cheng, H. In vitro cytotoxicity of self-adhesive dual-cured resin cement polymerized beneath three different cusp inclinations of zirconia. *BioMed Res. Int.* **2019**, *2019*, 7404038. [CrossRef] [PubMed]
121. Wataha, J.C. Predicting clinical biological responses to dental materials. *Dent. Mater.* **2012**, *28*, 23–40. [CrossRef] [PubMed]

polymers

MDPI

Article

Comprehensive Development of a Cellulose Acetate and Soy Protein-Based Scaffold for Nerve Regeneration

Brandon Gutiérrez [1], María Eugenia González-Quijón [2,3], Paulina Martínez-Rodríguez [4], Josefa Alarcón-Apablaza [5,6], Karina Godoy [3], Diego Pulzatto Cury [7,8], María Florencia Lezcano [9], Daniel Vargas-Chávez [6,10] and Fernando José Dias [4,11,*]

1 Master Program in Dental Sciences, Dental School, Universidad de La Frontera, Temuco 4780000, Chile; brandon.gutierrez@ufrontera.cl
2 Department of Chemical Engineering, Universidad de La Frontera, Temuco 4780000, Chile; mariaeugenia.gonzalez@ufrontera.cl
3 Scientific and Technological Bioresource Nucleus (BIOREN), Universidad de La Frontera, Temuco 4780000, Chile; karina.godoy@ufrontera.cl
4 Oral Biology Research Centre (CIBO-UFRO), Dental School, Universidad de La Frontera, Temuco 4780000, Chile; paulinaconstanza.martinez@ufrontera.cl
5 Research Centre in Dental Sciences (CICO-UFRO), Dental School, Universidad de La Frontera, Temuco 4780000, Chile; josefa.alarcon@ufrontera.cl
6 Doctoral Program in Morphological Sciences, Faculty of Medicine, Universidad de La Frontera, Temuco 4780000, Chile; daniel.vargas@umayor.cl
7 Department of Anatomy, Institute of Biomedical Sciences, Universidade de São Paulo (ICB-USP), São Paulo 05508-000, Brazil; diegocury@usp.br
8 Department of Cellular Biology and the Development, Institute of Biomedical Sciences, Universidade de São Paulo (ICB-USP), São Paulo 05508-000, Brazil
9 Departamento de Bioingeniería, Facultad de Ingeniería, Universidad Nacional de Entre Ríos, Oro Verde 3100, Argentina; lezcano.f@gmail.com
10 Facultad de Medicina y Ciencias de la Salud, Universidad Mayor, Escuela Medicina Veterinaria, Temuco 4780000, Chile
11 Department of Integral Adults Dentistry, Dental School, Universidad de La Frontera, Temuco 4780000, Chile
* Correspondence: fernando.dias@ufrontera.cl; Tel.: +56-452596399

Abstract: Background: The elaboration of biocompatible nerve guide conduits (NGCs) has been studied in recent years as a treatment for total nerve rupture lesions (axonotmesis). Different natural polymers have been used in these studies, including cellulose associated with soy protein. The purpose of this report was to describe manufacturing NGCs suitable for nerve regeneration using the method of dip coating and evaporation of solvent with cellulose acetate (CA) functionalized with soy protein acid hydrolysate (SPAH). Methods: The manufacturing method and bacterial control precautions for the CA/SPAH NGCs were described. The structure of the NGCs was analyzed under a scanning electron microscope (SEM); porosity was analyzed with a degassing method using a porosimeter. Schwann cell (SCL 4.1/F7) biocompatibility of cell-seeded nerve guide conduits was evaluated with the MTT assay. Results: The method employed allowed an easy elaboration and customization of NGCs, free of bacteria, with pores in the internal surface, and the uniform wall thickness allowed manipulation, which showed flexibility; additionally, the sample was suturable. The NGCs showed initial biocompatibility with Schwann cells, revealing cells adhered to the NGC structure after 5 days. Conclusions: The fabricated CA/SPAH NGCs showed adequate features to be used for peripheral nerve regeneration studies. Future reports are necessary to discuss the ideal concentration of CA and SPAH and the mechanical and physicochemical properties of this biomaterial.

Keywords: nerve injury; biomaterial; regenerative biology; nerve guide conductors; biocompatibility

1. Introduction

Injuries of the peripheral nerves can occur due to ischemic, chemical, mechanical, and thermal factors or during surgery [1], which leads to the partial or complete loss of sensory, motor, and autonomous functions and the generation of neuropathic pain, which can negatively affect quality of life [1,2]. Many people are affected by peripheral nerve injuries; according to recent studies, the mean incidence of this type of injury in England is 11.2 events per 100,000 persons per year [1] and 7.8 events per 100,000 persons per year in South Korea [2]. In developed countries, 13 to 23 of 100,000 persons/year are affected by nerve injuries [3] and more than 100,000 patients undergo nerve repair surgeries in the United States and Europe every year [4].

Nerve injuries are divided into neuropraxia, axonotmesis, and neurotmesis [1,5,6]. There is no loss of continuity in the nerve trunk in neuropraxia and axonotmesis, so it can regenerate spontaneously [7,8]; in the presence of neurotmesis, the loss of continuity of the nerve trunk occurs [1,5,6], and the length of the nerve gap or the distance between the proximal and distal ends affects the capacity for functional recovery [5,9,10]; in these cases, functional recovery is rarely restored [7,8], therefore patients do not recover normal motor control and fine sensitivity [11].

Treatment methods for nerve injuries can be categorized into surgical and nonsurgical treatments, depending on the type of nerve injury [5]. Surgical techniques are most frequently used in the treatment of neurotmesis [1], and the "gold standard" in the treatment of neurotmesis is autologous grafts [1–3,8,12–15]; however, the success rate of this therapy does not always result in correct nerve regeneration [14,16] and is associated with loss of function [1,3,10,14] and potential neuroma formation at the donor site [14,15]. To overcome these disadvantages, the development of NGCs to unite defects in peripheral nerves has been an important field of research in recent years [2,3,14,15,17].

NGCs are tubular scaffolds through which the regeneration of the lost nervous tissue is stimulated and directed [9]. The requirements of an ideal NGC include an architecture of mechanical properties that provide structural support, permeability, conductivity, flexibility, and biodegradability [1,3,13]. The design of NGCs is varied, and the configuration of the fibers, presence of pores and grooves, and thickness of the walls are fundamental to achieve successful nerve regeneration [1,13,18].

Several manufacturing methods have been described for the elaboration of NGCs [1,19], namely, drying in the cold (freeze-drying), micropatterning, electrospinning, additive manufacturing, gas foaming, phase separation, and finally evaporation, for which there are three different methods, namely, the dip–coating method, molding method (solvent casting), and rolling method; the latter can be used both for the solvent evaporation method and micropatterning and electrospinning. Each one has advantages and disadvantages that need to be investigated [1]. The solvent evaporation technique is described as a simple, reproducible, and low-cost procedure, but its main drawbacks are the use of toxic chemical solvents and irregular porosity [1].

There is a large list of biomaterials used in the creation of an NGC applied to nerve regeneration [1,3,8,12–14]. Nondegradable synthetic materials, such as silicone, were the first materials to be used [1,14] but have been abandoned, as they can cause a chronic reaction to a foreign body and mechanical grip, which make functional recovery difficult and a second intervention for the removal of these materials necessary [20,21]. This is the reason why biodegradable polymeric materials have been used to manufacture NGCs [1,9,13,17,21].

Natural polymer materials such as collagen [1,15,22], chitosan [15,23], silk [15,22,24], cellulose, and soy protein [1,2,21,25] have demonstrated great efficiency in nerve regeneration comparable to autologous nerve grafts [21]. Among the natural polymers, cellulose is the most abundant [21,26], and its biocompatibility, biodegradability, and mechanical properties have already been studied and adapted to different fields of tissue engineering [2,12,21,25]. In recent years, the association of cellulose/isolated soy protein (ISP) has been shown to be effective in the creation of NGCs capable of promoting nerve regeneration in sciatic nerve defects measuring 10 mm in width in rats [21], and ISP scaffolds have revealed biodegradable and biocompatible properties [2,12,21,25,26] and cytocompatibility with Schwann cells [21,27]; in addition, soy protein provides a morphological functionalization of porosity to the scaffold [28], which provides permeability allowing the supply of nutrients and exchange of gases and could prevent scar tissue infiltration and the diffusion of growth factors outside the lumen [13,29].

The purpose of this report is to describe the details and precautions for manufacturing NGCs with morphological and structural characteristics suitable for nerve regeneration using the method of dip coating and evaporation of solvent with cellulose functionalized with soy protein. The NGC structure is analyzed via scanning electron microscopy; its porosity via porosimeter; and its biocompatibility with Schwann cells via MTT assay and scanning electron microscopy.

2. Materials and Methods

Cellulose acetate (CA) (Product Code: 419028, ~50 g/mol, Sigma-Aldrich, Saint Louis, MO, USA) and soy protein acid hydrolysate (SPAH) (Product Code: S1674, Sigma-Aldrich, Saint Louis, MO, USA) were used to manufacture the NGCs. Analytical grade acetone (Product Code: 270725 purity 99.9%, 58.08 g/mol, Sigma-Aldrich, Saint Louis, MO, USA) was used to dissolve the CA and SPAH.

Prior to use, the CA was sterilized in 0.5 g aliquots in 2 mL cryogenic tubes (Sorenson BioScience, Salt Lake City, UT, USA). Before use, the SPAH was subjected to ultraviolet radiation for 30 min inside the laminar flow cabinet of a vertical air blower (BBS-v800, Biobase, Tsinan, China).

The antisepsis of the work area was carried out with ethyl alcohol 70° (Pure Ethyl Alcohol, Sigma-Aldrich, Saint Louis, MO, USA). The necessary instruments were as follows: anatomical forceps, scalpel handle No. 3, disposable scalpel blade No. 15, 1.4 mm diameter galvanized wire, 6 cm straight scissors, disposable sterile field cloths (square area of 0.25 m^2) and sterile gauze (square area of 0.0016 m^2). The asepsis of the instruments and aliquots of CA was carried out in a 12-L semiautomatic autoclave (Dabi-Atlante, Ribeirão Preto, Brazil).

2.1. Protocol for the Elaboration of NGC Cellulose Acetate/Soy Protein Acid Hydrolysate

Prior to the elaboration of the NGC, the asepsis and antisepsis of the instruments, supplies, and work area were carried out; all metallic instruments, beakers, magnetic stirrers, and cellulose acetate were sterilized; the antisepsis of the work area was carried out with 70° alcohol, and it was subjected to UV radiation for 30 min in a vertical laminar flow chamber. The NGC manufacturing protocol is described below.

I. In a sterile beaker, CA was dissolved in acetone, at concentrations of 9.08, 11.3, and 14.4 wt%; the solution was mixed for 10 min with a sterile magnetic stirrer at room temperature until a clear and homogeneous solution was achieved.

II. In parallel, in a sterile beaker, the SPAH was dissolved in acetone at a concentration of 25 and 30 wt%; the mixture was stirred for 30 min at room temperature until a homogeneous solution was achieved under constant UV light irradiation.

III. The SPAH solution was slowly poured into the CA solution, stirring constantly for 15 min at room temperature under UV light irradiation until a homogeneous and bubble-free mixture was achieved. The final concentrations were as follows: S1, CA: 4.1 and SPAH: 16.3 wt%; S2, CA: 5.1 and SPAH: 13.6 wt%; and S3, CA: 6.5 and SPAH: 13.6 wt%.

IV. The solution was deposited in 15 mL conical-bottom tubes/graduated cylinder (Corning, Corning, NY, USA), and before it settled, a 1.4 mm diameter galvanized rod/wire was introduced for 5 s and slowly withdrawn while making rotational movements so the polymer excess dropped off. In this way, the tubular shape of the NGC was obtained (Figure 1A).

V. The above procedure was repeated 9 times, always letting it drain between each immersion. The new immersions were carried out 60 s apart to allow the complete evaporation of the acetone and integration of the layers.

VI. Finally, the rod with the impregnated polymers was placed in ultrapure distilled water for 5 min.

VII. With the help of scissors, the ends were cut, and the NGC was removed by pushing and rotating at the same time (Figure 1B,C).

VIII. Disinfection protocol: the NGCs were submerged in 70° alcohol in 2 cycles of 30 min each (Figure 1D).

IX. Subsequently, the NGCs were preserved in a 1% penicillin/streptomycin antibiotic solution.

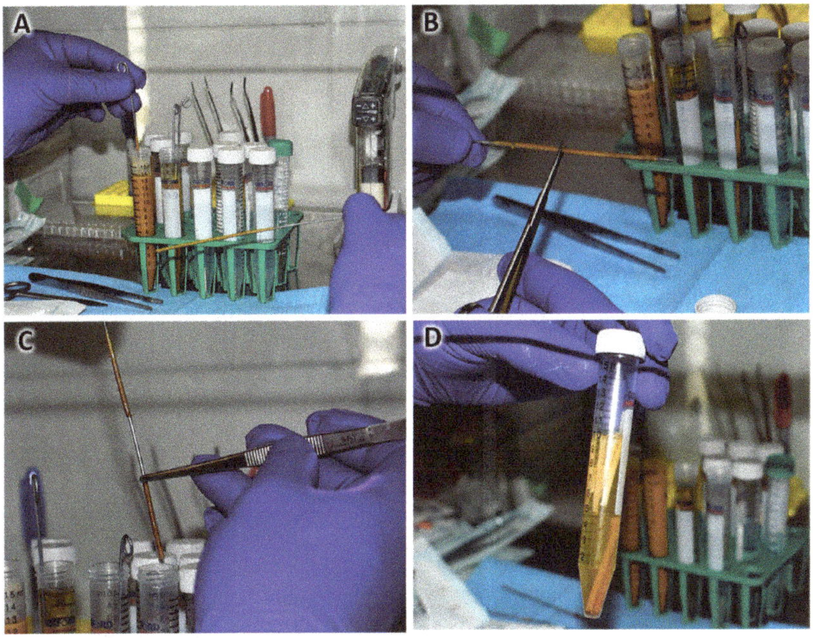

Figure 1. Steps in the method of preparing NGCs from cellulose and soy protein in a sterile environment (laminar flow chamber). (**A**) Immersions in 5.1% cellulose acetate solution and 13.6% soy protein acid hydrolysate. (**B**) NGC frame section and steel mold release. (**C**) Placement of prepared NGCs in preservation medium. (**D**) NGCs in the stock medium during soy protein release.

2.2. NGC Microbiological Growth Tests

To determine the adequate elaboration protocol described, it was necessary to test different methodologies to ensure the sterility of the NGCs. The NGCs were manufactured following the same manufacturing steps considering the following variables: the method of asepsis/antisepsis of CA and PS polymers and the medium for maintenance or cultivation.

To determine the adequate sterility protocol for preparing NGCs, a total of 16 NGCs were manufactured, of which 8 were made with autoclaved CA and sterilized SPAH under UV radiation for 30 min, and 8 were prepared with CA and PS sterilized under UV radiation for 30 min. Half of the samples from each group were cultured for 7 days at room temperature; additionally, half of the samples (n = 4) were cultured in 10 mL ultrapure distilled water (Corning), and the other half (n = 4) were cultured in 1% penicillin/streptomycin (Cytiva, Marlborough, MA, USA).

After 7 days, longitudinally cut 6 mm long NGC sections were cultured for 48 h in Ham's F12 culture medium (Cytiva) supplemented with 10% fetal bovine serum (Cytiva,

37 °C, 5% CO_2 in a culture chamber). Sample analysis was initially performed using an inverted microscope (EUROMEX), observing mainly the characteristics of the culture medium. After verifying the absence of supernatant material, the analysis of the NGC samples was completed using a scanning electron microscope (SEM-Hitachi SU3500, Hitachi, Tokyo, Japan).

2.3. NGC Structural Characterization—Macroscopic and Scanning Electron Microscopy Analyses

The morphological and morphometric analysis of the manufactured NGCs was performed with a scanning electron microscope (SEM-Hitachi SU3500, Hitachi, Tokyo, Japan) at a voltage of 10 kV. The samples were cut longitudinally and transversally. The samples were observed at different magnifications to analyze the external face, internal face, and cross-section. The morphometric structural characterization of the images was performed using the ImageJ v1.46r software (National Institutes of Health, Bethesda, MD, USA).

2.4. NGC Porosity and Surface Area—Nitrogen Adsorption and Degassing

The analysis of the surface area and pore size was performed using a porosimeter device (NOVA 1000e, Quantachrome instruments, Boynton Beach, FL, USA) according to a nitrogen adsorption and degassing method. Prior to analysis, the NGC samples were dried in a constant temperature incubator (Biobase, Tsinan, China) at 50° for 24 h. The results delivered were pore volume (g/mm^3), surface area (mean unit), and pore diameter (measured unit). The results were analyzed with Software Design Expert 6.0.6 (Stat-ease, Inc., Minneapolis, MN, USA).

2.5. Schwann Cell Viability Direct Biocompatibility—MTT Assay and Scanning Electron Microscopy

In vitro biocompatibility tests were carried out with a colorimetric assay to evaluate cellular metabolic activity according to the MTT assay.

Schwann cells (SCL 4.1/F7, ECACC 93031204) were grown in a T-75 bottle (Corning, Corning, NY, USA) with Ham's F12 culture medium (Cytiva) supplemented with 10% fetal calf serum (Cytiva) and 1% penicillin/streptomycin (Cytiva) at 37 °C, 5% CO_2 in a humidified environment (culture chamber). Cells were released from the plate with 2 mL of 0.25% Trypsin-EDTA (Cytiva) at 37 °C, 5% CO_2 in a humidified environment. After incubation, 4 mL of supplemented culture medium was added to inactivate the enzyme. Cells were collected in a 15 mL conical bottom tube (Corning) and centrifuged at 1000 rpm for 5 min. Cells were suspended in Ham's F12 medium supplemented at a concentration of 100,000 cells/10 µL.

Longitudinally cut, 6 mm long NGCs were placed in a 96-well plate (Corning) to which 100,000 cells were seeded on the inside of the NGC; the plates were incubated for 2 h at 37 °C, 5% CO_2 in a humidified environment. At the end of the incubation, 90 µL of supplemented culture medium was added to the wells until the next test.

The MTT test (Proliferation Kit I, Roche, Basel, Switzerland) was applied to assess each NGC at 24 and 48 h [30]. NGCs were transferred to a new 96-well plate supplemented with 100 µL of Ham's F12, and 10 µL of "MTT labeling reagent, 1×" (final concentration 0.5 mg/mL) was added to each NGC-containing well. The plates were incubated for 4 h at 37 °C, 5% CO_2. Once the incubation was finished, 100 µL of the "solubilization buffer, 1×" reagent was added, and the plates were incubated overnight at 37 °C, 5% CO_2. The following day, the NGCs were removed, and the reading was performed using Infinite50 equipment (Tecan, Männedorf, Switzerland) at 570 nm. The data were collected and analyzed in Excel (Office 365 Education—Microsoft, Redmond, WA, USA).

A total of 100,000 Schwann cells were seeded on the inside of longitudinally cut, 6 mm long NGCs, which were incubated for 24 and 72 h at 37 °C, 5% CO_2. As soon as the incubation time was over, the NGCs were washed twice with 1 mL of 1× PBS (Cytiva), and 2 mL of 2.5% glutaraldehyde in 0.1 M Sorensen's phosphate buffer (Electron Microscopy Sciences—EMS, Hatfield, PA, USA) was added. The samples were incubated for 30 min at room temperature. The fixed NGCs were washed 3 times with ultrapure distilled water (Corning) and analyzed under a scanning electron microscope (SEM-Hitachi SU3500, Hitachi, Tokyo, Japan) at a voltage of 10 kV.

2.6. Statistical Analysis

Statistical analysis of data was performed using SigmaPlot 12.0 software (Systat Software Inc., Palo Alto, CA, USA). Quantitative data are presented as the mean ± standard deviation. Statistical analysis was performed with a one-way analysis of variance (ANOVA), followed by a Bonferroni post-test. Values of $p < 0.05$ were considered statistically significant.

3. Results

3.1. NGC Microbiological Growth Tests

Observation of the Ham's F12 culture media under the inverted microscope after the 48 h NGC incubation indicated that bacteria were only found in the samples prepared with CA and PS irradiated with UV light and submerged in ultrapure distilled water for the previous 7 days. The other three preparation protocols did not reveal characteristics that would suggest contamination and bacterial proliferation in the prepared samples. Thus, the samples were analyzed using a scanning electron microscope to confirm the absence of bacteria (Figure 2A).

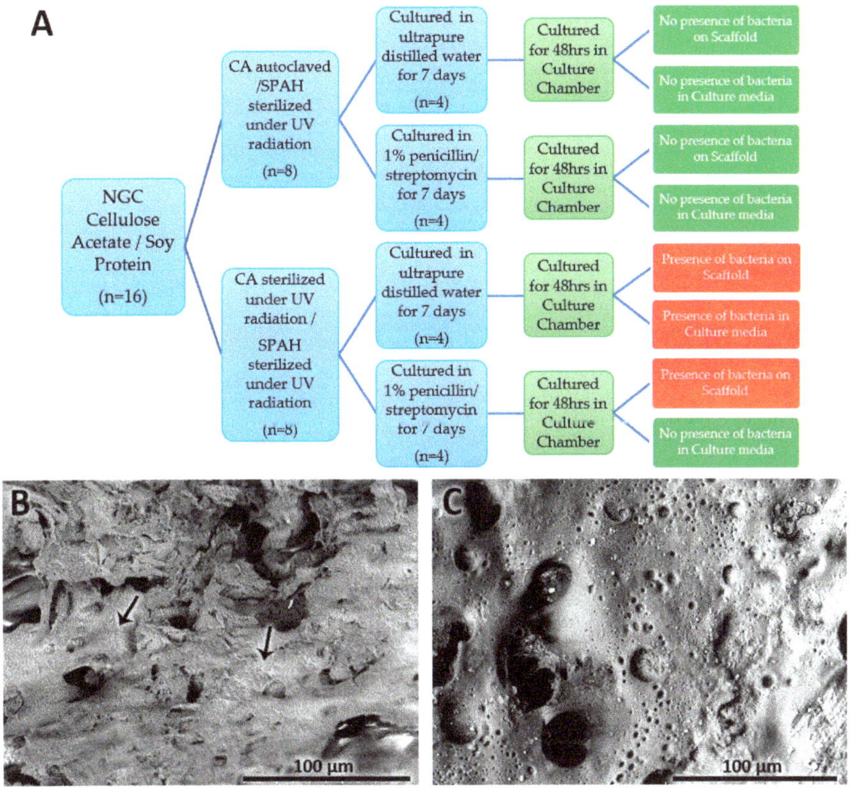

Figure 2. (**A**) Scheme of bacterial growth results from different manufacturing and storage protocols for cellulose and soy protein NGCs. (**B**) Cellulose and soy protein NGCs prepared without autoclaving cellulose showing the presence of bacteria (arrows) (Mag: ×500). (**C**) NGC prepared by autoclaving cellulose acetate and using UV light revealing the absence of bacteria (Mag: ×500).

The analysis of NGCs under SEM showed that bacteria adhered to the NGC structure of the samples prepared with CA and PS irradiated with UV light regardless of if they were immersed in ultrapure distilled water or 1% penicillin/streptomycin for 7 days; the strains found corresponded to diplococci and bacilli (Figure 2B). However, no bacteria were adhered to the NGC structures fabricated with autoclaved/sterile CA and PS irradiated with UV light, regardless of the medium in which they were immersed (Figure 2C).

3.2. NGC Structural Characterization

3.2.1. NGC Porosity and Surface Area—Liquid Nitrogen Adsorption and Degassing

The analysis of the surface area and porosity of NGC samples prepared with different CA and SPAH concentrations using the porosimeter with the liquid nitrogen adsorption and degassing method revealed the following data presented in Table 1.

Table 1. NGC porosity and surface area.

S1. CA 4.1 and SPAH 16.3 wt%	S2. CA 5.1 and SPAH 13.6 wt%	S3. CA 6.5 and SPAH 13.6 wt%
Pore Volume = 0.041 cc/g Pore Diameter Dv(d) = 3.251 nm Surface Area = 40.63 m^2/g	Pore Volume = 0.132 cc/g Pore Diameter Dv(d) = 3.249 nm Surface Area = 146.4 m^2/g	Pore Volume = 0.074 cc/g Pore Diameter Dv(d) = 3.28 nm Surface Area = 76.2 m^2/g

Considering the porosity assessment, it was determined that all following analyses were carried out on the NGC samples prepared with a concentration of CA5.1% and SPAH 13.6% wt, as they presented the largest surface area and pore volume.

3.2.2. Macroscopic Structural Characterization

Using the proposed manufacturing protocol, it was possible to manufacture NGCs up to 50 mm in length, which allowed cutting to different lengths according to different needs. These NGCs revealed consistency and flexibility similar to cardboard in samples that were kept moist (Figure 3A). The internal diameter remained constant at 1.4 mm, which corresponded to the diameter of the steel rod used as the mold (Figure 3B). This size was defined as ideal for implantation in the sciatic nerve of rats, but the size can also be customized depending on the nerve in which it is intended to be implanted.

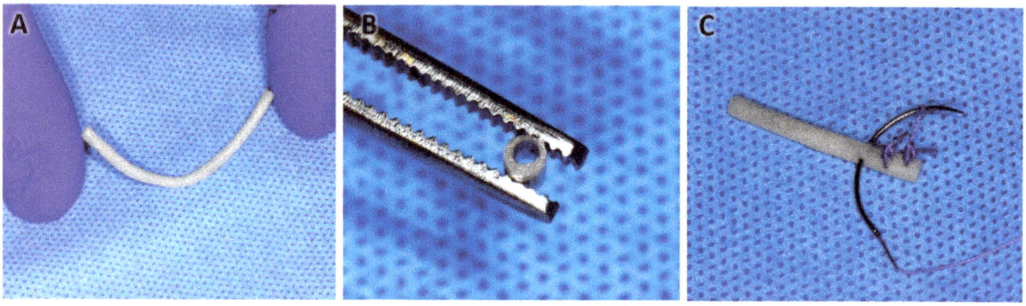

Figure 3. Macroscopic structure of the NGC. (**A**) Flexibility of the NGC while maintaining the tubular characteristic. (**B**) Cross-section of the NGC. (**C**) NGC structure sutured with silk and polyglactin.

The NGC structure obtained was suturable (Figure 3C), granting resistance to tearing using silk sutures and polyglactin 910. In addition, the tubular structure offered resistance to compression and some flexibility when pressed with the fingers and taken with anatomical forceps. This would allow its manipulation and implantation in future animal studies.

3.2.3. Microscopic Structural Characterization—Scanning Electron Microscopy

Analysis of CA and SPAH NGCs subjected to different sterilization and storage protocols revealed different morphological characteristics observed under SEM (Figure 4). It was noted that compared to the NGCs that were manufactured from cellulose subjected to the UV irradiation sterilization process, those manufactured from cellulose subjected to the autoclave sterilization process presented an internal surface with a rougher structure (Figure 5F,H) and better distribution of an apparently greater number of pores of different sizes (Figure 5B,D). The medium in which the samples were stored, regardless of the sterilization method, did not seem to alter the structure of the manufactured tubular scaffolds. In the four variations, the wall of the manufactured NGCs did not appear as a compact structure, revealing spaces interspersed with a trabecular structure.

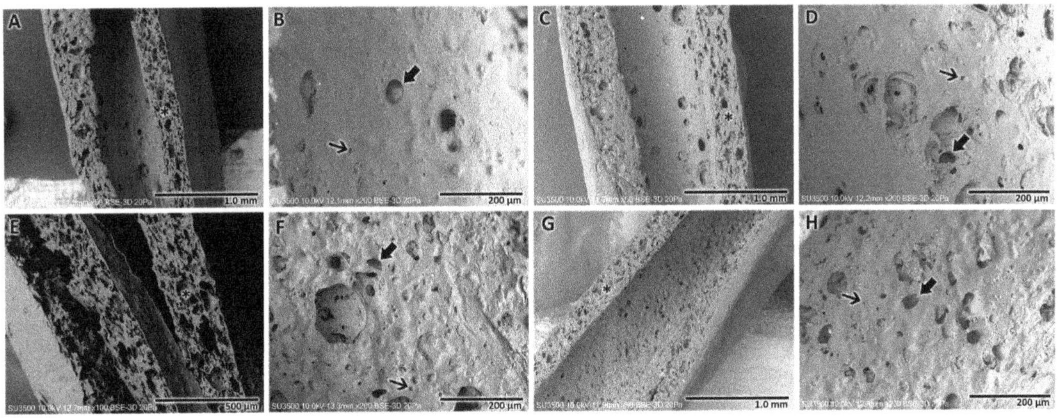

Figure 4. Scanning electron microscopy analysis of NGCs after different forms of sterilization and storage. UV-irradiated CA and SPAH NGC stored in ultrapure water. (**A**) Cross-section of the tubular structure revealing a noncompact wall (*) of the tubular structure (Mag.: ×50). (**B**) Internal surface revealing few pores of different sizes, major (arrow), minor (thin arrow), and a smoother surface (Mag.: ×200). UV-irradiated CA and SPAH NGC stored in 1% penicillin/streptomycin. (**C**) Cross-section of the NGC revealing the noncompact wall (*) of the structure (Mag.: ×50). (**D**) Inner surface of NGC, revealing major (arrow) and minor (thin arrow) pores (Mag.: ×200). Autoclaved CA and UV-irradiated SPAH NGC stored in ultrapure water. (**E**) Note the noncompact wall of the tubular NGC (*) (Mag.: ×100). (**F**) Internal surface of the NGC in which a rougher wall and a greater number of pores can be seen (major—arrow and minor pores—thin arrow) (Mag.: ×200). Autoclaved CA and UV-irradiated SPAH NGC stored in 1% penicillin/streptomycin. (**G**) Cross-section of the NGC revealing the wall with spaces and trabecular structure (*) (Mag.: ×50). (**H**) NGC internal surface with the presence of pores of different sizes (major—arrow; minor—thin arrow) distributed over the entire area of the structure (Mag.: ×200).

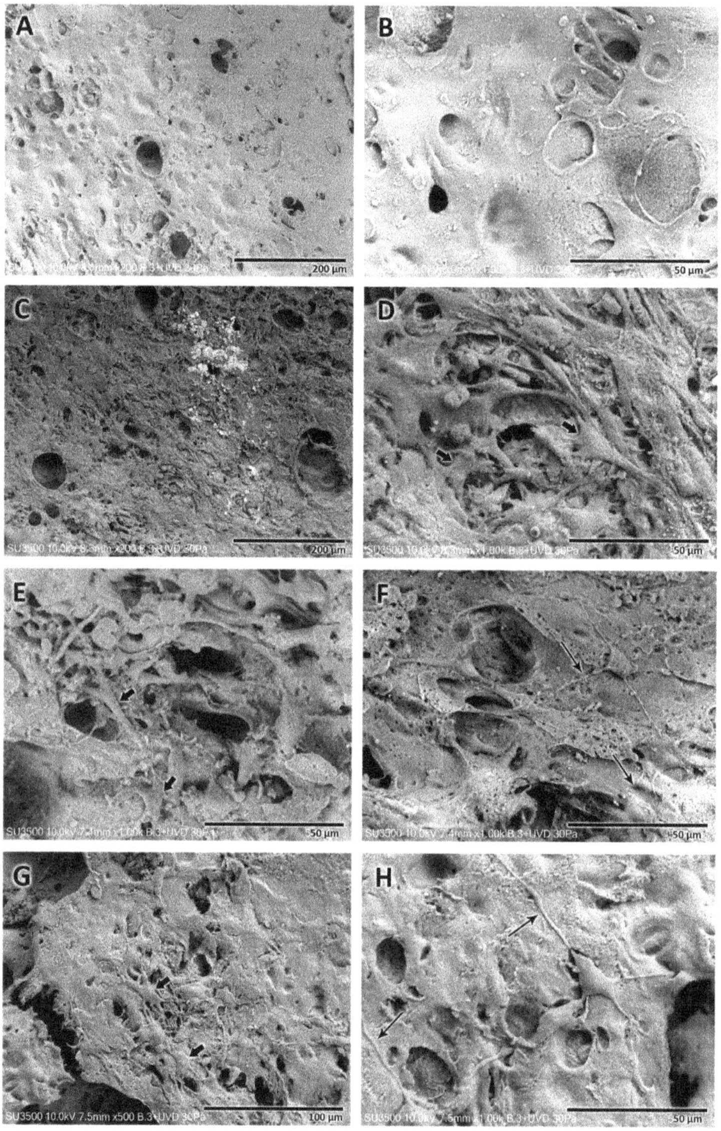

Figure 5. Direct cytocompatibility analysis of Schwann cells seeded on cellulose NGC functionalized with soy protein. (**A**) (Mag.: ×200) and (**B**) (Mag.: ×1000). Internal surface of the NGC without Schwann cell seeding; note the surface with the presence of pores and the absence of bacteria. (**C**) (Mag.: ×200) and (**D**) (Mag.: ×1000). On the inner surface of the NGC after 1 day of Schwann cell seeding, a high cell presence is observed forming a monolayer adhered to a star-shaped structure (arrow). (**E**) (Mag.: ×1000) and (**F**) (Mag.: ×1000). Internal surface of the NGC after 3 days of Schwann cell seeding. Note the presence of adhered Schwann cells (arrow) with a star-shaped shape in smaller quantities compared to the previous period and long cytoplasmic prolongations (long arrow). (**G**) (Mag.: ×500) and (**H**) (Mag.: ×1000). Internal surface of the NGC after 5 days of Schwann cell seeding. The presence of cells with characteristics similar to those observed after 3 days is observed, with the presence of cells with stellate morphology (arrow) and long cytoplasmic extensions (long arrow).

3.2.4. Microscopic NGC Morphometrics—Scanning Electron Microscopy

The morphometric characterization of the structure of the NGCs manufactured with the four sterilization and storage protocols was carried out using the SEM images obtained for the samples. The thickness of the walls and the diameter of the pores observed on the inner surface were analyzed and compared. In order to improve the analysis, the evaluated pores were divided into two types: major pores (>25 µm) and minor pores (<25 µm).

Tables 2–4 reveal the data for the NGC wall thickness and the major/minor internal pores, respectively.

Table 2. NGC wall width after different sterilization processes and storage methods.

Material (Sterilization)	Storage Method	Wall Width (µm, Mean ± SD)	Range (µm)	
			Min.	Max.
CA (UV rad.); SPAH (UV rad.)	ultrapure water	256.4 ± 17.1	237.5	278.1
CA (UV rad.); SPAH (UV rad.)	1% penicillin/streptomycin	267 ± 6.6	260.4	275.8
CA (autoclave); SPAH (UV rad.)	ultrapure water	248.5 ± 1.8	246.7	250.9
CA (autoclave); SPAH (UV rad.)	1% penicillin/streptomycin	259.6 ± 12.1	249	273.6

Table 3. NGC internal major pores (>25 µm) after different sterilization processes and storage methods.

Material (Sterilization)	Storage Method	Major Pore Size (µm, Mean ± SD)	Range (µm)	
			Min.	Max.
CA (UV rad.); SPAH (UV rad.)	ultrapure water	61.1 ± 5.4 a	54	66.6
CA (UV rad.); SPAH (UV rad.)	1% penicillin/streptomycin	32.8 ± 3.7 b	28.3	37.7
CA (autoclave); SPAH (UV rad.)	ultrapure water	42.1 ± 4.8	33.5	48.5
CA (autoclave); SPAH (UV rad.)	1% penicillin/streptomycin	40.7 ± 10.1	27.4	60.1

a = CA (UV rad.); SPAH (UV rad.) in ultrapure water larger than other 3 samples. b = CA (UV rad.); SPAH (UV rad.) in 1% penicillin/streptomycin smaller than CA (autoclave); SPAH (UV rad.) in ultrapure water.

Table 4. NGC internal minor pores (<25 µm) after different sterilization processes and storage methods.

Material (Sterilization)	Storage Method	Minor Pore Size (µm, Mean ± SD)	Range (µm)	
			Min.	Max.
CA (UV rad.); SPAH (UV rad.)	ultrapure water	11 ± 4.4	5.2	22.1
CA (UV rad.); SPAH (UV rad.)	1% penicillin/streptomycin	9.2 ± 5.5	1.9	23.5
CA (autoclave); SPAH (UV rad.)	ultrapure water	8.5 ± 4.3	2	23.3
CA (autoclave); SPAH (UV rad.)	1% penicillin/streptomycin	9.2 ± 4.8	2.5	23.3

Morphometric analysis of the NGCs manufactured with the four different sterilization and storage protocols revealed no significant differences ($p > 0.05$) in relation to the thickness of the walls of the tubular structure. The NGC wall thicknesses ranged between 237.5 and 275.8 µm.

Analysis of observable pores revealed significant differences in major pores. NGCs prepared with CA and SPAH irradiated with UV and stored in ultrapure water revealed the largest major pores (61.1 ± 5.4 µm, $p < 0.05$) compared to all other sterilization and storage protocols. Furthermore, samples of NGCs prepared with CA and SPAH irradiated with UV and stored in 1% penicillin/streptomycin revealed significantly smaller major pores (32.8 ± 3.7 µm, $p < 0.05$) than NGCs prepared with autoclaved CA and UV-irradiated SPAH stored in ultrapure water.

The minor pores also did not reveal significant differences ($p > 0.05$) among the four sterilization and storage protocols. The minor pore values for the NGCs ranged between 2 and 23.5 µm.

3.3. MTT Assay—Schwann Cell Viability/Biocompatibility

Tables 5 and 6 show the results of the first and second replicates of the cell viability assay, respectively.

Table 5. MTT assay for Schwann cell biocompatibility replicate 1.

Replicate 1 24–25 May 2023	Observance 24 h (Arbitrary Units)	Observance 48 h (Arbitrary Units)	Viability (obs 24 h/obs 48 h)
Sample 1	0.0291	0.0216	74.23%
Sample 2	0.0062	0.0055	88.71%
Sample 3	0.0415	0.0257	61.93%
Mean	0.026	0.018	68.75%

Table 6. MTT assay for Schwann cell biocompatibility replicate 2.

Replicate 2 31 May–1 June 2023	Observance 24 h (Arbitrary Units)	Observance 48 h (Arbitrary Units)	Viability (obs 24 h/obs 48 h)
Sample 1	0.0479	0.0367	76.62%
Sample 2	0.0374	0.0304	81.28%
Sample 3	0.0912	0.0556	60.96%
Mean	0.059	0.041	69.52%

Biocompatibility analysis using the MTT assay revealed a similar Schwann cell viability after 48 h in the two replicates performed on NGCs prepared with cellulose sterilized by autoclaving and the soy protein/cellulose mixture treated with UV light and stored in medium with 1% penicillin/streptomycin. Cell viability values ranged from 88.7% to 60.96%, with the average of the three samples from the first replication being 68.75% and that for the second replication being 69.52%.

3.4. NGC Direct Cytocompatibility of Schwann Cells

Samples of cellulose NGC functionalized with soy protein, analyzed using a scanning electron microscope without the seeded Schwann cells, revealed a porous structure without the presence of bacteria, as previously described.

The direct cytocompatibility analysis of Schwann cells was performed for three different periods, namely, 1, 3, and 5 days after cell seeding on the prepared NGCs (Figure 5). The different time points revealed different characteristics.

After 1 day of Schwann cell seeding, a monolayer of Schwann cells was observed that proliferated and adhered to the internal face of the tubular scaffold. These Schwann cells revealed cytoplasm with multiple projections with starry morphology occupying most of the surface, except for areas with pores larger than 60 µm; in these spots, the cells were not capable of forming a uniform layer.

In this period of analysis, some spherical structures were also observed, which suggested the presence of cells not adhered to the structure of the NGC because they were on top of other adhered cells, a fact that reinforced the formation of a monolayer of cells adhered to the internal surface of the manufactured cellulose NGCs.

After 3 and 5 days of cell seeding, the presence of Schwann cells adhered to the internal surface of the NGCs was still observed. Nevertheless, a reduction in the number of cells and the area occupied by the monolayer on the inner surface of the tubular scaffold was evident. The Schwann cells observed in these periods were adhered to the scaffolds, showing long cytoplasmic extensions and, consequently, maintaining the star-like shape.

4. Discussion

There are currently several methods for preparing NGCs [1,19]; however, the detailed manufacturing protocols, which would allow the exact reproducibility of these scaffolds, are normally not published to safeguard the copyright information.

Nevertheless, this report aimed to explain the entire manufacturing process of an NGC in a simplified and understandable manner, favoring its reproducibility and exposing some problems and solutions found during its development, such as the tube molding process and bacterial contamination.

The NGC tube molding was achieved after trying three variations of the solvent evaporation technique: 1. the solvent casting method, 2. the rolling layer method, and 3. the dip-coating method [1,21].

I. Solvent casting (molding method) used cellulose/soy protein dissolved in urea/NaOH solution and poured into a polyurethane mold with a central glass rod; the coagulation was carried out with the application of acetic acid [23,31]. Another study used hydroxyethyl cellulose and isolated soy protein dissolved in NaOH solution, and the solution was poured into molds with 1.3 or 1.7 metal rods; through a freeze-drying process, the final tubular shape was achieved [28].

In our methodological proposal, it was not possible to achieve an adequate resistant tubular structure; coagulation at room temperature results in a gel form that is complex to demold a tubular structure. The cellulose acetate was insoluble in a solution of urea/NaOH; thus, an attempt was made to coagulate the polymer solution with 5% acetic acid without obtaining positive results. In addition, the tubular structure adhered to the mold, which prevented its demolding. These inconveniences led us to discard this evaporation method.

An attempt was made to use petroleum jelly on the surfaces of the mold and central rod as well as to accelerate the acetone evaporation process in a constant temperature incubator (Biobase, Tsinan, China) at 25 °C for 5–10 and 15 min. The result was an amorphous and fragile NGC due to syneresis by polymer baking. In addition, it was suggested that the petroleum jelly adhered to the NGC could interfere with the cell adhesion results, and no literature was found using petroleum jelly or methods for its removal from the tubular structure (Figure 6A). These inconveniences led us to discard this evaporation method.

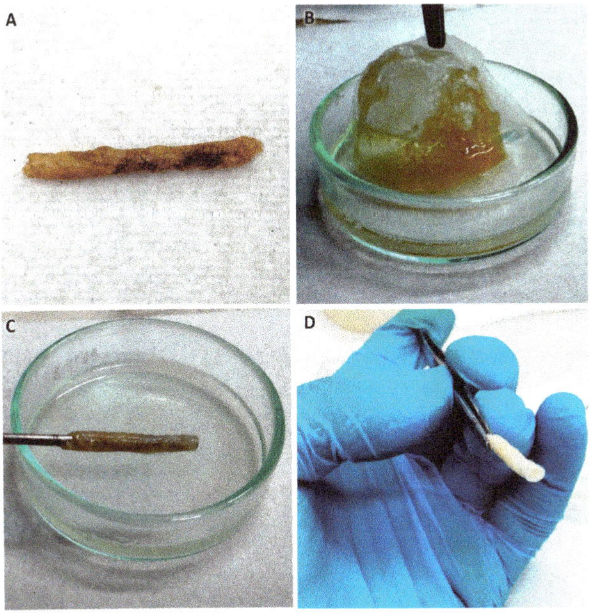

Figure 6. (**A**) NGC manufactured using cellulose acetate and soy protein by solvent casting method. (**B–D**) NGC manufactured using cellulose acetate and soy protein using the rolling layer method.

II. Rolling layer method (coiled membrane, Figure 6B–D): This method was used before to manufacture cellulose/soy protein membranes [31]. To give it a tubular shape, the floating membrane in distilled water was taken with tweezers, and with the help of a metallic rod with a 1.4 mm diameter, it was rolled by pressing on a sterile gauze or gauze pad; the rolling method is complex, and three disadvantages were detected:
    1. The membrane, still wet, is fragile, and therefore cracks and ruptures of the membrane occur, which are difficult to prevent during handling.
    2. The thickness of the floating membrane cannot be estimated, so it is difficult to predict the thickness of the resulting NGC walls; in addition, water and acetone are trapped between each membrane twist.
    3. As the soy protein is hydrophilic and disintegrates into islets it negatively affects the final morphology and porosity of the NGCs.

A second way to use this method of membrane elaboration is to let the polymers rest with the solvent in a Petri dish (Sorenson BioScience, Salt Lake City, UT, USA) for 24 h at room temperature [26] achieving the evaporation of the acetone without adding water. The results are not positive, possibly due to the different molecular weights of the polymers, which generate the decantation of the soy protein, obtaining a cellulose acetate membrane with a plastic aspect that is not flexible, is brittle, and is impossible to roll up. The disadvantages raised led us to discard this method.

III. Dip-coating method: After the failure of the aforementioned methods, the idea of granting the tubular shape by introducing a rod into the polymer solution [21] arose. When it was withdrawn intermittently from the solution, the evaporation of the solvent occurred, leaving the mixed polymers adhered to the rod; successively, in this way, the thickness of the NGC can be increased in a controlled manner. The process is practical and simple, allowing us to develop an NGC with a determined thickness and internal diameter; in addition, the resulting NGC presents relative flexibility, an ideal morphological characteristic that an NGC should possess [15]. Currently, the dip-coating manufacturing method has already been used in the manufacture of NGCs [21]; however, this process has not been reported in the literature when used with cellulose acetate functionalized with soy protein. A systematic review [21] reported that a disadvantage of the dip-coating method for the manufacture of NGCs is the impossibility of creating pores for the transport of nutrients. Our study, however, showed that the manufactured NGCs had pores on the inner face, in addition to a porous structure on the wall of the tubular structure.

The solvent evaporation method eliminated the disadvantages of the two aforementioned methods. When the rod was introduced into the polymers and slowly removed, a thin layer of cellulose/soybean protein was adhered, which allowed controlling the final thickness of the NGC; no oven was required to accelerate the evaporation of the solvent; the tubular structure formed instantaneously without requiring rolling; by not letting the polymers rest, there is no decantation and/or separation due to their different molecular weights; and the tubular structure was adhered only by the walls of the lumen facilitating the removal of the metal rod.

4.1. *Concentrations of Cellulose Acetate and Soy Protein for NGC Manufacturing*

The literature reveals that the concentrations of cellulose acetate and soy protein suitable for the manufacture of NGCs have not yet been defined since studies have reported a great variety of concentrations. Luo et al. [31] and Gan et al. [23] used a concentration of 3.5% soy protein and 10% cellulose; Kim et al. [32] used between 5 and 20 wt% of soy protein associated with poly(D,L-lactide-co-glycolide) to prepare microcapsules. Wang et al. [26] in chitosan/SPI membranes used a concentration of 3 wt% chitosan and varied between 0 and 30 wt% soy protein.

It has been determined that the SPAH concentration is the variable that determines the integrity and porosity of NGCs [26,31].

In our research, by mixing the independent solutions of each polymer, three different solutions, namely, CA 4.1 wt% and SPAH 16.3 wt%, CA 5.1 wt% and SPAH 13.6 wt%, and CA 6.5 wt% and SPAH 13.6 wt%, were obtained. It was noted that the different concentrations of CA and SPAH generated different porosities, in which both the increase in the CA and SPAH values and the reduction in the CA concentration revealed a reduction in the porosity of the samples obtained. For this reason, the sample with intermediate values of CA (5.1 wt%) and SPAH (13.6 wt%) was chosen, which revealed better porosity to carry out structural and biocompatibility analyses with Schwann cells. Furthermore, mixtures with CA concentrations above 6.45 wt% and SPAH above 17 wt% did not enable the manufacture of NGCs.

4.2. Bacterial Control of NGCs

NGC sterility is essential for biocompatibility with Schwann cells and subsequent implantation in living beings. After observing the first manufactured NGC samples under SEM, we found the presence of bacteria adhered to the structure.

Sarhane et al. [33] performed sterilization using ethanol solution baths (70° and 95° for 30 min) in scaffolds prepared by electrospinning poly-ϵ-caprolactone. In our case, despite carrying out this process, it was not possible to eliminate the bacteria that adhered to the NGCs. Romo-Velera et al. [34] used UV light for sterilization of scaffolds made of collagen/soy protein, irradiating the membrane for 20 min. In the same way, Oh et al. [35] sterilized the NGCs of PLGA with ultraviolet irradiation for 30 min, and Varshney et al. [36] sterilized silk/soy protein membranes with UV light for 30 min. The drawback of adopting this sterilization medium is that the tubular structure of the NGC makes it difficult for UV radiation to penetrate the lumen.

Therefore, the sterilization of NGCs with steam in an autoclave was proposed by Luo et al. [31] for cellulose membranes and soy protein and by Kundu et al. [27] for silk fiber NGCs. Since our tubular structure of the CA/SPAH suffers syneresis, modifying its physical properties, this method of sterilization was also discarded.

Considering the previous background, it was decided to carry out the autoclave sterilization methods of the CA and PS polymers dispensed in 1 g aliquots, which were used in the manufacture of the NGCs.

In our experiments, it was possible to sterilize the CA using moist heat in the autoclave cycle, without syneresis or imbibition of the material. On the other hand, SPAH underwent water imbibition in the autoclave sterilization process, which transformed the material into a compact mass that was not possible to subsequently dissolve in acetone to generate the final mixture with CA.

Thus, it was decided to sterilize the SPAH with UV radiation for 45 min during its mixing phase with acetone. In addition, a storage medium with ultrapure distilled water with 1% penicillin/streptomycin was used, and an aqueous medium was used in cell culture [34,37]. With these precautions, the sterility of the developed CA/SPAH NGCs was achieved.

4.3. Structural Characteristics of NGCs

The elaboration method proposed in the present study generated a structure with the presence of visible pores on the inner face and in the cross-section of the NGC. The internal diameter obtained was variable according to the diameter of the metal rod used.

Our NGC wall widths ranged from 237.5 to 278.1 μm, values considered adequate [15] that would allow correct handling and relative flexibility. In addition, our samples proved to be suturable with polyglactin 910.

The internal face porosity analyzed under SEM of the NGC samples was different from NGC images obtained by Gan et al. [23], in which it was not possible to observe internal pores. The addition of SPI caused microprotrusions on the surface that increased the structural surface area, as in our study, the release of SPAH from the polymeric matrix of cellulose acetate caused porosity to the structure and increased its surface area [38]. An immersion in a soluble liquid for soy protein for at least 24 h was necessary to form a porous tubular structure, similar to that reported by Luo et al. [29].

The concentration of 13.6 wt% of SPI used generates pores between 2 and 66 μm in diameter, with the average of "smaller pores" being 10 μm in diameter, similar to that reported by Gan et al. [23]. With the dip-coating method, it was possible to manufacture NGCs up to 6 cm in length, obtaining a uniform wall thickness; this is essential for repairing peripheral nerve injuries of greater extension.

Zhao et al.'s [28] scaffold elaborated with hydroxyethyl-cellulose/SPI reportedly had porosity with diameters from 20 to 200 μm in diameter, pores much larger than those reported in our study. A smaller pore range, as observed in our samples, could ensure the correct diffusion of nutrients [2] reducing the risk of infiltrating scar tissue [15].

4.4. Biocompatibility Evaluation of the Tubular Scaffold

Finally, the biocompatibility of Schwann cells was analyzed on the developed CA and SPAH NGCs.

Schwann cells were chosen for this analysis because they are recognized as fundamental glial cells in peripheral nerve regeneration. These cells contribute to neuronal survival and were important in the discovery of the axonal pathway, and they orchestrate the structure of the nerve with blood vessels and connective layers, endo, peri, and epineurium, during development. SCs exhibit remarkable plasticity in the context of nerve injury, wherein they differentiate into reparative cells and command a regenerative response that promotes nerve repair [39].

In our study, we have demonstrated a cellular viability of approximately 70% of SCs seeded on NGCs after 48 h. The viability values we obtained were similar to those reported by other authors using other cell types: 72% viability of human corneal epithelial cells and fibroblasts on ISP/glycerol membranes at 48 h [34] and 82.1% viability of fibroblasts on silk/ISP membranes over 4 days of culture [38]. Zhao et al. [38] evaluated the viability of fibroblasts seeded on hydroxyethylcellulose/SPI membranes around 80% at 72 h, concluding that the polymers pose low toxicity and high viability. Finally, Luo et al. [31] evaluated the cytocompatibility of SCs on cellulose/SPI membranes, reporting values of 120% viability at 72 h, as it seems that the high concentration of SPI at 30 wt% could be the factor that benefited this result.

Cellular cytocompatibility analysis under SEM was carried out on polymer membranes associated with soy protein. Wang et al. [40] in chitosan/ISP membranes concluded that ISP hydrolysis was beneficial for the proliferation, growth, and adherence of L929 cells. Zhao et al. [38] evaluated the cell arrangement of fibroblasts on membranes of hydroxyethyl-cellulose/ISP, reporting the presence of uniformly distributed cells. These results were similar to those observed in our investigation, namely, that the Schwann cells adhered and proliferated in the structure in a single layer. Specifically, regarding SCs, Luo et al. [31] analyzed their direct compatibility on cellulose/SPI membranes, showing good adhesion and proliferation 72 h after seeding. Our results were similar, revealing the presence of adhered cells up to 5 days after seeding on the developed AC/SPAH NGCs.

The NGCs manufactured with the proposed method were a first step towards the generation of a biomaterial that will allow the optimization and future complementation of this material with other nerve regeneration strategies. Our initial results suggest that this biomaterial could be enriched with cells and therapeutic agents as well as the combination of other biocompatible materials and neurotrophic growth factors that could generate an effective microenvironment that will favor cell growth and the regeneration of nervous tissue, including peripheral nerve tissue.

Among the study limitations and opportunities for improvement, the authors mention the possibility of refining the manufacture of NGCs with the evaporation of solvent/dip-coating method, mainly regarding the concentrations of CA and SPAH, which could improve the physicochemical properties of NGCs. These improvements could allow future implantation in different animal models for the study of functional recovery associated with nerve regeneration. Furthermore, there is still a lack of biomechanical analyses of this biomaterial and other more detailed cell proliferation and imaging analyses, such as confocal microscopy, which will help to understand the 3D interaction of cells with this biomaterial. Our research group will continue working to better understand and evaluate the material obtained, in addition to seeking to implement improvements in these NGCs.

5. Conclusions

The solvent evaporation/dip-coating method allowed an easy elaboration and customization of NGCs of cellulose acetate supplemented with soy protein. It was demonstrated in this report that a tubular structure could be fabricated, with different dimensions in length and internal diameter as well as presenting a uniform wall thickness that allowed its manipulation, which showed flexibility and was suturable.

The presence of pores on the surface and structural porosity was essential to allow the diffusion of nutrients. Initial biocompatibility analyses of NGCs with Schwann cells were promising. However, more cell proliferation analyses will be necessary to prove this biocompatibility.

In subsequent studies, it will be essential to find the ideal concentration of polymers to allow a higher concentration of SPAH since this is the polymer that provides porosity and increases biocompatibility. This study demonstrates that the NGC produced is safe and free of bacteria and that in the future it can be used for peripheral nerve regeneration studies as well as analyses of the mechanical and physicochemical properties of this biomaterial.

Author Contributions: Conceptualization, F.J.D., B.G., K.G. and M.E.G.-Q.; methodology, F.J.D., B.G., D.P.C., J.A.-A., D.V.-C. and P.M.-R.; investigation, B.G., F.J.D., D.P.C., M.F.L., D.V.-C. and P.M.-R.; formal analysis, F.J.D., B.G., K.G. and M.E.G.-Q.; writing—original draft preparation, B.G. and F.J.D.; supervision, F.J.D.; project administration, F.J.D.; and funding acquisition/resources, F.J.D. All authors have read and agreed to the published version of the manuscript.

Funding: This research was funded by ANID-Agencia Nacional de Investigación y Desarrollo de Chile. Fondo Nacional de Desarrollo Científico y Tecnológico—grant number: FONDECYT Iniciación 11190300. National ANID Master's Scholarship 2022, No. 4464-2022.

Institutional Review Board Statement: The in vitro study protocol was approved by the Scientific Ethics Committee of the Universidad de La Frontera (CEC-UFRO), under protocol number 096_22.

Informed Consent Statement: Not applicable.

Data Availability Statement: The data that support the findings of this report are available from the corresponding author [F.J.D.] upon reasonable request.

Conflicts of Interest: The authors declare no conflicts of interest. The funders had no role in the design of the study; in the collection, analyses, or interpretation of data; in the writing of the manuscript; or in the decision to publish the results.

References

1. Vijayavenkataraman, S. Nerve guide conduits for peripheral nerve injury repair: A review on design, materials and fabrication methods. *Acta Biomater.* **2020**, *106*, 54–69. [CrossRef] [PubMed]
2. Wu, P.; Tong, Z.; Luo, L.; Zhao, Y.; Chen, F.; Li, Y.; Huselstein, C.; Ye, Q.; Ye, Q.; Chen, Y. Comprehensive strategy of conduit guidance combined with VEGF producing Schwann cells accelerates peripheral nerve repair. *Bioact. Mater.* **2021**, *6*, 3515–3527. [CrossRef]
3. Murphy, R.N.A.; de Schoulepnikoff, C.; Chen, J.H.C.; Columb, M.O.; Bedford, J.; Wong, J.K.; Reid, A.J. The incidence and management of peripheral nerve injury in England (2005–2020). *J. Plast. Reconstr. Aesthet. Surg.* **2023**, *80*, 75–85. [CrossRef] [PubMed]
4. Kim, S.J.; Kwon, Y.M.; Ahn, S.M.; Lee, J.H.; Lee, C.H. Epidemiology of upper extremity peripheral nerve injury in South Korea, 2008 to 2018. *Medicine* **2022**, *101*, e31655. [CrossRef] [PubMed]
5. Li, R.; Liu, Z.; Pan, Y.; Chen, L.; Zhang, Z.; Lu, L. Peripheral Nerve Injuries Treatment: A Systematic Review. *Cell Biochem. Biophys.* **2014**, *68*, 449–454. [CrossRef] [PubMed]
6. Martinez de Albornoz, P.; Delgado, P.J.; Forriol, F.; Maffulli, N. Non-surgical therapies for peripheral nerve injury. *Br. Med. Bull.* **2011**, *100*, 73–100. [CrossRef]

7. Campbell, W.W. Evaluation and management of peripheral nerve injury. *Clin. Neurophysiol.* **2008**, *119*, 1951–1965. [CrossRef]
8. Allodi, I.; Udina, E.; Navarro, X. Specificity of peripheral nerve regeneration: Interactions at the axon level. *Prog. Neurobiol.* **2012**, *98*, 16–37. [CrossRef]
9. Raimondo, S.; Fornaro, M.; Tos, P.; Battiston, B.; Giacobini-Robecchi, M.G.; Geuna, S. Perspectives in regeneration and tissue engineering of peripheral nerves. *Ann. Anat.-Anat. Anz.* **2011**, *193*, 334–340. [CrossRef]
10. Ray, W.Z.; Mackinnon, S.E. Management of nerve gaps: Autografts, allografts, nerve transfers, and end-to-side neurorrhaphy. *Exp. Neurol.* **2010**, *223*, 77–85. [CrossRef]
11. Hasirci, V.; Arslantunali, D.; Dursun, T.; Yucel, D.; Hasirci, N. Peripheral nerve conduits: Technology update. *Med. Devices Evid. Res.* **2014**, *7*, 405–424. [CrossRef] [PubMed]
12. Grinsell, D.; Keating, C.P. Peripheral Nerve Reconstruction after Injury: A Review of Clinical and Experimental Therapies. *BioMed Res. Int.* **2014**, *2014*, 698256. [CrossRef] [PubMed]
13. Ferguson, T.A.; Son, Y.J. Extrinsic and intrinsic determinants of nerve regeneration. *J. Tissue Eng.* **2011**, *2*, 204173141141839. [CrossRef] [PubMed]
14. Jiang, X.; Lim, S.H.; Mao, H.Q.; Chew, S.Y. Current applications and future perspectives of artificial nerve conduits. *Exp. Neurol.* **2010**, *223*, 86–101. [CrossRef] [PubMed]
15. Apablaza, J.A.; Lezcano, M.F.; Lopez Marquez, A.; Godoy Sánchez, K.; Oporto, G.H.; Dias, F.J. Main Morphological Characteristics of Tubular Polymeric Scaffolds to Promote Peripheral Nerve Regeneration—A Scoping Review. *Polymers* **2021**, *13*, 2563. [CrossRef]
16. Pfister, L.A.; Papaloïzos, M.; Merkle, H.P.; Gander, B. Nerve conduits and growth factor delivery in peripheral nerve repair. *J. Peripher. Nerv. Syst.* **2007**, *12*, 65–82. [CrossRef]
17. Madduri, S.; Gander, B. Growth factor delivery systems and repair strategies for damaged peripheral nerves. *J. Controlled Release* **2012**, *161*, 274–282. [CrossRef]
18. Lin, Y.C.; Marra, K.G. Injectable systems and implantable conduits for peripheral nerve repair. *Biomed. Mater.* **2012**, *7*, 024102. [CrossRef]
19. Pierucci, A.; de Duek, E.A.R.; de Oliveira, A.L.R. Peripheral Nerve Regeneration through Biodegradable Conduits Prepared Using Solvent Evaporation. *Tissue Eng. Part A* **2008**, *14*, 595–606. [CrossRef]
20. Oh, S.H.; Kim, J.R.; Kwon, G.B.; Namgung, U.; Song, K.S.; Lee, J.H. Effect of Surface Pore Structure of Nerve Guide Conduit on Peripheral Nerve Regeneration. *Tissue Eng. Part C Methods* **2013**, *19*, 233–243. [CrossRef]
21. Kang, N.U.; Lee, S.J.; Gwak, S.J. Fabrication Techniques of Nerve Guidance Conduits for Nerve Regeneration. *Yonsei Med. J.* **2022**, *63*, 114. [CrossRef]
22. Merle, M.; Dellon, A.L.; Campbell, J.N.; Chang, P.S. Complications from silicon-polymer intubulation of nerves. *Microsurgery* **1989**, *10*, 130–133. [CrossRef] [PubMed]
23. Gan, L.; Zhao, L.; Zhao, Y.; Li, K.; Tong, Z.; Yi, L.; Wang, X.; Li, Y.; Tian, W.; He, X.; et al. Cellulose/soy protein composite-based nerve guidance conduits with designed microstructure for peripheral nerve regeneration. *J. Neural Eng.* **2016**, *13*, 056019. [CrossRef]
24. Catrina, S.; Gander, B.; Madduri, S. Nerve conduit scaffolds for discrete delivery of two neurotrophic factors. *Eur. J. Pharm. Biopharm.* **2013**, *85*, 139–142. [CrossRef] [PubMed]
25. Samrot, A.V.; Sathiyasree, M.; Rahim, S.B.A.; Renitta, R.E.; Kasipandian, K.; Krithika Shree, S.; Rajalakshmi, D.; Shobana, N.; Dhiva, S.; Abirami, S.; et al. Scaffold Using Chitosan, Agarose, Cellulose, Dextran and Protein for Tissue Engineering—A Review. *Polymers* **2023**, *15*, 1525. [CrossRef]
26. Wang, W.; Meng, Q.; Li, Q.; Liu, J.; Zhou, M.; Jin, Z.; Zhao, K. Chitosan Derivatives and Their Application in Biomedicine. *Int. J. Mol. Sci.* **2020**, *21*, 487. [CrossRef] [PubMed]
27. Kundu, B.; Kurland, N.E.; Bano, S.; Patra, C.; Engel, F.B.; Yadavalli, V.K.; Kundu, S.C. Silk proteins for biomedical applications: Bioengineering perspectives. *Prog. Polym. Sci.* **2014**, *39*, 251–267. [CrossRef]
28. Zhao, Y.; Zhang, Q.; Zhao, L.; Gan, L.; Yi, L.; Zhao, Y.; Xue, J.; Luo, L.; Du, Q.; Geng, R.; et al. Enhanced Peripheral Nerve Regeneration by a High Surface Area to Volume Ratio of Nerve Conduits Fabricated from Hydroxyethyl Cellulose/Soy Protein Composite Sponges. *ACS Omega* **2017**, *2*, 7471–7481. [CrossRef]
29. Luo, L.H.; Zhang, Y.F.; Wang, X.M.; Wan, Y.; Chang, P.R.; Anderson, D.P.; Chen, Y. Preparation, Characterization, and In Vitro and In Vivo Evaluation of Cellulose/Soy Protein Isolate Composite Sponges. *J. Biomater. Appl.* **2010**, *24*, 503–526. [CrossRef]
30. Thurzo, A.; Gálfiová, P.; Nováková, Z.V.; Polák, Š.; Varga, I.; Strunga, M.; Urban, R.; Surovková, J.; Leško, Ľ.; Hajdúchová, Z.; et al. Fabrication and In Vitro Characterization of Novel Hydroxyapatite Scaffolds 3D Printed Using Polyvinyl Alcohol as a Thermoplastic Binder. *Int. J. Mol. Sci.* **2022**, *23*, 14870. [CrossRef]
31. Luo, L.; Gong, W.; Zhou, Y.; Yang, L.; Li, D.; Huselstein, C.; Wang, X.; He, X.; Li, Y.; Chen, Y. Cellulose/soy protein isolate composite membranes: Evaluations of in vitro cytocompatibility with Schwann cells and in vivo toxicity to animals. *Biomed. Mater. Eng.* **2015**, *25*, 57–64. [CrossRef] [PubMed]
32. Kim, J.R.; Netravali, A.N. Self-Healing Properties of Protein Resin with Soy Protein Isolate-Loaded Poly(D,L -lactide- co -glycolide) Microcapsules. *Adv. Funct. Mater.* **2016**, *26*, 4786–4796. [CrossRef]
33. Sarhane, K.A.; Ibrahim, Z.; Martin, R.; Krick, K.; Cashman, C.R.; Tuffaha, S.H.; Broyles, J.M.; Prasad, N.; Yao, Z.C.; Cooney, D.S.; et al. Macroporous nanofiber wraps promote axonal regeneration and functional recovery in nerve repair by limiting fibrosis. *Acta Biomater.* **2019**, *88*, 332–345. [CrossRef]

34. Romo-Valera, C.; Guerrero, P.; Arluzea, J.; Etxebarria, J.; de la Caba, K.; Andollo, N. Cytocompatibility and Suitability of Protein-Based Biomaterials as Potential Candidates for Corneal Tissue Engineering. *Int. J. Mol. Sci.* **2021**, *22*, 3648. [CrossRef]
35. Oh, S.H.; Kim, J.H.; Song, K.S.; Jeon, B.H.; Yoon, J.H.; Seo, T.B.; Namgung, U.; Lee, I.W.; Lee, J.H. Peripheral nerve regeneration within an asymmetrically porous PLGA/Pluronic F127 nerve guide conduit. *Biomaterials* **2008**, *29*, 1601–1609. [CrossRef] [PubMed]
36. Varshney, N.; Sahi, A.K.; Poddar, S.; Mahto, S.K. Soy protein isolate supplemented silk fibroin nanofibers for skin tissue regeneration: Fabrication and characterization. *Int. J. Biol. Macromol.* **2020**, *160*, 112–127. [CrossRef]
37. Luo, L.H.; Wang, X.M.; Zhang, Y.F.; Liu, Y.M.; Chang, P.R.; Wang, Y.; Chen, Y. Physical properties and biocompatibility of cellulose/soy protein isolate membranes coagulated from acetic aqueous solution. *J. Biomater. Sci. Polym. Ed.* **2008**, *19*, 479–496. [CrossRef]
38. Zhao, Y.; He, M.; Zhao, L.; Wang, S.; Li, Y.; Gan, L.; Li, M.; Xu, L.; Chang, P.R.; Anderson, D.P.; et al. Epichlorohydrin-Cross-linked Hydroxyethyl Cellulose/Soy Protein Isolate Composite Films as Biocompatible and Biodegradable Implants for Tissue Engineering. *ACS Appl. Mater. Interfaces* **2016**, *8*, 2781–2795. [CrossRef]
39. Bosch-Queralt, M.; Fledrich, R.; Stassart, R.M. Schwann cell functions in peripheral nerve development and repair. *Neurobiol. Dis.* **2023**, *176*, 105952. [CrossRef]
40. Wang, X.; Hu, L.; Li, C.; Gan, L.; He, M.; He, X.; Tian, W.; Li, M.; Xu, L.; Li, Y.; et al. Improvement in physical and biological properties of chitosan/soy protein films by surface grafted heparin. *Int. J. Biol. Macromol.* **2016**, *83*, 19–29. [CrossRef]

polymers

MDPI

Review

Bioadhesive and Injectable Hydrogels and Their Correlation with Mesenchymal Stem Cells Differentiation for Cartilage Repair: A Mini-Review

Ján Kováč [1,2], Petra Priščáková [1,2], Helena Gbelcová [1,2], Abolfazl Heydari [1,3] and Stanislav Žiaran [1,4,*]

1 Medical Vision, Záhradnícka 55, 821 08 Bratislava, Slovakia; jan.kovac@nurch.sk (J.K.); petra.priscakova@fmed.uniba.sk (P.P.); helena.gbelcova@fmed.uniba.sk (H.G.); abolfazl.heydari@savba.sk (A.H.)
2 Institute of Medical Biology, Genetics and Clinical Genetics, Faculty of Medicine, Comenius University, 811 08 Bratislava, Slovakia
3 Polymer Institute of the Slovak Academy of Sciences, Dúbravská Cesta 9, 845 41 Bratislava, Slovakia
4 Department of Urology, Faculty of Medicine, Comenius University, Limbová 5, 833 05 Bratislava, Slovakia
* Correspondence: stanoziaran@gmail.com

Abstract: Injectable bioadhesive hydrogels, known for their capacity to carry substances and adaptability in processing, offer great potential across various biomedical applications. They are especially promising in minimally invasive stem cell-based therapies for treating cartilage damage. This approach harnesses readily available mesenchymal stem cells (MSCs) to differentiate into chondrocytes for cartilage regeneration. In this review, we investigate the relationship between bioadhesion and MSC differentiation. We summarize the fundamental principles of bioadhesion and discuss recent trends in bioadhesive hydrogels. Furthermore, we highlight their specific applications in conjunction with stem cells, particularly in the context of cartilage repair. The review also encompasses a discussion on testing methods for bioadhesive hydrogels and direct techniques for differentiating MSCs into hyaline cartilage chondrocytes. These approaches are explored within both clinical and laboratory settings, including the use of genetic tools. While this review offers valuable insights into the interconnected aspects of these topics, it underscores the need for further research to fully grasp the complexities of their relationship.

Keywords: bioadhesive; injectable; hydrogel; stem cell; differentiation

1. Introduction

Hydrogels, which are 3D cross-linked natural or synthetic polymer networks with high water-absorbing capacity and versatile fabrication characteristics, have wide-ranging applications, particularly in the fields of tissue engineering (TE) and regenerative medicine [1].

Injectable hydrogels specially offer potential advantages in minimally invasive local drug delivery, precise and site-specific implantation, as well as targeted delivery to hard-to-reach tissue sites and interface tissues. The phase transition in a polymer solution, from liquid to solid at a critical point, is known as the sol–gel transition state. Injectable hydrogels, including in situ forming and shear-thinning hydrogels, undergo a rapid sol–gel phase transition, which allows the matrix an easy taking of the shape of the cavity, providing a suitable fit and interface in tissues [2–4]. In this light, the adhesivity of applied hydrogel is one of the crucial properties for hydrogels in biomedicine.

Bioadhesive hydrogels have emerged as pivotal materials in the realm of cell therapy research, owing to their exceptional attributes. These attributes, including desired biocompatibility, biodegradability, tissue and cellular adhesion capabilities, as well as mechanical properties conducive to the emulation of the extracellular matrix (ECM), play a pivotal role in fostering critical cellular processes such as proliferation, wound healing, and tissue regeneration [5–7]. Drawing upon the information presented thus far and the observed experimental outcomes, it can be cautiously inferred that hydrogels exhibit favorable attributes as a potential material for biomedical applications, notably hinting at their potential suitability as a conducive environment for the proliferation of stem cells [8,9].

In the current scientific landscape, it is notable that there is a lack of recent reviews specifically focusing on the interplay between the bioadhesive properties of hydrogels and their role in the induction of MSCs differentiation. This absence of literature motivates our study for this significant correlation, which deserves attention. While the extant literature does contain a number of reviews that discuss hydrogels in the context of stem cells [10,11], none of them describes the relationship we explore in this review. We intend to examine the existing knowledge regarding hydrogel bioadhesiveness and its correlation to MSC differentiation, with a particular emphasis on elucidating the underlying mechanisms of bioadhesion and MSC proliferation and differentiation. This work is the first in this field due to its unique focus on revealing the multifaceted relationship between bioadhesiveness in hydrogel materials and its implications for the induction of MSC proliferation.

2. Bioadhesion

By definition, bioadhesion is the phenomenon in which natural and synthetic materials adhere to biological surfaces. This may or may not be associated with the use of adhesives to bond the material to the biological surface. Bioadhesion also refers to the incorporation of a biomaterial into the body, manifested by the formation of a biofilm on the biomaterial. Xiong at al. divided bioadhesion into three aspects: mucosal adhesion, cell adhesion and bioadhesives [7]. Mucoadhesion is a specific type of bioadhesion in which a layer of mucus gel forms on the surface of the biological surface during the adhesion process [12]. Cell adhesion is a complex phenomenon where aside from morphology, the chemical composition of the biomaterial surface interacts with surface molecules on cells [13]. Bioadhesives derived either from synthetic or biological source are highly biocompatible and biodegradable polymers, which are used to join two surfaces where at least one of them is a living tissue [14]. There are also other approaches to classify bioadhesion. Chopra et al. proposed three types: Type 1: adhesion between two biological phases; Type 2: adhesion of a biological phase to an artificial substrate; Type 3: adhesion of an artificial material to a biological substrate [15]. Overall, hydrogel adhesion involves a complex interplay of chemistry, topology, and mechanics, as various types of bonds are introduced (Table 1 or Figure 1). Hydrogels can manifest robust adhesion through the involvement of both covalent and noncovalent bonds. Covalent bonds contribute their inherent strength individually, while noncovalent bonds, through the synergistic interplay of polymer chains, collectively impart substantial adhesive properties [7,16,17]. The nature of bonds present within hydrogels significantly influences the process of cross-linking, which subsequently impacts their adhesive properties. Notably, heightened cross-linking levels tend to diminish the adhesive capacity of hydrogels. This reduction is attributed to the constrained mobility resulting from increased cross-linking, thereby impeding functional groups along polymer chains from accessing the hydrogel surface and establishing interactions with the substrate for adhesion [18].

Table 1. Overview of representative chemistry bonds that link hydrogel to biological surfaces known as bioadhesiveness [16].

Bond Types	Representative Bonds
Permanent covalent bonds	Carbon–carbon, Siloxane, Amide, Carbon–nitrogen
Dynamic covalent bonds	Disulfide, Imine, Boronate ester complexations
Non-covalent bonds	Ionic interactions, Hydrogen bonds, Hydrophobic interaction, Dipole–dipole interaction, π–π interaction

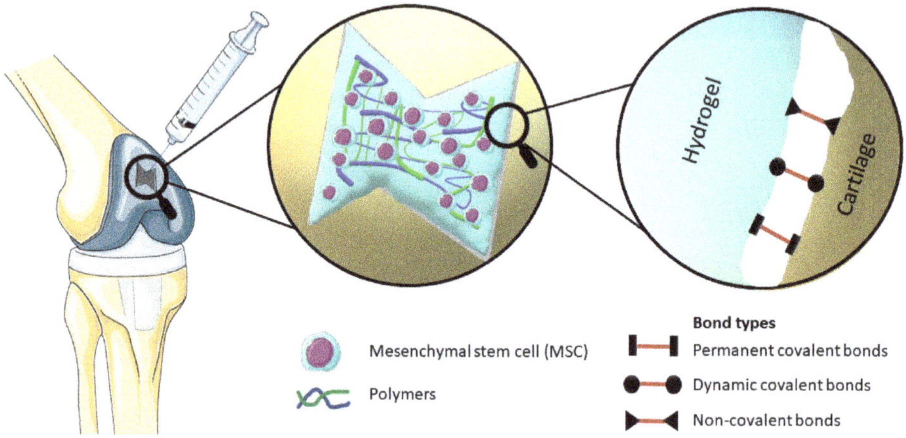

Figure 1. Elucidation of the mechanism of an injectable bioadhesive hydrogel with incorporated stem cells, which effectively occupies the defect in the cartilage structure, provides a visual representation of the different types of bonds involved in bioadhesion, namely Permanent covalent bonds, Dynamic covalent bonds and Non-covalent bonds.

Generally, bioadhesives can be classified into three basic categories: I. Wound closure, II. Sealing leakage, III. Immobilization [19]. An ideal bioadhesive polymer is characterized by the following criteria [15]:

1. the polymer and its degradation products must be non-toxic, biodegradable and non-absorbable;
2. it should have the ability to establish robust bonds with mucus or other biological surfaces;
3. rapid and strong adhesion to surfaces should be achievable;
4. it should offer ease of formulation with drugs without impacting drug release patterns.

An overview of recent research developments in the realm of bioadhesive hydrogels (Table 2) collectively offers insights into a variety of applications and material enhancements, underscoring the ongoing progress in the field of bioadhesive hydrogel technology. Then, a more narrowly focused overview of recent research in bioadhesive and bioadhesive injectable hydrogels in cartilage applications is presented in Table 3.

Table 2. An overview in recent research developments in bioadhesive hydrogels.

Materials	Adhesiveness Origins	Application	Ref.
Acrylic acid and -N,N′-methylenebisacrylamide (MBA) as a cross-linking agent	Phenol groups	Acceleration of oral wound healing	[20]
Vonoprazan fumarate (VF) and acidic fibroblast growth factor (AFGF)	Nano-silica	Hydrogel for Stomach Perforation Repair	[21]
Thiolated γ-polyglutamic acid (PGA-Cys)	Disulfide	For local delivery of keratinocyte growth factor (KGF)	[22]
Poly(acrylic acid) grafted with N-hydroxysuccinimide ester (PAAc-NHS ester) and gelatin	-	Spherical hydrogel network inhalation for enhanced lung defence (SHIELD) against SARS-CoV-2	[23]
Tyramine-modified hyaluronic acid (HA-Tyr)	Tyrosine-containing extracellular matrix proteins	Cartilage repair	[24]
Ion-activated hydrogel (Natural corneal extracellular matrix and peptide-modified alginate)	-	Corneal regeneration	[25]
Polyvinyl alcohol (PVA), dextran (Dex) and borax	Phenol groups	Effectively activated wound healing	[26]
Poloxamer and thiolated γ-PGA polymer	Glycoproteins groups	Diabetic wound healing	[27]

Table 2. *Cont.*

Materials	Adhesiveness Origins	Application	Ref.
Nitrobenzene-modified hyaluronic acid (HA-NB) + methacrylated polyvinyl alcohol (PVA-MA), with lithium phenyl-2,4,6-trimethylbenzoylphosphinate (LAP) as the photoinitiator	Aldehyde groups	Repair of arterial bleeding	[28]
The protocatechuic aldehyde hybridized collagen-based all-natural hydrogel (FGMA/FG/PA)	Protocatechuic aldehyde (PA)	Promotes Angiogenesis and Diabetic Wound Healing	[29]
Cathechol-conjungated chitosan (CHI-C)	Catechol groups	Hemostasis and bone regeneration	[30]
Catechol-grafted carbon-quantum-dot DA-CQD@Pd hydrogel	Catechol groups	Cancer immunotherapy	[31]
Elastic protein-based hydrogel—grafted glycidyl methacrylate on the gelatin backbone (GELGYM)	-	Ocular tissue engineering	[5]
GTT-3 hydrogel: Tannic acid modified gelatin (Gel-TA) with transglutaminase (TG)	Hydrogen bonding, imine linking, and acyl-transfer reaction	Tissue wound hemostasis	[32]
Methacrylate-hybridizedpoly(3,4-ethylenedioxythiophene)nanoparticle(dPEDOTNP)-incorporated hydrogel	Catechol groups	Brain–machine interference (BMI)	[33]
Chondroitin sulfate (CS), a cartilage-derived sulfated glycosaminoglycan (GAG) as the backbone + catechol moities 3,4-dihydroxyphenylalanine (DOPA)	Catechol groups	Multi-functional bioadhesive	[34]
Platelet-rich plasma (PRP)-laden GelMA hydrogel	-	Hydrogel Contact Lens for the Treatment of Ocular Surface Chemical Injuries	[35]
Silk fibroin (SF) + Poly(ethylene glycol) (PEG) hydrogel	Catechol groups	Wound closure	[36]
Methacryloylated gelatin (GelMA), Pluronic F127 diacrylate (F127DA) & Aldehyded Pluronic F127 (AF127) co-assembled bi-functional micelles and collagen type I (COL I) hydrogel	Aldehyde groups	Corneal patch for in situ sutureless corneal repair	[37]
Cephalexin NT and chitosane hydrogel	Positively charged groups	Antibacterial drug delivery	[38]
Hybrid double-network polydopamine–hyaluronic acid hydrogel	-	Disposable wound dressing	[39]
Gelatin and glycerine-based hydrogel	-	Site-specific drug release	[40]
Methacrylate gelatine + methacrylated silk fibroin + Pluronic F127 diacrylate	Amine-MA and thiol-MA groups plus intermolecular non-covalent bonds	Bladder injury repair	[41]

Table 3. An overview in recent research developments in bioadhesive and bioadhesive-injectable hydrogels in cartilage repair.

Materials	Adhesiveness Origins	Ref.
poly(ethyleneglycol)(PEG)–basedhydrogel	Sulfate and acrylate groups.	[42]
Polydopamine–chondroitin sulfate–polyacrylamide (PDA–CS–PAM)	Catechol group	[43]
Catechol-functionalized Chondoritin sulfate (CS-CA) hydrogel	Catechol group	[44]
Polydopamine/gelatin-poly(acrylic acid) (PDA/Gel-PAA) composite hydrogel	Catechol groups	[45]
Oxidized hyaluronic acid (OHA) and N-(2-hydroxypropyl)-3-trimethylammonium chitosan chloride (HTCC) methacrylate (HTCCMA) hydrogel	Aldehyde groups	[46]
Hyaluronic acid-transglutaminase (HA-TG) hydrogel	Covalent bonds	[47]
Self-cross-linked oxidized alginate/gelatin hydrogel	Aldehyde groups	[48]
Doublle cross-linked Hyalourinoc acid hydrogel (Deial Alder clic reaction ad phenyl-boronate ester bond)	Catechol groups	[49]
Tyramine-modified hyaluronic acid (HA-Tyr) hydrogel	Adhesion to cartilage due to hyaluronic acid	[24]
Sulfhydryl chondroitin sulfate and polydopamine (CS-PDA)	Catechol groups	[50]
Gelatine–Methacryloyl(GelMA)–glycol chitosan hydrogel	Protein and methacrylate groups	[51]
Eznymytically cross-linked alginate—dopamine, chondroitin sulfate, and regenerated silk fibroin (AD/CS/RSF)	Aldehyde and dopamine groups	[52]

3. Testing of Bioadhesion

Adhesion represents a multifaceted phenomenon governed by intricate interactions involving chemical, topological, and mechanical factors. The comprehensive evaluation of adhesion typically encompasses four distinct mechanical assessments. Notably, within this set of tests, the probe-pull and lap-shear methodologies serve to quantitatively assess adhesion strength by specifically gauging the maximum force per unit area. Meanwhile, the peel and bilayer-stretch tests are employed to assess adhesion toughness, quantifying the energy necessary for separation per unit area. These four tests serve to investigate and differentiate various facets of adhesion properties [16,53].

The majority of adhesion and bioadhesion tests are typically mechanical tests conducted usually ex vivo. Peel tests are a type of mechanical tests used to assess the strength of adhesive bonds, particularly for flexible adherents [54]. There are more variants based on the peeling angle, and all are standardized protocols; e.g., Wei at al. followed standard protocol for peeling adhesion test ASTM F2256-05 [55,56], sometimes minorly modified by research teams; e.g., Jeon et al. utilized a 90° peel test with a porcine skin substrate [57].

One of the most commonly employed assessments for evaluating the adhesive properties of bioadhesive hydrogels involves the utilization of the lap-shear test, also referred to as bulk adhesion testing. The test assesses shear strength, with cohesive failure occurring within the adhesive, while the adhesive failure depends on the adherend's interface properties [58]. It is a standardized method (ASTM F2255:2005 [59]) that research teams modify [60]; usually, the test undergoes ex vivo utilizing porcine skin [20,61], and the test could be also performed in vitro [41].

The form of a bilayer stretch test methodology can be applied to assess extensional adhesion, wherein the adhesion energy is quantified when the hydrogels are either in their unextended or extended states [62]. Moreover, novel perspectives on adhesion measurement are emerging; e.g., Dehene et al. recently introduced a straightforward and replicable supplement method in viable tissues [63]. Ultimately, scientific teams frequently quantify adhesion in a straightforward manner by using weights and increasing tensile loading till adhesion failure [64,65].

In addition to mechanical tests, biocompatibility tests are an important part of hydrogel bioadhesiveness tests. One such test is the ISO-10993-11 [66] medical device rules and standards. Thanusha at al. evaluated biocompatibility for the developed hydrogel wound dressing [67].

The evaluation of bioadhesive hydrogels also includes clinical trials. For instance, as reported in the study by Øvrebø et al., the transition of hydrogels from laboratory development to clinical application necessitates adherence to an extensive array of protocols and regulatory standards, as well as the establishment of post-market surveillance measures [68].

4. Application of Bioadhesive Injectable Hydrogels in Cartilage Regeneration

Bioadhesive injectable hydrogels have garnered substantial interest in recent years due to their remarkable properties. The diverse applications of these hydrogels, ranging from wound healing and tissue repair to cell adhesion and wearable sensors, are discussed, underscoring their promising role in biomedicine and offering valuable insights for future research [7]. As an illustration of the increasing prominence of bioadhesive injectable hydrogels in medicine, various studies stand out. These studies encompass adhesive hydrogels for delivering mesenchymal stem cell-derived exosomes to treat spinal cord injuries [69], an innovative approach using hypoxia-stimulated exosomes within a peptide-modified adhesive hydrogel for spinal cord injury treatment [70], the GelMA–dopamine–EV hydrogel for enhanced MSC-EV function in diabetic wound healing [71], and an adhesive hydrogel integrated with placental mesenchymal stem cell-conditioned medium (CM) to prevent uterine adhesions and improve patient outcomes [72]. Additionally, a PEG-based hydrogel shows potential for muscle regeneration [73], and Col/APG hydrogels incorporating umbilical cord stem cell factor (SCF) offer effective therapeutic treatment for diabetic wounds [12] as well as for diabetic ulcer treatment [74]. The bioadhesive injectable hydrogel with a phenolic nanozyme (SAN) and a CpGODN adjuvant holds promise for localized immunomodulation and catalytic immunotherapy in the tumor microenvironment [31]. Inspired by mussel adhesive proteins, a dopamine-modified poly(α,β-aspartic acid) derivative (PDAEA) forms an injectable bioadhesive hydrogel with strong adhesion and drug delivery potential [75]. An innovative dynamic cross-linked photothermal hydrogel adhesive exhibits photothermal effects and on-demand removability, suitable for wound closure and healing, including MRSA-infected wounds [76]. A novel injectable acacia gum (AG) hydrogel with rapid gelation, self-healing, and effective bioadhesion holds promise for future biomedical applications as a wound-healing agent carrier [77]. Lastly, a composite hydrogel designed for bladder injuries shows potential for tissue engineering and bladder tissue regeneration [41], and a Tetra-PEG hydrogel bioadhesive (SS) offers sutureless repair of GI defects with controlled inflammation and tissue regeneration [78].

Articular cartilage has limited regenerative capacity. MSC-based approaches have emerged as a promising alternative in the treatment of cartilage defects and osteoarthritis. MSCs are a promising source of therapeutically relevant cells for hyaline cartilage regeneration due to their capacity to differentiate into the chondrogenic lineage. However, experimental evidence suggests that after a while, intra-articularly injected MSCs tend to differentiate into transient cartilage that is transformed into bone by the endochondral ossification rather than hyaline articular cartilage. This process leads to decreased effectiveness of the treatment. Similarly, the stratified ultrastructure and spatial organization of native hyaline cartilage disappear [79]. At the same time, the most MSCs injected intra-articularly fail to attach to the damaged cartilage layer, and it is possible that they quickly spread into systemic circulation due to the rapid turnover of synovial capillaries and lymphatic vessels [80]. Consequently, for the optimization of clinical strategies in the domain of cell-based cartilage engineering, it becomes imperative to establish a conducive 3D microenvironment. This microenvironment should comprise a tailored amalgamation of biomaterials and bioactive factors, aimed at further augmenting the differentiation of MSCs into chondrocytes.

Mesenchymal stem cells (MSC) are a promising source of therapeutically relevant cells for hyaline cartilage regeneration due to their capacity to differentiate into the chondrogenic lineage. The aim of the targeted differentiation is to obtain an artificial cartilage tissue with biomechanical properties similar to that of native hyaline cartilage (hyperelastic and dissipative properties, smoothness, toughness, wear resistance, resistance to compressive, tensile, and shear forces). In addition to the MSC differentiation into chondrocytes, the enhancement of the synthesis of the proteins of the hyaline cartilage extracellular matrix including fibronectin, collagens, glycosaminoglycans, proteoglycans, cytokines, and growth factors involved in the functioning of cartilage [81,82] is necessary.

MSCs can be applied to a suitable scaffold without prior induction of differentiation. Then, the so-called indirect method of differentiation is carried out, and its success depends on the properties of the scaffold. Another method is the in vitro targeted direct differentiation of MSCs into the chondrocytes that are subsequently applied to the scaffold (Figure 2). Clinical trials with MSC therapies for the regeneration of hyaline cartilage are summarized by Carneiro et al. [83]. Most experimental methods of hyaline cartilage regeneration, which have already been introduced in clinical practice, use direct modification techniques.

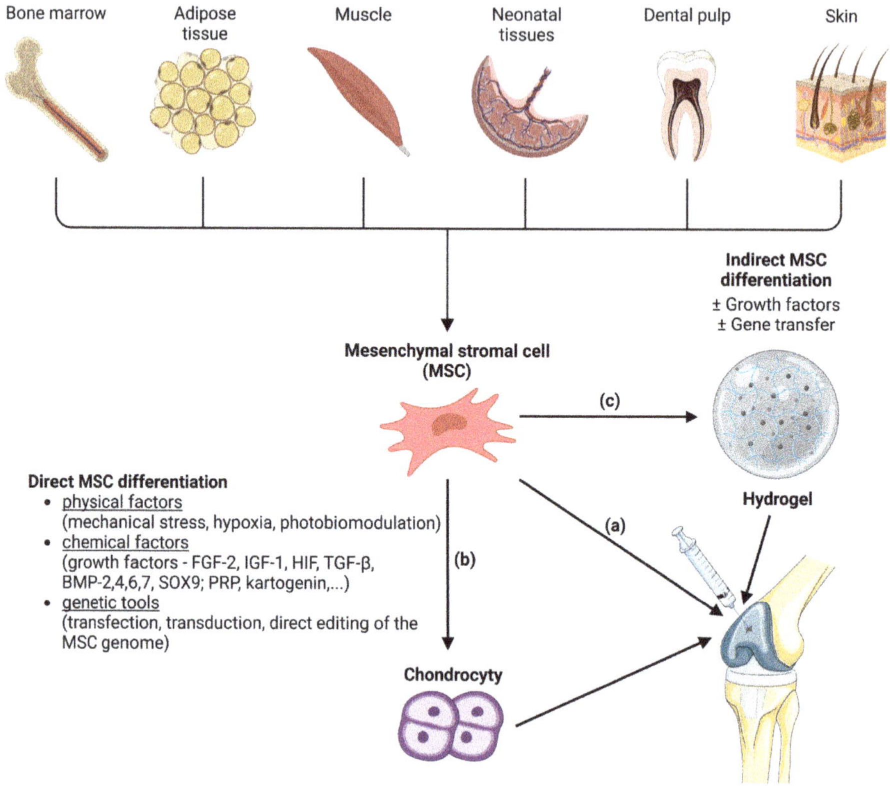

Figure 2. Methods of mesenchymal stem cell (MSC) differentiation into the hyaline cartilage´s chondrocytes. MSCs are a source of the therapeutically relevant cells for cartilage regeneration. MSCs can be harvested from various tissues (bone marrow, adipose tissue, muscle, neonatal tissues, dental pulp and skin). MSCs can be injected directly into joint (**a**). Another approach is induction of MSC differentiation into chondrocytes before microinjection into the joint. Induction of differentiation can be direct (**b**) or indirect (**c**). Direct MSC differentiation can be induced by various physical, chemical and genetical factors. In the case of indirect MSC differentiation, the MSCs are applied to a suitable scaffold (like hydrogels) where the differentiation is induced with or without the presence of growth factors or gene transfer. After that, hydrogel with attached cells is microinjected into the joint. Created with BioRender.com.

5. The Clinically Used Direct Methods of the MSC Differentiation into the Hyaline Cartilage's Chondrocytes

Various physical and chemical factors can be used to induce differentiation of the MSCs into the hyaline cartilage's chondrocytes. The main physical factors affecting the proliferation of chondrogenic cells are periodic mechanical stress, like cyclic strain or fluid shear stress [84–86], hypoxia [87], and electromagnetic radiation, so-called photobiomodulation [88–91]. Subsequent transformation of the physical signal into a biochemical signal is mediated by integrins and focal adhesion [92]. The potential of feasible hydrostatic pressure to effectively promote the proliferation and chondrogenic differentiation of mesenchymal stem cells was demonstrated in vitro [93], and the effect of photobiomodulation was successfully preclinically [94] as well as clinically [95] tested. Moreover, the combination of the physical stimuli with a scaffold that mimics the native cartilage microenvironment has been found to enhance chondrogenesis for cartilage repair [96].

From the chemical factors, growth factors are an integral part of forming the real conditions of the microenvironment and play a key role in the processes of cell development, including chondrogenic differentiation. Modification of MSCs using recombinant growth factors is one of the simplest, safest, and experimentally and clinically proven approaches to induce differentiation and changes in cell proliferation. The most studied factors with a direct influence on the chondrogenic differentiation of mesenchymal stem cells are fibroblast growth factors FGF-2 [97], insulin-like growth factor IGF-1 [98], hypoxia factors HIFs, cytokines of the transforming growth factor superfamily TGF-β [99,100], and associated bone morphogenetic proteins BMP-2,4,6,7 [101], as well as the SOX9 transcription factor [91,102,103]. A preclinical study on a rabbit cartilage injury model demonstrated the ability of recombinant SOX9 protein to induce reparative tissue formation with features of hyaline cartilage when administered at the site of microfracture [104]. The clinical study focused on the safety and tolerability as well as the dose-limiting toxicity and the maximally tolerated dose of intra-articular BMP-7 finished in 2010, and provided support for the continued development for the treatment of osteoarthritis [105]. However, the results from the phase II studies were not published, and no further studies have been proposed. Currently, there are no BMP-2 products in clinical trials for the treatment of osteoarthritis [106].

The application of platelet-rich plasma (PRP) is another therapeutic approach. Activated platelets (by thrombin, calcium, or collagen) release growth factors that induce chondrogenic differentiation of MSCs that leads to cartilage regeneration in rabbits [107]. Phase I and phase II clinical trials to assess the effectiveness of PRP in the treatment of knee osteoarthritis are ongoing (ClinicalTrials.gov ID: NCT05579665, ClinicalTrials.gov ID NCT02118519).

Small molecule drugs, such as kartogenin (KGN), curcumin, and resveratrol, show promising advantages over frequently used growth factors. They are too small to induce the immune response and are more affordable [108]. KGN is one of the most common small molecules used as a chondrogenic factor, and it is also able to improve the production of chondrogenesis-related proteins of MSCs, including collagen type II and aggrecan [109,110].

6. Genetic Tools Used for the Direct MSC Differentiation into the Hyaline Cartilage's Chondrocytes In Vivo and In Vitro

By genetic modification of cells, which includes transfection, transduction, and direct editing of the genome, it is possible to control chondrogenic differentiation as well as the production of ECM proteins. All three methods have been used to repair cartilage in animals, but have not yet been used in clinical practice. For the application of genetic engineering tools, it is necessary to know in detail the molecular mechanisms of chondrogenic differentiation of MSCs. The topic of signalling pathways and cytokines in chondrogenic differentiation of MSCs is summarized by Yang X et al. [111]. Transgenic growth factors are used for induction of chondrogenic differentiation, transgenic transcription factors are used for the regulation of differentiation, and oligonucleotide delivery can direct MSC differentiation through post-transcriptional gene regulation. Unlike transgenic proteins and RNA, DNA must be carried into the nucleus. From a number of existing methods, those suitable for the modification of a larger population of cells should be selected. Similar to chemical factors, genetic tools allow the use of direct and indirect methods of MSC differentiation; therefore, it is necessary to decide whether the MSC differentiation should take place prior to scaffold seeding, or whether scaffold containing plasmid DNA (pDNA) should be used.

In the process of transduction adenoviruses, retroviruses, herpes simplex virus, adeno-associated viruses, and lentiviruses have been used as vectors to induce chondrogenic differentiation of MSCs [112]. Through transduction, genetic material (vector and transgene) can be delivered directly into the joint (in vivo procedure) or into the explants of cells taken from the joint (ex vivo procedure), whose safety is checked before redelivering into the joint [113,114]. The scientific literature contains many successful examples of MSC transduction to enhance chondrogenesis by increasing the expression of genes responsible for hyaline cartilage proliferation performed either in vitro or in vivo [115]. Although there have been few successful clinical studies in terms of the using transduction for cartilage regeneration [116–118], there have also been several clinical cases of the development of tumorigenesis, namely leukemia, after transduction [119]. Therefore, other methods of gene therapy for cartilage regeneration are being investigated for clinical practice.

Primary cell lines, including MSCs and chondrocytes, can be transfected by electroporation [120], or nanocarrier materials, including lipofectamine [119,121–123]. The delivery of nucleic acid directly into the nucleus is called nucleofection [124]. Nucleofection is possible by microinjection or by nanocarriers if a nuclear localization sequence is introduced together [125]. Microinjection and nanocarriers are not suitable for large populations of cells [126]. For the elimination of intracellular inflammation and transgene silencing, it has been proven suitable to deliver the minicircle DNA (mcDNA) prepared from pDNA by removing bacterial sequences [127].

The advantage of plasmids is simple preparation and chemical stability, the disadvantage is the necessity of their transcription. In contrast, mRNA and oligonucleotide (small interfering siRNA and micro miRNA) transfection does not require nuclear transport and transcription [128,129]. Similar to DNA and mRNA, siRNA and miRNA can be delivered via nanocarriers [128]. Preclinical and clinical studies on gene therapy for the repair of articular cartilage are reviewed by Bellavia et al. [130].

Another approach of the gene modification of MSCs to enhance chondrogenesis is the knockout of genes in combination with miRNAs that affect the expression of ECM genes; namely, the overexpression of miR-140, miR-21 and miR-675 can stimulate chondrogenesis in MSC cells [131]. New possibilities for the effective treatment of osteoarthritis and other degenerative joint diseases are provided by the relatively new genome editing method CRISPR/Cas [132–134]. Through knockout, the CRISPR/Cas technology ensures the activation of a specific gene expression. This approach was successfully applied to the production of ECM proteins in hADSC cells [135] and activation-mediated synthesis of type II collagen and aggrecan in hMSC cells [136].

The use of viral transduction of MSCs for clinical use requires experience in working with viruses and cell cultures and increases the cost of the technique. Transfection, although allowing limited clinical application, leads to low efficiency and high cytotoxicity of MSCs. The application of CRISPR/Cas-mediated gene therapy in clinical practice is limited, in addition to ethical problems, by the possibility of introducing additional, unwanted mutations into the genome. Aspects of genetic modification of MSCs for hyaline cartilage repair are reviewed by Le H et al. [109].

Treatment approaches that incorporate MSCs (even genetically modified), growth factors, and growth-promoting substrates into biocompatible scaffolds can help improve cartilage regeneration [137]. MSCs (or other cells) and growth factors can be applied to the defect site by inserting a three-dimensional (3D) scaffold that promotes growth factor release, cell adhesion, proliferation, and differentiation [138,139]. There are even several promising studies combining the effect of physical (hypoxia) and chemical (growing factors) factors with genetic tools (RNA interference) on MSCs in a 3D scaffold [140,141]. The importance of appropriate stiffness and adhesion of the gene-activated scaffolds that can enhance transfection and chondrogenesis of MSCs has also been confirmed [142].

In the context of employing bioadhesive hydrogels for cartilage injury treatment, innovative approaches include a hybrid photo-cross-linkable (HPC) hydrogel, showcasing fast gelation, robust strength, and tissue adhesion for arthroscopic cartilage repair, with potential as an autologous chondrocyte implantation scaffold [143]. Another study explores an injectable, highly adhesive hydrogel with exosomes for early-stage osteoarthritis cartilage defects, offering potential for minimal cartilage defect treatment and stem cell-based repair [52]. Further research focuses on triple network (TN) hydrogels mimicking cartilage properties, relevant for stem cell-based cartilage repair [144]. Additionally, a mussel-inspired adhesive hydrogel (PDA/Gel-PAA) exhibits high mechanical strength, adhesion, and stem cell-friendly cartilage repair [45]. Lastly, a double cross-linked (DC) hydrogel shows promise for enhancing stem cell-based cartilage repair with improved mechanical properties and anti-degradation characteristics [49]. Building upon the information provided in the preceding referenced studies and publications, it can be reasonably postulated that bioadhesion plays a significant role in the application of hydrogels within the realm of biomedical and stem cell-related research.

7. Correlation between Bioadhesion and Stem Cell Differentiation

Cells respond to mechanical cues provided by materials serving as scaffolds. For instance, stiffer hydrogels promote faster growth of neural stem cells [145], while hydrogels with weaker cross-linking facilitate cell migration, favoring regeneration. Adhesion in such environments enhances the exchange of necessary molecules, ultimately improving regeneration [146].

Also, pluripotent stem cell heterogeneity is influenced by mechanical factors related to cell–cell and cell–matrix adhesion. Researchers have experimentally manipulated the spatial polarization of cell–cell adhesion, identifying E-cadherin as one of as key regulators in the differentiation process [147]. The above findings point to a mechanistic role of adhesive materials and of hydrogels in controlling stem cell proliferation and differentiation.

Materials with functional groups can impact cell behavior and stem cell differentiation, often associated with cell spreading. A study employed a unique surface patterning technique to explore the effects of various functional groups (-CH3, -OH, -COOH, -NH2) on mesenchymal stem cells during chondrogenic induction. It was observed that the type of functional groups had an indirect influence on cell differentiation, primarily through protein adsorption, non-specific cell adhesion, and subsequent cell spreading [148].

Surface chemistry of biomaterials has been widely recognized for its role in modulating human mesenchymal stem cell (hMSC) differentiation along specific lineages [149,150]. In an innovative approach, researchers introduced functional groups (acrylic acid and phosphates) onto silk surfaces to direct hMSC differentiation into chondrocytes and osteocytes. Notably, this method does not rely on the addition of growth factors or external signals; differentiation is initiated by the distinct surface functional groups [151].

In addition, controlling cell–polymer interactions is a crucial aspect of scaffold development for tissue engineering. One study investigated the influence of adhesion ligand spacer arm length, such as the RGD peptide coupled to alginate hydrogels, on stem cell behavior. It was found that the spacer arm's length played a key role in regulating stem cell proliferation and differentiation within polymer scaffolds, holding significance for tissue engineering applications [152]. These findings collectively underscore the complex mechanisms by which adhesion/bioadhesion influence stem cell proliferation and differentiation in tissue engineering contexts.

8. Future Perspective

By 2023, 15 randomized controlled clinical trials and 11 nonrandomized RCTs using culture-expanded MSCs in the treatment of knee OA were finished [153]. A total of 179 are ongoing [154], and they suggest net positive effects of MSCs on mitigating pain and symptoms and on cartilage protection and/or repair. In 2023, there are 14 finished or ongoing clinical trials using hydrogels as potential therapies for the management of OA [154]. Adhesive hydrogels improve therapeutic outcomes through offering stable integration between tissue and implants. A total of 133 clinical studies are currently testing effect of platelet-rich plasma and 24 explore the usage of gene therapy in the treatment of OA. Application of hydrogel, MSCs or chemical factors can induce cartilage regeneration, but only to some extent. Many deal with a short half-life and easy clearance from the intended site, leading to low bioavailability. Hydrogels have been shown to influence cell functionality in both chondrocytes and MSCs promoting chondrogenesis and the formation of hyaline cartilage-like ECM. The addition of specific growth factors or gene transfer to hydrogel-encapsulated MSCs may influence cell functionality in cartilage regeneration.

Thus, the combination of adhesive hydrogels, MSC differentiation inductors (growth factors, gene transfer) and MSCs may improve chondrogenic differentiation, maintain chondrogenic phenotype and decrease hypertrophic differentiation in seeded cells while providing optimal biocompatibility, stability, and biodegradability. The combination with bioactive substances and MSCs may help to overcome the hurdle of rapid degradation in natural hydrogels, which often takes place faster than hydrogels can be replaced by de novo ECM [155]. To our knowledge, the combined effectiveness of MSC-laden hydrogels with growth factors/gene transfer has been tested only on animal models (summarized in [155]), but so far with promising outcomes in cartilage regenerations. Next essential steps, as clinical trials, are much needed.

9. Conclusions

In conclusion, our comprehensive review underscores the potential correlation between bioadhesive injectable hydrogels and MSC differentiation in the context of cartilage repair. The reviewed literature strongly suggests that these hydrogels hold promise as a promising platform to enhance the regenerative capacity of MSCs for cartilage tissue regeneration. However, further research and clinical studies are imperative to validate and optimize this innovative approach for effective clinical applications.

Author Contributions: Conceptualization, S.Ž. and A.H.; writing—original draft preparation, J.K., P.P. and H.G.; writing—review and editing, J.K., P.P., H.G., A.H. and S.Ž.; funding, S.Ž. All authors have read and agreed to the published version of the manuscript.

Funding: This research was the result of the project implementation: Advanced bioactive hydrogel materials for regenerative medicine (ABSCARM); ITMS2014+: 313011BWL6 supported by the Operational Programme Integrated infrastructure funded by the European Regional Development Fund.

Institutional Review Board Statement: Not applicable.

Informed Consent Statement: Not applicable.

Data Availability Statement: Not applicable.

Acknowledgments: Parts of the figure were drawn using pictures from Servier Medical Art. Servier Medical Art by Servier is licensed under a Creative Commons Attribution 3.0 Unported License (https://creativecommons.org/licenses/by/3.0/).

Conflicts of Interest: The authors declare no conflict of interest.

References

1. Wang, W.; Narain, R.; Zeng, H. Hydrogels. In *Polymer Science and Nanotechnology*; Elsevier: Amsterdam, The Netherlands, 2020; pp. 203–244.
2. Guvendiren, M.; Lu, H.D.; Burdick, J.A. Shear-thinning hydrogels for biomedical applications. *Soft Matter* **2011**, *8*, 260–272. [CrossRef]
3. Rizzo, F.; Kehr, N.S. Recent Advances in Injectable Hydrogels for Controlled and Local Drug Delivery. *Adv. Health Mater.* **2020**, *10*, e2001341. [CrossRef] [PubMed]
4. Sun, Y.; Nan, D.; Jin, H.; Qu, X. Recent advances of injectable hydrogels for drug delivery and tissue engineering applications. *Polym. Test.* **2019**, *81*, 106283. [CrossRef]
5. Sharifi, S.; Islam, M.M.; Sharifi, H.; Islam, R.; Koza, D.; Reyes-Ortega, F.; Alba-Molina, D.; Nilsson, P.H.; Dohlman, C.H.; Mollnes, T.E.; et al. Tuning gelatin-based hydrogel towards bioadhesive ocular tissue engineering applications. *Bioact. Mater.* **2021**, *6*, 3947–3961. [CrossRef] [PubMed]
6. Chang, M.; Liu, X.; Wang, X.; Peng, F.; Ren, J. Mussel-inspired adhesive hydrogels based on biomass-derived xylan and tannic acid cross-linked with acrylic acid with antioxidant and antibacterial properties. *J. Mater. Sci.* **2021**, *56*, 14729–14740. [CrossRef]
7. Xiong, Y.; Zhang, X.; Ma, X.; Wang, W.; Yan, F.; Zhao, X.; Chu, X.; Xu, W.; Sun, C. A review of the properties and applications of bioadhesive hydrogels. *Polym. Chem.* **2021**, *12*, 3721–3739. [CrossRef]
8. Stan, D.; Tanase, C.; Avram, M.; Apetrei, R.; Mincu, N.B.; Mateescu, A.L.; Stan, D. Wound healing applications of creams and 'smart' hydrogels. *Exp. Dermatol.* **2021**, *30*, 1218–1232. [CrossRef]
9. Dey, K.; Agnelli, S.; Serzanti, M.; Ginestra, P.; Scarì, G.; Dell'Era, P.; Sartore, L. Preparation and properties of high performance gelatin-based hydrogels with chitosan or hydroxyethyl cellulose for tissue engineering applications. *Int. J. Polym. Mater. Polym. Biomater.* **2019**, *68*, 183–192. [CrossRef]

10. Yazdani, M.; Shahdadfar, A.; Jackson, C.J.; Utheim, T.P. Hyaluronan-Based Hydrogel Scaffolds for Limbal Stem Cell Transplantation: A Review. *Cells* **2019**, *8*, 245. [CrossRef]
11. Li, Z.; Yue, M.; Liu, Y.; Zhang, P.; Qing, J.; Liu, H.; Zhou, Y. Advances of Engineered Hydrogel Organoids within the Stem Cell Field: A Systematic Review. *Gels* **2022**, *8*, 379. [CrossRef]
12. Kheyraddini Mousavi, A.; Leseman, Z.C.; Palacio, M.L.B. Bioadhesion. In *Encyclopedia of Nanotechnology*; Springer: Dordrecht, The Netherlands, 2012; pp. 186–193.
13. Palacio, M.L.B.; Bhushan, B. Bioadhesion: A review of concepts and applications. *Philos. Trans. R. Soc. A Math. Phys. Eng. Sci.* **2012**, *370*, 2321–2347. [CrossRef] [PubMed]
14. Uma, K. Bioadhesives for clinical applications—A mini review. *Mater Adv.* **2023**, *4*, 2062–2069.
15. Chopra, H.; Kumar, S.; Singh, I. Bioadhesive Hydrogels and Their Applications. In *Bioadhesives in Drug Delivery*; Wiley: Hoboken, NJ, USA, 2020; pp. 147–170.
16. Yang, J.; Bai, R.; Chen, B.; Suo, Z. Hydrogel Adhesion: A Supramolecular Synergy of Chemistry, Topology, and Mechanics. *Adv. Funct. Mater.* **2020**, *30*, 1901693. [CrossRef]
17. Yao, H.; Wu, M.; Lin, L.; Wu, Z.; Bae, M.; Park, S.; Wang, S.; Zhang, W.; Gao, J.; Wang, D.; et al. Design strategies for adhesive hydrogels with natural antibacterial agents as wound dressings: Status and trends. *Mater. Today Bio.* **2022**, *16*, 100429. [CrossRef] [PubMed]
18. Li, Z.; Yu, C.; Kumar, H.; He, X.; Lu, Q.; Bai, H.; Kim, K.; Hu, J. The Effect of Crosslinking Degree of Hydrogels on Hydrogel Adhesion. *Gels* **2022**, *8*, 682. [CrossRef]
19. Duan, W.; Bian, X.; Bu, Y. Applications of Bioadhesives: A Mini Review. *Front. Bioeng. Biotechnol.* **2021**, *9*, 716035. [CrossRef]
20. Sun, J.; Chen, T.; Zhao, B.; Fan, W.; Shen, Y.; Wei, H.; Zhang, M.; Zheng, W.; Peng, J.; Wang, J.; et al. Acceleration of Oral Wound Healing under Diabetes Mellitus Conditions Using Bioadhesive Hydrogel. *ACS Appl. Mater. Interfaces* **2023**, *15*, 416–431. [CrossRef]
21. Liu, S.; Luan, Z.; Wang, T.; Xu, K.; Luo, Q.; Ye, S.; Wang, W.; Dan, R.; Shu, Z.; Huang, Y.; et al. Endoscopy Deliverable and Mushroom-Cap-Inspired Hyperboloid-Shaped Drug-Laden Bioadhesive Hydrogel for Stomach Perforation Repair. *ACS Nano* **2023**, *17*, 111–126. [CrossRef]
22. Xu, H.L.; Tong, M.Q.; Wang, L.F.; Chen, R.; Li, X.Z.; Sohawon, Y.; Zhao, Y.Z. Thiolated γ-polyglutamic acid as a bioadhesive hydrogel-forming material: Evaluation of gelation, bioadhesive properties and sustained release of KGF in the repair of injured corneas. *Biomater. Sci.* **2019**, *7*, 2582–2599. [CrossRef]
23. Mei, X.; Li, J.; Wang, Z.; Zhu, D.; Huang, K.; Hu, S.; Cheng, K. An inhaled bioadhesive hydrogel to shield non-human primates from SARS-CoV-2 infection. *Nat. Mater.* **2023**, *22*, 903–912. [CrossRef]
24. Behrendt, P.; Ladner, Y.; Stoddart, M.J.; Lippross, S.; Alini, M.; Eglin, D.; Armiento, A.R. Articular Joint-Simulating Mechanical Load Activates Endogenous TGF-β in a Highly Cellularized Bioadhesive Hydrogel for Cartilage Repair. *Am. J. Sports Med.* **2020**, *48*, 210–221. [CrossRef] [PubMed]
25. Zhao, L.; Shi, Z.; Sun, X.; Yu, Y.; Wang, X.; Wang, H.; Li, T.; Zhang, H.; Zhang, X.; Wang, F.; et al. Natural Dual-Crosslinking Bioadhesive Hydrogel for Corneal Regeneration in Large-Size Defects. *Adv. Healthc. Mater.* **2022**, *11*, e2201576. [CrossRef] [PubMed]
26. Xie, X.; Lei, Y.; Li, Y.; Zhang, M.; Sun, J.; Zhu, M.-Q.; Wang, J. Dual-crosslinked bioadhesive hydrogel as NIR/pH stimulus-responsiveness platform for effectively accelerating wound healing. *J. Colloid Interface Sci.* **2023**, *637*, 20–32. [CrossRef] [PubMed]
27. Yao, Q.; Shi, Y.; Xia, X.; Tang, Y.; Jiang, X.; Zheng, Y.-W.; Zhang, H.; Chen, R.; Kou, L. Bioadhesive hydrogel comprising bilirubin/β-cyclodextrin inclusion complexes promote diabetic wound healing. *Pharm. Biol.* **2021**, *59*, 1137–1147. [CrossRef]
28. Luo, S.; Yang, L.; Zou, Q.; Yuan, D.; Xu, S.; Zhao, Y.; Wu, X.; Wang, Z.; Ye, C. Rapid suture-free repair of arterial bleeding: A novel approach with ultra-thin bioadhesive hydrogel membrane. *Chem. Eng. J.* **2023**, *472*, 144865. [CrossRef]
29. Fu, Y.; Shi, Y.; Wang, L.; Zhao, Y.; Wang, R.; Li, K.; Zhang, S.; Zha, X.; Wang, W.; Zhao, X.; et al. All-Natural Immunomodulatory Bioadhesive Hydrogel Promotes Angiogenesis and Diabetic Wound Healing by Regulating Macrophage Heterogeneity. *Adv. Sci.* **2023**, *10*, 2206771. [CrossRef]
30. Huang, W.; Cheng, S.; Wang, X.; Zhang, Y.; Chen, L.; Zhang, L. Noncompressible Hemostasis and Bone Regeneration Induced by an Absorbable Bioadhesive Self-Healing Hydrogel. *Adv. Funct. Mater.* **2021**, *31*, 2009189. [CrossRef]
31. He, H.; Fei, Z.; Guo, T.; Hou, Y.; Li, D.; Wang, K.; Ren, F.; Fan, K.; Zhou, D.; Xie, C.; et al. Bioadhesive injectable hydrogel with phenolic carbon quantum dot supported Pd single atom nanozymes as a localized immunomodulation niche for cancer catalytic immunotherapy. *Biomaterials* **2022**, *280*, 121272. [CrossRef]
32. Zhou, J.; Wu, Y.; Zhang, X.; Lai, J.; Li, Y.; Xing, J.; Teng, L.; Chen, J. Enzyme Catalyzed Hydrogel as Versatile Bioadhesive for Tissue Wound Hemostasis, Bonding, and Continuous Repair. *Biomacromolecules* **2021**, *22*, 1346–1356. [CrossRef]
33. Wang, X.; Sun, X.; Gan, D.; Soubrier, M.; Chiang, H.-Y.; Yan, L.; Li, Y.; Li, J.; Yu, S.; Xia, Y.; et al. Bioadhesive and conductive hydrogel-integrated brain-machine interfaces for conformal and immune-evasive contact with brain tissue. *Matter* **2022**, *5*, 1204–1223. [CrossRef]
34. Zhu, W.; Iqbal, J.; Wang, D.-A. A DOPA-functionalized chondroitin sulfate-based adhesive hydrogel as a promising multi-functional bioadhesive. *J. Mater. Chem. B* **2018**, *7*, 1741–1752. [CrossRef] [PubMed]

35. Soykan, M.N.; Altug, B.; Bas, H.; Ghorbanpoor, H.; Avci, H.; Eroglu, S.; Sengel, S.B.; Sariboyaci, A.E.; Bagis, S.G.; Uysal, O.; et al. Developing a Novel Platelet-Rich Plasma-Laden Bioadhesive Hydrogel Contact Lens for the Treatment of Ocular Surface Chemical Injuries. *Macromol. Biosci.* 2023, *early view*.
36. Liu, H.; Yuan, M.; Sonamuthu, J.; Yan, S.; Huang, W.; Cai, Y.; Yao, J. A dopamine-functionalized aqueous-based silk protein hydrogel bioadhesive for biomedical wound closure. *New J. Chem.* **2020**, *44*, 884–891. [CrossRef]
37. Li, M.; Wei, R.; Liu, C.; Fang, H.; Yang, W.; Wang, Y.; Zhou, X. A 'T.E.S.T.' hydrogel bioadhesive assisted by corneal cross-linking for in situ sutureless corneal repair. *Bioact. Mater.* **2023**, *25*, 333–346. [CrossRef] [PubMed]
38. Salatin, S.; Jelvehgari, M. Desirability function approach for development of a thermosensitive and bioadhesive nanotransfersome–hydrogel hybrid system for enhanced skin bioavailability and antibacterial activity of cephalexin. *Drug Dev. Ind. Pharm.* **2020**, *46*, 1318–1333. [CrossRef] [PubMed]
39. Gao, Y.-M.; Li, Z.-Y.; Zhang, X.-J.; Zhang, J.; Li, Q.-F.; Zhou, S.-B. One-Pot Synthesis of Bioadhesive Double-Network Hydrogel Patch as Disposable Wound Dressing. *ACS Appl. Mater. Interfaces* **2023**, *15*, 11496–11506. [CrossRef] [PubMed]
40. Cassano, R.; Curcio, F.; Mandracchia, D.; Trapani, A.; Trombino, S. Gelatin and Glycerine-Based Bioadhesive Vaginal Hydrogel. *Curr. Drug Deliv.* **2020**, *17*, 303–311. [CrossRef]
41. Fu, Z.; Xiao, S.; Wang, P.; Zhao, J.; Ling, Z.; An, Z.; Shao, J.; Fu, W. Injectable, stretchable, toughened, bioadhesive composite hydrogel for bladder injury repair. *RSC Adv.* **2023**, *13*, 10903–10913. [CrossRef]
42. Sharma, B.; Fermanian, S.; Gibson, M.; Unterman, S.; Herzka, D.A.; Cascio, B.; Coburn, J.; Hui, A.Y.; Marcus, N.; Gold, G.E.; et al. Human Cartilage Repair with a Photoreactive Adhesive-Hydrogel Composite. *Sci. Transl. Med.* **2013**, *5*, 167ra6. [CrossRef]
43. Han, L.; Wang, M.; Li, P.; Gan, D.; Yan, L.; Xu, J.; Wang, K.; Fang, L.; Chan, C.W.; Zhang, H.; et al. Mussel-Inspired Tissue-Adhesive Hydrogel Based on the Polydopamine–Chondroitin Sulfate Complex for Growth-Factor-Free Cartilage Regeneration. *ACS Appl. Mater. Interfaces* **2018**, *10*, 28015–28026. [CrossRef]
44. Shin, J.; Kang, E.H.; Choi, S.; Jeon, E.J.; Cho, J.H.; Kang, D.; Lee, H.; Yun, I.S.; Cho, S.-W. Tissue-Adhesive Chondroitin Sulfate Hydrogel for Cartilage Reconstruction. *ACS Biomater. Sci. Eng.* **2021**, *7*, 4230–4243. [CrossRef]
45. Yan, L.; Zhou, T.; Ni, R.; Jia, Z.; Jiang, Y.; Guo, T.; Wang, K.; Chen, X.; Han, L.; Lu, X. Adhesive Gelatin-Catechol Complex Reinforced Poly(Acrylic Acid) Hydrogel with Enhanced Toughness and Cell Affinity for Cartilage Regeneration. *ACS Appl. Bio Mater.* **2022**, *5*, 4366–4377. [CrossRef]
46. Qiu, H.; Deng, J.; Wei, R.; Wu, X.; Chen, S.; Yang, Y.; Gong, C.; Cui, L.; Si, Z.; Zhu, Y.; et al. A lubricant and adhesive hydrogel cross-linked from hyaluronic acid and chitosan for articular cartilage regeneration. *Int. J. Biol. Macromol.* **2023**, *243*, 125249. [CrossRef]
47. Levinson, C.; Cavalli, E.; von Rechenberg, B.; Zenobi-Wong, M.; Darwiche, S.E. Combination of a Collagen Scaffold and an Adhesive Hyaluronan-Based Hydrogel for Cartilage Regeneration: A Proof of Concept in an Ovine Model. *Cartilage* **2021**, *13*, 636S–649S. [CrossRef]
48. Balakrishnan, B.; Joshi, N.; Jayakrishnan, A.; Banerjee, R. Self-crosslinked oxidized alginate/gelatin hydrogel as injectable, adhesive biomimetic scaffolds for cartilage regeneration. *Acta Biomater.* **2014**, *10*, 3650–3663. [CrossRef]
49. Yu, C.; Gao, H.; Li, Q.; Cao, X. Injectable dual cross-linked adhesive hyaluronic acid multifunctional hydrogel scaffolds for potential applications in cartilage repair. *Polym. Chem.* **2020**, *11*, 3169–3178. [CrossRef]
50. Li, D.; Li, J.; Chen, T.; Qin, X.; Pan, L.; Lin, X.; Liang, W.; Wang, Q. Injectable Bioadhesive Hydrogels Scavenging ROS and Restoring Mucosal Barrier for Enhanced Ulcerative Colitis Therapy. *ACS Appl. Mater. Interfaces* **2023**, *15*, 38273–38284. [CrossRef]
51. Paul, S.; Schrobback, K.; Tran, P.A.; Meinert, C.; Davern, J.W.; Weekes, A.; Klein, T.J. Photo-Cross-Linkable, Injectable, and Highly Adhesive GelMA-Glycol Chitosan Hydrogels for Cartilage Repair. *Adv. Healthc. Mater.* 2023; *early view*.
52. Zhang, F.-X.; Liu, P.; Ding, W.; Meng, Q.-B.; Su, D.-H.; Zhang, Q.-C.; Lian, R.-X.; Yu, B.-Q.; Zhao, M.-D.; Dong, J.; et al. Injectable Mussel-Inspired highly adhesive hydrogel with exosomes for endogenous cell recruitment and cartilage defect regeneration. *Biomaterials* **2021**, *278*, 121169. [CrossRef]
53. Wang, M.; Wang, C. Bulk Properties of Biomaterials and Testing Techniques. In *Encyclopedia of Biomedical Engineering*; Elsevier: Amsterdam, The Netherlands, 2019; pp. 53–64.
54. Ebnesajjad, S.; Landrock, A.H. Testing of Adhesive Bonds. In *Adhesives Technology Handbook*; Elsevier: Amsterdam, The Netherlands, 2015; pp. 339–352.
55. Wei, K.; Senturk, B.; Matter, M.T.; Wu, X.; Herrmann, I.K.; Rottmar, M.; Toncelli, C. Mussel-Inspired Injectable Hydrogel Adhesive Formed under Mild Conditions Features Near-Native Tissue Properties. *ACS Appl. Mater. Interfaces* **2019**, *11*, 47707–47719. [CrossRef]
56. *ASTM F2256-05*; Standard Test Method for Strength Properties of Tissue Adhesives in T-Peel by Tension Loading. ASTM International: West Conshehoken, PA, USA, 2015.
57. Jeon, E.Y.; Joo, K.I.; Cha, H.J. Body temperature-activated protein-based injectable adhesive hydrogel incorporated with decellularized adipose extracellular matrix for tissue-specific regenerative stem cell therapy. *Acta Biomater.* **2020**, *114*, 244–255. [CrossRef]
58. Mathias, J.-D.; Grédiac, M.; Michaud, P. Bio-based adhesives. In *Biopolymers and Biotech Admixtures for Eco-Efficient Construction Materials*; Elsevier: Amsterdam, The Netherlands, 2016; pp. 369–385.
59. *ASTM F2255:2005*; Standard Test Method for Strength Properties of Tissue Adhesives in Lap-Shear by Tension Loading. ASTM International: West Conshehoken, PA, USA, 2005.

60. Jung, H.Y.; Le Thi, P.; HwangBo, K.-H.; Bae, J.W.; Park, K.D. Tunable and high tissue adhesive properties of injectable chitosan based hydrogels through polymer architecture modulation. *Carbohydr. Polym.* **2021**, *261*, 117810. [CrossRef]
61. Sigen, A.; Xu, Q.; Johnson, M.; Creagh-Flynn, J.; Venet, M.; Zhou, D.; Wang, W. An injectable multi-responsive hydrogel as self-healable and on-demand dissolution tissue adhesive. *Appl. Mater. Today* **2021**, *22*, 100967.
62. Yang, J.; Steck, J.; Bai, R.; Suo, Z. Topological adhesion II. Stretchable adhesion. *Extreme. Mech. Lett.* **2020**, *40*, 100891. [CrossRef]
63. Dehne, T.; Zehbe, R.; Krüger, J.P.; Petrova, A.; Valbuena, R.; Sittinger, M.; Schubert, H.; Ringe, J. A method to screen and evaluate tissue adhesives for joint repair applications. *BMC Musculoskelet. Disord.* **2012**, *13*, 175. [CrossRef] [PubMed]
64. Zhu, F.; Wang, C.; Yang, S.; Wang, Q.; Liang, F.; Liu, C.; Qiu, D.; Qu, X.; Hu, Z.; Yang, Z. Injectable tissue adhesive composite hydrogel with fibroblasts for treating skin defects. *J. Mater. Chem. B* **2017**, *5*, 2416–2424. [CrossRef]
65. Chen, S.; Tomov, M.L.; Ning, L.; Gil, C.J.; Hwang, B.; Bauser-Heaton, H.; Chen, H.; Serpooshan, V. Extrusion-Based 3D Bioprinting of Adhesive Tissue Engineering Scaffolds Using Hybrid Functionalized Hydrogel Bioinks. *Adv. Biol.* **2023**, *7*, e2300124. [CrossRef]
66. *ISO-10993-11*; Biological Evaluation of Medical Devices. Part 11: Tests for Systemic Toxicity. ISO: Geneva, Switzerland, 2017.
67. Thanusha, A.V.; Koul, V. Biocompatibility evaluation for the developed hydrogel wound dressing—ISO-10993-11 standards—In vitro and in vivo study. *Biomed. Phys. Eng. Express* **2022**, *8*, 015010. [CrossRef]
68. Øvrebø, Ø.; Perale, G.; Wojciechowski, J.P.; Echalier, C.; Jeffers, J.R.; Stevens, M.M.; Rossi, F. Design and clinical application of injectable hydrogels for musculoskeletal therapy. *Bioeng. Transl. Med.* **2022**, *7*, e10295. [CrossRef]
69. Li, L.; Zhang, Y.; Mu, J.; Chen, J.; Zhang, C.; Cao, H.; Gao, J. Transplantation of Human Mesenchymal Stem-Cell-Derived Exosomes Immobilized in an Adhesive Hydrogel for Effective Treatment of Spinal Cord Injury. *Nano Lett.* **2020**, *20*, 4298–4305. [CrossRef]
70. Mu, J.; Li, L.; Wu, J.; Huang, T.; Zhang, Y.; Cao, J.; Ma, T.; Chen, J.; Zhang, C.; Zhang, X.; et al. Hypoxia-stimulated mesenchymal stem cell-derived exosomes loaded by adhesive hydrogel for effective angiogenic treatment of spinal cord injury. *Biomater. Sci.* **2022**, *10*, 1803–1811. [CrossRef]
71. Wang, Y.; Song, P.; Wu, L.; Su, Z.; Gui, X.; Gao, C.; Zhou, C. In situ photo-crosslinked adhesive hydrogel loaded with mesenchymal stem cell-derived extracellular vesicles promotes diabetic wound healing. *J. Mater. Chem. B* **2023**, *11*, 837–851. [CrossRef]
72. Zhu, Y.; Wang, T.; Bao, M.-J.; Qu, X.-H.; Li, Z.-M. Effect of stem cell conditional medium-loading adhesive hydrogel on TGF-β1-induced endometrial stromal cell fibrosis. *Front. Bioeng. Biotechnol.* **2023**, *11*, 1168136. [CrossRef] [PubMed]
73. Han, W.M.; Mohiuddin, M.; Anderson, S.E.; García, A.J.; Jang, Y.C. Co-delivery of Wnt7a and muscle stem cells using synthetic bioadhesive hydrogel enhances murine muscle regeneration and cell migration during engraftment. *Acta Biomater.* **2019**, *94*, 243–252. [CrossRef] [PubMed]
74. Zhang, L.; Zhou, Y.; Su, D.; Wu, S.; Zhou, J.; Chen, J. Injectable, self-healing and pH responsive stem cell factor loaded collagen hydrogel as a dynamic bioadhesive dressing for diabetic wound repair. *J. Mater. Chem. B* **2021**, *9*, 5887–5897. [CrossRef] [PubMed]
75. Gong, C.; Lu, C.; Li, B.; Shan, M.; Wu, G. Injectable dopamine-modified poly(α,β-aspartic acid) nanocomposite hydrogel as bioadhesive drug delivery system. *J. Biomed. Mater. Res. A* **2017**, *105*, 1000–1008. [CrossRef]
76. Kang, X.; Guan, P.; Xiao, C.; Liu, C.; Guan, Y.; Lin, Y.; Tian, Y.; Ren, K.; Huang, Y.; Fu, R.; et al. Injectable Intrinsic Photothermal Hydrogel Bioadhesive with On-Demand Removability for Wound Closure and MRSA-Infected Wound Healing. *Adv. Health Mater.* **2023**, *12*, e2203306. [CrossRef]
77. Ahmadian, Z.; Jelodar, M.Z.; Rashidipour, M.; Dadkhah, M.; Adhami, V.; Sefareshi, S.; Ebrahimi, H.A.; Ghasemian, M.; Adeli, M. A self-healable and bioadhesive acacia gum polysaccharide-based injectable hydrogel for wound healing acceleration. *DARU J. Pharm. Sci.* **2023**, 1–15. [CrossRef]
78. Li, S.; Xian, Y.; He, G.; Chen, L.; Chen, Z.; Hong, Y.; Wu, D. In situ Injectable Tetra-PEG Hydrogel Bioadhesive for Sutureless Repair of Gastrointestinal Perforation. *Chin. J. Chem.* 2023; *early view*.
79. Somoza, R.A.; Welter, J.F.; Correa, D.; Caplan, A.I. Chondrogenic Differentiation of Mesenchymal Stem Cells: Challenges and Unfulfilled Expectations. *Tissue Eng. Part B Rev.* **2014**, *20*, 596–608. [CrossRef]
80. Huang, J.; Liu, Q.; Xia, J.; Chen, X.; Xiong, J.; Yang, L.; Liang, Y. Modification of mesenchymal stem cells for cartilage-targeted therapy. *J. Transl. Med.* **2022**, *20*, 515. [CrossRef]
81. Fuentes-Mera, L.; Camacho, A.; Moncada-Saucedo, N.K.; Peña-Martínez, V. Current Applications of Mesenchymal Stem Cells for Cartilage Tissue Engineering. In *Mesenchymal Stem Cells—Isolation, Characterization and Applications*; InTech: Houston, TX, USA, 2017.
82. Danišovič, L'.; Boháč, M.; Zamborský, R.; Oravcová, L.; Provazníková, Z.; Csöbönyeiová, M.; Varga, I. Comparative analysis of mesenchymal stromal cells from different tissue sources in respect to articular cartilage tissue engineering. *Gen. Physiol. Biophys.* **2016**, *35*, 207–214. [CrossRef]
83. Carneiro, D.D.C.; Araújo, L.T.D.; Santos, G.C.; Damasceno, P.K.F.; Vieira, J.L.; Santos, R.R.D.; Soares, M.B.P. Clinical Trials with Mesenchymal Stem Cell Therapies for Osteoarthritis: Challenges in the Regeneration of Articular Cartilage. *Int. J. Mol. Sci.* **2023**, *24*, 9939. [CrossRef]
84. Huang, X.; Das, R.; Patel, A.; Duc Nguyen, T. Physical Stimulations for Bone and Cartilage Regeneration. *Regen. Eng. Transl. Med.* **2018**, *4*, 216–237. [CrossRef] [PubMed]
85. Shahmoradi, S.R.; Kabir Salmani, M.; Soleimanpour, H.R.; Tavakoli, A.H.; Hosaini, K.; Haghighipour, N.; Bonakdar, S. Induction of Chondrogenic Differentiation in Human Mesenchymal Stem Cells Cultured on Human Demineralized Bone Matrix Scaffold under Hydrostatic Pressure. *Tissue Eng. Regen. Med.* **2019**, *16*, 69–80. [CrossRef] [PubMed]

86. Dusfour, G.; Maumus, M.; Cañadas, P.; Ambard, D.; Jorgensen, C.; Noël, D.; Le Floc'H, S. Mesenchymal stem cells-derived cartilage micropellets: A relevant in vitro model for biomechanical and mechanobiological studies of cartilage growth. *Mater. Sci. Eng. C* **2020**, *112*, 110808. [CrossRef]
87. Gómez-Leduc, T.; Desancé, M.; Hervieu, M.; Legendre, F.; Ollitrault, D.; de Vienne, C.; Herlicoviez, M.; Galéra, P.; Demoor, M. Hypoxia Is a Critical Parameter for Chondrogenic Differentiation of Human Umbilical Cord Blood Mesenchymal Stem Cells in Type I/III Collagen Sponges. *Int. J. Mol. Sci.* **2017**, *18*, 1933. [CrossRef] [PubMed]
88. Parate, D.; Franco-Obregón, A.; Fröhlich, J.; Beyer, C.; Abbas, A.A.; Kamarul, T.; Hui, J.H.P.; Yang, Z. Enhancement of mesenchymal stem cell chondrogenesis with short-term low intensity pulsed electromagnetic fields. *Sci. Rep.* **2017**, *7*, 9421. [CrossRef]
89. Marks, R. Articular Cartilage Degradation and Photobiomodulation Therapy. *CPQ Orthop.* **2021**, *5*, 1–20.
90. Schneider, C.; Dungel, P.; Priglinger, E.; Danzer, M.; Schädl, B.; Nürnberger, S. The impact of photobiomodulation on the chondrogenic potential of adipose-derived stromal/stem cells. *J. Photochem. Photobiol. B Biol.* **2021**, *221*, 112243. [CrossRef]
91. Bozhokin, M.S.; Vcherashnii, D.B.; Yastrebov, S.G.; Beilinson, L.L.; Zherebtsova Ju, V.; Khotin, M.G. Low-intensity photobiomodulation at 632.8 nm increases tgfβ3, col2a1, and sox9 gene expression in rat bone marrow mesenchymal stem cells in vitro. *Lasers Med. Sci.* **2022**, *37*, 435–441. [CrossRef]
92. Potter, C.M.F.; Lao, K.H.; Zeng, L.; Xu, Q. Role of Biomechanical Forces in Stem Cell Vascular Lineage Differentiation. *Arter. Thromb. Vasc. Biol.* **2014**, *34*, 2184–2190. [CrossRef]
93. Cheng, B.; Tu, T.; Shi, X.; Liu, Y.; Zhao, Y.; Zhao, Y.; Li, Y.; Chen, H.; Chen, Y.; Zhang, M. A novel construct with biomechanical flexibility for articular cartilage regeneration. *Stem Cell Res. Ther.* **2019**, *10*, 298. [CrossRef]
94. Fekrazad, R.; Eslaminejad, M.B.; Shayan, A.M.; Kalhori, K.A.; Abbas, F.M.; Taghiyar, L.; Pedram, M.S.; Ghuchani, M.S. Effects of Photobiomodulation and Mesenchymal Stem Cells on Articular Cartilage Defects in a Rabbit Model. *Photomed. Laser Surg.* **2016**, *34*, 543–549. [CrossRef]
95. Ferraresi, C.; Freire, F.; Hamblin, M. Photobiomodulation in Cartilage: In vitro, in vivo, and Clinical Trials. In *Low-Level Light Therapy: Photobiomodulation*; SPIE: Bellingham, WA, USA, 2018; pp. 183–194.
96. Sawatjui, N.; Limpaiboon, T.; Schrobback, K.; Klein, T. Biomimetic scaffolds and dynamic compression enhance the properties of chondrocyte- and MSC -based tissue-engineered cartilage. *J. Tissue Eng. Regen. Med.* **2018**, *12*, 1220–1229. [CrossRef]
97. Khan, W.S.; Tew, S.R.; Adesida, A.B.; Hardingham, T.E. Human infrapatellar fat pad-derived stem cells express the pericyte marker 3G5 and show enhanced chondrogenesis after expansion in fibroblast growth factor-2. *Thromb. Haemost.* **2008**, *10*, R74. [CrossRef]
98. Armakolas, N.; Dimakakos, A.; Armakolas, A.; Antonopoulos, A.; Koutsilieris, M. Possible role of the Ec peptide of IGF-1Ec in cartilage repair. *Mol. Med. Rep.* **2016**, *14*, 3066–3072. [CrossRef] [PubMed]
99. Jia, Z.; Wang, S.; Liang, Y.; Liu, Q. Combination of kartogenin and transforming growth factor-β3 supports synovial fluid-derived mesenchymal stem cell-based cartilage regeneration. *Am. J. Transl. Res.* **2019**, *11*, 2056–2069. [PubMed]
100. Monaco, G.; El Haj, A.J.; Alini, M.; Stoddart, M.J. Sodium Hyaluronate Supplemented Culture Media as a New hMSC Chondrogenic Differentiation Media-Model for in vitro/ex vivo Screening of Potential Cartilage Repair Therapies. *Front. Bioeng. Biotechnol.* **2020**, *8*, 243. [CrossRef] [PubMed]
101. Fu, H.; Wang, H.; Li, D. BMP-7 accelerates the differentiation of rabbit mesenchymal stem cells into cartilage through the Wnt/β-catenin pathway. *Exp. Ther. Med.* **2017**, *14*, 5424–5428. [CrossRef]
102. Kozhemyakina, E.; Lassar, A.B.; Zelzer, E. A pathway to bone: Signaling molecules and transcription factors involved in chondrocyte development and maturation. *Development* **2015**, *142*, 817–831. [CrossRef]
103. Gibson, J.D.; O'sullivan, M.B.; Alaee, F.; Paglia, D.N.; Yoshida, R.; Guzzo, R.M.; Drissi, H. Regeneration of Articular Cartilage by Human ESC-Derived Mesenchymal Progenitors Treated Sequentially with BMP-2 and Wnt5a. *Stem Cells Transl. Med.* **2017**, *6*, 40–50. [CrossRef]
104. Zhang, X.; Wu, S.; Naccarato, T.; Prakash-Damani, M.; Chou, Y.; Chu, C.-Q.; Zhu, Y. Regeneration of hyaline-like cartilage in situ with SOX9 stimulation of bone marrow-derived mesenchymal stem cells. *PLoS ONE* **2017**, *12*, e0180138. [CrossRef]
105. Hunter, D.J.; Pike, M.C.; Jonas, B.L.; Kissin, E.; Krop, J.; McAlindon, T. Phase 1 safety and tolerability study of BMP-7 in symptomatic knee osteoarthritis. *BMC Musculoskelet. Disord.* **2010**, *11*, 232. [CrossRef] [PubMed]
106. Whitty, C.; Pernstich, C.; Marris, C.; McCaskie, A.; Jones, M.; Henson, F. Sustained delivery of the bone morphogenetic proteins BMP-2 and BMP-7 for cartilage repair and regeneration in osteoarthritis. *Osteoarthr. Cartil. Open* **2022**, *4*, 100240. [CrossRef] [PubMed]
107. Chen, Y.-C.; Hsu, Y.-M.; Tan, K.P.; Fang, H.-W.; Chang, C.-H. Intraarticular injection for rabbit knee osteoarthritis: Effectiveness among hyaluronic acid, platelet-rich plasma, and mesenchymal stem cells. *J. Taiwan Inst. Chem. Eng.* **2018**, *91*, 138–145. [CrossRef]
108. Feng, X.; Li, J.; Zhang, X.; Liu, T.; Ding, J.; Chen, X. Electrospun polymer micro/nanofibers as pharmaceutical repositories for healthcare. *J. Control. Release* **2019**, *302*, 19–41. [CrossRef]
109. Le, H.; Xu, W.; Zhuang, X.; Chang, F.; Wang, Y.; Ding, J. Mesenchymal stem cells for cartilage regeneration. *J. Tissue Eng.* **2020**, *11*, 204173142094383. [CrossRef]
110. Spakova, T.; Plsikova, J.; Harvanova, D.; Lacko, M.; Stolfa, S.; Rosocha, J. Influence of Kartogenin on Chondrogenic Differentiation of Human Bone Marrow-Derived MSCs in 2D Culture and in Co-Cultivation with OA Osteochondral Explant. *Molecules* **2018**, *23*, 181. [CrossRef] [PubMed]

111. Yang, X.; Tian, S.; Fan, L.; Niu, R.; Yan, M.; Chen, S.; Zheng, M.; Zhang, S. Integrated regulation of chondrogenic differentiation in mesenchymal stem cells and differentiation of cancer cells. *Cancer Cell Int.* **2022**, *22*, 169. [CrossRef]
112. Chen, F.H.; Rousche, K.T.; Tuan, R.S. Technology Insight: Adult stem cells in cartilage regeneration and tissue engineering. *Nat. Clin. Pract. Rheumatol.* **2006**, *2*, 373–382. [CrossRef]
113. Steinert, A.F.; Nöth, U.; Tuan, R.S. Concepts in gene therapy for cartilage repair. *Injury* **2008**, *39*, 97–113. [CrossRef]
114. Li, K.-C.; Hu, Y.-C. Cartilage Tissue Engineering: Recent Advances and Perspectives from Gene Regulation/Therapy. *Adv. Healthc. Mater.* **2015**, *4*, 948–968. [CrossRef]
115. Rodriguez-Merchan, E.C.; Valentino, L.A. The Role of Gene Therapy in Cartilage Repair. *Arch. Bone Jt. Surg.* **2019**, *7*, 79–90.
116. Ha, C.-W.; Noh, M.J.; Choi, K.B.; Lee, K.H. Initial phase I safety of retrovirally transduced human chondrocytes expressing transforming growth factor-beta-1 in degenerative arthritis patients. *Cytotherapy* **2012**, *14*, 247–256. [CrossRef]
117. Kim, M.K.; Ha, C.W.; In, Y.; Cho, S.D.; Choi, E.S.; Ha, J.K.; Lee, M.C. A Multicenter, Double-Blind, Phase III Clinical Trial to Evaluate the Efficacy and Safety of a Cell and Gene Therapy in Knee Osteoarthritis Patients. *Hum. Gene Ther. Clin. Dev.* **2018**, *29*, 48–59. [CrossRef]
118. Lee, B.; Parvizi, J.; Bramlet, D.; Romness, D.W.; Guermazi, A.; Noh, M.; Sodhi, N.; Khlopas, A.; Mont, M.A. Results of a Phase II Study to Determine the Efficacy and Safety of Genetically Engineered Allogeneic Human Chondrocytes Expressing TGF-β1. *J. Knee Surg.* **2020**, *33*, 167–172. [CrossRef]
119. Hamann, A.; Nguyen, A.; Pannier, A.K. Nucleic acid delivery to mesenchymal stem cells: A review of nonviral methods and applications. *J. Biol. Eng.* **2019**, *13*, 7. [CrossRef]
120. Bucher, C.; Gazdhar, A.; Benneker, L.M.; Geiser, T.; Gantenbein-Ritter, B. Nonviral Gene Delivery of Growth and Differentiation Factor 5 to Human Mesenchymal Stem Cells Injected into a 3D Bovine Intervertebral Disc Organ Culture System. *Stem Cells Int.* **2013**, *2013*, 326828. [CrossRef]
121. Kelly, A.M.; Plautz, S.A.; Zempleni, J.; Pannier, A.K. Glucocorticoid Cell Priming Enhances Transfection Outcomes in Adult Human Mesenchymal Stem Cells. *Mol. Ther.* **2016**, *24*, 331–341. [CrossRef]
122. Park, J.S.; Yi, S.W.; Kim, H.J.; Kim, S.M.; Kim, J.-H.; Park, K.-H. Construction of PLGA Nanoparticles Coated with Polycistronic *SOX5*, *SOX6*, and *SOX9* Genes for Chondrogenesis of Human Mesenchymal Stem Cells. *ACS Appl. Mater. Interfaces* **2017**, *9*, 1361–1372. [CrossRef]
123. de Carvalho, T.G.; Pellenz, F.M.; Laureano, A.; Silla, L.M.d.R.; Giugliani, R.; Baldo, G.; Matte, U. A simple protocol for transfecting human mesenchymal stem cells. *Biotechnol. Lett.* **2018**, *40*, 617–622. [CrossRef]
124. Nakashima, S.; Matsuyama, Y.; Nitta, A.; Sakai, Y.; Ishiguro, N. Highly Efficient Transfection of Human Marrow Stromal Cells by Nucleofection. *Transplant. Proc.* **2005**, *37*, 2290–2292. [CrossRef]
125. Hoare, M.; Greiser, U.; Schu, S.; Mashayekhi, K.; Aydogan, E.; Murphy, M.; Barry, F.; Ritter, T.; O'Brien, T. Enhanced lipoplex-mediated gene expression in mesenchymal stem cells using reiterated nuclear localization sequence peptides. *J. Gene Med.* **2010**, *12*, 207–218. [CrossRef]
126. Han, S.-W.; Nakamura, C.; Kotobuki, N.; Obataya, I.; Ohgushi, H.; Nagamune, T.; Miyake, J. High-efficiency DNA injection into a single human mesenchymal stem cell using a nanoneedle and atomic force microscopy. *Nanomedicine* **2008**, *4*, 215–225. [CrossRef]
127. Mun, J.-Y.; Shin, K.K.; Kwon, O.; Lim, Y.T.; Oh, D.-B. Minicircle microporation-based non-viral gene delivery improved the targeting of mesenchymal stem cells to an injury site. *Biomaterials* **2016**, *101*, 310–320. [CrossRef]
128. Benoit, D.S.W.; Boutin, M.E. Controlling Mesenchymal Stem Cell Gene Expression Using Polymer-Mediated Delivery of siRNA. *Biomacromolecules* **2012**, *13*, 3841–3849. [CrossRef]
129. Levy, O.; Zhao, W.; Mortensen, L.J.; LeBlanc, S.; Tsang, K.; Fu, M.; Karp, J.M. mRNA-engineered mesenchymal stem cells for targeted delivery of interleukin-10 to sites of inflammation. *Blood* **2013**, *122*, e23–e32. [CrossRef]
130. Bellavia, D.; Veronesi, F.; Carina, V.; Costa, V.; Raimondi, L.; De Luca, A.; Giavaresi, G. Gene therapy for chondral and osteochondral regeneration: Is the future now? *Cell. Mol. Life Sci.* **2018**, *75*, 649–667. [CrossRef]
131. Gurusinghe, S.; Strappe, P. Gene Modification of Mesenchymal Stem Cells and Articular Chondrocytes to Enhance Chondrogenesis. *BioMed Res. Int.* **2014**, *2014*, 1–10. [CrossRef]
132. Adkar, S.S.; Brunger, J.M.; Willard, V.P.; Wu, C.-L.; Gersbach, C.A.; Guilak, F. Genome Engineering for Personalized Arthritis Therapeutics. *Trends Mol. Med.* **2017**, *23*, 917–931. [CrossRef]
133. Nishimasu, H.; Shi, X.; Ishiguro, S.; Gao, L.; Hirano, S.; Okazaki, S.; Noda, T.; Abudayyeh, O.O.; Gootenberg, J.S.; Mori, H.; et al. Engineered CRISPR-Cas9 nuclease with expanded targeting space. *Science* **2018**, *361*, 1259–1262. [CrossRef]
134. Tanikella, A.S.; Hardy, M.J.; Frahs, S.M.; Cormier, A.G.; Gibbons, K.D.; Fitzpatrick, C.K.; Oxford, J.T. Emerging Gene-Editing Modalities for Osteoarthritis. *Int. J. Mol. Sci.* **2020**, *21*, 6046. [CrossRef]
135. Farhang, N.; Davis, B.; Weston, J.; Ginley-Hidinger, M.; Gertz, J.; Bowles, R.D. Synergistic CRISPRa-Regulated Chondrogenic Extracellular Matrix Deposition Without Exogenous Growth Factors. *Tissue Eng. Part A* **2020**, *26*, 1169–1179. [CrossRef] [PubMed]
136. Huynh, N.P.; Gloss, C.C.; Lorentz, J.; Tang, R.; Brunger, J.M.; McAlinden, A.; Guilak, F. Long non-coding RNA GRASLND enhances chondrogenesis via suppression of the interferon type II signaling pathway. *Elife* **2020**, *9*, e49558. [CrossRef]
137. Huang, H.; Hu, X.; Zhang, X.; Duan, X.; Zhang, J.; Fu, X.; Dai, L.; Yuan, L.; Zhou, C.; Ao, Y. Codelivery of Synovium-Derived Mesenchymal Stem Cells and TGF-β by a Hybrid Scaffold for Cartilage Regeneration. *ACS Biomater. Sci. Eng.* **2018**, *5*, 805–816. [CrossRef] [PubMed]

138. Tsuzuki, N.; Seo, J.P.; Yamada, K.; Haneda, S.; Furuoka, H.; Tabata, Y.; Sasaki, N. The effect of a gelatin β-tricalcium phosphate sponge loaded with mesenchymal stem cells (MSC), bone morphogenic protein-2, and platelet-rich plasma (PRP) on equine articular cartilage defect. *Can. Vet. J.* **2013**, *54*, 573–580. [PubMed]
139. Huynh, N.P.T.; Brunger, J.M.; Gloss, C.C.; Moutos, F.T.; Gersbach, C.A.; Guilak, F. Genetic Engineering of Mesenchymal Stem Cells for Differential Matrix Deposition on 3D Woven Scaffolds. *Tissue Eng. Part A* **2018**, *24*, 1531–1544. [CrossRef]
140. Legendre, F.; Ollitrault, D.; Gomez-Leduc, T.; Bouyoucef, M.; Hervieu, M.; Gruchy, N.; Galéra, P. Enhanced chondrogenesis of bone marrow-derived stem cells by using a combinatory cell therapy strategy with BMP-2/TGF-β1, hypoxia, and COL1A1/HtrA1 siRNAs. *Sci. Rep.* **2017**, *7*, 3406. [CrossRef]
141. Branly, T.; Bertoni, L.; Contentin, R.; Rakic, R.; Gomez-Leduc, T.; Desancé, M.; Galéra, P. Characterization and use of Equine Bone Marrow Mesenchymal Stem Cells in Equine Cartilage Engineering. Study of their Hyaline Cartilage Forming Potential when Cultured under Hypoxia within a Biomaterial in the Presence of BMP-2 and TGF-ß1. *Stem Cell Rev. Rep.* **2017**, *13*, 611–630. [CrossRef]
142. Ledo, A.M.; Vining, K.H.; Alonso, M.J.; Garcia-Fuentes, M.; Mooney, D.J. Extracellular matrix mechanics regulate transfection and SOX9-directed differentiation of mesenchymal stem cells. *Acta Biomater.* **2020**, *110*, 153–163. [CrossRef]
143. Hua, Y.; Xia, H.; Jia, L.; Zhao, J.; Zhao, D.; Yan, X.; Zhang, Y.; Tang, S.; Zhou, G.; Zhu, L.; et al. Ultrafast, tough, and adhesive hydrogel based on hybrid photocrosslinking for articular cartilage repair in water-filled arthroscopy. *Sci. Adv.* **2021**, *7*, eabg0628. [CrossRef]
144. Demott, C.J.; Jones, M.R.; Chesney, C.D.; Grunlan, M.A. Adhesive Hydrogel Building Blocks to Reconstruct Complex Cartilage Tissues. *ACS Biomater. Sci. Eng.* **2023**, *9*, 1952–1960. [CrossRef]
145. Liang, Y.; Li, S.; Li, Y.; Li, M.; Sun, X.; An, J.; Xu, Q.; Chen, Z.; Wang, Y. Impact of hydrogel stiffness on the induced neural stem cells modulation. *Ann. Transl. Med.* **2021**, *9*, 1784. [CrossRef] [PubMed]
146. Feng, Q.; Wei, K.; Lin, S.; Xu, Z.; Sun, Y.; Shi, P.; Li, G.; Bian, L. Mechanically resilient, injectable, and bioadhesive supramolecular gelatin hydrogels crosslinked by weak host-guest interactions assist cell infiltration and in situ tissue regeneration. *Biomaterials* **2016**, *101*, 217–228. [CrossRef] [PubMed]
147. Toh, Y.-C.; Xing, J.; Yu, H. Modulation of integrin and E-cadherin-mediated adhesions to spatially control heterogeneity in human pluripotent stem cell differentiation. *Biomaterials* **2015**, *50*, 87–97. [CrossRef] [PubMed]
148. Cao, B.; Peng, Y.; Liu, X.; Ding, J. Effects of Functional Groups of Materials on Nonspecific Adhesion and Chondrogenic Induction of Mesenchymal Stem Cells on Free and Micropatterned Surfaces. *ACS Appl. Mater. Interfaces* **2017**, *9*, 23574–23585. [CrossRef] [PubMed]
149. Amadori, S.; Torricelli, P.; Panzavolta, S.; Parrilli, A.; Fini, M.; Bigi, A. Multi-Layered Scaffolds for Osteochondral Tissue Engineering: In Vitro Response of Co-Cultured Human Mesenchymal Stem Cells. *Macromol. Biosci.* **2015**, *15*, 1535–1545. [CrossRef]
150. Dong, L.; Cheng, K.; Zhou, Y.; Yu, M.; Gong, J.; Lin, Y.; Luo, Q.; Wang, Q.; Weng, W.; Wang, H. Surface Atomic Structure Directs the Fate of Human Mesenchymal Stem Cells. *ACS Appl. Mater. Interfaces* **2017**, *9*, 15274–15285. [CrossRef]
151. Patil, S.; Singh, N. Spatially controlled functional group grafting of silk films to induce osteogenic and chondrogenic differentiation of human mesenchymal stem cells. *Mater. Sci. Eng. C* **2018**, *91*, 796–805. [CrossRef]
152. Lee, J.W.; Kim, H.; Lee, K.Y. Effect of spacer arm length between adhesion ligand and alginate hydrogel on stem cell differentiation. *Carbohydr. Polym.* **2016**, *139*, 82–89. [CrossRef]
153. Copp, G.; Robb, K.P.; Viswanathan, S. Culture-expanded mesenchymal stromal cell therapy: Does it work in knee osteoarthritis? A pathway to clinical success. *Cell. Mol. Immunol.* **2023**, *20*, 626–650. [CrossRef]
154. Duan, W.-L.; Zhang, L.-N.; Bohara, R.; Martin-Saldaña, S.; Yang, F.; Zhao, Y.-Y.; Xie, Y.; Bu, Y.-Z.; Pandit, A. Adhesive hydrogels in osteoarthritis: From design to application. *Mil. Med. Res.* **2023**, *10*, 1–26. [CrossRef]
155. Wagenbrenner, M.; Mayer-Wagner, S.; Rudert, M.; Holzapfel, B.M.; Weissenberger, M. Combinations of Hydrogels and Mesenchymal Stromal Cells (MSCs) for Cartilage Tissue Engineering—A Review of the Literature. *Gels* **2021**, *7*, 217. [CrossRef] [PubMed]

polymers

MDPI

Article

Mevacor/Poly(vinyl acetate/2-hydroxyethyl methacrylate) as Solid Solution: Preparation, Solubility Enhancement and Drug Delivery

Mohammed Alassaf [1], Saad Mohammed Alqahtani [1], Rana Salem Al Khulaifi [1], Waseem Sharaf Saeed [1], Faisal S. Alsubaie [1], Abdelhabib Semlali [2] and Taieb Aouak [1,*]

1 Department of Chemistry, College of Science, King Saud University, Riyadh 11451, Saudi Arabia; salqahtani2@ksu.edu.sa (S.M.A.); ranaalkhulaifi@gmail.com (R.S.A.K.); wsaeed@ksu.edu.sa (W.S.S.)
2 Groupe de Recherche en Écologie Buccale, Faculté de Médecine Dentaire, Université Laval, Quebec City, QC G1V 0A6, Canada
* Correspondence: taouak@ksu.edu.sa

Abstract: Mevacor/Poly(vinyl acetate-co-2-hydroxyethyl methacrylate) drug carrier systems (MVR/VAC-HEMA) containing different Mevacor (MVR) contents were prepared in one pot by free radical copolymerization of vinyl acetate with 2-hydroxyethyl methacrylate using an LED lamp light in the presence of camphorquinone as a photoinitiator and Mevacor as a drug filler. The prepared material was characterized by FTIR, ^{1}H NMR, DSC, SEM and XRD methods. Different parameters influencing the efficiency in the Mecvacor-water solubility and the drug delivery of this system, such as the swelling capacity of the carrier, the amount of Mevacor loaded and the pH medium have been widely investigated. The results obtained revealed that the Mevacor particles were uniformly dispersed in their molecular state in the copolymer matrix forming a solid solution; the cell toxicity of the virgin poly(vinyl acetate-co-2-hydroxy ethyl methacrylate) (VAC-HEMA) and MVR/VAC-HEMA drug carrier system exhibited no significant effect on their viability when between 0.25 and 2.00 wt% was loaded in these materials; the average swelling capacity of VAC-HEMA material in water was found to be 45.16 wt%, which was practically unaffected by the pH medium and the solubility of MVR deduced from the release process reached more than 22 and 37 times that of the powder dissolved directly in pH 1 and 7 media, respectively. The in vitro MVR release kinetic study revealed that the MVR/VAC-HEMA system containing 0.5 wt% MVR exhibited the best performance in the short gastrointestinal transit (GITT), while that containing 2.0 wt% is for the long transit as they were able to considerably reduce the minimum release of this drug in the stomach (pH1).

Keywords: Mevacor/poly(vinyl acetate-co-2-hydroxyl methyl methacrylate); solid solution; photo-copolymerization; LED light; Mevacor delivery; solubility enhancement

1. Introduction

Today, polymer materials occupy practically all sectors of industry. In the biomedical field, these materials are now involved in several formulations, particularly in the drug delivery domain in order to obtain desirable pharmacokinetics and therapeutic indices. Controlling the amount of drug released over time and its rate are key elements in the development of drug delivery systems—the rapid release of medication can have adverse effects on the body, and too little can prevent this drug from achieving the desired effectiveness. Therefore, drug carriers must be designed in a fashion that allows them to deliver the optimal dosage for the optimal duration of time in order to render drugs more effective [1–3].

Hydrogel-based drug delivery systems provide an alternative to the conventional drug formulations. They have been utilized in recent years to tackle the inadequate bioavailability of drugs. Hydrogels are highly water-absorbent mesh networks that are formed naturally or synthetically. The chains of polymers in these systems are crosslinked physically by means of many intermolecular forces (e.g., hydrogen bonding, dispersion interaction) or bound to each other covalently [4]. In many studies, researchers made use of polymer hydrogels as carriers due to their biocompatibility and physicochemical tolerability [5,6]. Among these studies, several investigations have been conducted in which polymer hydrogels were employed to enhance bioavailability and controlled the amount of drugs being released.

Mevacor (MVR) (Scheme 1), sold under the brand name "Lovastatin" or "Altoprev" among others, belongs to a class of drugs called HMG-CoA reductase inhibitors, also known as statins. This medication is used to treat high cholesterol and reduce the risk of cardiovascular disease [7]. This medication which is orally administered is recommended with lifestyle changes. As side effects, this medication can cause serious muscle damage, among other things, and also increases the risk of liver disease [8]. MVR is poorly soluble in water; according to the Biopharmaceutics Classification System (BCS) its solubility in water is 0.4 $\mu g\ mL^{-1}$ [9]. It is highly permeable, and the rate of its oral absorption is often controlled by the dissolution in the gastrointestinal rate. When drugs are orally administered, their water solubility is one of the factors most affecting their efficacy [10], because a limited solubility of an active ingredient in water considerably reduces the dynamics of the drug absorption. Therefore, an enhancement in its dissolution in water leads to the improving of its bioavailability. Among the techniques used to improve the solubility of the lovastatin in water, solid dispersion has been shown to be an effective and practical method by different researchers in this field [11–13].

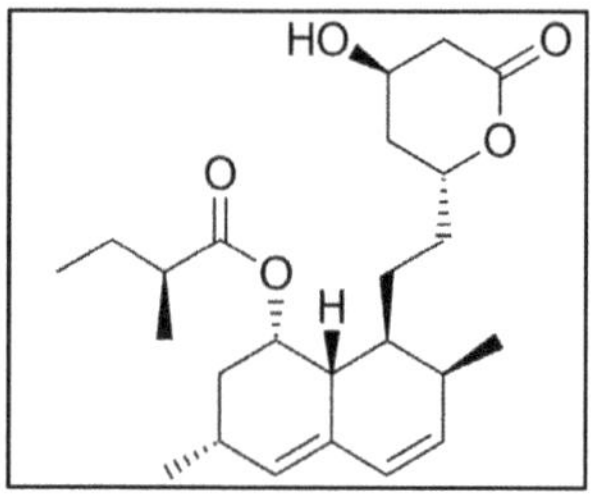

Scheme 1. Chemical structure of Mevacor (MVR).

Poly(2-hydroxyethyl methacrylate) (PHEMA) swells but does not dissolve in water when its polymerization degree, $\overline{Dp}$, is greater than 20 [14,15] and its swelling capacity is moderate (~40 wt%) [16]. With a $\overline{Dp}$ lower than 20, this polymer is soluble at all temperatures. Those with $\overline{Dp}$ localized between 20 and 45 exhibit cloud points. According to Armes et al. [15], the insolubility of PHEMA in water is mainly due to intra- and inter-chain H-bonding between polymeric hydroxy groups. The short chains may H-bond less efficiently, leaving a greater fraction of hydroxy groups to interact with water, or the H-bonding may simply not be sufficient to overcome the increased entropy of the short chains. As long as this polymer is soluble in other solvents such as dimethylformamide and dimethyl sulfoxide, this proves that the PHEMA is crosslinked physically (formation of inter-chain hydrogen bonds) and not chemically (formation of interchain covalent bonds).

PHEMA hydrogel was employed for the first time in the biological field by Wichterle and Lim [17]. Due to its safe tolerance, good biocompatibility, non-toxicity and non-antigenic properties, this hydrophilic material is widely studied in biomedical applications [18]. For example, this hydrogel was also developed as a carrier for water-soluble anticancer drugs, such as 5-fluorouracil [19], topical mitomycin-C [20], and cytarabine [21].

Poly(vinyl acetate) (PVAc) belongs to the family of shape memory polymers (SMP) which are smart materials that can be temporarily modeled and return to their original shape when an external stimulus is applied to them. PVAc is a hydrophobic and inexpensive polymer. It can be easily transformed into a hydrophilic biopolymer by hydrolysis in an acidic or basic medium by removing the molecules of acetic acid as a by-product giving poly(vinyl alcohol). This polymer, similarly to PHEMA, which is synthesized mainly by the free radical polymerization route, is biocompatible, biodegradable, non-toxic and non-carcinogenic [22–24]. However, its application as a shape memory polymer in the biomedical field is limited due to its weak mechanical properties such as the resistance modulus [22,25]. Due to its biological nature which respects the environment in which it is applied, it is used in various medical fields including cardiovascular devices, implants of artificial organs, cartilage skin and contact lenses. It is also used in wound dressings and various drug delivery applications [23,24].

In order to improve its mechanical properties and broaden its application in the medical field, PVAc is blended with others polymers having desirable properties [23,25–27] or employed as copolymers [28–30].

Several studies were performed to improve the dissolution properties and bioavailability of MVR, thereby increasing its drug efficacy. For example, Lin et al. [31] prepared nanoparticles involving poly(d, l-lactide-co-glycolide acid) (PLGA) and Mevacor using the double emulsion technique for its application in the dentistry field in the direct capping of pulp. It was found that these systems released MVR over a period of 44 days. In another study published by Tarafder et al. [32] on the release behavior of this same molecule from a coating of poly (ε-caprolactone) (PCL) on β-tricalcium phosphate (β-TCP), it was revealed that the hydrophilic-hydrophobic and hydrophobic-hydrophobic interactions that exist between MVR and PCL were the key factors controlling the diffusion dominated release kinetics of this medication from PCL coating. Recently, Madhavi et al. [33] studied the improvement of MVR absorption and its oral bioavailability. These researchers attempted to formulate floating drug delivery systems using Mevacor as a drug candidate using Methocel of different grades, e.g., K4 and K5. According to the results obtained, the F6 formulation containing 20 mg of MVR, xanthum gum was revealed as the optimized formulation and the result obtained was a release of more than 98.9% of this drug in 12 h.

To reduce the undesirable side effects of this very beneficial medication mainly used to treat high cholesterol and reduce the risk of cardiovascular disease, controlling the amount of MVR released, its uniformity, location, speed and duration of release must be well controlled.

In this work, to control the fraction of Mevacor released in a targeted medium over time, the reduction of the swelling rate of the PHEMA support and the improvement of its flexibility must be considered. To achieve this goal, a small amount of vinyl acetate unit (hydrophobic monomer) was incorporated into these PHEMA chains by copolymerization of HEMA with Vac comonomers. A series of Mevacor/poly(vinyl acetate-co-2-hydroxyethyl methacrylate) (MVR/VAC-HEMA) solid solutions with different MVR contents was prepared in one pot by the free radical photo-copolymerization reaction in the presence of this medication using camphorquinone as a photo-initiator under LED light. The structures of copolymer and solid solutions obtained were highlighted by FTIR and ^{1}H NMR spectrometry and the distribution of MVR particles in the VAC-HEMA matrix was investigated by DSC, XRD and SEM techniques. The "in vitro" release kinetics were carried out in different pH media and the percentage of Mevacor released was evaluated by the UV-visible method. Different parameters influencing the release dynamics of this medication were widely studied such as the swellability of the copolymer material, the amount of the MVR loaded and the pH medium. Among others, the solubility enhancement of Mecavor in acidic and neutral pH media and the estimated distribution of the cumulative MVR released from MVR/VAC-HEMA hydrogel systems on the principal digestive organs timed according to Belzer et al.'s approach [34] were also investigated in this work.

2. Materials and Methods

2.1. Chemicals

Camphorquinone (CQ) (purity, 97%), vinyl acetate (VAc) (purity, 99%), and 2-hydroxyethyl methacrylate (HEMA) (purity, 99%) were supplied by Sigma Aldrich (KGaA, Germany).

Mevacor (MVR) was provided from TCI America (Portland, OR, USA). Before usage, monomers were held under nitrogen gas after being purified by distillation at a reduced pressure. Photo-initiator (CQ) was used without purification. The human epithelial cell (HEC) line, GMSM-K prepared by Gilchrist et al. (2000) was offered by Dr Grenier of Laval Universitéy (Quebec, QC, Canada). Dulbecco's Modified Eagle's Medium (DMEM) was provided from Corning, Manassas, VA, USA. Fetal bovine serum (FBS) was purchased from Gibco (New York, NY, USA) and 1% of penicillin/streptomycin solution was supplied from Sigma Aldrich (St. Louis, MO, USA). A lactate dehydrogenase (LDH) detection kit was provided by BioVision (BioVision, Milpitas, CA, USA).

2.2. Preparation of PHEMA and PVAc Homopolymers

PHEMA and PVAc homopolymers were prepared by free radical polymerization of HEMA and Vac, respectively, under a lamp light led (6 × 5 watts) using CQ as photo-initiator. Initially, 5.35 g (0.041 mol) of HEMA, 3.530 g (0.041 mol) of VAc and 0.008 g of CQ were mixed in a beaker by stirring at an ambient temperature (~25 °C) until complete dissolution of the photo-initiator. The obtained solution was placed in a cylindrical mold made of Teflon and exposed to the LED lamp light during the polymerization reaction. The polymerization reaction was carried out under a moderated flux of nitrogen gas (3 mL·min^{-1}) passing through the reactor fabricated as described by Alqahtani et al. [35,36]. The polymer was easily detached from the molder and dried in the open air for 12 h followed by the same duration in a vacuum oven at 40 °C. Both the resulting polymeric materials are a cylindrical sheet that is thin, transparent and flexible.

2.3. Preparation of MVR/VAC-HEMA Solid Solution

MVR/VAC-HEMA solid solution was prepared in one pot by free radical copolymerization of Vac with HEMA in presence of MVR as filler and CQ as photo-initiator. First, 0.041 mol (5.35 g) of HEMA and 0.007 mol (0.62 g) of VAc were mixed in a beaker by stirring at an ambient temperature (~25 °C), then 0.008 g of CQ was added to the reaction mixture. The copolymerization reaction was carried out as shown in Scheme 2 in the same conditions as described in the Section 2.2. As in the previous case, the copolymer was easily detached from the molder and dried in the open air for 12 h followed by a same duration in a vacuum oven at 40 °C. The resulting polymeric material is a cylindrical sheet that is thin, transparent and flexible. A series of MVR/VAC-HEMA containing 0.5 wt% (0.03 g), 1.0 wt% (0.06 g), 1.5 wt% (0.09 g) and 2.0 wt% (0.12 g) MVR contents was prepared by this same procedure.

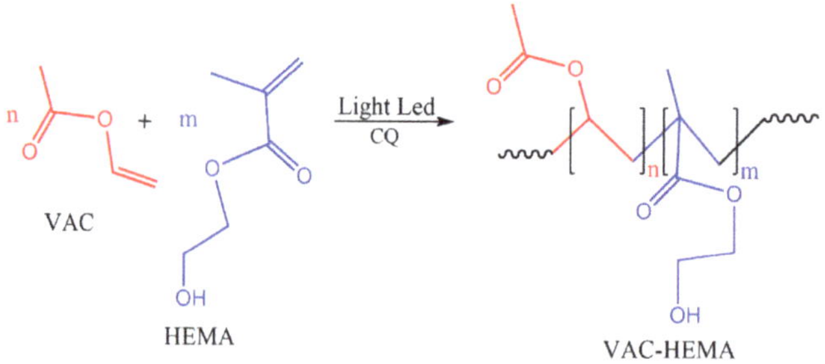

Scheme 2. Free radical copolymerization reaction of HEMA with VAC.

2.4. Characterization

2.4.1. FTIR Analysis

A Perkin Elmer Spectrum GX FTIR spectrometer (Waltham, MA, USA) operating in attenuated total reflection mode, in a wave number range of 4000 to 650 cm^{-1}. Thirty-two (32) cycles of scanning, and a resolution of 2 cm^{-1} was used to record the Fourier transform infrared (FTIR) spectra of pure MVR, PVAc, PHEMA homopolymers, and VAC-HEMA copolymer. The samples were analyzed in the form of thin transparent films except that of MVR which was analyzed as powder diluted in dehydrated KBr.

2.4.2. HNMR Analysis

A JEOL FX 90 Q NMR spectrometer, (Tokyo, Japan) at 500 MHz was used to characterize the ^{1}HNMR copolymer and their corresponding homopolymers were dissolved in DMSO-d_6.

2.4.3. SEC Analysis

The SEC chromatogram of VAC-HEMA was performed on a Varian apparatus equipped with JASCO type 880-PU HPLC pump, refractive index, UV detectors and TSK gel columns calibrated with polystyrene standards. Dimethylformamide (DMF) was used as an eluent at 30 °C.

2.4.4. DSC Analysis

The DSC thermograms of MVR, virgin VAC-HEMA and MVR/VAC-HEMA solid solutions containing different MVR contents were plotted on a DSC device (Shimadsu DSC 60, Tokyo, Japan) previously calibrated with indium. Between 10 and 12 mg of samples were packaged in aluminum DSC capsules before being placed in the DSC cell. Samples were scanned by heating from −25 to 250 °C with a heating rate of 10 °C·min^{-1}. The glass transition temperatures (Tgs) of homopolymers, copolymer and copolymers containing medication with different contents were taken from the inflection point of the thermal curves, which characterizes the sliding of about 50% of the polymer chains. The melting temperature of MVR powder was determined from the top of the endothermic peak corresponding to its enthalpy of fusion.

2.4.5. XR-Diffraction Analysis

An advanced BRUKER D8 diffractometer (Germany) was used to perform the samples X-ray spectrographs of pure and mixed materials. Patterns were run with Cu Kα radiation at 40 mA and 40 kV with a 2θ scan rate of 2° min^{-1}.

2.4.6. SEM Analysis

Before and after the release procedure, the micrographs of MVR powder, virgin VAC-HEMA, and MVR/VAC-HEMA dried films of various MVR contents coated with gold grid were investigated by scanning electron microscope (SEM) using a Hitachi S4700 Field emission (Tokyo, Japan).

2.4.7. Cell Toxicity and Cell Adhesion

According to Semlali et al. [37], the LDH assay was used to assess the cellular cytotoxicity of all specimens. On 24-well plates for 24 h, 2×10^5 cells containing a specimen were seeded. Fifty (50) µL of duplicate LDH mix solution was added to an equal volume of each supernatant. These solutions were placed in a 96-well plate and incubated for 20 min at room temperature in the dark until the yellow color appeared. The reading was taken at 490 nm with an iMark Microplate Absorbance Spectrophotometer (Bio-Rad, Mississauga, ON, Canada). A percentage was used to represent the cell toxicity. Next, 2×10^5 cells treated with 1% of Triton X-100 were used as a positive control for LDH and which corresponds to 100% cytotoxicity. This experiment was replicated three times for each specimen.

According to our previous works, the MTT test was used to measure cell adhesion for each specimen placed on 24-well plates [38,39]. Briefly, 2×10^5 GMSMK cells/specimen were seeded in 24-well plate, cultured for adhesion overnight. Each cell culture was then given a 1/10 volume addition of a 5 mg·mL^{-1} MTT solution for 3 h of incubation at 37 °C in the dark before using 300 µL of isopropanol 0.05N of HCL solution to lyse the cells. Next, 2×100 µL of lysed solution was poured into a 96-well microplate to assess absorbance at 550 nm by an iMark reader (Bio-Rad, Mississauga, ON, Canada). The percentage cell viability was used to assess the percentage of viable, proliferating cells. In addition, 0% adhesion corresponds to Absorbance (A) at 550 nm for a cell-free specimen.

2.4.8. Swellability

At 37 °C, distilled water was used to test the copolymer hydrogel's swelling degree until saturation (swelling equilibrium). A square film sample of copolymer measuring 3 cm × 3 cm × 3 mm was completely immersed into a beaker that had been filled with 100 mL of distillate water. The entire system was kept at a temperature of 37 °C with moderate agitation. After being withdrawn from the beaker at regular intervals, the specimen was gently wiped of the droplets that had been left on the film specimen's two surfaces with absorbent paper before being weighed on a precision balance. The identical conditions were used to conduct this experiment three times, and the swelling degree, *SD* (*wt*%), was calculated from the mean arithmetic values obtained using Equation (1)

$$SD(wt\%) = \frac{m_t - m_o}{m_o} \times 100, \tag{1}$$

where m_o and m_t are the masses of the film specimen weighted before and at *t* time of the swelling process, respectively.

2.4.9. In Vitro Release Dynamic

MVR/VAC-HEMA film samples of different MVR contents were immersed separately in 20 mL of bi-distilled water maintained at pH levels of 1, 3, 5, and 7 at body temperature (37 °C) for 72 h with a stirring rate of 100 rpm. To monitor the amount of MVR released, aliquots of 0.5 mL are taken at time intervals and immediately returned to the media after analysis. This operation allows for a constant volume of media during the release process. The concentration of MVR released during a given period time was determined by UV-visible analysis using a double beam U-2910 Hitachi spectrophotometer(New York, NY, USA).

The cumulative mass of MVR released was obtained from the calibration curve indicating the variation in medication concentration versus absorbance. A buffer solution does not need to be added to the water in this situation since the small amount of MVR produced (pKa = 13.49) [36] during the release process has practically no impact on the stability of the pH medium.

3. Results and Discussion

3.1. Characterization

3.1.1. FTIR Analysis

The FTIR spectra of pure MVR, virgin VAC-HEMA and MVR/VAC-HEMA solid solution with different MVR contents are gathered in Figure 1. The comparison of the spectrum of VAC-HEMA with those of PHEMA and PVAc reveals the presence of the combination of the signals attributed to these two homopolymers. The absorption band of the signals attributed to the carbonyl groups of the HEMA and VAc units is observed at 1704.70 cm^{-1} and that of the hydroxyl group at 3365.04 cm^{-1}. Given the small amount of MVR loaded into the MVR/VAC-HEMA system, the spectra of the solid solution containing 0.5 and 1.0 wt% of this medication do not show additional signals characterizing the structure of this drug, such as that of the signal at 801 cm^{-1} which is assigned to the δ(CH) rocking mode [40]. The comparison of these spectra also reveals shifts in the carbonyl groups from 1704.70 to 1716.08 cm^{-1} and another shift in the hydroxyl groups of the copolymer from 3365 to 3358.27 cm^{-1}, thus revealing the presence of hydrogen bonds between the hydroxyl and the carbonyl groups of the HEMA unit and those of MVR medication. This confirms the uniform dispersion of this medication in its molecule level in the copolymer matrix.

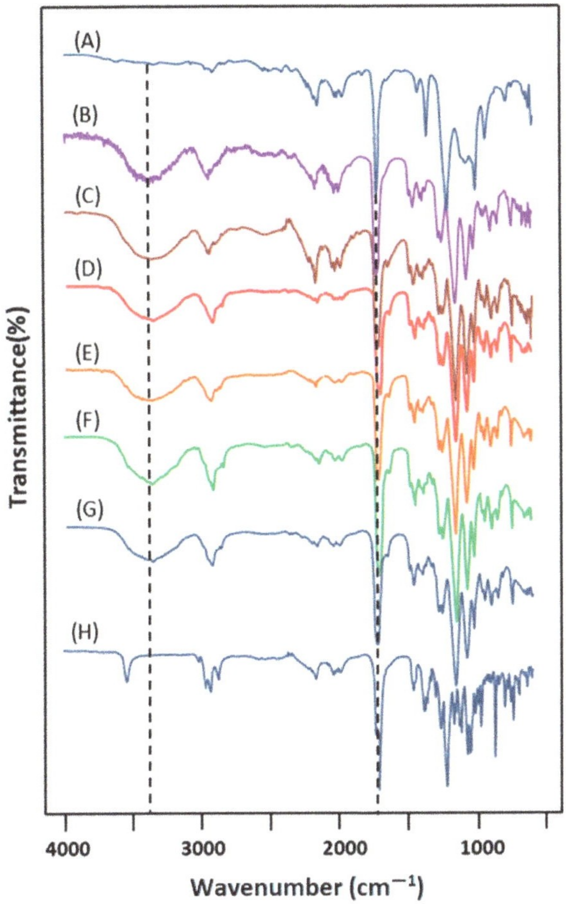

Figure 1. FTIR spectra of PVAc (A), PHEMA (B) and VAC-HEMA containing 0 (C), 0.5 (D), 1.0 (E), 1.5 (F), 2.0 (G) wt% MVR and (H) pure MVR. (---) Reference bands compared to those of pure PHEMA.

3.1.2. NMR Analysis

The structure of the VAC-HEMA was demonstrated by ^{1}H NMR analysis in the light of the comparison of its spectrum with those of their corresponding PHEMA and PVAc homopolymers as shown in Figure 2. Indeed, the spectrum of VAC-HEMA overlap the signals of the two different homopolymers, thus indicating the presence of the two monomers involved in the copolymerization reaction. The multiplets that appear between 0.35 and 1.0 ppm are assigned to the methyl group of the HEMA sequence isomers in the copolymer chain, and those between 1.3 and 2.1 ppm are attributed to the ethylenyl groups of the isomers of the two different units involved in the copolymer [41]. The percentage of the VAc and HEMA monomeric units in the VAC-HEMA copolymer was determined on the basis of the areas of the signals attributed to the three protons of the methylene group (a) and those of the signals of the ethynyl group ($d + d'$) common to the two different monomeric units using Equation (2),

$$HEMA(mol\%) = \frac{2\delta_{CH_3(a)}}{3\delta_{CH_2(d+d')}} \times 100 \quad (2)$$

where $\delta_{CH_3(a)}$ and $\delta_{CH_2(d+d')}$ are the surface areas of the three protons of $-CH_3$ (a) and the two protons of $-CH_2$-$(d + d')$ common to the two monomeric units, respectively. The determination of the composition in HEMA units in the VAC-HEMA copolymer indicates 84.22% by mole taken as the arithmetic mean of three spectra for this same copolymer. This composition is desirable in order to maintain the affinity of the MVR with the resulting material (miscibility) while limiting its swelling to a rate allowing it to be used as an adequate carrier in the ACR/VAC-HEMA drug carrier system.

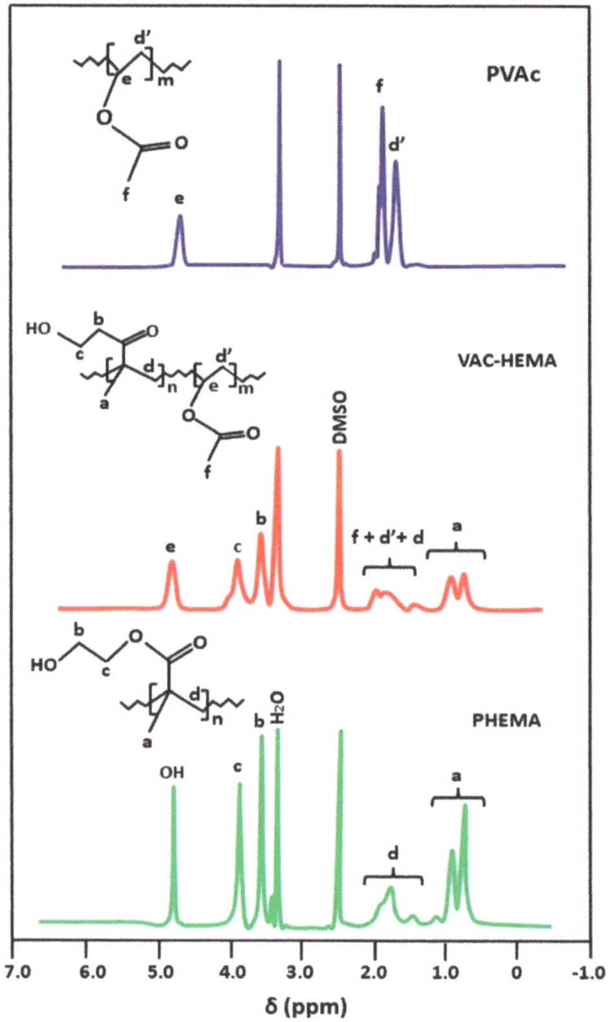

Figure 2. ^{1}HNMR spectra of PHEMA, PVAc and VAC-HEMA copolymer.

The deconvolution of these signals between 0.5 and 2.3 ppm in Laurentzien curves, as shown in Figure 3, permitted to determine the average tacticity (%) of the triad, tetrahedral and pentahedral sequences distributed in the VAC-HEMA chains. The results obtained revealed that this copolymer has a mainly heterotactic microstructure with atactic pentaeds *mrrr* (31.26%), *rmrr* (15.72%) and *mmmr* (10.14%), and the rest are mainly syndiotactic tetraheds *rrr* (21.42%) and atactics *rmr* + *mmr* (11.62%).

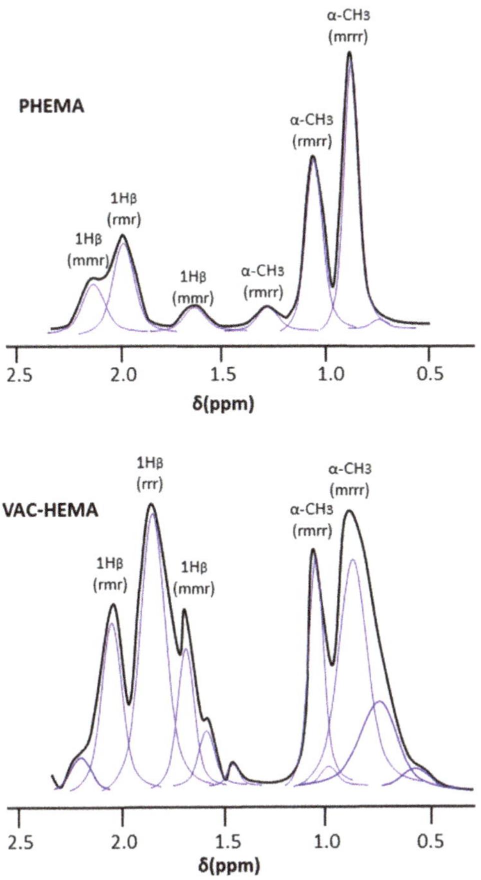

Figure 3. Decomposition in Lorentzian curves of the ^{1}HNMR signals of PHEMA and VAC-HEMA between 0.5 and 2.3 ppm.

These values were taken from the calculations of triplicate spectra for each sample, and the results obtained revealed that the microstructure of VAC-HEMA copolymer obtained practically reflects that of the PHEMA homopolymer which is mainly heterotactic as shown in this figure. These results are not surprising, because it is well known that in general, the polymerization of vinyl or acrylic monomers by the free radical route leads to mainly heterotactic microstructures due to the uncontrollable free radicals resulted during the reaction.

3.1.3. XR-Diffraction

The XR-Diffraction spectra of MVR powder, virgin VAC-HEMA and MVR/VAC-HEMA solid solutions are grouped for comparison in Figure 4. The spectrograph of the virgin VAC-HEMA shows an amorphous structure by the absence of sharp signals characterizing a crystalline structure. The spectrum of MVR powder shows a crystalline structure through the appearance of sharp signals at 7.90, 9.31, 11.35, 15.00, 16.80, 19.21 and 26.50° 2 theta, which agree with the literature [12]. The disappearance of these signals in the spectra of MVR/VAC-HEMA systems indicates that the MVR are homogeneously dispersed in the VAC-HEMA matrix in its molecular level. Noting that, the presence of aggregated particles in an amorphous polymer is evidenced by the appearance of signals specific to the crystalline structure of the filler in the spectrum of the MVR/VAC-HEMA solid solution or by their shift in the event of modification of the geometry of the resulted crystals.

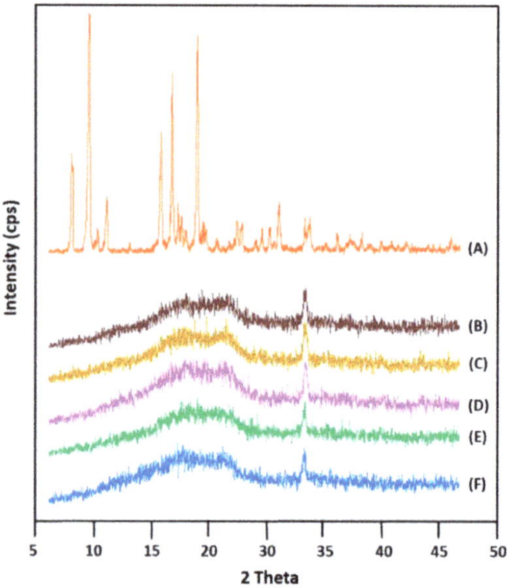

Figure 4. XR-Diffraction patterns of: (A) MVR powder; (B) VAC-HEMA; (C) MVR/VAC-HEMA0.5; (D) MVR/VAC-HEMA1.0; (E) MVR/VAC-HEMA1.5 and (F) MVR/VAC-HEMA2.0.

3.1.4. SEC Analysis

The SEC chromatogram of VAC-HEMA in Figure 5 indicates a relatively high molecular weight ($\overline{Mn}$ =4.12 × 10^5 g·mol^{-1} and $\overline{Mw}$ = 6.25 × 10^5 g·mol^{-1}) characterized by a relatively narrow molecular distribution, in which the polydispersity index (PDI) was 1.52. This indicates that there was an important reduction in the transfer reactions during the copolymerization reaction. Note that a large PDI indicates the presence of a small fraction of low macromolecular weight in the polymers. According to different authors [15,16], PHEMA chains having degrees of polymerization lower than 20 are soluble in water at all temperatures. This is not desirable as a carrier for use in the drug release fields. The incorporation of 15.78 mol% of VAC hydrophobic units in the PHEMA chains considerably reduce the water solubility of this copolymer.

3.1.5. DSC Analysis

The DSC curves of the MVR/VAC-HEMA solid solutions and their pure components are grouped together in Figure 6. As can be observed from the thermogram of pure MVR, this medication shows a melting temperature at 173 °C, which agrees with the literature (173.4 °C) [11] and the profile of the thermal curve of the virgin VAC-HEMA shows a Tg of 113 °C for this copolymer. On the other hand, when 0.5 wt% of MVR was incorporated in the VAC-HEMA matrix, a drop in the Tg of 25 °C was observed which passed from 113 to 88 °C. As the content of MVR incorporated into the copolymer becomes greater, a significant increase in the Tg of the copolymer is noted reaching a maximum of 145 °C when 2.0 wt% of this drug was loaded. Regarding the MVR/VAC-HEMA systems, the total disappearance of the endothermic peak characterizing the fusion of MVR in their thermograms indicates that this drug is indeed uniformly dispersed in the copolymer matrix in its molecular level. This confirms the formation of a solid solution involving the MVR as solute and the VAC-HEMA as solvent.

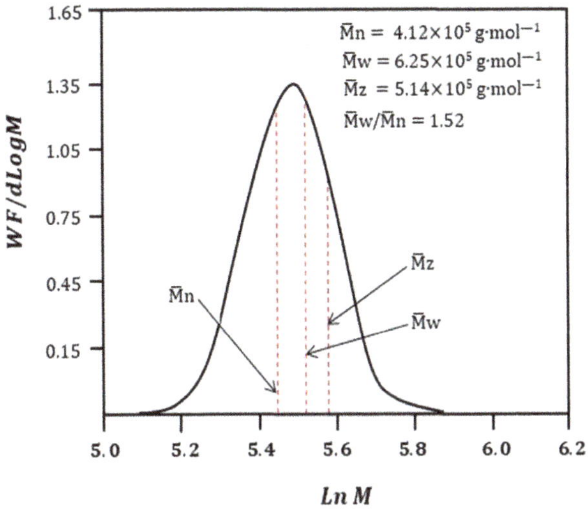

Figure 5. Chromatogram SEC of VAC-HEMA obtained at 30 °C in DMF used as eluent at 30 °C.

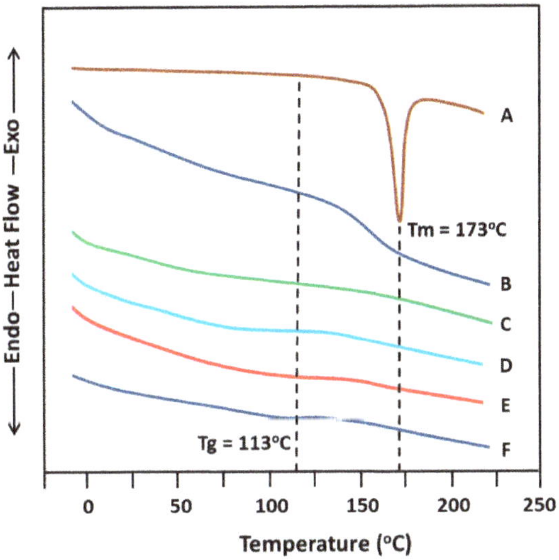

Figure 6. DSC thermograms of MVR (A), virgin VAC-HEMA (B) and MVR/VAC-HEMA containing 0.5 wt% (C), 1.0 wt% (D), 1.5 wt% (E) and 2.0 wt% (F) MVR.

3.1.6. SEM Analysis

The surface morphologies of the MVR/VAC-HEMA drug carrier systems and their pure components were examined by the scanning electron microscopy and the images of pure MVR, virgin VAC-HEMA and their mixtures containing 0.5 and 2.0 wt% medication are shown as examples in Figure 7. The image of the pure MVR (A) shows smooth rectangular rod-shaped particles on the corners of different dimensions. A similar photo of this medication was also obtained by Guan et al. [42]. The micrograph of the virgin VAC-HEMA (B) shows a smooth, slightly rugged surface devoid of any particles deposited on the surface, resembling a snowy mountain descent. Before the release process, the image of the MVR/VAC-HEMA0.5 system (C) shows a slightly undulating surface resembling muddy ground. As can also be seen in this photo, this surface is devoid of any particles deposited or embedded on the copolymer material. This indicates that the Mevacor particles are well dispersed homogeneously in their molecular level in the VAC-HEMA matrix. This confirms the results obtained by DSC and XR-D analysis. This same film sample, after the release process, shows a surface containing a high density of well-dispersed circular and oval-shaped pores with an average diameter varying between 0.3 and 16 μm, testifying to a massive release of MVR medication. On the other hand, before the release process, the image of MVR/VAC-HEMA2.0 drug carrier system, which contains the maximum amount of MVR (D), shows a smooth surface containing plaice formed during the vacuum drying of the sample. After release, this same sample, as in the case of the MVR/VAC-HEMA0.5 specimen, shows a high density of pores, but sometimes of a much larger size testifying to the release of a significant amount of drug.

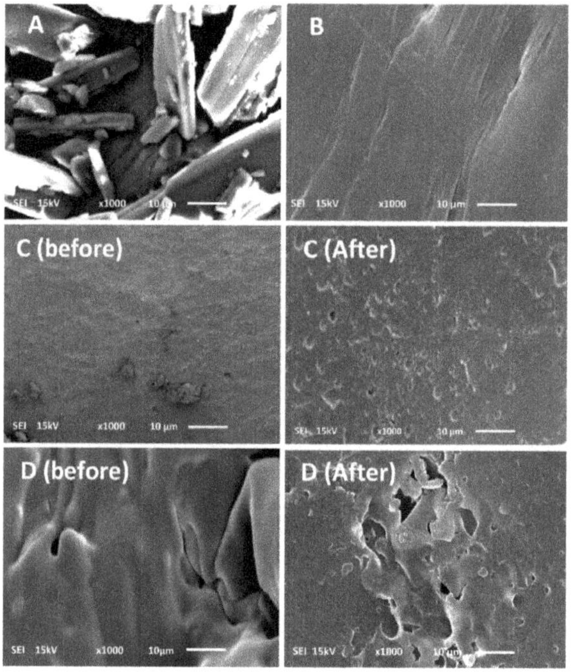

Figure 7. SEM micrographs of MVR (**A**), VAC-HEMA (**B**) and MVR/VAC-HEMA drug carrier systems containing: 0.5 wt% (**C**) and 2.0 wt% (**D**) MVR before and after the release process.

3.1.7. Swellability

To study the dynamics of a drug released from a drug carrier system in an aqueous media, knowledge of the swellability of a material used as a carrier is a key factor that must be taken into account. According to the literature, at body temperature (37 °C), this parameter is influenced by different factors such as the nature of the used carrier, the crosslinking degree of the polymer [43,44] and the pH medium in which the drug will be released [45,46]. In this work, the variation in the swelling degree of VAC-HEMA copolymer versus time, carried out at 37 °C during one week in different pH media, is plotted in Figure 8. As can be observed from the profiles of these curves obtained after triplicate experiments, usual exponential functions are obtained and the values of the swelling capacity of the specimen in pH media 1, 3, 5 and 7 varied between 42.2 ± 0.3 and 45.8 ± 0.5 wt%, which reveal no significant change in the swelling capacity of VAC-HEMA material in this pH range. This result was predicted, because the nature of the copolymer in question does not a seat of acid-base interaction with the pH medium. Based on this principle, the release dynamics of the MVR will depend only on a synergy involving the properties of VAC-HEMA copolymer and those of the medication loaded and not on the pH medium.

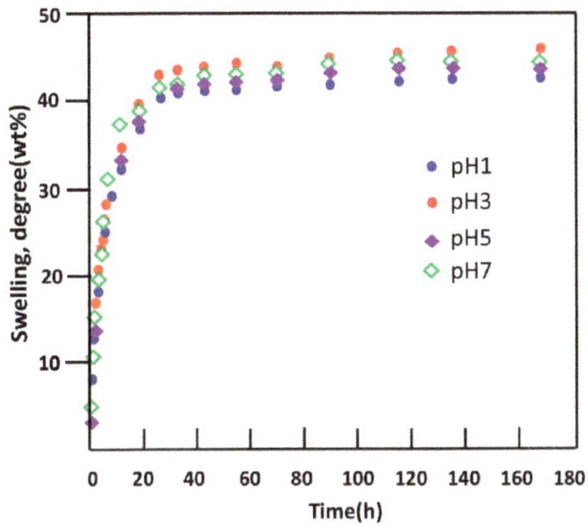

Figure 8. Swelling behavior of virgin VAC-HEMA material in water at 37 °C.

3.1.8. Water Diffusion

According to Comyn [47], the diffusion of small molecules such as water through a polymer material for a short period of diffusion and when the wt/w_∞ is less than 0.5 is expressed by the following relationship:

$$\frac{w_t}{w_\infty} = 2\left(\frac{D}{\pi l^2} \times t\right)^{0.5}, \tag{3}$$

where w_t and w_∞ are the weight of the sorbed molecules during a t time and the total weight absorbed, respectively. l and D are the polymer film thickness and the diffusion coefficient of the molecules sorbed, respectively. The linearization of Equation (4) leads to Equation (5)

$$Ln\left(\frac{w_t}{w_\infty}\right) = lnk + nLnt, \text{ with } k = \frac{2}{l}\left(\frac{D}{\pi}\right)^{0.5}, \tag{4}$$

The slope of the linear portion of the curve indicating the change in the $Ln(w_t/w_\infty)$ as function of the square root of time (Figure 9) give the D value of water molecules through the VAC-HEMA hydrogel. Indeed, the profile of the curves obtained reveals straight line for this hydrogel in all pH media investigated. This reveals that the diffusion of water molecules through the VAC-HEMA matrix obeys a Fickian model. The basic equation allowing to obtain the mass of molecules absorbed by a polymeric material is given by the following equation [48],

$$\frac{w_t}{w_\infty} = kt^n, \tag{5}$$

where n and k are the type of the diffusion mechanism and a constant relating the swelling rate, which depends on the film thickness and the diffusion coefficient, respectively. In this work, the n and k values are deducted from the slope and the intercept of these linear curves and the results obtained are gathered in Table 1.

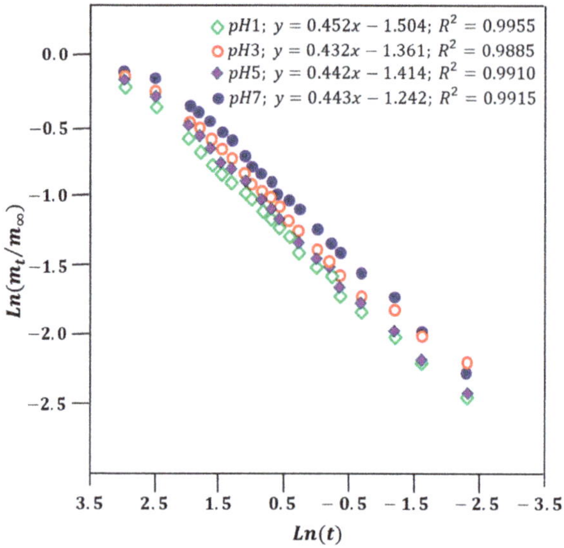

Figure 9. Variation in $Ln(m_t/m_\infty)$ of virgin VAC-HEMA hydrogel versus *Ln*(*t*) at 37 °C and in different pH media.

Table 1. Kinetics parameters deducted from the straight lines of the curves $Ln(m_t/m_\infty)$ of virgin VAC-HEMA hydrogel versus *Ln*(*t*) at 37 °C obtained in different pH media.

	VAC-HEMA Hydrogel			
pH Medium	**1**	**3**	**5**	**7**
n	0.488 ± 0.087	0.492 ± 0.086	0.502 ± 0.073	0.493 ± 0.072
$k(\text{h})^{-1}$	0.222 ± 0.120	0.256 ± 0.111	0.243 ± 0.121	0.289 ± 0.118
$D{\cdot}10^{-3}(\text{mm}^2{\cdot}\text{h}^{-1})$	1.901 ± 0.054	2.273 ± 0.034	2.435 ± 0.028	2.792 ± 0.025

In general, the values of n are close to those of the exponent in Equation (6), thus characterizing the order 0.5 of the diffusion of water molecules in the VAC-HEMA carrier. On the other hand, the values of k and D slowly increase with the pH of the medium. This indicates that, during the swelling process, the water diffusion rate increased with the pH of the medium and reached about 1.5 times that in the neutral medium. This seems to indicate that in a neutral medium the hydrogen bonds between the molecules of water and those of the copolymer are more favorable. This can be interpreted as a small reduction in the hydrophilicity of the copolymer due to a reduction in the density of the inter- and intra-chain hydroxyl bonds of the copolymer in acidic media. This is probably due to intra- and inter-chain etherification reactions involving two adjacent hydroxyl groups of the HEMA units that result in the elimination of water molecules by forming hydrophobic ether bonds. Consequently, this leads to a reduction in the diffusion rate of water molecules inside the copolymer matrix.

3.1.9. MVR Solubility Enhancement

The maximum solubility of MVR in media pH 1 and 7 was deduced from the maximum amount of this medication released from the MVR/VAC-HEMA drug carrier systems over time using the shake flask method [49]. The solubility of MVR powder in water at pH 1 and 7 was measured at saturation characterized by the beginning of the appearance of turbidity in which the reflectance in the UV-visible range increased sharply indicating a supersaturation of the resulted solution. The supersaturated solution is centrifuged at ambient temperature (~25 °C) and the liquid supernatant is filtered. To avoid crystallization of MVR in the medium, 0.5 mL of the supernatant is taken and then diluted with the same medium. The measurements of the absorbance and reflectance were performed at 37 °C and the solubility at equilibrium of MVR in neutral pH and pH medium 1 were taken from the arithmetic average of three experiments. The maximum solubility deducted from the maximum release of MVR from the MVR/VAC-HEMA drug carrier systems in these same media are gathered with that dissolved directly as powder in Table 2. As can be seen from these data, the maximum solubility of MVR increased significantly when dispersed at its molecular level in the VAC-HEMA matrix (Scheme 3). Indeed, the solubility of this same medication obtained by this method increased more than 22 and 37 times that dissolved as powder in media pH 1 and pH 7, respectively. This is due to a significant increase in the contact surface between the dispersed MVR molecules and those of the aqueous medium inside the carrier matrix. According to the results obtained on the physico-chemical and thermal characterization of MVR/VAC-HEMA systems by XRD (Section 3.1.3) and DSC (Section 3.1.4), it was revealed that the MVR was effectively homogeneously distributed in the carrier matrix in its molecular state, this implies that the solvation of this drug was in its maximum state.

Table 2. Comparison between the maximum solubility (mg·mL^{-1}) of MVR released from the MVR/VAC-HEMA system with that directly dissolved as powder at 37 °C in pH media 1 and 7.

	MVR (mg·mL^{-1})	
pH	**1**	**7**
Powder	0.0060 ± 0.0013	0.0038 ± 0.0015
MVR/VAC-HEMA1.5	0.1313 ± 0.0010	0.1407 ± 0.0012

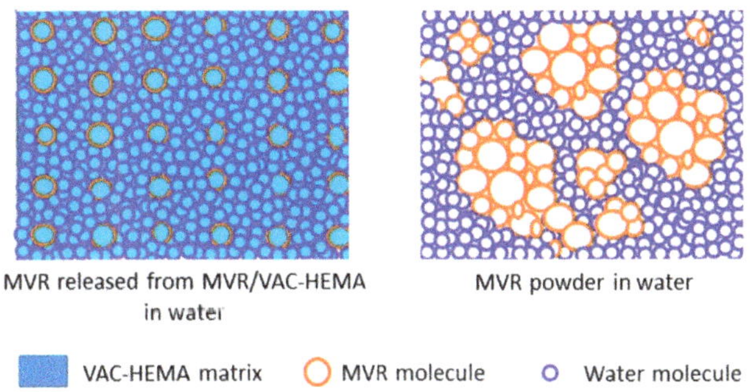

Scheme 3. Solubility enhancement of MVR.

3.1.10. Cytotoxicity

The cell toxicity of VAC-HEMA and MVR/VAC-HEMA materials was measured at 490 nm from triplicated tests and the percentage toxicity taken from their average values is visualized in a histogram in Figure 10A. As can be observed from these results, the normal HEC appear to be unaffected by the virgin specimen (VAC-HEMA) and no significant effect on their viability is observed when between 0.25 and 2.00 wt% was loaded in this material. Although the cytotoxicity of this material was not significant, increasing the MVR content in the MVR/VAC-HEMA system from 0.25 to 2.00 wt% leads to increase the cell cytotoxicity from 19.2 to 58.3%. This appears to be due to the effect of this drug as an antiviral agent on healthy cells.

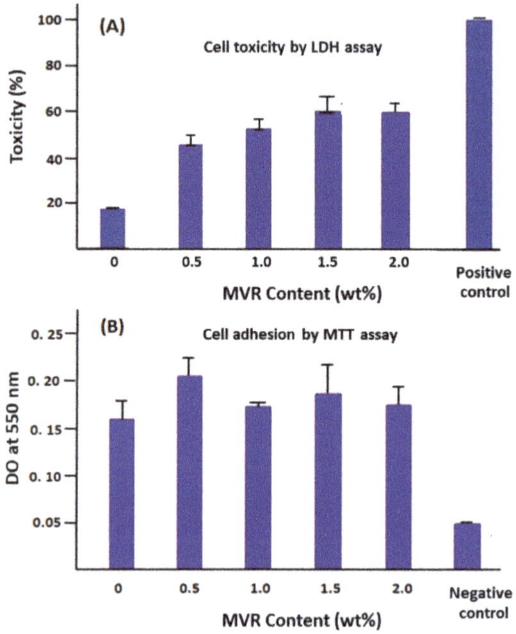

Figure 10. (**A**) The effect of MVR/VAC-HEMA materials on the HEC toxicity examined by the LDH assay and (**B**) HEC adhesion on the virgin VAC-HEMA and MVR/VAC-HEMA specimens. The positive control used in this experiment is the collagen membrane (collapse) giving 100% adhesion.

3.1.11. Cell Adhesion

The cell adhesion testing on virgin VAC-HEMA and MVR/VAC-HEMA drug carrier systems was performed by the standard MTT assay and the results obtained were taken from triplicated tests and the average percentages of cell adhesion for each sample are presented in the histogram of Figure 10B. These data, compared to the negative control, indicate a significant cell adhesion rate for all prepared samples. These data also reveal that the absorbance of the virgin VAC-HEMA increased by more than three fold that without cells, however no significant difference is noted when this hydrogel is loaded with 0.25 to 2.00 wt% of the MVR content.

3.2. *In Vitro Release Dynamic*

For non-porous (dense) and asymmetric drug-carrier systems, the mechanism of a drug released in aqueous medium involves three principal phenomena: (i) the penetration of the release medium into the drug-support system takes place under the action of the osmosis phenomenon (the water passes from the least concentrated medium to the most concentrated medium), and its rate of diffusion depends mainly on the affinity between the release medium and the drug-carrier system, (ii) the dissolution of the drug inside the drug-carrier system and this depends mainly on its solubility in this medium, (iii) the release of the drug which is performed in solution under the action of a thermodynamic imbalance caused by the large difference between the free enthalpies of the mixture inside and outside the drug-carrier system due to the large difference between the drug concentration inside and outside the system.

In this present work, the release of MVR from the MVR/VAC-HEMA drug carrier systems was carried out at 37 °C during 72 h in media pH 1, 3, 5 and 7, separately, and the results obtained are plotted in Figure 11. For all samples, the release dynamics of this drug follows a logarithmic pattern with a variable slope depending on the MVR content in the drug carrier system. These data reveal, for any pH medium, that the maximum cumulative of MVR released is reached with the MVR/VAC-HEMA0.5 specimen containing the lower percentage of MVR initially loaded (0.5 wt%) in the copolymer matrix. In terms of the amount released, the maximum drug release is obtained with the system that contains the most MVR loaded in the VAC-HEMA carrier (2.0 wt%). The decrease in the percentage of the drug released when the initial amount of MVR loaded in the VAC-HEMA matrix seem to be due to the limit solubility of MVR in water inside the copolymer matrix, because a small amount of medication is much easier to dissolve than a larger amount. As can be seen from these results, for all MVR/VAC-HEMA compositions, the release percentage of this medication decreased with the pH of the medium. The explanation of this phenomenon will be amply presented in Section 3.3.

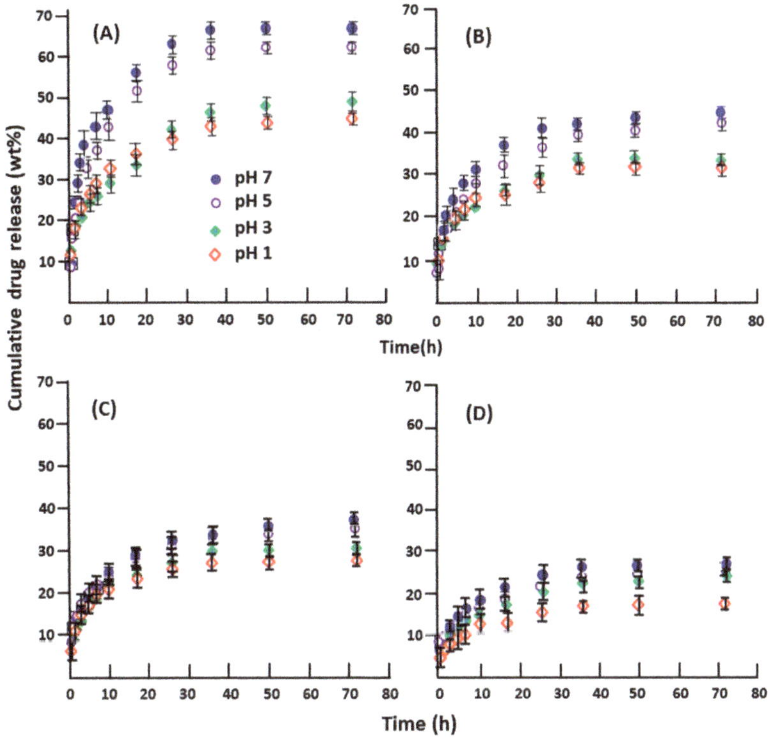

Figure 11. Change in the cumulative MVR released from the MVR/VAC-HEMA drug carrier systems containing 0.5 (**A**), 1.0 (**B**), 1.5 (**C**) and 2.0 (**D**) wt% of MVR versus time.

3.2.1. Diffusion Behavior of MVR

In order to understand the release dynamics of MVR over time, the diffusion behavior of this medication through MVR/VAC-HEMA drug carrier systems was investigated. The MVR released from these materials in different pH media does not exceed 60 wt% of this medication initially loaded in these drug carrier systems. Therefore, the Fickian model is applicable to describe the diffusion mechanism of MVR through the VAC-HEMA material. According to Masaro et al. [48], the relationship between the amount of drug released as a function of time resulting from the Fick model is given by Equation (6)

$$\frac{m_t}{m_o} = k'\sqrt{t} \tag{6}$$

where m_t/m_0 and k' are the weight ratio of MVR released at t time of the release process and a constant related to the drug carrier system, respectively.

If the drug molecules diffuse through the MVR/VAC-HEMA system according to the Fick model, the graph indicating the variation in the drug fraction released m_t/m_0 as a function of the square root of time would give a straight line with a slope corresponding to the value of the k' constant. Once the k' for a drug carrier system is determined, the D' value is easily calculated using Equation (7):

$$k' = 4\sqrt{\frac{D'}{\pi l^2}}, \tag{7}$$

where l is the thickness of the MVR/VAC-HEMA film sample. As shown in Figure 12, the variation in the weight fraction of the MVR (m_t/m_0) released as function of the square root of time describes a straight line and the k' and D' values, deducted for all MVR/VAC-HEMA drug-carrier systems, are gathered for comparison in Table 3. As can be seen from these data, the D' value of the MVR molecules through the MVR/VAC-HEMA drug carrier system increased slowly with the MVR content.

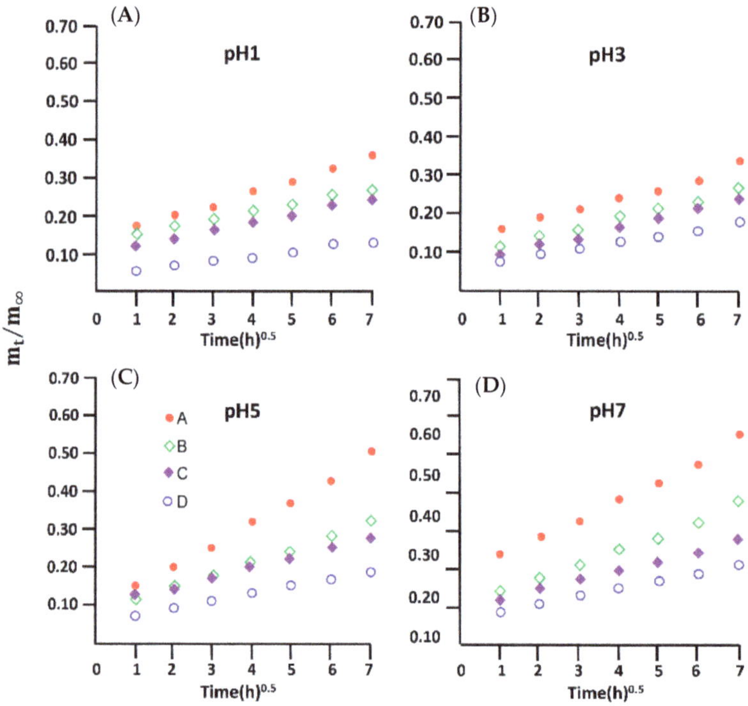

Figure 12. Variation in the drug fraction released from MVR/VAC-HEMA systems containing 0.5 (**A**), 1.0 (**B**), 1.5 (**C**) and 2.0 (**D**) wt% of MVR in different pH media versus the square root of time.

Table 3. k' and D' values deducted from the kinetic curves indicating the variation in (m_t/m_0) versus the square root of time.

System	pH	$k' \cdot 10^2$ ($h^{-0.5}$)	R^2	$D' \cdot 10^{-3}$ ($mm^2 \cdot h^{-1}$)	System	$k' \cdot 10^2$ ($h^{-0.5}$)	R^2	$D' \cdot 10^{-3}$ ($mm^2 \cdot h^{-1}$)
I	1	3.29	0.9966	3.394	III	2.00	0.9900	0.788
	3	3.58	0.9454	4.473		2.59	0.9969	1.979
	5	6.11	0.9874	9.221		2.64	0.9961	1.741
	7	5.57	0.9818	7.793		2.81	0.9881	1.961
II	1	2.11	0.9883	1.348	IV	1.46	0.9839	0.629
	3	2.79	0.9844	2.203		1.85	0.9785	0.984
	5	3.65	0.9952	3.850		2.05	0.9909	1.266
	7	3.79	0.9898	4.130		2.17	0.9913	1.298

I: MVR/VAC-HEMA0.5; II: MVR/VAC-HEMA1.0; III: MVR/VAC-HEMA1.5; IV: MVR/VAC-HEMA2.0.

3.2.2. Effect of the Initial Amount of MVR

The effect of the MVR amount initially loaded in the VAC-HEMA carrier was carried out over 3 days in different pH media and the results obtained are presented in Figure 13. The profile of the curves obtained shows a significant decrease in the MVK released in any pH medium when this medication initially loaded in the VAC-HEMA increased. This is undoubtedly due to the limited solubility of MVR in water inside the VAC-HEMA matrix, which was found in this investigation to equal 0.131 and 0.141 mg·mL^{-1} in pH media 1 and 7, respectively. Indeed, as demonstrated in Section 3.1.6, the swelling capacity of the VAC-HEMA carrier remains unchanged regardless of the pH medium. Therefore, the concentration of the dissolved drug inside the VAC-HEMA matrix also remains unchanged regardless of the amount of MVK initially loaded into this material. Therefore, an excess of this medication leads to decrease the percentage released. This can also be interpreted by the diffusion coefficient of this drug in these different media inside the copolymer matrix (Table 3). Indeed, this parameter was faster in the case of drug carrier systems containing 0.5 and 1.0 wt% MVR which varies between 3.394 and 7.793 mm^2·h^{-1} for the MVR/VAC-HEMA0.5 system and between 1.348 and 4.130 mm^2·h^{-1} for the MVR/VAC-HEMA1.0 system, depending on the pH medium. On the other hand, this parameter is less rapid in the case of the two other systems containing 1.5 and 2.0 wt% MVR with diffusion coefficients varying between 0.788 and 1.961 mm^2·h^{-1} for the MVR/VAC-HEMA1.5 system and between 0.629 and 1.298 mm^2·h^{-1} for the MVR/VAC-HEMA system.

3.3. *Effect of pH Medium*

The effect of the pH medium on the release dynamic of MVR from the MVR/VAC-HEMA systems was studied over a period of 3 days of the release process and the results obtained are presented out in Figure 14. These curves reveal an increase in the percentage of the cumulative MVR released with the pH of medium, passing through an acceleration step between pH 3 and 5, which increased with the amount of MVR loaded except that of the drug carrier system containing the maximum amount of MVR (2.0 wt%) which stabilized in this pH range.

As stated earlier in Section 3.1.6, the swelling capacity of the VAC-HEMA carrier is virtually unaffected by the pH of the absorbed medium. In this case, the variation in the percentage of MVR released depends mainly on the interactions between pH and its solubility in the medium where it will be released. Indeed, the solubility of MVR in pH media 1 and 7 obtained in this work (Section 3.1.8) was found equal to 0.1313 and 0.1407 mg·mL^{-1}, respectively. This seems to explain why the release of MVR increased with the pH of medium increased from 1 to 7.

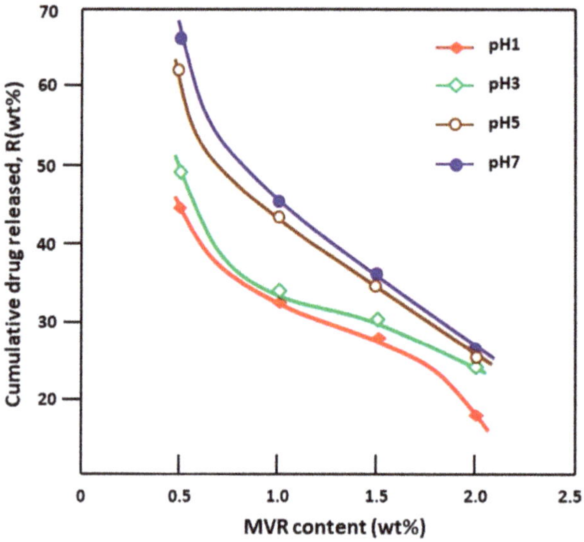

Figure 13. Cumulative MVR released form the MVR/VAC-HEMA drug carrier system versus the MVR content loaded in the VAC-HEMA matrix obtained over 3 days of the release process.

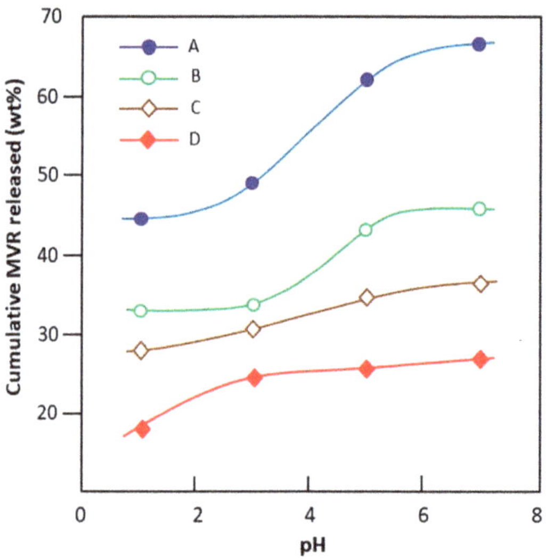

Figure 14. The change in the cumulative MVR released from MVR/VAC-HEMA0.5 (A), MVR/VAC-HEMA1.0 (B), MVR/VAC-HEMA1.5 (C) and MVR/VAC-HEMA2.0 (D) systems versus the pH of the medium taken at 3 days of the release time.

The presence of MVR in the VAC-HEMA matrix made it possible to reduce the percentage of MVR released in media pH acid. This can be explained by a reduction in the swelling capacity of the MVR/VAC-HEMA system due to a rearrangement of the inter-chain hydrogen bonds of the hydroxyl-hydroxyl and hydroxyl-carbonyl type in this medium, thus reducing the affinity between VAC-HEMA and water molecules in favor of that between MVR and the copolymer. In this situation, fewer water molecules penetrate the copolymer matrix due to the hydrophobicity of the MVR physically linked to this material, resulting in less solvation of the drug and consequently less release in this environment.

3.4. Performance of the MVR/VAC-HEMA Drug Carrier System

The study of the effectiveness of the MVR/VAC-HEMA drug carrier system in the delivery of MVR in the different PH media investigated was carried out in this work on the basis of three essential factors: (i) the cumulative amount of MVR released, (ii) the stability of the release rate and iii) the duration of the release process. In general, the most effective system is the one that releases the greatest amount of drug in a neutral pH medium (intestinal) and little in an acidic pH medium (stomach) with a uniform release rate and for a long period. The profile of each kinetic curve in Figure 10 reveals three main pseudo-linear steps, the durations of which depend on the pH of the medium and the amount of drug initially loaded into the VAC-HEMA carrier.

The percentage of MVR released and its instantaneous release velocity for each period and each system are deduced from the coordinates and slope of each corresponding pseudo-linear curve, respectively, and the results obtained are gathered for comparison in Table 4. The first step, which is the fastest and the shortest, occurs during the first 2 to 5 h of the release process is characterized by a significant release percent of MVR in these different media (8.0–37.7 wt%) with a stable release rate mainly of 6.0 wt%$\cdot$h^{-1}. The second stage, which comes just after, is medium and goes from 8 to 31 h of the release process. During this period, between 6.0 and 26.5 wt% of MVR were uniformly released with a release rate varying between 0.34 and 1.54 wt%$\cdot$h^{-1}, depending on the pH medium and the MVR initially loaded in the VAC-HEMA carrier. The third stage of the release process is the longest (36 to 59 h) during which a small percent of MVR are released (0.3–8 wt%) with a stable release rate oscillating between 0.10 and 0.72 wt%$\cdot$h^{-1}, depending on the pH of the medium and the MVR initially loaded in the carrier. The performance of the MVR/VAC-HEMA drug carrier systems was estimated according to the criteria cited above which are: (i) an adequate fraction of drug released in the intestinal transit (neutral pH medium), (ii) a minimum fraction of drug released in the stomach (acidic pH medium) and iii) a stability in the drug release rate over a long period of time. Considering these criteria, it was found that the MVR/VAC-HEMA systems containing 0.5 and 1.0 wt% MVR contents seemed to give the best performances with 62.2 and 45.0 wt% of this drug released in neutral pH medium and 36 and 28.3% by weight in the acidic media (pH 1 and 3), respectively.

According to the literature [34], it was found that the mean gastrointestinal transit time (GITT) which depends on the age, sex and weight of the person is between 53 and 88 h distributed over three main steps: (i) gastric transition time between 1 and 4 h (pH 1.5–3.5), (ii) intestinal transit time between 4 and 12 h (pH 7–9) and (iii) colon transit time between 48 and 72 h (pH 5–7 h). Taking into account the release of the MVR in different pH media of Table 4 and the GITT distribution on the different digestive organs approached by Beltzer et al. [34], it was possible to approximately estimate the percentages of cumulative MVR released from the MVR/VAC-HEMA drug carrier systems in different organs. The average stomach/digestive organs rate (*SDO*) was evaluated, disregarding the effects of microorganisms and enzymes, using Equation (8) and the data obtained are illustrated in Table 5.

$$SDO(wt\%) = \frac{r_{St}}{r_{st} + r_{sint} + r_{col}} \times 100, \quad (8)$$

where r_{st}, r_{sint} and r_{col} are the MVK in wt% released in the stomach, small intestine and colon, respectively, during a certain transit time.

Table 4. Distribution of the cumulative MVR released from the MVR/VAC-HEMA on the different pseudo stable zones.

System	pH	Pseudo-Stable Zone (h)	Cumulative MVR Released (wt%)	Release Rate (wt%/h)	System	Pseudo-Stable Zone (h)	Cumulative MVR Released (wt%)	Release Rate (wt%/h)
I	1	0–3	22.0 ± 0.3	7.33 ± 0.10	III	0–3	18.0 ± 0.5	06.0 ± 0.17
		5–36	12.0 ± 0.3	0.41 ± 0.01		7–36	10.0 ± 0.4	0.34 ± 0.01
		36–72	02.0 ± 0.2	0.07 ± 0.01		36–72	0.3 ± 0.2	0.01 ± 0.01
	3	0–5	24.0 ± 0.3	4.80 ± 0.06		0–3	18.0 ± 0.5	06.0 ± 0.17
		5–36	26.5 ± 0.6	0.85 ± 0.02		3–35	15.0 ± 0.5	0.47 ± 0.02
		36–72	06.5 ± 0.3	0.18 ± 0.01		36–72	01.0 ± 0.2	0.03 ± 0.01
	5	0–5	32.0 ± 0.4	6.40 ± 0.08		0–3	19.0 ± 0.5	6.33 ± 0.17
		5–22	25.0 ± 0.3	1.47 ± 0.02		3–16	13.0 ± 0.4	01.0 ± 0.03
		26–72	08.0 ± 0.2	0.17 ± 0.01		16–72	06.0 ± 0.3	0.11 ± 0.01
	7	0–5	32.7 ± 0.5	6.54 ± 0.10		0–3	19.0 ± 0.6	6.33 ± 0.02
		5–26	22.5 ± 0.3	1.07 ± 0.01		3–16	20.0 ± 0.6	1.54 ± 0.05
		26–72	07.0 ± 0.3	0.15 ± 0.01		16–72	06.0 ± 0.4	0.11 ± 0.01
II	1	0–2	10.0 ± 0.3	5.00 ± 0.15	IV	0–2	05.0 ± 0.2	2.50 ± 0.10
		2–12	11.3 ± 0.3	1.13 ± 0.03		2–10	08.0 ± 0.3	1.00 ± 0.04
		17–72	4.0 ± 0.2	0.72 ± 0.01		10–72	08.0 ± 0.4	0.13 ± 0.01
	3	0–2	11.0 ± 0.4	5.50 ± 0.20		0–2	08.0 ± 0.4	4.00 ± 0.20
		2–14	15.0 ± 0.5	1.25 ± 0.04		2–16	10.0 ± 0.4	0.71 ±0.03
		14–72	3.0 ± 0.2	0.05 ± 0.01		16–72	06.0 ± 0.4	0.11 ± 0.01
	5	0–2	13.0 ± 0.3	06.5 ± 0.15		0–2	09.0 ± 0.2	4.50 ± 0.10
		2–17	15.0 ± 0.4	01.0 ± 0.03		2–11	09.0 ± 0.4	1.00 ± 0.04
		19–72	7.0 ± 0.3	0.13 ± 0.01		13–72	06.0 ± 0.3	0.10 ± 0.01
	7	0–2	14.0 ± 0.2	07.0 ± 0.10		0–2	10.0 ± 0.2	5.00 ± 0.10
		2–19	6.0 ± 0.4	0.35 ± 0.02		2–13	11.5 ± 0.3	1.06 ± 0.03
		19–72	6.0 ± 0.3	0.11 ± 0.01		13–72	06.5 ± 0.3	0.11 ± 0.01

I: MVR/VAC-HEMA0.5; II: MVR/VAC-HEMA1.0; III: MVR/VAC-HEMA1.5; IV: MVR/VAC-HEMA2.0.

Table 5. Distribution of the total MVR released from MVR/VAC-HEMA drug carrier systems on the principal digestive organs timed according to the Belzer approach.

MVR/VAC-HEMA System	Stomach Transit (wt%)		Small Intestine Transit (wt%)		Colon Transit (wt%)		SDO (wt%)	
Transit Time	Min (1 h)	Max (4 h)	Min (4 h)	Max (12 h)	Min (48 h)	Max (72)	Min (48 h)	Max (72)
MVR/VAC-HEMA0.5	6.07	13.37	26.16	40.19	33.06	36.66	10.86	14.81
MVR/VAC-HEMA1.0	6.00	18.41	20.00	28.10	42.52	45.82	8.76	19.90
MVR/VAC-HEMA1.5	5.25	10.13	14.70	17.50	23.19	25.83	12.17	18.95
MVR/VAC-HEMA2.0	6.25	8.21	12.12	20.60	25.35	28.00	14.38	14.45

For the short gastro intestinal transit, these data reveal that the MVR/VAC-HEMA drug system delivers in the stomach between 8.76 and 14.38 wt% of the initial MVR loaded and between 14.81 and 19.00 wt% for the long gastro intestinal transit. The systems containing 0.5 and 1.0 wt% of this medication show the best performance because they were able to reduce the minimum percentage release of MVR in the stomach to 8.76 and 14.38 wt% for MVR/VAC-HEMA0.5 and MVR/VAC-HEMA1.0, respectively. This is beneficial because among the goals in the drug delivery field is their delivery in the intestines, which are the main seat of absorption and not the stomach which is the seat of the chemical and mechanical crushing of food. Sometimes, when a drug is administered directly orally, a considerable amount will be destroyed in the stomach under the effect of enzymes and the acidity of the environment and therefore only a part of the drug will reach its objective (the intestines). In addition, the products resulting from the breakdown of this drug in the stomach can cause several side effects. In this case, the smaller the amount of drug released into the stomach, the more these side effects are reduced. On the other hand, the incorporation of medication in the copolymer used as carrier such as VAC-HEMA also allows the reduction in the prescribed dose for the patient, and this is performed by subtracting a destroyed part of the drug in the stomach. The dosage of MVR prescribed for a patient according to sex, age, weight and health can be easily controlled by the amount of MVR/VAC-HEMA drug carrier system administered.

4. Conclusions

The objectives of this work were successfully achieved. Indeed, the photo-copolymerization of 2-hydroxyethyl methyl methacrylate with vinyl acetate in a desired percentage (~15 wt% vinyl acetate) in the presence of desired ratios of AVR was successfully performed under LED light. The uniform dispersion of the Mevacor particles at their molecular level in the VAC-HEMA matrix, confirming the obtaining of a solid solution, has been demonstrated by the DSC and XRD and SEM analysis methods. The swelling investigation of VAC-HEMA material in pH medium ranged between 1 and 7 demonstrated that the swellability of this material was 45.20 ± 0.82 wt% which is practically unaffected by the pH of the medium. No significant cytotoxicity of virgin VAC-HEMA and MVR/VAC-HEMA drug carrier systems was revealed by the LDH test and the percentage of the cell adhesion on these systems examined by MTT assay indicated much higher than the negative control. The "in vitro" of the Mevacor released from these systems revealed a maximum increase of 22- and 37-fold that dissolved as powder under the same temperature and media. These data also indicate that the highest Mevacor release occurs when the drug carrier system initially contained 0.5% MVR. In any pH medium, the incorporation of MVR into the copolymer matrix drastically decreased the percentage of the drug released. It was also found from these results that the cumulative MVR released increased with the pH medium. The MVR/VAC-HEMA drug delivery system containing 0.5 and 1.0 wt% MVR appears to provide the best performance with 62.2 and 45.0 wt% of this drug evenly released in neutral pH medium, respectively, and 36 and 28.3 wt% in pH medium 1, respectively.

Utilizing the Belzer approach, the results of average stomach/digestive organs rate indicated that the MVR/VAC-HEMA drug carrier system delivers between 8.76 and 14.38 wt% of Mevacor in the stomach during the short GITT and between 14.81 and 19.00 wt% during the long GITT. The drug carrier systems including 0.5 and 1.0 wt% MVR demonstrate the highest performance, reducing the minimum release of this medication in the stomach to 8.76 and 14.38 wt%, respectively. This is advantageous, because one of the intended targets in the field of drug delivery is their release directly into the intestines, which is the main site of food and drug absorption.

Author Contributions: Data curation, M.A. and R.S.A.K.; Formal analysis, M.A., S.M.A., R.S.A.K., F.S.A., A.S. and T.A.; Funding acquisition, S.M.A. and T.A.; Investigation, S.M.A. and W.S.S.; Methodology, M.A., S.M.A., R.S.A.K., F.S.A., A.S. and T.A.; Project administration, S.M.A. and T.A.; Resources, M.A. and W.S.S.; Software, W.S.S.; Supervision, S.M.A.; Visualization, M.A.; Writing—original draft, S.M.A. and T.A. All authors have read and agreed to the published version of the manuscript.

Funding: The authors extend their appreciation to Researchers Supporting Project number (RSP2023R475), King Saud University, Riyadh, Saudi Arabia.

Institutional Review Board Statement: Not applicable.

Data Availability Statement: Not applicable.

Acknowledgments: The authors extend their appreciation to Researchers Supporting Project number (RSP2023R475), King Saud University, Riyadh, Saudi Arabia.

Conflicts of Interest: The authors declare no conflict of interest.

References

1. Akelah, A.; Rehab, A. Controlled release systems based on polyacrylates and polyacrylates/clay nanocomposites of atenolol drug. *SPE Polym.* **2022**, *3*, 118–127. [CrossRef]
2. Lee, S.; Na, K. Oleic acid conjugated polymeric photosensitizer for metastatic cancer targeting in photodynamic therapy. *Biomater. Res.* **2020**, *24*, 1. [CrossRef] [PubMed]
3. Tiwari, S.; Marathe, S.; Chandola, M.; Thakur, N.; Rathore, R.; Garg, S. Synthesis and characterization of nanocomposites for drug delivery applications. *J. Polym. Texfile Eng.* **2016**, *3*, 1–7.
4. Vigata, M.; Meinert, C.; Hutmacher, D.W.; Bock, N. Hydrogels as drug delivery systems: A review of current characterization and evaluation techniques. *Pharmaceutics* **2020**, *12*, 1188. [CrossRef] [PubMed]
5. Naahidi, S.; Jafari, M.; Logan, M.; Wang, Y.; Yuan, Y.; Bae, H.; Dixon, B.; Chen, P. Biocompatibility of hydrogel-based scaffolds for tissue engineering applications. *Biotechnol. Adv.* **2017**, *35*, 530–544. [CrossRef] [PubMed]
6. Loessner, D.; Meinert, C.; Kaemmerer, E.; Martine, L.C.; Yue, K.; Levett, P.A.; Klein, T.J.; Melchels, F.P.; Khademhosseini, A.; Hutmacher, D.W. Functionalization, preparation and use of cell-laden gelatin methacryloyl–based hydrogels as modular tissue culture platforms. *Nat. Protoc.* **2016**, *11*, 727–746. [CrossRef] [PubMed]
7. Gotto, A.M., Jr. Statins: Powerful drugs for lowering cholesterol: Advice for patients. *Circulation* **2002**, *105*, 1514–1516. [CrossRef] [PubMed]
8. Tobert, J.A. Efficacy and long-term adverse effect pattern of lovastatin. *Am. J. Cardiol.* **1988**, *62*, J28–J34. [CrossRef]
9. Lovastatin Monograph for Professionals. Available online: https://www.drugs.com/monograph/lovastatin.html (accessed on 1 January 2022).
10. Fink, C.; Sun, D.; Wagner, K.; Schneider, M.; Bauer, H.; Dolgos, H.; Mäder, K.; Peters, S.A. Evaluating the role of solubility in oral absorption of poorly water-soluble drugs using physiologically-based pharmacokinetic modeling. *Clin. Pharmacol. Ther.* **2020**, *107*, 650–661. [CrossRef]
11. Górniak, A.; Gajda, M.; Pluta, J.; Czapor-Irzabek, H.; Karolewicz, B. Thermal, spectroscopic and dissolution studies of lovastatin solid dispersions with acetylsalicylic acid. *J. Therm. Anal. Calorim.* **2016**, *125*, 777–784. [CrossRef]
12. UmakantVerma, J.; Mokale, V. Preparation of freeze-dried solid dispersion powder using mannitol to enhance solubility of lovastatin and development of sustained release tablet dosage form. *Am. J. Pharm. Sci. Nanotechnol.* **2014**, *1*, 11–26.
13. Patel, R.P.; Patel, D.J.; Bhimani, D.B.; Patel, J.K. Physicochemical characterization and dissolution study of solid dispersions of furosemide with polyethylene glycol 6000 and polyvinylpyrrolidone K30. *Dissolut. Technol.* **2008**, *15*, 17–25. [CrossRef]
14. Mochizuki, A.; Ogawa, H.; Nishimori, Y. Water structure in poly (2-hydroxyethyl methacrylate): Effect of molecular weight of poly (2-hydroxyethyl methacrylate) on its property related to water. *J. Appl. Polym. Sci.* **2012**, *125*, 53–60. [CrossRef]
15. Weaver, J.; Bannister, I.; Robinson, K.; Bories-Azeau, X.; Armes, S.; Smallridge, M.; McKenna, P. Stimulus-responsive water-soluble polymers based on 2-hydroxyethyl methacrylate. *Macromolecules* **2004**, *37*, 2395–2403. [CrossRef]
16. Macret, M.; Hild, G. Hydroxyalkyl methacrylates: Hydrogel formation based on the radical copolymerization of 2-hydroxyethylmethacrylate and 2, 3-dihydroxypropylmethacrylate. *Polymer* **1982**, *23*, 748–753. [CrossRef]
17. Wichterle, O.; Lim, D. Hydrophilic gels for biological use. *Nature* **1960**, *185*, 117–118. [CrossRef]
18. Jeyanthi, R.; Rao, K.P. In vivo biocompatibility of collagenpoly (hydroxyethyl methacrylate) hydrogels. *Biomaterials* **1990**, *11*, 238–243. [CrossRef] [PubMed]
19. Denizli, A.; Kiremitci, M.; Pişkin, E. Subcutaneous polymeric matrix system p (HEMA-BGA) for controlled release of an anticancer drug (5-fluorouracil): II: Release kinetics. *Biomaterials* **1988**, *9*, 363–366. [CrossRef]
20. Kiremitçi-Gümüşderelioglu, M.; Gökçe, M.; Fikret Akata, R. A novel MMC-loaded pHEMA drainage device for the treatment of glaucoma: In vitro and in vivo studies. *J. Biomater. Sci. Polym. Ed.* **1996**, *7*, 857–869. [CrossRef]
21. Trigo, R.; Blanco, M.; Teijon, J.; Sastre, R. Anticancer drug, ara-C, release from pHEMA hydrogels. *Biomaterials* **1994**, *15*, 1181–1186. [CrossRef]
22. Vinyl acetate, polyvinyl acetate and polyvinyl alcohol. *IARC Monogr. Eval. Carcinog. Risk Chem. Hum.* **1979**, *19*, 341–366.
23. Sivalingam, G.; Chattopadhyay, S.; Madras, G. Enzymatic degradation of poly (ε-caprolactone), poly (vinyl acetate) and their blends by lipases. *Chem. Eng. Sci.* **2003**, *58*, 2911–2919. [CrossRef]
24. Novoa, G.A.G.; Heinämäki, J.; Mirza, S.; Antikainen, O.; Colarte, A.I.; Paz, A.S.; Yliruusi, J. Physical solid-state properties and dissolution of sustained-release matrices of polyvinylacetate. *Eur. J. Pharm. Biopharm.* **2005**, *59*, 343–350. [CrossRef] [PubMed]
25. Babaahmadi, M.; Sabzi, M.; Ghasemi, I. Mechanical and Shape Memory Properties of Poly (vinyl acetate)/Poly (lactic acid) Blends. *Iran. J. Polym. Sci. Technol.* **2018**, *31*, 57–68.
26. Jannesari, M.; Varshosaz, J.; Morshed, M.; Zamani, M. Composite poly (vinyl alcohol)/poly (vinyl acetate) electrospun nanofibrous mats as a novel wound dressing matrix for controlled release of drugs. *Int. J. Nanomed.* **2011**, *6*, 993–1003.
27. Sivalingam, G.; Karthik, R.; Madras, G. Blends of poly (ε-caprolactone) and poly (vinyl acetate): Mechanical properties and thermal degradation. *Polym. Degrad. Stab.* **2004**, *84*, 345–351. [CrossRef]
28. Bailly, N.; Thomas, M.; Klumperman, B. Poly (N-vinylpyrrolidone)-block-poly (vinyl acetate) as a drug delivery vehicle for hydrophobic drugs. *Biomacromolecules* **2012**, *13*, 4109–4117. [CrossRef]
29. Osman, A.F.; Hamid, A.R.A.; Fitri, T.F.M.; Fauzi, A.A.A.; Halim, K.A.A. Poly (ethylene-co-vinylacetate) copolymer based nanocomposites: A review. *IOP Conf. Ser. Mater. Sci. Eng.* **2020**, *864*, 012121. [CrossRef]

30. Akl, M.; Abd Elrazek, H.; Abdel Bary, E. Poly (Ethylene-Co-Vinyl Acetate) Blends for Controlled Drug Release. *Am. J. Adv. Drug Deliv.* **2018**, *6*, 052–060. [CrossRef]
31. Lin, H.-P.; Tu, H.-P.; Hsieh, Y.-P.; Lee, B.-S. Controlled release of lovastatin from poly (lactic-co-glycolic acid) nanoparticles for direct pulp capping in rat teeth. *Int. J. Nanomed.* **2017**, *12*, 5473. [CrossRef]
32. Tarafder, S.; Nansen, K.; Bose, S. Lovastatin release from polycaprolactone coated β-tricalcium phosphate: Effects of pH, concentration and drug–polymer interactions. *Mater. Sci. Eng. C* **2013**, *33*, 3121–3128. [CrossRef] [PubMed]
33. Madhavi, V. Formulation and evaluation of lovstatin controlled release tablets. *IJPPA* **2017**, *1*, 79–84.
34. Belzer, C.; De Vos, W.M. Microbes inside—From diversity to function: The case of Akkermansia. *ISME J.* **2012**, *6*, 1449–1458. [CrossRef] [PubMed]
35. Alqahtani, S.M.; Al Khulaifi, R.S.; Alassaf, M.; Saeed, W.S.; Bedja, I.; Aldarwesh, A.; Aljubailah, A.; Semlali, A.; Aouak, T. Preparation and Characterization of Poly (vinyl acetate-co-2-hydroxyethyl methacrylate) and In Vitro Application as Contact Lens for Acyclovir Delivery. *Int. J. Mol. Sci.* **2023**, *24*, 5483. [CrossRef] [PubMed]
36. Chuong, M.C.; Choy, E.; Douk, K.; Duong, L.H.; Hoang, S.K.; Le, N.; Lim, M.; Poirier, B.; Prasad, D.; Radke, M. The development of delayed-then-extended-release lovastatin tablet. *Int. J. Appl. Pharm.* **2013**, *5*, 11–18.
37. Semlali, A.; Jacques, E.; Rouabhia, M.; Milot, J.; Laviolette, M.; Chakir, J. Regulation of epithelial cell proliferation by bronchial fibroblasts obtained from mild asthmatic subjects. *Allergy* **2010**, *65*, 1438–1445. [CrossRef] [PubMed]
38. Alghamdi, A.A.; Alsolami, A.; Saeed, W.S.; Al-Odayni, A.-B.M.; Semlali, A.; Aouak, T. Miscibility of poly (acrylic acid)/poly (methyl vinyl ketone) blend and in vitro application as drug carrier system. *Des. Monomers Polym.* **2018**, *21*, 145–162. [CrossRef]
39. Semlali, A.; Papadakos, S.; Contant, C.; Zouaoui, I.; Rouabhia, M. Rapamycin inhibits oral cancer cell growth by promoting oxidative stress and suppressing ERK1/2, NF-κB and beta-catenin pathways. *Front. Oncol.* **2022**, *12*, 873447. [CrossRef]
40. Hakiem, A.F.A.; Mohamed, N.A.; Ali, H.R. FTIR spectroscopic study of two isostructural statins: Simvastatin and Lovastatin as authentic and in pharmaceuticals. *Spectrochim. Acta Part A Mol. Biomol. Spectrosc.* **2021**, *261*, 120045.
41. Pham, Q.T.; Petiaud, R.; Llauro-Darricades, M.-F.; Waton, H. *Proton & Carbon NMR Spectra of Polymers*; CRC Press: Boca Raton, FL, USA, 2019.
42. Guan, J.; Liu, Q.; Zhang, X.; Zhang, Y.; Chokshi, R.; Wu, H.; Mao, S. Alginate as a potential diphase solid dispersion carrier with enhanced drug dissolution and improved storage stability. *Eur. J. Pharm. Sci.* **2018**, *114*, 346–355. [CrossRef]
43. Khan, S.; Ranjha, N.M. Effect of degree of cross-linking on swelling and on drug release of low viscous chitosan/poly (vinyl alcohol) hydrogels. *Polym. Bull.* **2014**, *71*, 2133–2158. [CrossRef]
44. Numata, I.; Cochrane, M.A.; Souza, C.M., Jr.; Sales, M.H. Carbon emissions from deforestation and forest fragmentation in the Brazilian Amazon. *Environ. Res. Lett.* **2011**, *6*, 044003. [CrossRef]
45. Akula, P.; PK, L. Effect of pH on weakly acidic and basic model drugs and determination of their ex vivo transdermal permeation routes. *Braz. J. Pharm. Sci.* **2018**, *54*, 1–8. [CrossRef]
46. Aljubailah, A.; Alharbi, W.N.O.; Haidyrah, A.S.; Al-Garni, T.S.; Saeed, W.S.; Semlali, A.; Alqahtani, S.; Al-Owais, A.A.; Karami, A.M.; Aouak, T. Copolymer Involving 2-Hydroxyethyl Methacrylate and 2-Chloroquinyl Methacrylate: Synthesis, Characterization and In Vitro 2-Hydroxychloroquine Delivery Application. *Polymers* **2021**, *13*, 4072. [CrossRef]
47. Comyn, J. Introduction to polymer permeability and the mathematics of diffusion. In *Polymer Permeability*; Springer: Dordrecht, The Netherlands, 1985; pp. 1–10.
48. Masaro, L.; Zhu, X. Physical models of diffusion for polymer solutions, gels and solids. *Prog. Polym. Sci.* **1999**, *24*, 731–775. [CrossRef]
49. Glomme, A.; März, J.; Dressman, J.B. Comparison of a miniaturized shake-flask solubility method with automated potentiometric acid/base titrations and calculated solubilities. *J. Pharm. Sci.* **2005**, *94*, 1–16. [CrossRef] [PubMed]

polymers

MDPI

Article

Degradation of Poly(ε-caprolactone) Resorbable Multifilament Yarn under Physiological Conditions

Monica V. Deshpande [1,*], Arjunsing Girase [1] and Martin W. King [1,2,*]

[1] Wilson College of Textiles, North Carolina State University, Raleigh, NC 27606, USA; aggirase@ncsu.edu
[2] College of Textiles, Donghua University, Shanghai 201620, China
* Correspondence: mdeshpande4293@gmail.com (M.V.D.); mwking2@ncsu.edu (M.W.K.)

Abstract: Poly(ε-caprolactone) (PCL) is a hydrophobic, resorbable aliphatic polymer recognized for its low tenacity and extensive elongation at break, making it a popular choice for fabricating biodegradable tissue engineering scaffolds. PCL's slow degradation rate typically results in a complete resorption period of 2 to 3 years. While numerous studies have examined the degradation of PCL in various forms such as films and webs, no study to date has investigated its physiological degradation in multifilament yarn form. In this study, we subjected PCL multifilament yarn samples to physiological conditions in phosphate-buffered saline (PBS) maintained at a consistent temperature of 37 ± 2 °C and agitated at 45 rpm for a period of 32 weeks. We retrieved samples at five different intervals to analyze the degradation profile of the multifilament yarn. This allowed us to estimate the complete resorption time and rate under these in vitro conditions. Over the 32-week period, the multifilament yarn's mass decreased by 4.8%, its elongation at break declined by 42%, the tenacity dropped by 40%, and the peak load at break fell by 46.5%. Based on these findings, we predict that a scaffold structure incorporating PCL multifilament yarn would undergo complete resorption in approximately 14 months under physiological conditions, such as in PBS solution at a pH of approximately 7 and a temperature of 37 °C.

Keywords: Poly(ε-caprolactone); biodegradable; physiological degradation; biomedical textiles; tissue engineering scaffold

1. Introduction

Poly(ε-caprolactone) (PCL) is a semicrystalline, highly elastomeric, biodegradable, and bioresorbable polymer characterized by a glass transition at −60 °C and a melting point at +60 °C [1]. The polymer is synthesized from the ring-opening polymerization of ε-caprolactone. PCL is biocompatible, approved by the FDA as the polymer for fabricating certain implantable devices, and distinguished by its hydrophobicity, low tensile strength (approximately 23 MPa), and high elongation at break (700%). It is a slow-degrading polymer, typically requiring 2 to 3 years for complete degradation. Hence, it is often used as a copolymer with other degradable polymers such as polylactic acid (PLA) and polyglycolic acid (PGA) so as to reduce the level of crystallinity and increase the rate of degradation. It is also often blended with biological materials, like hydroxyapatite and collagen. Its ease of manufacturing allows PCL to be extensively utilized in fabricating biodegradable scaffolds for craniofacial tissue regeneration using electrospinning and 3D printing techniques [2,3].

Degradation studies are critical for evaluating the degradation profile of polymeric structures and estimating the time required for complete degradation or resorption following implantation in the body as a tissue engineering scaffold. Polymeric biomaterials can degrade via several mechanisms, including hydrolytic or oxidative degradation [4]. The 'functional time', or the period the structure maintains its desirable function, and the 'disappearance time', which is the period for the structure to completely resorb and lose its mass, can be used to determine the degradability of biocompatible polymers [4]. Various degradation studies have been undertaken, both in vitro and in vivo, over different periods of time so as to determine the suitability of such polymers for specific applications in tissue engineering or drug delivery [4].

For instance, Yoon et al. carried out a 180-day degradation study, both in vitro and in vivo, to assess the biocompatibility of a biodegradable cylindrical PLLA mesh [5]. They subjected PLLA biomaterial to degradation in PBS solution in the presence of lysozyme to observe changes in mass and tensile strength in vitro. For in vivo degradation, they implanted PLLA mesh in rats for 180 days and retrieved them at various time points to analyze the in vivo degradation profile through changes in mass and tensile strength [5].

In a different study, Yeo et al. evaluated the in vitro degradation of an 80/20 poly(ε-caprolactone)/tricalcium phosphate (PCL–TCP) scaffold in Dulbecco's modified Eagle media (DMEM), measuring changes in porosity, pH, physical state, and mechanical properties [6].

As Labet et al. noted, the hydrolytic degradation profile of PCL based on its molecular weight distribution initially involves the degradation of the amorphous phase, which causes the level of crystallinity to increase while maintaining the molecular weight [7]. This is followed by the cleavage of the ester bonds, resulting in the loss of mass. At higher temperatures, the polymer degrades through a different mechanism of end-chain scission [7].

Previous research has examined PCL degradation in various forms, including films, monofilaments, copolymer structures, and polymer meshes [6–11]. However, the degradation profile of PCL in its multifilament form remains unexplored. Multifilament yarns offer distinct mechanical and surgical advantages, including high flexibility, ease of knotting, and improved knot security, over their monofilament counterparts. These properties make them more suitable for certain biomedical applications, such as the production of bioresorbable sutures and woven or knitted scaffolds for tissue engineering. It is, therefore, critical to understand the degradation behavior of PCL in multifilament yarn form to better predict its performance and lifespan in these applications.

Recognizing this knowledge gap, our study was designed to monitor the degradation profile of PCL multifilament yarns under physiological conditions for a period of 8 months. Through tracking the loss in mass and the changes in ultimate strength and elongation at break, our research aims to provide valuable insights into how PCL multifilament yarns can be used in tissue engineering and other biomedical applications. These data can serve as a baseline for predicting the lifespan of future biomaterials incorporating PCL multifilaments, ultimately improving their design and functionality in clinical applications.

2. Materials and Methods

2.1. Procurement of Yarn and Sample Preparation

The primary material used in this study was a bobbin of bioresorbable, 100% poly(ε-caprolactone) (PCL) multifilament yarn. This was sourced from Guangdong Zhuhai Adhesive Products Co., Ltd. (Zhuhai, Guangdong, China). This PCL yarn was spun from a semicrystalline polymer with a high amorphous content. This contributes to its high elongation at break, which is a critical property for the purpose of this research. The physical properties of the PCL multifilament yarn, outlined in Table 1, underscore its relevance to the study's objectives. The chemical structure of PCL is shown in Figure 1.

For experimental consistency, the PCL multifilament yarn was processed into individual test specimens. Each yarn specimen was cut precisely to a 5-inch length. This uniform length ensured that the surface area exposed to the degrading solution would be constant across all samples, thereby allowing for more accurate comparisons.

To optimize the yarn's contact with the surrounding degradation solution during the study, each 5-inch specimen was carefully wound onto a styrene frame. This ensured that the entire surface area of the yarn was exposed to the solution, which enhanced the accuracy and precision of the degradation study. The specimens were weighed prior to being wound onto the frames. Their initial mass was recorded and used for comparison between samples and over time during the degradation study.

Table 1. Physical properties of poly(ε-caprolactone) multifilament yarn as received from the manufacturer.

Polymer	100% Poly(ε-caprolactone)
Approximate average Mw	80 kDa
Denier	160
Number of filaments	36
Elongation at break at room temperature	>150%
Glass transition temperature (Tg)	−60 °C
Melting temperature (Tm)	60 °C
Cross-section	Circular
Storage condition	4 °C

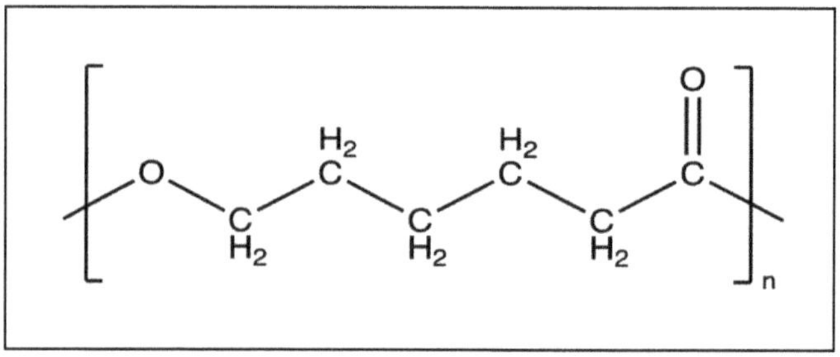

Figure 1. Chemical structure of PCL polymer.

2.2. PCL Yarn Degradation Study

The styrene frames carrying the prepared PCL yarn samples were submerged individually in glass beakers. Each beaker was filled with 20 mL of Dulbecco's phosphate-buffered saline (dPBS) solution, sourced from Gibco Life Technologies (Gaithersburg, Maryland). The beakers were then sealed with parafilm to prevent contamination and evaporation during the experiment.

For the degradation study, the sealed beakers were placed in a Boekel Grant ORS 200 shaker bath (Boekel Scientific, Feasterville, Pennsylvania). The bath was set to maintain a consistent shaking speed of 45 rpm and a temperature of 37 ± 2 °C. This setup ensured an even distribution of forces acting on the surface of the multifilament yarn samples.

To maintain the pH of the dPBS solution at a consistent level of 7.0 ± 0.5, the solution in each beaker was replaced with fresh dPBS every two weeks. This step was crucial in mimicking physiological conditions for the degradation study.

Five specimens were collected at each of four different time points, namely 8, 16, 24, and 32 weeks, and the degradation profile of the PCL yarn samples was determined. The mass, tensile strength, and elongation at the break of each specimen were measured at each retrieval time.

Upon removal from the dPBS solution, the specimens underwent a cleaning process to remove residual solution. They were thoroughly rinsed with deionized water and then air-dried under a hood for 30 min. This ensured that the measured properties were inherent to the yarn samples and not influenced by any remaining surface contamination.

2.3. Evaluation of Tensile Strength and Elongation at Break

Prior to mechanical testing, each yarn specimen was weighed in milligrams. This step was vital to ascertain the linear density or denier of the yarn, which is a standard unit of measurement in grams per 9000 m.

The mechanical properties, including the tensile strength or tenacity and percent strain at break, were evaluated using an MTS Q-test mechanical tester (MTS Systems Corp., Eden Prairie, MN, USA) with a 5-pound load cell, a 1-inch gauge length, and a crosshead speed of 300 cm/min. The peak load at break, expressed in gram-force (gf), and the percent elongation at break were recorded for each yarn specimen. The tenacity of the yarn samples was then calculated in terms of gram-force per denier. This was achieved through dividing the peak load at break (in gf) by the denier of the yarn. This analysis followed the standard test method ASTM D2256 and was conducted at a room temperature of 20 ± 2 °C.

For a robust statistical analysis, these tests were replicated on five specimens at each of the designated time points. The average and standard deviation of the recorded values were calculated to provide a clear indication of the yarn's mechanical properties during the degradation process.

2.4. Statistical Analysis

All data collected during the study are presented in terms of the mean ± standard deviation, derived from five individual data points. To compare the differences in data and ascertain their statistical significance, a student's t-test was applied using JMP Pro® statistical software (15.2.0, SAS Institute Inc., Cary, NC, USA). Statistical significance was pre-determined at a p-value threshold of less than 0.05. Any result yielding a p-value less than this threshold was deemed to be statistically significant.

3. Results and Discussions

The 5-inch-long PCL multifilament yarn samples were retrieved from the dPBS solution at each time point, namely 8, 16, 24, and 32 weeks. The mass of each 5-inch specimen was measured in milligrams, and the changes in tenacity, breaking force, and elongation at break were monitored over the course of the study. The rationale behind assessing the reduction rates of these properties was to estimate the timeline over which the filaments could maintain their mechanical strength and elongation at break.

The average mass in milligrams, the percentage elongation at break, the peak load in gf, and the tenacity in gf/den were measured and calculated at five time points. However, the measurements were converted into SI units for tenacity (cN/tex) and peak load (cN) as shown in Table 2. Figures 2–5 illustrate the changes in these properties plotted against the time of the degradation study in weeks.

Table 2. Average parameters such as mass, percent elongation at break, tenacity, and peak load of the PCL multifilament yarn during 32 weeks of hydrolytic degradation in dPBS.

Yarn Properties	0 Weeks	8 Weeks	16 Weeks	24 Weeks	32 Weeks
Average mass (mg)	2.1 (±0.1)	2.1 (±0.1)	2.0 (±0.2)	2.0 (±0.1)	2.0 (±0.1)
Percent elongation at break (%)	160.7 (±23.7)	105.5 (±28.4)	79.9 (±28.6)	116.2 (±15.8)	93.3 (±28.0)
Tenacity (cN/tex)	240.2 (±24.7)	197.0 (±45.9)	137.7 (±38.8)	174.1 (±24.7)	144.8 (±24.7)
Peak load (cN)	426.4 (±43.7)	319.0 (±74.6)	219.2 (±61.9)	274.6 (±39.3)	228.1 (±38.9)

Equation (1) was derived from extrapolating the rate of reduction in total mass. This suggests that the PCL filament yarn will take approximately 55.1 weeks to dissolve completely under similar conditions; namely, aqueous solution at pH 7, shaken continuously at 45 rpm, and maintained at 37 ± 2 °C.

$$y = -0.039x + 2.149 \tag{1}$$

Comparisons were drawn between this study's results and similar degradation studies on alternative forms of PCL as described in the literature. Over the 32-week period, we observed an approximately 6% loss in mass, a 42% decrease in elongation at break, and about a 40% loss in tenacity. These results are consistent with the findings reported by Lam et al. [8], who studied the changes in mass, maximum stress, crystallinity, and molecular weight during the degradation of PCL in dPBS in a physiological environment.

Figure 2 illustrates the rate of reduction in the total mass of the polymeric filament yarn specimens during the degradation study. During the initial 8 weeks, no significant loss in mass was observed. However, a linear decrease in mass was observed from Week 8 to Week 24, after which the rate of reduction again slowed down. This pattern may reflect the faster degradation of the amorphous regions on the surface of the filaments during the initial stage, followed by a slowed degradation rate as the water molecules took longer to diffuse into the more crystalline structure. This degradation process mirrors that described by Sun et al., who recorded similar stages of PCL capsule degradation in vivo [9].

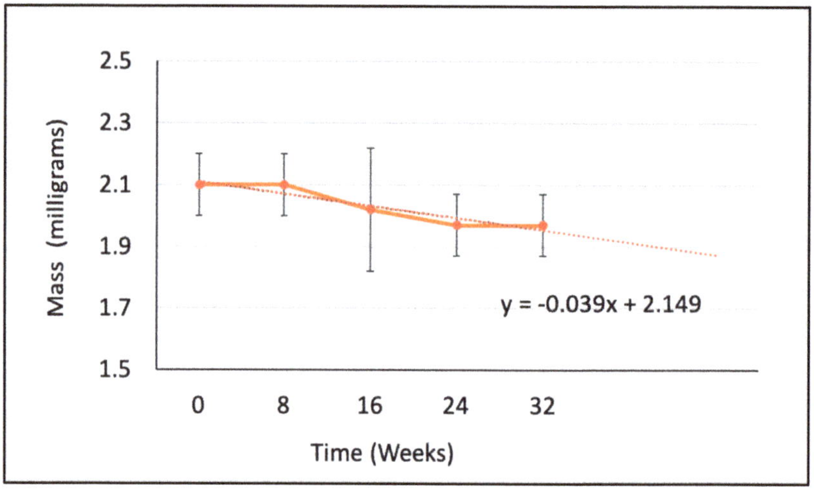

Figure 2. The change in mass in milligrams of the PCL multifilament yarn at each time point over 32 weeks of hydrolytic degradation in dPBS (n = 5).

Figure 3 shows the changes in tenacity of the PCL multifilament yarn during the degradation study. At the start of the study and continuing for the first 16 weeks, the PCL yarn's tenacity decreased linearly by over 40%. This suggests that the material's ability to withstand stress was diminishing during this period, likely due to degradation occurring primarily in the less structured, amorphous regions of the polymer. However, between the 16th and 24th weeks, the trend changed direction. Instead of the tenacity continuing to decrease, the yarn experienced an approximate 30% increase in strength. This unexpected enhancement in tenacity could be attributed to the complete degradation of the amorphous regions of the PCL, leaving behind a more crystalline, well-ordered, and aligned structure. A similar trend was observed in the percent elongation at break (Figure 4) and the changes in the peak load at break (Figure 5). The more ordered packing of the polymer chains in the crystalline regions generally imparts a higher tensile strength, which is likely to explain the observed temporary increase in yarn tenacity. After Week 24, the tenacity of the PCL yarn started to decline once more. This final decrease suggests that even the more resistant crystalline regions of the polymer began to degrade under the sustained hydrolytic experimental conditions.

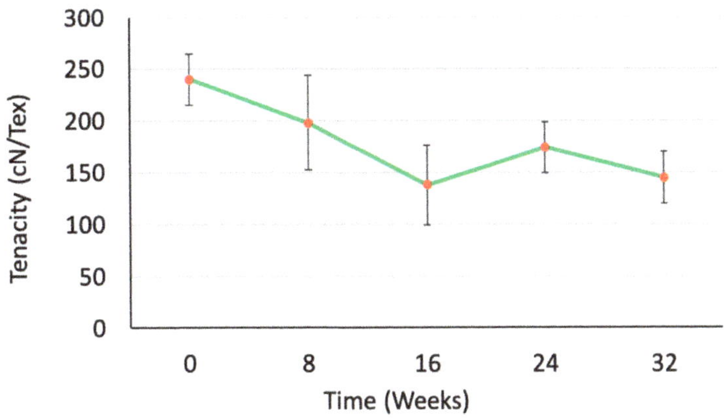

Figure 3. Changes in tenacity in cN per tex of the PCL multifilament yarn at each time point over 32 weeks of hydrolytic degradation in dPBS (n = 5).

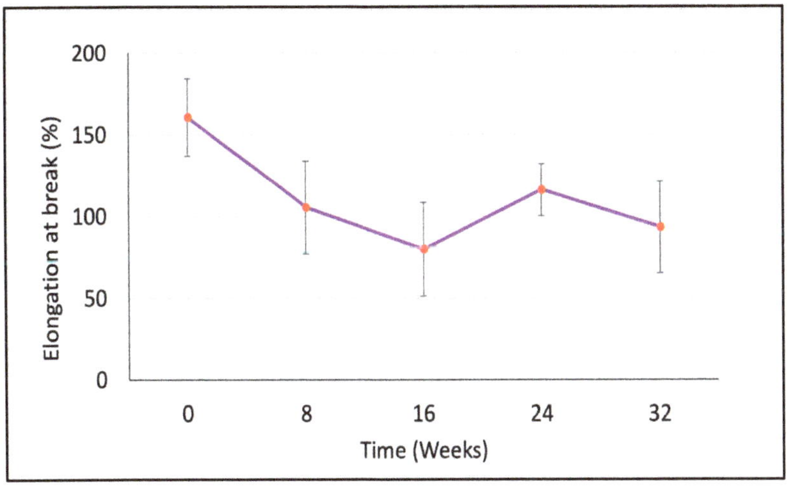

Figure 4. Changes in percent elongation at break of the PCL multifilament yarn at each time point over 32 weeks of hydrolytic degradation in dPBS (n = 5).

In in vitro and in vivo degradation studies of a 50:50 elastomeric poly(L-lactic acid)/poly(ε-caprolactone) copolymer in the form of an extruded porous tubular scaffold, Jeong et al. observed that the PCL component of the copolymer degraded faster than the PLLA component due to a higher proportion of amorphous regions available for hydrolytic reaction, despite PCL's greater hydrophobicity [10]. This might explain why the PCL multifilament yarn did not show any structural disintegration throughout the 32-week study and why the total loss in mass was no more than 6%.

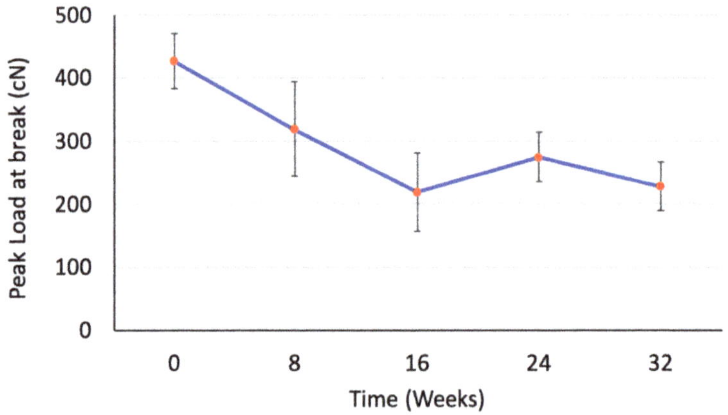

Figure 5. Change in peak load at break in cN of the PCL multifilament yarn at each time point over 32 weeks of hydrolytic degradation in dPBS (n = 5).

PCL, or poly(ε-caprolactone), is a semicrystalline polymer, which means it consists of both amorphous and crystalline regions. The amorphous regions are those where the polymer chains do not have a well-defined, ordered structure, while the crystalline regions are areas where the polymer chains are neatly arranged in a regular, repeating pattern. In this degradation study, PCL multifilament yarn was subjected to hydrolytic conditions for a prolonged period of 32 weeks. As the study progressed, the polymer started to degrade, and the effect of this degradation on the mass, strength, and elasticity of the PCL yarn was measured at different time points. The initial stages of degradation occurred in the amorphous regions of the PCL yarn. Water molecules infiltrate these less-ordered regions more readily, breaking the polymer chains and leading to an initial drop in mass and tensile strength. This explains why the study observed a linear decrease in mass from Week 8 to Week 24.

Once the amorphous regions had substantially degraded, the remaining PCL yarn was more crystalline in nature. At this point, a temporary increase in strength was observed, which was most likely due to the higher tensile strength associated with the ordered crystalline polymer structures. The crystalline regions are more resistant to degradation, hence, the slower rate of mass reduction after 24 weeks. However, eventually, the degradation process also affected the crystalline regions, leading to a further loss in mass and strength over the remaining period of degradation. This slower degradation of the crystalline regions explains why the PCL filament yarn was calculated to take approximately 55.1 weeks to be completely resorbed under the experimental conditions. Overall, both the amorphous and crystalline regions in PCL yarn play a crucial role in determining its degradation profile, which in turn affects its potential application in bioresorbable medical devices.

4. Conclusions

This 32-week degradation study provided valuable insights into the behavior of poly(ε-caprolactone) (PCL) multifilament yarn under conditions resembling the physiological environment. Notably, we identified a two-stage degradation process attributable to the semicrystalline nature of PCL. In the initial stage, the amorphous regions of the polymer were rapidly absorbed, followed by a slower degradation of the more ordered crystalline regions.

The study quantitatively documented a linear decrease of over 40% in the tenacity of the yarn up until Week 16, followed by an approximate 30% increase from Weeks 16 to 24, before the tenacity began to decrease again. This fluctuation can be attributed to the increase in the total crystallinity of the yarn as the amorphous regions were fully degraded. The total loss in mass observed during the study was no more than 6%, indicating the material's resistance to complete degradation during the 32-week experiment.

Extrapolation of the obtained data suggests that full resorption of the PCL multifilament yarn under these conditions will take approximately 14 months. The specific rate and extent of degradation can vary depending on factors such as the exact physiological conditions and the specific design and structure of the implantable device made from the resorbable PCL yarn.

For a more comprehensive understanding of the degradation process, future studies could investigate changes in the average molecular weight and molecular weight distribution of the PCL multifilament yarn during different stages of degradation. This would provide additional evidence to support and possibly refine the degradation profile derived from the present study.

The findings of this study have substantial implications for the design of implantable devices. Understanding the time-dependent mechanical properties and degradation profile of PCL yarn will enable engineers and researchers to adjust their designs so as to predict the in vivo performance and lifespan of implants made from this material. Despite the controlled nature of the experiment, it is acknowledged that actual in vivo conditions may present additional complexities. Therefore, in vitro degradation studies such as this one are pivotal in providing baseline data and should be complemented with in vivo trials for a comprehensive evaluation of the material's suitability for specific medical applications. The evolving nature of the biomedical industry demands more rigorous, in-depth, and detailed analyses of biodegradable materials. The degradation profile of poly(ε-caprolactone) (PCL) multifilament yarn, as presented in this study, is a pivotal step towards this direction. Future endeavors might shift focus towards understanding the nano-structural transformations and enzymatic interplays that affect the degradation behavior of PCL. This can be further investigated in the blend of materials science and biology to ensure safer, more predictable, and adaptable medical implants for future healthcare needs.

Author Contributions: Conceptualization, M.V.D. and M.W.K.; methodology, M.V.D. and M.W.K.; validation, M.V.D. and M.W.K.; formal analysis, M.W.K.; investigation, M.V.D.; resources, M.W.K.; data curation, M.W.K.; writing, M.V.D. and A.G.; writing—review and editing, M.V.D. and A.G.; supervision, M.W.K.; project administration, M.W.K.; funding acquisition, M.W.K. All authors have read and agreed to the published version of the manuscript.

Funding: This research received no external funding.

Institutional Review Board Statement: Not applicable.

Informed Consent Statement: Not applicable.

Data Availability Statement: All the data supporting the findings of the study are available within the article.

Acknowledgments: The authors would like to thank Lu Wang and Ting Zou for providing us with the PCL yarn sample.

Conflicts of Interest: The authors declare that the research was conducted in the absence of any commercial or financial relationships that could be construed as a potential conflict of interest.

References

1. Arif, Z.U.; Khalid, M.Y.; Noroozi, R.; Sadeghianmaryan, A.; Jalalvand, M.; Hossain, M. Recent advances in 3D-printed polylactide and polycaprolactone-based biomaterials for tissue engineering applications. *Int. J. Biol. Macromol.* **2022**, *218*, 930–968. [CrossRef] [PubMed]
2. Borrelli, M.R.M.; Hu, M.S.; Longaker, M.T.M.; Lorenz, H.P.M. Tissue Engineering and Regenerative Medicine in Craniofacial Reconstruction and Facial Aesthetics. *J. Craniofacial Surg.* **2020**, *31*, 15–27. [CrossRef] [PubMed]
3. Zhang, W.; Yelick, P.C. Craniofacial Tissue Engineering. *Cold Spring Harb. Perspect. Med.* **2017**, *8*, a025775. [CrossRef] [PubMed]
4. Lyu, S.; Untereker, D. Degradability of Polymers for Implantable Biomedical Devices. *Int. J. Mol. Sci.* **2009**, *10*, 4033–4065. [CrossRef] [PubMed]

5. Yoon, S.-D.; Kwon, Y.-S.; Lee, K.-S. Biodegradation and Biocompatibility of Poly L-lactic Acid Implantable Mesh. *Int. Neurourol. J.* **2017**, *21*, S48–S54. [CrossRef] [PubMed]
6. Yeo, A.; Rai, B.; Sju, E.; Cheong, J.; Teoh, S. The degradation profile of novel, bioresorbable PCL–TCP scaffolds: Anin vitro andin vivo study. *J. Biomed. Mater. Res. Part A* **2008**, *84A*, 208–218. [CrossRef] [PubMed]
7. Labet, M.; Thielemans, W. Synthesis of polycaprolactone: A review. *Chem. Soc. Rev.* **2009**, *38*, 3484–3504. [CrossRef] [PubMed]
8. Lam, C.X.F.; Savalani, M.M.; Teoh, S.-H.; Hutmacher, D.W. Dynamics of in vitro polymer degradation of polycaprolactone-based scaffolds: Accelerated versus simulated physiological conditions. *Biomed. Mater.* **2008**, *3*, 034108. [CrossRef] [PubMed]
9. Sun, H.; Mei, L.; Song, C.; Cui, X.; Wang, P. The in vivo degradation, absorption and excretion of PCL-based implant. *Biomaterials* **2006**, *27*, 1735–1740. [CrossRef] [PubMed]
10. Jeong, S.I.; Kim, B.-S.; Lee, Y.M.; Ihn, K.J.; Kim, S.H.; Kim, Y.H. Morphology of Elastic Poly(l-lactide-*co*-ε-caprolactone) Copolymers and in Vitro and in Vivo Degradation Behavior of Their Scaffolds. *Biomacromolecules* **2004**, *5*, 1303–1309. [CrossRef] [PubMed]
11. Abid, Z.; Mosgaard, M.D.; Manfroni, G.; Petersen, R.S.; Nielsen, L.H.; Müllertz, A.; Boisen, A.; Keller, S.S. Investigation of Mucoadhesion and Degradation of PCL and PLGA Microcontainers for Oral Drug Delivery. *Polymers* **2019**, *11*, 1828. [CrossRef] [PubMed]

polymers

MDPI

Article

Tuning of Silver Content on the Antibacterial and Biological Properties of Poly(ε-caprolactone)/Biphasic Calcium Phosphate 3D-Scaffolds for Bone Tissue Engineering

Francesca Menotti [1], Sara Scutera [1], Bartolomeo Coppola [2], Fabio Longo [1], Narcisa Mandras [1], Lorenza Cavallo [1], Sara Comini [1], Rosaria Sparti [1], Elisa Fiume [2], Anna Maria Cuffini [1], Giuliana Banche [1,*], Paola Palmero [2,†] and Valeria Allizond [1,†]

1 Department of Public Health and Pediatrics, University of Torino, 10126 Turin, Italy; francesca.menotti@unito.it (F.M.); sara.scutera@unito.it (S.S.); f.longo@unito.it (F.L.); narcisa.mandras@unito.it (N.M.); lorenza.cavallo@unito.it (L.C.); rosaria.sparti@unito.it (R.S.); annamaria.cuffini@unito.it (A.M.C.); valeria.allizond@unito.it (V.A.)

2 Department of Applied Science and Technology, Politecnico di Torino, 10129 Turin, Italy; bartolomeo.coppola@polito.it (B.C.); paola.palmero@polito.it (P.P.)

* Correspondence: giuliana.banche@unito.it; Tel.: +39-011-670-5627

† These authors contributed equally to this work.

Abstract: There is a growing interest in tissue engineering, in which biomaterials play a pivotal role in promoting bone regeneration. Furthermore, smart functionalization can provide biomaterials with the additional role of preventing orthopedic infections. Due to the growing microbial resistance to antimicrobials used to treat those infections, metal ions, such as silver, thanks to their known wide range of bactericidal properties, are believed to be promising additives in developing antibacterial biomaterials. In this work, novel poly(ε-caprolactone) (PCL)-based 3D scaffolds have been designed and developed, where the polymer matrix was modified with both silver (Ag), to supply antibacterial behavior, and calcium phosphates (biphasic calcium phosphate, BCP) particles to impart bioactive/bioresorbable properties. The microstructural analysis showed that constructs were characterized by square-shaped macropores, in line with the morphology and size of the templating salts used as pore formers. Degradation tests demonstrated the important role of calcium phosphates in improving PCL hydrophilicity, leading to a higher degradation degree for BCP/PCL composites compared to the neat polymer after 18 days of soaking. The appearance of an inhibition halo around the silver-functionalized PCL scaffolds for assayed microorganisms and a significant ($p < 0.05$) decrease in both adherent and planktonic bacteria demonstrate the Ag^+ release from the 3D constructs. Furthermore, the PCL scaffolds enriched with the lowest silver percentages did not hamper the viability and proliferation of Saos-2 cells. A synergic combination of antimicrobial, osteoproliferative and biodegradable features provided to 3D scaffolds the required potential for bone tissue engineering, beside anti-microbial properties for reduction in prosthetic joints infections.

Keywords: poly(ε-caprolactone)-based biomaterial; calcium phosphates; silver; *Staphylococcus aureus*; *S. epidermidis*; *Escherichia coli*; anti-adhesive/antibacterial properties; Saos-2 cells' cell viability/proliferation

1. Introduction

Bone tissue diseases are a growing concern due to the aging population. In fact, adverse clinical conditions, attributable to trauma, pathologies, tumors, or previous surgical procedures, may affect the population, especially those over 50 [1]. While tissue regeneration is a natural process, certain clinical cases may benefit from tissue engineering to boost the production of new bone tissue. Self-regeneration is a multifaceted biological process that includes molecular signaling and many different types of cells. Osteoblasts, as bone synthesizing cells, produce the bone matrix, and, subsequently, the remodeling of new bone tissue occurs to finalize mature bone [2,3]. However, these physiological healing processes decrease in older people; therefore, tissue engineering can offer a new approach for bone restoration and the improvement of medical procedures [1,4].

Engineered bone graft substitutes are designed to restore impaired bone tissue and its mechanical properties, allowing, at the same time, natural bone regeneration and healing progression. Deficient tissue can be repaired and substituted by a three-dimensional (3D) porous construct known as a "scaffold", which has been identified as an innovative therapeutic answer for cavity filling in bone pathological alterations or for the healing of fractured bone [1]. Such bone biocompatible 3D scaffolds may be obtained from natural or synthetic sources. The latter comprise polymers such as polycaprolactone (PCL), a nontoxic aliphatic polyester with FDA-approved clinical applicability thanks to its biodegradability and biocompatibility properties [5,6]. In medical applications, PCL is investigated for wound dressing [7,8], fixation implants [9], drug delivery [10], or in bone-tissue engineering for the manufacturing of long-term implantable devices [3,5,6,11,12].

In order to better match the compositional, physical, and mechanical properties of natural bone, synthetic polymers can be added with ceramic particles, mainly hydroxyapatite (HA), α/β-tricalcium phosphate (α/β-TCP) or biphasic calcium phosphate (BCP, a mixture of HA/β-TCP), known as the mineral components of the natural bone [1,6,13,14]. The incorporation of calcium phosphates (CaPs), as the bioactive component, into PCL provides the 3D scaffold with the required multi-functionalities, merging mechanical features, bioactivity, biodegradability, and the ability to stimulate, promote, and induce osteogenic cell differentiation [3,15].

Furthermore, it is pivotal that these 3D scaffolds can be additionally tuned to let the controlled—fast or prolonged—release of antibacterial agents, allowing the development of novel personalized tissue engineering applications with the ability to act as a targeted delivery system, achieving a high local concentration of loaded molecules with a dose-dependent effect [13,16]. PCL itself could be the cargo of antimicrobial substances such as copper, silver, zinc oxide, or graphene [15,17]. Notably, silver displays a wide spectrum of antimicrobial activity against different pathogens, including those resistant to various antibiotics. In addition, due to its low toxicity to mammalian cells [8,16], the use of silver in orthopedics appears to be a strategic choice.

In one of our previous studies [11], with the aim of developing antibacterial and anti-biofilm biomaterials for tissue regeneration, we focused on the design and manufacturing of BCP/PCL composites functionalized with silver at 1.67wt% with respect to the polymer matrix. These 3D scaffolds showed a highly porous structure, suitable mechanical properties, and relevant antibacterial/anti-adhesive activity against *Staphylococcus aureus* as well. On the other hand, despite the fact that the silver content was relatively low, they displayed cytotoxic behavior towards eukaryotic cells, specifically human osteoblasts.

For these reasons, in the present research, the silver content in the BCP/PCL porous architecture was tuned in the range ~0.8–1.2%, featuring a 3D scaffold characterized by both antibacterial/anti-adhesive and osteogenic properties. This research presents a comparative study of the antibacterial properties of the 3D scaffolds made of BCP/PCL with the addition of varying silver content against three typical human pathogens—*S. aureus*, *S. epidermidis*, and *Escherichia coli*—involved in orthopedic infections.

2. Materials and Methods

2.1. BCP/PCL-Based Scaffold Preparation and Characterization

Poly(ε-caprolactone) (PCL, Merck KGaA, Milan, Italy) pellets were solubilized in acetone (20 wt%) at 40 °C for 24 h. In order to fabricate 3D porous structures, two types of inorganic salts, with sizes in the range 125–355 µm, were used as pore formers, precisely NaCl and $NaNO_3$ (Sigma Aldrich, St. Louis, MO, USA, >99.5% purity). Salt granules were mixed with the solubilized polymer (NaCl or $NaNO_3$:PCL weight ratio 90:10); once homogenized, the suspension was cast into cylindrical plastic molds ($\varnothing$ = 20 mm, h = 10 mm). Once dried and demolded, samples were soaked in deionized water for 4 days, renewing the water each day, to solubilize the salt and generate the required porosity. To fabricate the porous poly(ε-caprolactone)/biphasic calcium phosphate (BCP/PCL) composites, first HA (Captal S BM192, Plasma Biotal Limited, Buxton, UK) and β-TCP (Captal R, Plasma Biotal Limited) powders were mixed under dry conditions in a 70:30 weight ratio to provide the BCP mixture. BCP composite powder was then added to acetone and stirred for 12 h, to which PCL was finally added (BCP:PCL weight ratio, 40:60). Further details on the experimental process can be found in Comini et al. 2021 [11]. Finally, for the (Ag)-doped materials, a variable concentration of silver nitrate ($AgNO_3$, Grade AR, Sigma Aldrich) was mixed with acetone until complete dissolution. In particular, the 1 wt% or 1.2 wt%, and the 0.79 wt% or 1 wt% (as respect to PCL) were introduced for the NaCl or $NaNO_3$ 3D-scaffolds, respectively. These silver concentrations to be added to the polymer were determined by preliminary tests aimed at providing an antibacterial action against the three different microorganisms while preserving the viability of the human osteosarcoma (SaoS)-2 cells. These preliminary results were obtained by testing—using the micro-dilution assays [18]—the direct effect of silver on both bacteria and eukaryotic cells.

Details of all the obtained specimens are reported in Table 1.

Table 1. 3D PCL-based constructs obtained for the present research.

Types of 3D Scaffolds	Salt Used as Pore Former	Composition of the 3D Scaffolds
pure PCL	NaCl	poly(ε-caprolactone)
pure BCP/PCL	NaCl	biphasic calcium phosphates/poly(ε-caprolactone)
PCL + Ag 1%	NaCl	poly(ε-caprolactone) + 1% of silver
PCL + Ag 1.2%	NaCl	poly(ε-caprolactone) + 1.2% of silver
BCP/PCL + Ag 1%	NaCl	biphasic calcium phosphates/poly(ε-caprolactone) + 1% of silver
BCP/PCL + Ag 1.2%	NaCl	biphasic calcium phosphates/poly(ε-caprolactone) + 1.2% of silver
pure PCL	$NaNO_3$	poly(ε-caprolactone)
pure BCP/PCL	$NaNO_3$	biphasic calcium phosphates/poly(ε-caprolactone)
PCL + Ag 0.79%	$NaNO_3$	poly(ε-caprolactone) + 0.79% of silver
PCL + Ag 1%	$NaNO_3$	poly(ε-caprolactone) + 1% of silver
BCP/PCL + Ag 0.79%	$NaNO_3$	biphasic calcium phosphates/poly(ε-caprolactone) + 0.79% of silver
BCP/PCL + Ag 1%	$NaNO_3$	biphasic calcium phosphates/poly(ε-caprolactone) + 1% of silver

The microstructural characterization was carried out by field emission scanning electron microscopy (FESEM, Zeiss Supra 40, Jena, Germany).

The phase composition was investigated with X-ray diffraction (XDR, Philips PW 1710, Eindhoren, The Netherlands) analysis. In particular, the relative ratio between the HA and β-TCP phases was determined by the following relationship [19]:

$$\%HA = \frac{I_{100}(HA)}{I_{100}(HA) + I_{100}(TCP)} \times 100 \quad (1)$$

where I_{100} (HA) and I_{100} (TCP) denote the relative peak intensities of the hydroxyapatite and β-TCP phases, respectively.

Simultaneous Thermogravimetry-Differential thermal analysis (TG-DTA, LabSys evo machine, Setaram, Caluire, France) was performed on both PCL and BCP/PCL samples to assess the influence of the ceramic filler on the thermal behavior of the polymer.

2.2. PCL-Based and BCP/PCL-Based 3D Scaffold Biodegradability Test

The biodegradability tests were performed as previously reported in detail [12]. Briefly, the BCP/PCL-based construct —prepared with different silver concentrations—was immersed in Dulbecco's modified eagle medium (DMEM; Merck KGaA) solution [12,20–22] at

37 °C during 3, 6, 12, and 18 days of incubation. After each soaking time, the scaffolds were removed from the DMEM and dried to reach a constant mass before weighing. The weight loss values of the pure PCL and composite BCP/PCL samples were obtained following this formula:

$$\text{weight loss}(\%) = \frac{m_0 - m_x}{m_0} \times 100 \quad (2)$$

where

m_0 = initial mass of the sample
m_x= mass of the dried sample after immersion at time x

2.3. Cell Viability Assays by Direct-Contact Assay

The in vitro cytotoxicity test was performed, as previously detailed [11,12], on human Saos-2 (American Type Culture Collection®, ATCC®, HTB-85, Manassas, Virginia, VA, USA), an osteosarcoma cell line characterized by an osteoblastic phenotype. These eukaryotic cells were cultured in DMEM high in glucose (Merck KGaA) with phenol red (plus sodium bicarbonate, 10% fetal bovine serum, and 1% penicillin–streptomycin), and incubated at 37 °C in a 5% CO_2 atmosphere.

The sterile BCP/PCL 3D scaffolds were cut into cylinders of 5 mm diameter and 5 mm height and put in 96-well plates; thereafter, their surface was covered by 2×10^4 Saos-2 cells and incubated in culture medium for different incubation times, specifically 0, 3, 6, and 12 days. All the BCP/PCL-based samples, with or without silver, were assayed in triplicate, and the medium was replaced every 2–3 days. At each time point, the cell viability was determined by the 3-(4,5-Dimethylthiazol-2-yl)-2,5-Diphenyltetrazolium Bromide (MTT) assay (Merck KGaA), and the optical density (OD) was measured at 570 nm using a microplate reader (VICTOR3TM, PerkinElmer, Waltham, MA, USA). The results were reported as OD values, subtracting the OD of the PCL or BCP/PCL-based scaffolds without Saos-2 cells from the OD of the PCL or BCP/PCL-based 3D scaffolds with Saos-2 cells. Differences among the obtained OD values on the various constructs with and without silver were determined and statistically analyzed.

2.4. In Vitro Antibacterial Assays

The in vitro antibacterial assays were performed by testing different pathogens involved in orthopedic infections [23–25], specifically *S. aureus* (ATCC® 29213), *S. epidermidis* (ATCC® 35984), and *E. coli* (ATCC® 25922). To investigate the activity of silver blending to PCL- or BCP/PCL-based 3D constructs, the inhibition halo (manual v 9.0; https://www.eucast.org/ast_of_bacteria/disk_diffusion_methodology, accessed on 30 July 2023) and the bacterial adhesion experiments were conducted on the three bacterial strains as recently reported in our research [11,12].

Briefly, for the inhibition halo test, 0.5 McFarland (1–2 $\times 10^8$ colony-forming units, CFU/mL) suspensions of each bacterium were uniformly spread on Mueller Hinton Agar (MHA, Becton Dickinson and Company, BD, Franklin Lakes, NJ, USA), then sterile both PCL- or BCP/PCL-pure scaffolds and the correspondingly silver-added ones were placed on agar. The silver release from functionalized specimens and its effect on bacterial development were evaluated after an incubation for 24 h at 35 ± 2 °C by measuring (mm) the inhibition halo [11,12].

Whereas, for the microbial adhesion test, as previously detailed [11,12], the bacteria were cultured for 18–24 h at 35 ± 2 °C in Mueller Hinton Broth (MHB, BD), then centrifuged and diluted in MHB to obtain a 10^4 CFU/mL inoculum. The sterile PCL- or BCP/PCL-based 3D constructs, with and without different silver contents, were surrounded by 7 mL of bacterial inoculum in a 6-well culture plate and incubated at 35 ± 2 °C by shaking for 24 h to permit in vitro bacterial bonding to the biomaterials. After incubation, the constructs were subjected to sonication for 30 min at room temperature in 10 mL of NaCl 0.9% solution (Bieffe Medital S.p.A., Grosotto, Italy) to detach bacteria, which were strongly bound to the samples themselves and quantified by a plating count on MHA. Planktonic bacteria were also counted as CFU/mL. The adhesion tests were conducted in triple for each 3D construct type and performed at least three times.

2.5. Statistical Analysis

The GraphPad Prism 9 software (San Diego, CA, USA) was employed to analyze the morphological parameters, the weight loss percentages, the microbiological (CFU/mL), and the MTT (OD, 570 nm) data by descriptive statistics (means and standard error of the means). An unpaired Student's *t*-test was used to find significant differences ($p < 0.05$) between the various tested samples.

3. Results and Discussion

The present study was aimed at designing and developing 3D scaffolds based on pure PCL or BCP/PCL functionalized biomaterials for orthopedic bone tissue engineering that would, in parallel, ensure bone regenerative potential through the presence of BCP and an antimicrobial feature by adding $AgNO_3$, released as Ag^+. The available literature pertains to PCL as fibers or printed constructs, but only when added with metal ions (i.e., silver, copper, zinc, etc.) [5,13–16,26,27], whereas no existing research has studied the combined use of calcium phosphates and silver, especially on 3D constructs fabricated by the salt leaching/polymer casting method.

3.1. Characterization of PCL- and BCP/PCL-Based Biomaterials

The dimensions of all the PCL- and BCP/PCL-based biomaterials are reported in Table 2. In brief, the 3D scaffolds exhibited a morphological cylinder-shaped geometry with comparable dimensions, specifically diameter (mm) and height (mm). Composite samples (BCP/PCL) presented a slightly larger diameter compared to pure PCL, indicating a lower shrinkage during drying of the former specimens, imputable to the constraining effect of the ceramic particles. Density values showed almost superimposable values among the same type of samples but obtained with different salts (NaCl or $NaNO_3$). Besides the negligible role of salts on density, these results highlight the high reproducibility of the process employed.

Table 2. Morphological characteristics (reported as mean ± standard error of the mean) of the pure PCL- or BCP/PCL-based 3D scaffolds, functionalized with low silver concentrations, and pored with NaCl (A) or $NaNO_3$ (B).

	Morphological Parameters			**Statistical Analysis**
A	**Diameter (mm)**	**Height (mm)**	**Density (mg/mm³)**	**Student's *t*-Test**
Scaffold Type				
PCL	18.31 ± 0.11	11.27 ± 0.16	0.126 ± 0.003	weight and density PCL vs. BCP/PCL $p < 0.001$
BCP/PCL	18.98 ± 0.10	10.80 ± 0.29	0.204 ± 0.005	
PCL + Ag 1%	18.31 ± 0.21	10.58 ± 0.49	0.133 ± 0.009	
PCL + Ag 1.2%	18.38 ± 0.22	11.18 ± 0.49	0.132 ± 0.005	
BCP/PCL + Ag 1%	18.73 ± 0.12	11.68 ± 0.13	0.213 ± 0.004	
BCP/PCL + Ag 1.2%	18.92 ± 0.13	12.15 ± 0.16	0.221 ± 0.003	
B				
PCL	18.10 ± 0.13	10.10 ± 0.39	0.127 ± 0.003	weight and density PCL vs. BCP/PCL $p < 0.001$
BCP/PCL	18.74 ± 0.10	9.87 ± 0.41	0.205 ± 0.005	
PCL + Ag 0.79%	18.61 ± 0.13	11.11 ± 0.65	0.133 ± 0.009	
PCL + Ag 1%	18.42 ± 0.22	11.81 ± 0.46	0.132 ± 0.005	
BCP/PCL + Ag 0.79%	18.63 ± 0.11	9.64 ± 0.25	0.213 ± 0.004	
BCP/PCL + Ag 1%	18.99 ± 0.07	11.25 ± 0.16	0.220 ± 0.003	

Abbreviations: PCL—poly(ε-caprolactone); BCP—biphasic calcium phosphates; Ag—silver.

In addition, considering the theoretical density of neat PCL (1.145 g/cm^3), the density of pure PCL specimens (0.126 ± 0.003 g/cm^3 and 0.127 ± 0.003 g/cm^3 for the 3D scaffolds pored with NaCl or $NaNO_3$, respectively) provided a total porosity of ~89%, in fair agreement with the expected nominal porosity of 90%. When the BCP was added to obtain the BCP/PCL composite constructs, the density increased to about 0.22 g/cm^3. This augmentation can be explained by considering the theoretical density of the ceramic phases (3.16 g/cm^3 for HA and 3.07 g/cm^3 for β-TCP), which provided a nominal density for the BCP/PCL composite of 1.53 g/cm^3 (as determined by the rule of mixture). Therefore, in this case, the overall porosity was ~86%, just slightly lower than the nominal 90% volume.

Notably, the addition of the different percentages of silver to the 3D scaffolds, either pure PCL or BCP/PCL, did not significantly modify neither the cylindrical geometry nor the dimensions, while a slight increase in density was observed (Table 2).

In Figure 1, the XRD patterns of neat PCL (a) and of BCP composite powder (b)—both used as references—and the pattern related to BCP/PCL composite scaffold (c) are presented. Concerning the neat polymer (a), the presence of two sharp peaks at 21.65 (°2θ) and 23.92 (°2θ), corresponding respectively to the (110) and (200) planes, denotes the presence of polycaprolactone with a semi-crystalline structure, in agreement with literature data [28,29]. In (b), the peaks of the HA and β-TCP phases were recognized and indexed through the JCPD files n. 00-009-0432 and 00-009-0169, respectively. By applying Equation (1), the HA: β-TCP weight ratio was 68:32. The fair agreement between the experimental and nominal ratios (70:30) strengthens the use of the above equation for the semi-quantification of the calcium phosphate phase's ratio. In (c), the PCL main signals have been determined, along with all the most intense peaks of the HA and β-TCP phases. These results suggest a negligible role for calcium phosphate particles in modifying the crystallinity of the polymer matrix. Equation (1) provides that the HA and β-TCP phases are present in the polymer matrix with a weight ratio of 72:28, still in very good agreement with the nominal ratio.

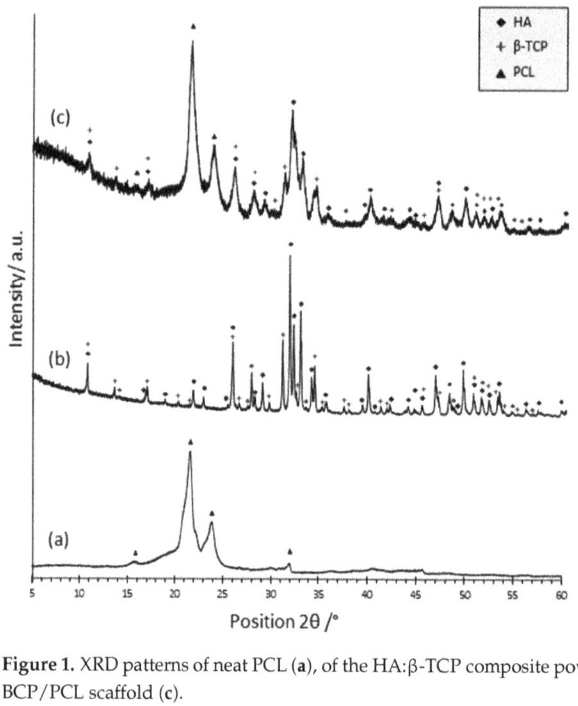

Figure 1. XRD patterns of neat PCL (**a**), of the HA:β-TCP composite powder (**b**), and of the composite BCP/PCL scaffold (**c**).

In addition, in order to quantify the PCL:BCP ratio, DTA-TG analysis was carried out on the composite scaffold. After thermal decomposition at 650 °C, a residual mass equal to 43% was determined, corresponding to the calcium phosphate particles, in agreement with the nominal content (40 wt% with respect to PCL).

Joint results concerning the calcium phosphate content in the polymer (by DTA-TG) and on the HA:β-TCP ratio (by XRD) clearly demonstrate that the long water soaking time of the samples (4 days), necessary to leach out the NaCl and $NaNO_3$ salts, did not induce any significant or preferential dissolution of the calcium phosphate phases. Although the above results were determined on samples prepared by using NaCl as a pore former, similar results were determined with $NaNO_3$ and therefore were not repeated here.

FESEM micrographs of neat PCL scaffolds as well as silver-added PCL and BCP/PCL composite constructs are illustrated in the following: In Figure 2, the morphology of the PCL scaffolds pored with NaCl (A) and $NaNO_3$ (D) salts is compared. While in the former case the pores are characterized by well-defined geometrical shapes due to regular-shaped NaCl granules [11,12], a less regular pore shape can be recognized in $NaNO_3$-derived materials. This difference in pore morphology is maintained in the micrographs of the silver-added PCL scaffolds, where squared pores characterized the NaCl-based sample (B,C), contrary to the $NaNO_3$-one (E,F). In spite of this, all the structures were characterized by a very high porosity and were open and interconnected (as evidenced by the higher magnification images, C and F), which is extremely required in bone tissue engineering. Notably, the addition of silver did not alter the pore morphology or the microstructures, as the same features can be observed in neat and silver-added PCL specimens.

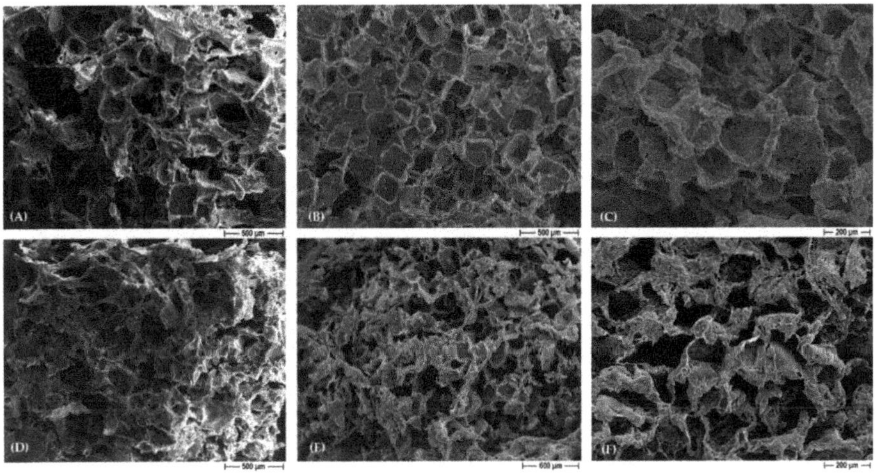

Figure 2. Representative FESEM micrographs of neat PCL scaffolds (**A**,**D**) and silver-added PCL (**B**,**C**,**E**,**F**) obtained by using NaCl (**A**–**C**) and $NaNO_3$ (**D**–**F**) as pore formers. For silver-added PCL, lower (**B**,**E**) and higher (**C**,**F**) magnification images are depicted.

Figure 3 shows the FESEM micrograph of BCP/PCL scaffolds obtained by using NaCl as a pore former, as indicated by the geometrically shaped pores. The lower magnification image (A) shows again the high porosity degree and the pore interconnection by means of tine pores in the cell walls; the higher magnification one (B) shows the homogeneous dispersion of the HA/β-TCP particles, well embedded into the polymer matrix. Similar microstructural features were observed in the $NaNO_3$-based samples and are thus not depicted here.

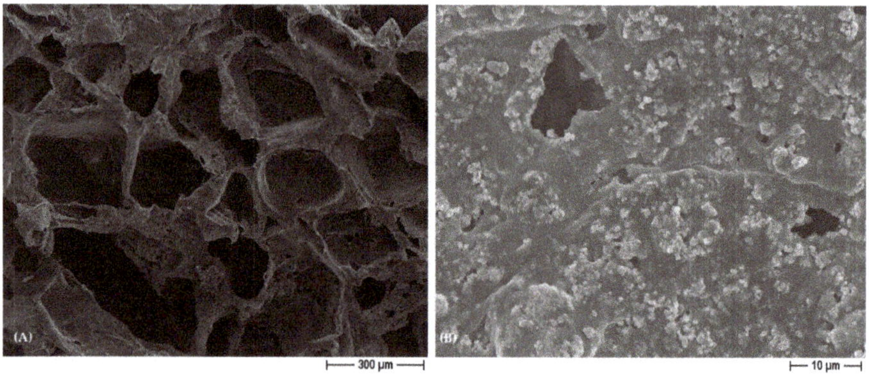

Figure 3. Representative FESEM micrographs of BCP/PCL (NaCl) scaffolds at lower (**A**) and higher (**B**) magnification.

3.2. PCL- and BCP/PCL-Based 3D Scaffold Biodegradability Degree

In Figure 4, the weight loss of the 3D scaffold as a function of the immersion time in DMEM is depicted.

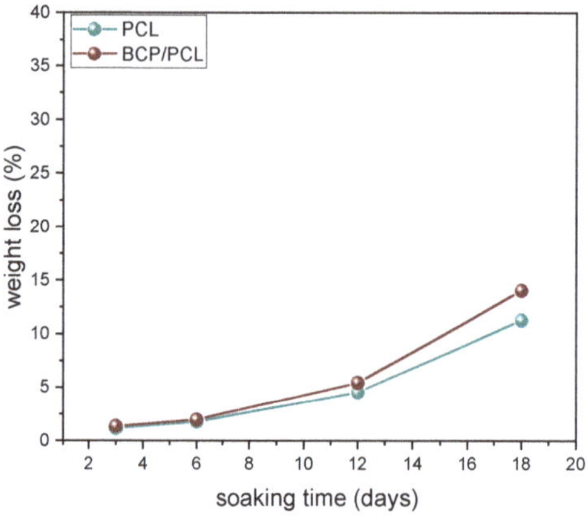

Figure 4. Weight loss of PCL and BCP/PCL scaffolds (by using NaCl as pore former salt) at different time points (3, 6, 12, and 18 days) after the immersion in DMEM.

For both PCL and BCP/PCL scaffolds, a gradual increase in mass loss was determined, with close values for the two types of samples up to 12 days (4.6% and 5.5%, respectively). After 18 days, a more significant mass loss was determined for the composite sample, reaching 14% of the weight loss compared to the neat polymer scaffold (11%).

These results are in poor agreement with literature data since, generally, a lower degradation degree for PCL-based materials is reported [30,31]. However, a clear comparison of results is difficult due to the various experimental conditions used to test the materials (such as soaking medium, time and temperature, degradation mechanism, etc.), as well as the different material features (such as porosity amount and size, crystallinity, thickness of the samples, etc.). With specific reference to biodegradability tests performed in DMEM, Lu et al. (2012) fabricated scaffolds by injecting the melt polymer or melt PCL added with β-TCP particles into a soluble porous mold [32]. The authors reported a weight loss of just 0.35% after 2 weeks of immersion, which increased to ~3% and 4% after 4 and 6 weeks, respectively. These values increased significantly for 10% and 20% β-TCP-containing materials, showing 2.4% and 2.8% weight loss, respectively, after 2 weeks of immersion.

A very low weight loss (0.7%) after 2 weeks of immersion of PCL scaffold, obtained by fused deposition modeling technique, was also reported by Hedayati et al. (2022) too [33]. Muhammad et al. (2012) compared the biodegradability of a commercial PCL scaffold with a self-crossing PCL-derived polymer, the polycaprolactone trifumarate, obtained by a salt-leaching method, similarly to our work but using a lower polymer: NaCl ratio of 1:1 [34]. While the former still presented a moderate weight loss after 2 weeks (~2.5%) but was comparable to the values achieved in this work after 12 days, the modified PCL achieved values close to 20%. Finally, Janarthanan et al. (2019) reported weight loss of ~8% and 10% for porous PCL and porous PCL/α-TCP, obtained by a solvent casting/salt leaching method similar to our process after 14 days of immersion, in good agreement with the present results [35]. The highly porous structure, characterized by open and well-interconnected pores and thin struts within the cells, is probably responsible for this high degradation degree and rate. A further explanation, when comparing the current results with tests carried out in other media, can be ascribed to the better wettability of PCL by DMEM than other mediums like physiological buffer solution (PBS), as determined by Musciacchio et al. (2022) [36].

In this work, to rule out a selective release of calcium phosphate particles during soaking in DMEM, XRD analyses were performed on BCP/PCL samples before and after each incubation time (Figure 5).

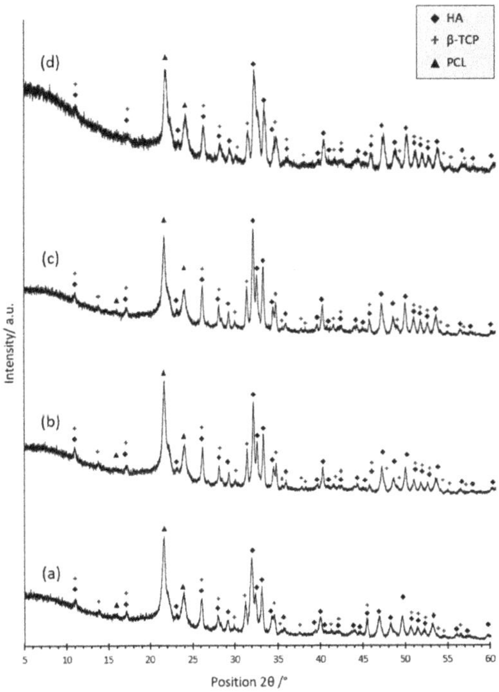

Figure 5. XRD patterns of BCP/PCL scaffolds after soaking for 3 (**a**), 6 (**b**), 12 (**c**), and 18 (**d**) days in DMEM.

Samples soaked in DMEM for different times showed very similar diffraction patterns in terms of the intensity of the signals associated with PCL, HA, and β-TCP phases. By applying Equation (1), the HA: β-TCP ratio was in fact 71.5, 70, 70, and 73 in samples incubated for 2, 6, 12, and 18 days, respectively, showing a very good match with the nominal ratio. These results rule out any significant dissolution of the ceramic particles into the incubation fluid; on the other hand, calcium phosphate particles can play a role in improving the hydrophilicity of PCL, thus explaining the highest degradation of BCP/PCL compared to the neat polymer after 18 days of soaking.

3.3. In Vitro Saos-2 Cell Viability/Proliferation Assay

Here, we evaluated Saos-2 cells since these eukaryotic cells are a useful in vitro model of typical osteoblast behavior and represent a mature osteoblast phenotype with high alkaline phosphatase activity and osteocalcin expression at similar levels as human primary osteoblasts [37–39].

In our previous paper, we demonstrated that the silver concentration of ~1.67%, even if it showed good efficacy against *S. aureus*, unfortunately impaired Saos-2 cell viability and proliferation [11]. Thus, here we tuned the silver amount into the 3D scaffolds to reach both antibacterial and non-cytotoxic behavior. Specifically, regarding samples pored with NaCl, they were added with 1% and 1.2%, whereas those pored with $NaNO_3$—for which we demonstrated a higher release of silver content into the medium [11]—were functionalized with 0.79% and 1%.

In Figure 6, the summary of all the results achieved by the MTT assays and expressed as optical density (OD 570 nm) concerning the viability and proliferation of Saos-2 cells in contact with the different specimens pored with NaCl (A) or $NaNO_3$ (B) within 12 days of incubation, blended or not with silver, is reported.

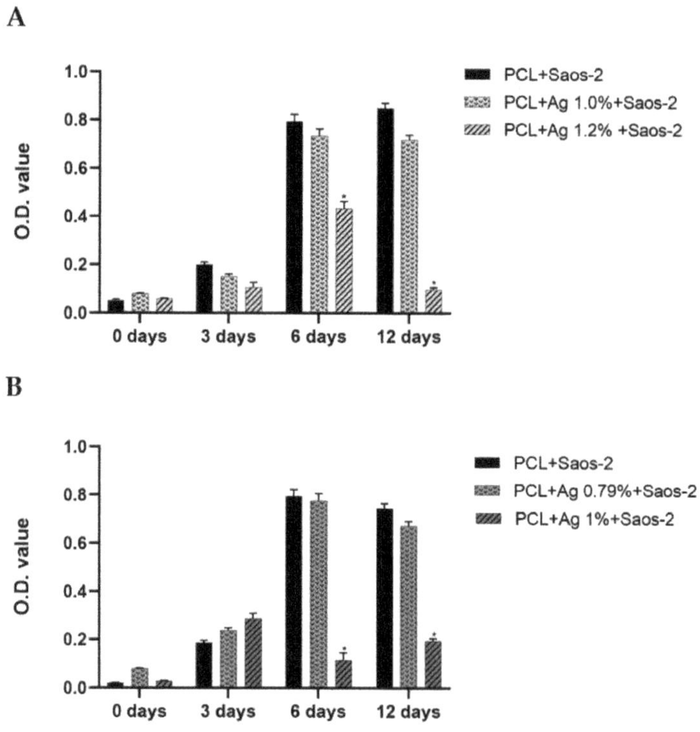

Figure 6. Cell viability (MTT analysis) of Saos-2 cells exposed to pure PCL-based biomaterial and to different concentrations of Ag-enriched PCL 3D scaffolds, fabricated with NaCl (**A**) or $NaNO_3$ (**B**) salts, expressed as optical density (OD) value at 570 nm. Results are means ± standard error of the mean (SEM) of at least three independent experiments; * $p < 0.001$ unpaired t-test.

Briefly, within 3 days of incubation, similar OD values were recorded for all the PCL-based 3D scaffolds, reporting no differences neither between controls and silver-functionalized specimens nor between the NaCl (A) and $NaNO_3$ (B) salts used to form the pores in the samples. At 6 days of incubation, while the OD for the pure PCL and the PCL + 1% of silver for NaCl constructs, or 0.79% of silver for the $NaNO_3$ ones, resulted in an additional increase in the OD values, the 3D scaffolds enriched with the higher silver concentrations—specifically 1.2% or 1% for NaCl and $NaNO_3$, respectively—highlighted a significant ($p < 0.001$) decrease in the OD values with respect to pure PCL (Figure 6). Notably, after 12 days of incubations, a further significant ($p < 0.001$) reduction in cell viability was evident for the 1.2% (NaCl pored) and the 1% ($NaNO_3$ pored) functionalized specimens, with respect to pure-PCL and to the ones added with the lower concentrations (1% or 0.79% for NaCl and $NaNO_3$ pored constructs, respectively). An analogous but not cytotoxic pattern pertaining to the 3D scaffolds obtained with BCP/PCL and with BCP/PCL added with the lower (1% for NaCl and 0.79% $NaNO_3$) silver concentrations was obtained, thus demonstrating that the functionalization with calcium phosphates did not impair eukaryotic cell viability.

Finally, for both pure-PCL and PCL enriched with the lower (1% or 0.79%) silver percentages, the preparation of the 3D scaffolds either with NaCl or with $NaNO_3$ promoted Saos-2 proliferation within 6 days, determining a confluence state and no additional growth at 12 days (Figure 6).

In the attempt to screen the existing literature on the effect of PCL enriched with silver on sarcoma Saos-2 cells viability, we faced the fact that very poor articles on the issue are available; therefore, in the following part of the paper, we compared the direct action of silver towards these types of eukaryotic cells. Research by Ashe S. et al. (2016) demonstrated that the presence of AgNPs (0.005–0.25 mM) ameliorate Saos-2 cell parameters such as morphology and restores native protein structure [40]. The same authors [41] prepared composite hydrogels encompassing AgNPs and determined that, with respect to pure biomaterials, a slight reduction in Saos-2 cells was revealed, and the initial concentration of AgNPs was 1 mg/mL. More recently, Rodriguez-Contreras et al. (2023) demonstrated both adhesion and proliferation of Saos-2 cells on silver and gallium-modified titanium surfaces in a dose-dependent manner [25]. Additionally, in current studies, the direct effect of AgNPs (3–250 μg/mL or 0.3125–10 ppm) on SaoS-2 cells was tested, and a dose-dependent cytotoxicity was shown [42,43]. These data are in good agreement with those here obtained, since a decrease in eukaryotic cell viability was highlighted only when higher silver concentrations were used, whereas when they were exposed to lower ones, no cytotoxic effects were demonstrated. The silver concentrations that displayed a toxic or nontoxic behaviour towards Saos-2 cells are very close; thus, we can speculate that the blending of silver into the PCL- and BCP/PCL-based biomaterials allowed a tailored amount that was consequently specifically released as Ag^+, demonstrating a cause/effect association between ions and cytotoxicity on eukaryotic cells [25,44].

In parallel, we further confirmed that the PCL polymer alone or blended with CaPs did not impair Saos-2 cell viability or proliferation [17,45,46].

3.4. Antibacterial Assays

A growing body of evidence indicates that the most common pathogens recovered from prosthetic joint infections (PJIs) in different clinical settings are Gram-positive bacteria belonging to the *Staphylococcus* species, mainly *S. aureus* and *S. epidermidis* [47,48]. Thereafter, other microorganisms, such as *Enterobacteriaceae* (i.e., *E. coli* or *Kleblsiella pneumoniae*), can be the causative bacterial species [48,49]. These bacteria can worsen the infectious process by both being resistant to the antimicrobial treatment and by producing a well-established biofilm [47,50,51].

For these reasons, three different pathogens—*S. aureus, S. epidermidis,* and *E. coli*—as demonstrative microorganisms involved in PJIs were used in the present research. The microbiological results were obtained by performing the inhibition halo (Figures 7–9 and Table 3) and the bacterial adhesion experiments (Tables 4 and 5). Even if only the lower concentration of silver (1% for NaCl and 0.79% for $NaNO_3$) did not display a reduction in Saos-2 cell viability and proliferation, being non-toxic for eukaryotic cells, all the microbiological assays were performed by using all the silver concentrations added to the PCL-based biomaterials. In fact, literature research also reported that close concentrations of silver are not toxic for eukaryotic cells but exert antibacterial activity as well [24,25,27,42].

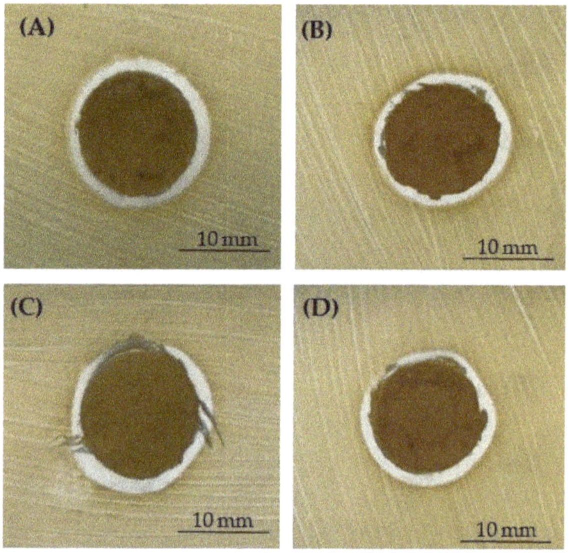

Figure 7. Representative images of the inhibition halo assay against *S. aureus* in the presence of the PCL-based samples, pored with NaCl, enriched with 1% (**A**) or 1.2% (**B**) of silver, or those pored with $NaNO_3$, enriched with 0.79% (**C**) or 1% (**D**) of silver.

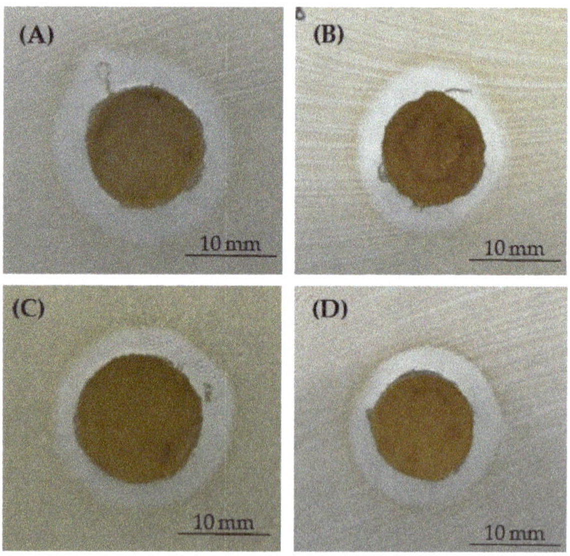

Figure 8. Representative images of the inhibition halo assay against *S. epidermidis* in the presence of the PCL-based samples, pored with NaCl, enriched with 1% (**A**) or 1.2% (**B**) of silver, or those pored with $NaNO_3$, enriched with 0.79% (**C**) or 1% (**D**) of silver.

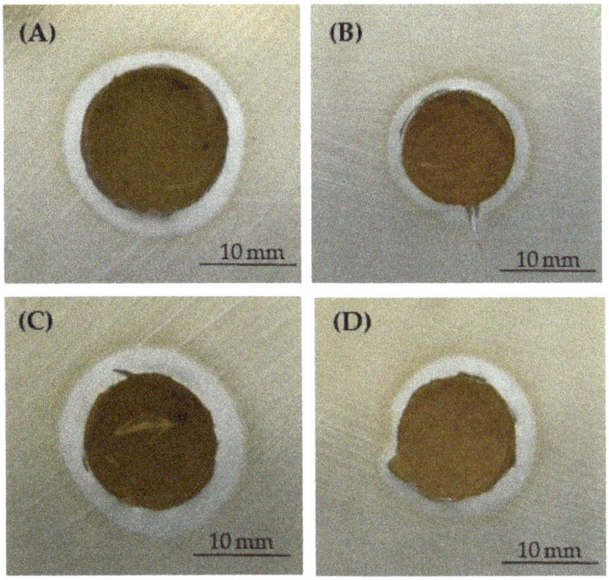

Figure 9. Representative images of the inhibition halo assay against *E. coli* in the presence of the PCL-based samples, pored with NaCl, enriched with 1% (**A**) or 1.2% (**B**) of silver, or those pored with $NaNO_3$, enriched with 0.79% (**C**) or 1% (**D**) of silver.

Table 3. Average diameters (reported as mean ± standard error of the mean) of the inhibition halo around the pure-PCL- or the BCP/PCL-based scaffolds functionalized with low silver concentrations, pored with NaCl (A) or $NaNO_3$ (B), towards the three assayed bacterial strains.

	Average Diameter ± SEM (mm)		
A	*S. aureus*	*S. epidermidis*	*E. coli*
Scaffold Type pored with NaCl			
PCL + Ag 1%	22.65 ± 0.16	27.41 ± 0.21	22.02 ± 0.36
PCL + Ag 1.2%	22.85 ± 0.32	30.82 ± 0.30	22.19 ± 0.13
BCP/PCL + Ag 1%	22.79 ± 0.24	28.13 ± 0.12	21.32 ± 0.16
BCP/PCL + Ag 1.2%	23.06 ± 0.11	31.65 ± 0.22	22.87 ± 0.41
B			
Scaffold Type pored with $NaNO_3$			
PCL + Ag 0.79%	23.12 ± 0.20	27.06 ± 0.38	21.11 ± 0.13
PCL + Ag 1%	23.13 ± 0.31	29.76 ± 0.18	22.29 ± 0.23
BCP/PCL + Ag 0.79%	24.09 ± 0.50	27.89 ± 0.22	21.51 ± 0.30
BCP/PCL + Ag 1%	24.23 ± 0.12	30.03 ± 0.47	22.77 ± 0.21

Table 4. Number of adherent staphylococci and *E. coli* (log_{10} colony-forming units, CFU/mL) on the PCL and BCP/PCL 3D scaffolds, pure or functionalized with low silver concentrations and pored with NaCl (A) or $NaNO_3$ (B), towards the three assayed bacterial strains.

	Number of Adhered Bacteria as log_{10} CFU/mL (Means ± Standard Error of the Means)			**Statistical Analysis**
A	*S. aureus*	*S. epidermidis*	*E. coli*	**Student's *t*-Test**
Scaffold Type pored with NaCl				
PCL	$2.16 \times 10^9 \pm 3.56 \times 10^8$	$1.55 \times 10^7 \pm 5.50 \times 10^6$	$1.45 \times 10^8 \pm 1.14 \times 10^6$	PCL and BCP/PCL vs. PCL + Ag and BCP/PCL + Ag $p < 0.001$
BCP/PCL	$3.05 \times 10^9 \pm 6.65 \times 10^8$	$2.15 \times 10^7 \pm 4.00 \times 10^6$	$1.24 \times 10^8 \pm 1.10 \times 10^6$	
PCL + Ag 1%	$2.36 \times 10^3 \pm 1.88 \times 10^2$	$4.67 \times 10^2 \pm 1.40 \times 10^1$	$2.43 \times 10^2 \pm 1.15 \times 10^1$	
PCL + Ag 1.2%	$2.01 \times 10^3 \pm 6.61 \times 10^1$	$2.27 \times 10^2 \pm 1.83 \times 10^1$	$2.86 \times 10^2 \pm 1.69 \times 10^1$	
BCP/PCL + Ag 1%	$2.66 \times 10^3 \pm 8.54 \times 10^1$	$3.32 \times 10^2 \pm 1.35 \times 10^1$	$2.52 \times 10^2 \pm 1.35 \times 10^1$	
BCP/PCL + Ag 1.2%	$1.55 \times 10^3 \pm 6.53 \times 10^1$	$2.42 \times 10^2 \pm 1.76 \times 10^1$	$1.42 \times 10^2 \pm 1.55 \times 10^1$	
B				
Scaffold Type pored with $NaNO_3$				
PCL	$1.48 \times 10^9 \pm 3.80 \times 10^8$	$2.66 \times 10^7 \pm 1.34 \times 10^6$	$2.19 \times 10^8 \pm 1.41 \times 10^6$	PCL and BCP/PCL vs. PCL + Ag and BCP/PCL + Ag $p < 0.001$
BCP/PCL	$2.82 \times 10^9 \pm 9.60 \times 10^7$	$2.65 \times 10^7 \pm 1.89 \times 10^6$	$1.78 \times 10^8 \pm 1.76 \times 10^6$	
PCL + Ag 0.79%	$3.43 \times 10^3 \pm 2.18 \times 10^2$	$2.30 \times 10^2 \pm 1.66 \times 10^1$	$3.12 \times 10^2 \pm 1.83 \times 10^1$	
PCL + Ag 1%	$2.31 \times 10^3 \pm 2.30 \times 10^2$	$2.01 \times 10^2 \pm 3.10 \times 10^1$	$1.06 \times 10^2 \pm 1.28 \times 10^1$	
BCP/PCL + Ag 0.79%	$6.18 \times 10^3 \pm 2.86 \times 10^2$	$2.52 \times 10^2 \pm 1.89 \times 10^1$	$2.89 \times 10^2 \pm 1.50 \times 10^1$	
BCP/PCL + Ag 1%	$4.96 \times 10^3 \pm 1.22 \times 10^2$	$2.44 \times 10^2 \pm 2.17 \times 10^1$	$1.60 \times 10^2 \pm 1.12 \times 10^1$	

Table 5. Number of planktonic staphylococci and *E. coli* (log_{10} colony-forming units, CFU/mL) in the presence of PCL and BCP/PCL 3D scaffolds, pure or functionalized with low silver concentrations and pored with NaCl (A) or $NaNO_3$ (B), towards the three assayed bacterial strains.

	Number of Planktonic Bacteria as log_{10} CFU/mL (Means ± Standard Error of the Means)			**Statistical Analysis**
A	*S. aureus*	*S. epidermidis*	*E. coli*	**Student's *t*-Test**
Scaffold Type Pored with NaCl				
PCL	$2.80 \times 10^9 \pm 2.82 \times 10^8$	$2.64 \times 10^8 \pm 6.62 \times 10^6$	$1.74 \times 10^9 \pm 3.13 \times 10^8$	PCL and BCP/PCL vs. PCL + Ag and BCP/PCL + Ag $p < 0.001$
BCP/PCL	$2.35 \times 10^9 \pm 7.25 \times 10^7$	$3.09 \times 10^8 \pm 1.21 \times 10^7$	$1.07 \times 10^9 \pm 9.35 \times 10^7$	
PCL + Ag 1%	$1.28 \times 10^5 \pm 1.18 \times 10^4$	$3.32 \times 10^4 \pm 1.28 \times 10^3$	$1.67 \times 10^3 \pm 1.14 \times 10^2$	
PCL + Ag 1.2%	$1.06 \times 10^5 \pm 5.53 \times 10^3$	$2.82 \times 10^4 \pm 1.80 \times 10^3$	$1.30 \times 10^3 \pm 1.60 \times 10^2$	
BCP/PCL + Ag 1%	$1.65 \times 10^5 \pm 1.64 \times 10^4$	$2.80 \times 10^4 \pm 1.55 \times 10^3$	$3.50 \times 10^3 \pm 2.51 \times 10^2$	
BCP/PCL + Ag 1.2%	$1.58 \times 10^5 \pm 2.40 \times 10^4$	$1.72 \times 10^4 \pm 1.39 \times 10^3$	$2.62 \times 10^3 \pm 1.64 \times 10^2$	
B				
Scaffold Type pored with $NaNO_3$				
PCL	$2.16 \times 10^9 \pm 5.81 \times 10^8$	$2.02 \times 10^8 \pm 2.00 \times 10^7$	$1.42 \times 10^9 \pm 1.56 \times 10^8$	PCL and BCP/PCL vs. PCL + Ag and BCP/PCL + Ag $p < 0.001$
BCP/PCL	$3.67 \times 10^9 \pm 3.22 \times 10^7$	$3.51 \times 10^8 \pm 2.75 \times 10^7$	$1.34 \times 10^9 \pm 4.33 \times 10^7$	
PCL + Ag 0.79%	$2.78 \times 10^5 \pm 2.76 \times 10^4$	$2.79 \times 10^4 \pm 2.13 \times 10^3$	$1.48 \times 10^3 \pm 1.07 \times 10^2$	
PCL + Ag 1%	$1.57 \times 10^5 \pm 1.46 \times 10^4$	$2.61 \times 10^4 \pm 2.03 \times 10^3$	$1.24 \times 10^3 \pm 1.33 \times 10^2$	
BCP/PCL + Ag 0.79%	$3.35 \times 10^5 \pm 8.26 \times 10^3$	$2.76 \times 10^4 \pm 1.02 \times 10^3$	$2.24 \times 10^3 \pm 8.05 \times 10^1$	
BCP/PCL + Ag 1%	$2.32 \times 10^5 \pm 4.21 \times 10^4$	$2.23 \times 10^4 \pm 5.80 \times 10^2$	$2.06 \times 10^3 \pm 1.00 \times 10^2$	

In Figures 7–9, representative images of the inhibition halo assay are depicted, and in Table 3, the average diameter (mm) of the inhibition growth around the 3D scaffolds is reported. These data established that the silver was released from the PCL- and BCP/PCL-based Ag-added samples and that it acted towards the three microorganisms tested, even if a more pronounced activity was revealed against *S. epidermidis* with respect to both *S. aureus* and *E. coli*. No inhibition halo was obtained around the pure-PCL and -BCP/PCL constructs, confirming that no antibacterial action is displayed by the polymer alone or by the polymer functionalized by CaPs. These results are in line with those of other researchers [13,16,52]. In a recently published article, the authors demonstrated that a growth-inhibition zone was generated around titanium samples doped with silver and gallium against both Gram-positive (*S. aureus* and *S. epidermidis*) and Gram-negative (*E. coli* and *Pseudomonas aeruginosa*) bacteria [25]. Additionally, a similar wide area of inhibition was obtained by testing *S. aureus* halo around poly-lactic-co-glycolic-acid/PCL scaffolds enriched with AgNPs [53] or by assaying *S. aureus* and *E. coli* halo around scaffolds containing silver-doped HA [27].

The microbiological results (expressed as $\log_{10}$ CFU/mL) of the adhered bacteria—precisely *S. aureus*, *S. epidermidis*, and *E. coli*—to the different PCL- and BCP/PCL-based biomaterials, produced with either NaCl or $NaNO_3$ salts, after 24 h of incubation are presented in Table 4.

Briefly, the three bacteria adhered to the pure PCL and BCP/PCL-based 3D scaffolds, used as controls, with different loads of 10^9 CFU/mL, 10^7 CFU/mL, and 10^8 CFU/mL, for *S. aureus*, *S. epidermidis*, and *E. coli*, respectively. Whereas, with respect to the controls, the silver-blend specimens were able to significantly ($p < 0.001$) reduce their bacterial adhesion, highlighting values at 10^3 CFU/mL for *S. aureus* and 10^2 CFU/mL for both *S. epidermidis* and *E. coli*. As expected, no differences between the two silver concentrations (1% and 1.2% for NaCl and 0.79% and 1% for $NaNO_3$) were revealed, neither for the Gram-positive nor for the Gram-negative bacteria (Table 4), suggesting that despite the variable silver amount in the 3D scaffolds, a reduction in the bacterial adhesion occurred anyway.

In Table 5, the growth of the three different microorganisms in the broths being in contact for 24 h with the specimens either controls—PCL and BCP/PCL—or silver-added ones, determined by the planktonic count as $\log_{10}$ CFU/mL, is shown.

These results confirmed that the silver was released by the enriched samples into the bacterial medium and exerted an antibacterial effect on their growth. In fact, a significant ($p < 0.001$) reduction in the planktonic *S. aureus*, *S. epidermidis*, and *E. coli* was detected, represented by a load of 10^5 CFU/mL, 10^4 CFU/mL, and 10^3 CFU/mL, respectively. Whereas, in agreement with other literature results, no innate antibacterial activity by the polymer alone or functionalized with CaPs was highlighted [6,16,27,52].

Thus, the silver presence in the 3D scaffolds revealed an anti-adhesive and antibacterial (Tables 4 and 5) action, and additionally to that, an anti-biofilm activity. In fact, it has to be highlighted that a well-structured biofilm was noted in the control materials for both the Gram-positive and the Gram-negative bacteria (Figure 10), whereas in the silver-functionalized PCL and BCP/PCL-biomaterials, only a few microorganisms were observed, and notably, they were altered in their usual morphology (Figure 11). Actually, the staphylococci lost their spherical shape, and *E. coli* displayed a more elongated-fusiform morphology; this effect was due to the direct effect of silver on the bacterial external structures. These data are in line with those of other scientific works that have proposed a destabilization of the biofilm exerted by silver [25,54]. Here, we can speculate that the silver presence is able to inhibit the biofilm's production rather than its dislocation.

All together, the microbiological results confirmed the anti-adhesive and anti-biofilm properties of the 3D scaffolds functionalized with both silver concentrations and, in parallel, antibacterial growth as well. In a study, the antimicrobial action of PCL/clay mineral vermiculite films doped or not with zinc was assayed, and the results demonstrated an antibacterial effect only against *E. coli* but not against *S. aureus* [5]. In addition, electrospun scaffolds containing silver-doped HA demonstrated a great antibacterial effect on *S. aureus* and *E. coli* growth over time [27]. Qian Y. et al. (2019) produced PLGA/PCL electrospun scaffolds, including AgNPs, and highlighted an inhibition in *S. aureus* growth [16].

Florea D.A. et al. (2022) evaluated the antibacterial performance of different coatings on magnesium phosphate-containing silver nanoparticles and demonstrated their strong anti-*S. aureus* and -*Ps. aeruginosa* efficiency [14].

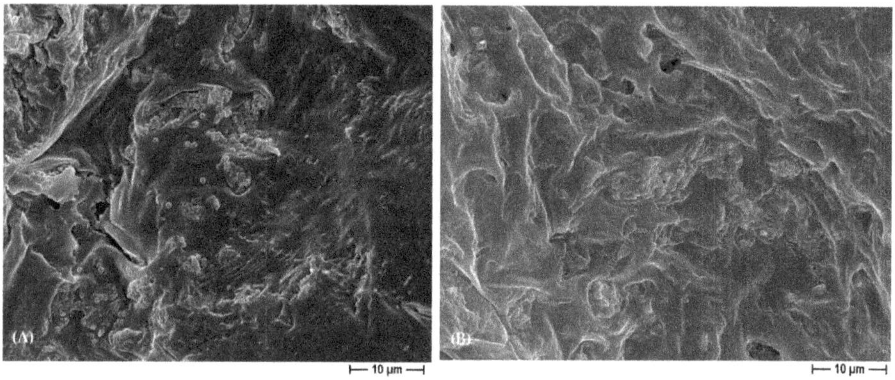

Figure 10. Representative FESEM micrograph of the control biomaterial, specifically BCP/PCL, non-sonicated in the presence of the Gram-positive (**A**) or the Gram-negative (**B**) bacteria, obtained by using NaCl salt as a template, at 5000× magnification, revealing a well-structured biofilm.

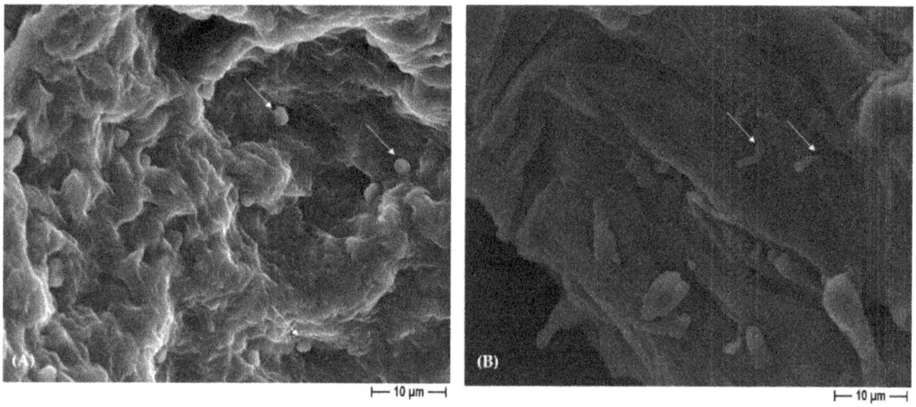

Figure 11. Representative FESEM micrographs reporting bacterial morphological modifications: *S. epidermidis* (**A**) has an oval-shaped morphology (white arrows) and *E. coli* (**B**) has an elongated morphology (white arrows) on PCL-based scaffolds added with 1% silver, obtained by using NaCl salt as a pore former (2000× magnification).

4. Conclusions

In natural conditions, during an adult's lifetime, bone regeneration and healing are well-orchestrated processes that involve different eukaryotic cells. However, the bone tissue remodeling due to different pathological events, such as bone fractures, tumors, skeletal abnormalities, and elderly diseases, requires a guiding substitute able to allow osteogenic cell proliferation and colonization. Notably, a relevant issue in the surgical implantation of a bone-tissue scaffold is the occurrence of infections that cause a shortcoming in the patient's quality of life. These infections might be exacerbated by the presence of antimicrobial drug-resistant and biofilm-producing bacteria, too. Recent research has tried to overcome these problems by adding metal ions instead of antibiotics to scaffolds designed for bone tissue engineering. In this context, the present study demonstrated that the 3D PCL-based constructs functionalized with calcium phosphates showed a highly interconnected porosity and, in parallel, revealed a good degradation behavior able to promote eukaryotic cell colonization. Moreover, the tuning of the silver content into the novel BCP/PCL-based construct of about 1% revealed sustained anti-adhesive, antibacterial, and anti-biofilm activity—towards a broad spectrum of microbial species—without affecting osteoblast viability and proliferation. Further studies with human primary osteoblasts aimed at deepening their integration and differentiation when in contact with these new scaffolds will be investigated to promote bone tissue regeneration. The use of the appropriate concentration of silver displays many advantages over conventional antibiotics, mainly the targeted release and minimizing the problem of antibiotic resistance by bacteria. In fact, silver killing activity against microorganisms is complex and involves its action on various external layers of the bacterial cell and on proteins, DNA, and other inner structures; thus, the risk of selecting resistance is low. Hence, the here designed and studied 3D scaffolds based on PCL blended with calcium phosphates and blended with silver for bone tissue engineering are an effective strategy to be applied to improve the healing of such important human tissue.

Author Contributions: Conceptualization, P.P. and V.A.; Data curation, R.S. and G.B.; Formal analysis, B.C., N.M. and E.F.; Funding acquisition, V.A.; Investigation, F.M., B.C., F.L., L.C., S.C., R.S. and E.F.; Methodology, F.M. and S.C.; Software, S.S. and N.M.; Supervision, P.P. and V.A.; Visualization, A.M.C.; Writing—original draft, G.B., P.P. and V.A.; Writing—review and editing, S.S. and A.M.C. All authors have read and agreed to the published version of the manuscript.

Funding: The research leading to some of these results has received funding Fondazione Cassa di Risparmio di Torino (RF = 2019.0588).

Institutional Review Board Statement: Not applicable.

Data Availability Statement: The source data underlying tables and figures are available from the authors upon request.

Acknowledgments: This work was presented in part at the 33rd European Congress of Clinical Microbiology and Infectious Diseases (ECCMID), Copenhagen 2023, P0685.

Conflicts of Interest: The authors declare no conflict of interest.

References

1. Jung, K.-J.; Sarkar, S.K.; Kim, W.-J.; Kim, B.-R.; Park, J.-S.; Lee, B.-T. Bone Regeneration by Multichannel Cylindrical Granular Bone Substitute for Regeneration of Bone in Cases of Tumor, Fracture, and Arthroplasty. *Int. J. Environ. Res. Public Health* **2022**, *19*, 8228. [CrossRef]
2. Safari, B.; Davaran, S.; Aghanejad, A. Osteogenic Potential of the Growth Factors and Bioactive Molecules in Bone Regeneration. *Int. J. Biol. Macromol.* **2021**, *175*, 544–557. [CrossRef] [PubMed]
3. Zhu, Y.; Goh, C.; Shrestha, A. Biomaterial Properties Modulating Bone Regeneration. *Macromol. Biosci.* **2021**, *21*, e2000365. [CrossRef] [PubMed]
4. Yudoh, K.; Sugishita, Y.; Suzuki-Takahashi, Y. Bone Development and Regeneration 2.0. *Int. J. Mol. Sci.* **2023**, *24*, 8761. [CrossRef]
5. Holešová, S.; Čech Barabaszová, K.; Hundáková, M.; Ščuková, M.; Hrabovská, K.; Joszko, K.; Antonowicz, M.; Gzik-Zroska, B. Development of Novel Thin Polycaprolactone (PCL)/Clay Nanocomposite Films with Antimicrobial Activity Promoted by the Study of Mechanical, Thermal, and Surface Properties. *Polymers* **2021**, *13*, 3193. [CrossRef]
6. Hajduga, M.B.; Bobinski, R.; Dutka, M.; Bujok, J.; Cwiertnia, M.; Pajak, C.; Kurowska, A.; Rajzer, I. The Influence of Graphene Content on the Antibacterial Properties of Polycaprolactone. *Int. J. Mol. Sci.* **2022**, *23*, 10899. [CrossRef] [PubMed]
7. Muwaffak, Z.; Goyanes, A.; Clark, V.; Basit, A.W.; Hilton, S.T.; Gaisford, S. Patient-Specific 3D Scanned and 3D Printed Antimicrobial Polycaprolactone Wound Dressings. *Int. J. Pharm.* **2017**, *527*, 161–170. [CrossRef]
8. Allizond, V.; Banche, G.; Salvoni, M.; Malandrino, M.; Cecone, C.; Cuffini, A.M.; Bracco, P. Facile One-Step Electrospinning Process to Prepare AgNPs-Loaded PLA and PLA/PEO Mats with Antibacterial Activity. *Polymers* **2023**, *15*, 1470. [CrossRef] [PubMed]
9. Bou-Francis, A.; Piercey, M.; Al-Qatami, O.; Mazzanti, G.; Khattab, R.; Ghanem, A. Polycaprolactone Blends for Fracture Fixation in Low Load-Bearing Applications. *J. Appl. Polym. Sci.* **2020**, *137*, 48940. [CrossRef]
10. Hou, Z.; Zhou, W.; Guo, X.; Zhong, R.; Wang, A.; Li, J.; Cen, Y.; You, C.; Tan, H.; Tian, M. Poly(ϵ-Caprolactone)-Methoxypolyethylene Glycol (PCL-MPEG)-Based Micelles for Drug-Delivery: The Effect of PCL Chain Length on Blood Components, Phagocytosis, and Biodistribution. *Int. J. Nanomed.* **2022**, *17*, 1613–1632. [CrossRef]

11. Comini, S.; Sparti, R.; Coppola, B.; Mohammadi, M.; Scutera, S.; Menotti, F.; Banche, G.; Cuffini, A.M.; Palmero, P.; Allizond, V. Novel Silver-Functionalized Poly(ε-Caprolactone)/Biphasic Calcium Phosphate Scaffolds Designed to Counteract Post-Surgical Infections in Orthopedic Applications. *Int. J. Mol. Sci.* **2021**, *22*, 10176. [CrossRef]
12. Comini, S.; Scutera, S.; Sparti, R.; Banche, G.; Coppola, B.; Bertea, C.M.; Bianco, G.; Gatti, N.; Cuffini, A.M.; Palmero, P.; et al. Combination of Poly(ε-Caprolactone) Biomaterials and Essential Oils to Achieve Anti-Bacterial and Osteo-Proliferative Properties for 3D-Scaffolds in Regenerative Medicine. *Pharmaceutics* **2022**, *14*, 1873. [CrossRef]
13. Timin, A.S.; Muslimov, A.R.; Zyuzin, M.V.; Peltek, O.O.; Karpov, T.E.; Sergeev, I.S.; Dotsenko, A.I.; Goncharenko, A.A.; Yolshin, N.D.; Sinelnik, A.; et al. Multifunctional Scaffolds with Improved Antimicrobial Properties and Osteogenicity Based on Piezoelectric Electrospun Fibers Decorated with Bioactive Composite Microcapsules. *ACS Appl. Mater. Interfaces* **2018**, *10*, 34849–34868. [CrossRef]
14. Florea, D.A.; Grumezescu, V.; Bîrcă, A.C.; Vasile, B.Ș.; Mușat, M.; Chircov, C.; Stan, M.S.; Grumezescu, A.M.; Andronescu, E.; Chifiriuc, M.C. Design, Characterization, and Antibacterial Performance of MAPLE-Deposited Coatings of Magnesium Phosphate-Containing Silver Nanoparticles in Biocompatible Concentrations. *Int. J. Mol. Sci.* **2022**, *23*, 7910. [CrossRef] [PubMed]
15. Afewerki, S.; Bassous, N.; Harb, S.; Palo-Nieto, C.; Ruiz-Esparza, G.U.; Marciano, F.R.; Webster, T.J.; Furtado, A.S.A.; Lobo, A.O. Advances in Dual Functional Antimicrobial and Osteoinductive Biomaterials for Orthopaedic Applications. *Nanomed. Nanotechnol. Biol. Med.* **2020**, *24*, 102143. [CrossRef] [PubMed]
16. Qian, Y.; Zhou, X.; Zhang, F.; Diekwisch, T.G.H.; Luan, X.; Yang, J. Triple PLGA/PCL Scaffold Modification Including Silver Impregnation, Collagen Coating, and Electrospinning Significantly Improve Biocompatibility, Antimicrobial, and Osteogenic Properties for Orofacial Tissue Regeneration. *ACS Appl. Mater. Interfaces* **2019**, *11*, 37381–37396. [CrossRef] [PubMed]
17. Dias, D.; Vale, A.C.; Cunha, E.P.F.; C Paiva, M.; Reis, R.L.; Vaquette, C.; Alves, N.M. 3D-Printed Cryomilled Poly(ε-Caprolactone)/Graphene Composite Scaffolds for Bone Tissue Regeneration. *J. Biomed. Mater. Res. B Appl. Biomater.* **2021**, *109*, 961–972. [CrossRef]
18. Tullio, V.; Mandras, N.; Scalas, D.; Allizond, V.; Banche, G.; Roana, J.; Greco, D.; Castagno, F.; Cuffini, A.M.; Carlone, N.A. Synergy of Caspofungin with Human Polymorphonuclear Granulocytes for Killing Candida Albicans. *Antimicrob. Agents Chemother.* **2010**, *54*, 3964–3966. [CrossRef] [PubMed]
19. Toth, J.M.; Hirthe, W.M.; Hubbard, W.G.; Brantley, W.A.; Lynch, K.L. Determination of the Ratio of HA/TCP Mixtures by x-Ray Diffraction. *J. Appl. Biomater.* **1991**, *2*, 37–40. [CrossRef]
20. Robu, A.; Antoniac, A.; Grosu, E.; Vasile, E.; Raiciu, A.D.; Iordache, F.; Antoniac, V.I.; Rau, J.V.; Yankova, V.G.; Ditu, L.M.; et al. Additives Imparting Antimicrobial Properties to Acrylic Bone Cements. *Materials* **2021**, *14*, 7031. [CrossRef] [PubMed]
21. Lee, J.T.Y.; Leng, Y.; Chow, K.L.; Ren, F.; Ge, X.; Wang, K.; Lu, X. Cell Culture Medium as an Alternative to Conventional Simulated Body Fluid. *Acta Biomater.* **2011**, *7*, 2615–2622. [CrossRef]
22. Petit, C.; Tulliani, J.-M.; Tadier, S.; Meille, S.; Chevalier, J.; Palmero, P. Novel Calcium Phosphate/PCL Graded Samples: Design and Development in View of Biomedical Applications. *Mater. Sci. Eng. C Mater. Biol. Appl.* **2019**, *97*, 336–346. [CrossRef]
23. Cazzola, M.; Ferraris, S.; Allizond, V.; Bertea, C.M.; Novara, C.; Cochis, A.; Geobaldo, F.; Bistolfi, A.; Cuffini, A.M.; Rimondini, L.; et al. Grafting of the Peppermint Essential Oil to a Chemically Treated Ti6Al4V Alloy to Counteract the Bacterial Adhesion. *Surf. Coat. Technol.* **2019**, *378*, 125011. [CrossRef]
24. Allizond, V.; Comini, S.; Cuffini, A.M.; Banche, G. Current Knowledge on Biomaterials for Orthopedic Applications Modified to Reduce Bacterial Adhesive Ability. *Antibiotics* **2022**, *11*, 529. [CrossRef] [PubMed]
25. Rodríguez-Contreras, A.; Torres, D.; Piñera-Avellaneda, D.; Pérez-Palou, L.; Ortiz-Hernández, M.; Ginebra, M.P.; Calero, J.A.; Manero, J.M.; Rupérez, E. Dual-Action Effect of Gallium and Silver Providing Osseointegration and Antibacterial Properties to Calcium Titanate Coatings on Porous Titanium Implants. *Int. J. Mol. Sci.* **2023**, *24*, 8762. [CrossRef] [PubMed]
26. Bharadwaz, A.; Jayasuriya, A.C. Recent Trends in the Application of Widely Used Natural and Synthetic Polymer Nanocomposites in Bone Tissue Regeneration. *Mater. Sci. Eng. C Mater. Biol. Appl.* **2020**, *110*, 110698. [CrossRef]
27. Paterson, T.E.; Shi, R.; Tian, J.; Harrison, C.J.; De Sousa Mendes, M.; Hatton, P.V.; Li, Z.; Ortega, I. Electrospun Scaffolds Containing Silver-Doped Hydroxyapatite with Antimicrobial Properties for Applications in Orthopedic and Dental Bone Surgery. *J. Funct. Biomater.* **2020**, *11*, 58. [CrossRef] [PubMed]
28. Longo, R.; Catauro, M.; Sorrentino, A.; Guadagno, L. Thermal and Mechanical Characterization of Complex Electrospun Systems Based on Polycaprolactone and Gelatin. *J. Therm. Anal. Calorim.* **2022**, *147*, 5391–5399. [CrossRef]
29. Fadaie, M.; Mirzaei, E. Nanofibrillated Chitosan/Polycaprolactone Bionanocomposite Scaffold with Improved Tensile Strength and Cellular Behavior. *Nanomed. J.* **2018**, *5*, 77–89. [CrossRef]
30. Armentano, I.; Dottori, M.; Fortunati, E.; Mattioli, S.; Kenny, J.M. Biodegradable Polymer Matrix Nanocomposites for Tissue Engineering: A Review. *Polym. Degrad. Stab.* **2010**, *95*, 2126–2146. [CrossRef]
31. Hutmacher, D.W. Scaffolds in Tissue Engineering Bone and Cartilage. *Biomaterials* **2000**, *21*, 2529–2543. [CrossRef] [PubMed]
32. Lu, L.; Zhang, Q.; Wootton, D.; Chiou, R.; Li, D.; Lu, B.; Lelkes, P.; Zhou, J. Biocompatibility and Biodegradation Studies of PCL/β-TCP Bone Tissue Scaffold Fabricated by Structural Porogen Method. *J. Mater. Sci. Mater. Med.* **2012**, *23*, 2217–2226. [CrossRef]
33. Hedayati, S.K.; Behravesh, A.H.; Hasannia, S.; Kordi, O.; Pourghaumi, M.; Saed, A.B.; Gashtasbi, F. Additive Manufacture of PCL/NHA Scaffolds Reinforced with Biodegradable Continuous Fibers: Mechanical Properties, in-Vitro Degradation Profile, and Cell Study. *Eur. Polym. J.* **2022**, *162*, 110876. [CrossRef]

34. Muhammad, K.B.; Abas, W.A.B.W.; Kim, K.H.; Pingguan-Murphy, B.; Zain, N.M.; Akram, H. In Vitro Comparative Study of White and Dark Polycaprolactone Trifumarate in Situ Cross-Linkable Scaffolds Seeded with Rat Bone Marrow Stromal Cells. *Clinics* **2012**, *67*, 629–637. [CrossRef] [PubMed]
35. Janarthanan, G.; Kim, I.G.; Chung, E.-J.; Noh, I. Comparative Studies on Thin Polycaprolactone-Tricalcium Phosphate Composite Scaffolds and Its Interaction with Mesenchymal Stem Cells. *Biomater. Res.* **2019**, *23*, 1. [CrossRef] [PubMed]
36. Musciacchio, L.; Mardirossian, M.; Guagnini, B.; Raffini, A.; Rizzo, M.; Trombetta, C.; Liguori, G.; Turco, G.; Porrelli, D. Rifampicin-Loaded Electrospun Polycaprolactone Membranes: Characterization of Stability, Antibacterial Effects and Urotheliocytes Proliferation. *Mater. Des.* **2022**, *224*, 111286. [CrossRef]
37. Czekanska, E.M.; Stoddart, M.J.; Ralphs, J.R.; Richards, R.G.; Hayes, J.S. A Phenotypic Comparison of Osteoblast Cell Lines versus Human Primary Osteoblasts for Biomaterials Testing. *J. Biomed. Mater. Res. Part A* **2014**, *102*, 2636–2643. [CrossRef]
38. Dvorakova, J.; Wiesnerova, L.; Chocholata, P.; Kulda, V.; Landsmann, L.; Cedikova, M.; Kripnerova, M.; Eberlova, L.; Babuska, V. Human Cells with Osteogenic Potential in Bone Tissue Research. *BioMed. Eng. OnLine* **2023**, *22*, 33. [CrossRef] [PubMed]
39. Wu, Y.; Fuh, J.; Ozbolat, I.T. Chapter 4—Bioprinting of Bone. In *3D Bioprinting in Tissue and Organ Regeneration*; Wu, Y., Fuh, J., Ozbolat, I.T., Eds.; Academic Press: Cambridge, MA, USA, 2023; pp. 95–118. ISBN 978-0-12-824291-9.
40. Ashe, S.; Nayak, D.; Kumari, M.; Nayak, B. Ameliorating Effects of Green Synthesized Silver Nanoparticles on Glycated End Product Induced Reactive Oxygen Species Production and Cellular Toxicity in Osteogenic Saos-2 Cells. *ACS Appl. Mater. Interfaces* **2016**, *8*, 30005–30016. [CrossRef]
41. Ashe, S.; Behera, S.; Dash, P.; Nayak, D.; Nayak, B. Gelatin Carrageenan Sericin Hydrogel Composites Improves Cell Viability of Cryopreserved SaOS-2 Cells. *Int. J. Biol. Macromol.* **2020**, *154*, 606–620. [CrossRef]
42. Sokołowska, P.; Siatkowska, M.; Białkowska, K.; Rosowski, M.; Komorowski, P.; Walkowiak, B. Osteosarcoma Cells in Early and Late Stages as Cancer in Vitro Progression Model for Assessing the Responsiveness of Cells to Silver Nanoparticles. *J. Biomed. Mater. Res. Part B Appl. Biomater.* **2022**, *110*, 1319–1334. [CrossRef]
43. Michalakis, K.; Bakopoulou, A.; Papachristou, E.; Vasilaki, D.; Tsouknidas, A.; Michailidis, N.; Johnstone, E. Evaluation of the Response of HOS and Saos-2 Osteosarcoma Cell Lines When Exposed to Different Sizes and Concentrations of Silver Nanoparticles. *BioMed Res. Int.* **2021**, *2021*, e5013065. [CrossRef]
44. De Matteis, V.; Malvindi, M.A.; Galeone, A.; Brunetti, V.; De Luca, E.; Kote, S.; Kshirsagar, P.; Sabella, S.; Bardi, G.; Pompa, P.P. Negligible Particle-Specific Toxicity Mechanism of Silver Nanoparticles: The Role of Ag+ Ion Release in the Cytosol. *Nanomed. Nanotechnol. Biol. Med.* **2015**, *11*, 731–739. [CrossRef] [PubMed]
45. de Andrade Pinto, S.A.; de Nadai Dias, F.J.; Brasil Camargo Cardoso, G.; Dos Santos Junior, A.R.; de Aro, A.A.; Pino, D.S.; Meneghetti, D.H.; Vitti, R.P.; Dos Santos, G.M.T.; de Carvalho Zavaglia, C.A. Polycaprolactone/Beta-Tricalcium Phosphate Scaffolds Obtained via Rotary Jet-Spinning: In Vitro and in Vivo Evaluation. *Cells Tissues Organs* **2021**, *211*, 477–491. [CrossRef] [PubMed]
46. Lama, S.; Luce, A.; Bitti, G.; Chacon-Millan, P.; Itro, A.; Ferranti, P.; D'Auria, G.; Cammarota, M.; Nicoletti, G.F.; Ferraro, G.A.; et al. Polydatin Incorporated in Polycaprolactone Nanofibers Improves Osteogenic Differentiation. *Pharmaceuticals* **2022**, *15*, 727. [CrossRef]
47. Siljander, M.P.; Sobh, A.H.; Baker, K.C.; Baker, E.A.; Kaplan, L.M. Multidrug-Resistant Organisms in the Setting of Periprosthetic Joint Infection—Diagnosis, Prevention, and Treatment. *J. Arthroplast.* **2018**, *33*, 185–194. [CrossRef] [PubMed]
48. Depypere, M.; Morgenstern, M.; Kuehl, R.; Senneville, E.; Moriarty, T.F.; Obremskey, W.T.; Zimmerli, W.; Trampuz, A.; Lagrou, K.; Metsemakers, W. J. Pathogenesis and Management of Fracture-Related Infection. *Clin. Microbiol. Infect.* **2020**, *26*, 572–578. [CrossRef] [PubMed]
49. Koepf, U.S.; Scheidt, S.; Hischebeth, G.T.R.; Strassburg, C.P.; Wirtz, D.C.; Randau, T.M.; Lutz, P. Increased Rate of Enteric Bacteria as Cause of Periprosthetic Joint Infections in Patients with Liver Cirrhosis. *BMC Infect. Dis.* **2022**, *22*, 389. [CrossRef]
50. Rudelli, B.A.; Giglio, P.N.; de Carvalho, V.C.; Pécora, J.R.; Gurgel, H.M.C.; Gobbi, R.G.; Vicente, J.R.N.; Lima, A.L.L.M.; Helito, C.P. Bacteria Drug Resistance Profile Affects Knee and Hip Periprosthetic Joint Infection Outcome with Debridement, Antibiotics and Implant Retention. *BMC Musculoskelet. Disord.* **2020**, *21*, 574. [CrossRef]
51. Imagama, T.; Seki, K.; Seki, T.; Matsuki, Y.; Yamazaki, K.; Sakai, T. Low Frequency of Local Findings in Periprosthetic Hip Infection Caused by Low-Virulent Bacteria Compared to Periprosthetic Knee Infection. *Sci. Rep.* **2021**, *11*, 11714. [CrossRef]
52. Alipour, M.; Pouya, B.; Aghazadeh, Z.; SamadiKafil, H.; Ghorbani, M.; Alizadeh, S.; Aghazadeh, M.; Dalir Abdolahinia, E. The Antimicrobial, Antioxidative, and Anti-Inflammatory Effects of Polycaprolactone/Gelatin Scaffolds Containing Chrysin for Regenerative Endodontic Purposes. *Stem Cells Int.* **2021**, *2021*, e3828777. [CrossRef] [PubMed]
53. Qian, Y.; Zhang, Z.; Zheng, L.; Song, R.; Zhao, Y. Fabrication and Characterization of Electrospun Polycaprolactone Blended with Chitosan-Gelatin Complex Nanofibrous Mats. *J. Nanomater.* **2014**, *2014*, e964621. [CrossRef]
54. Doherty, C.; Byrne, C.V.; Baqader, S.; El-Chami, C.; McBain, A.J.; Thomason, H.A. Anti-Biofilm Effects and Healing Promotion by Silver Oxynitrate-Based Dressings. *Sci. Rep.* **2023**, *13*, 2014. [CrossRef] [PubMed]

polymers

MDPI

Article

Design of Potent and Salt-Insensitive Antimicrobial Branched Peptides

Janet To, Xiaohong Zhang and James P. Tam *

Synzymes and Natural Products Center (SYNC), School of Biological Sciences, Nanyang Technological University, 60 Nanyang Drive, Singapore 637551, Singapore
* Correspondence: jptam@ntu.edu.sg

Abstract: Dendrimeric and branched peptides are polypeptides formed by diverse types of scaffolds to give them different forms. Previously, we reported a cascade-type, Lys-scaffolded antimicrobial peptide dendrimer D4R tethered with four RLYR tetrapeptides. Antimicrobial D4R is broad-spectrum, salt insensitive, and as potent as the natural-occurring tachyplesins, displaying minimum inhibitory concentrations (MIC) < 1 μM. However, the relationships between scaffolds and antimicrobial potency remain undefined. Here, we report the design of four novel types of peptide antimicrobials whose scaffolded backbones are lysine (Lys), iso-Lys, ornithine (Orn), or iso-Orn tethered with RLYR on their α- or sidechain-amines to give ε-, δ-, and their α-branched peptides. When assayed against ten microorganisms, the Lys-scaffolded α- and ε-branched peptides are broadly active, salt insensitive, and as potent as D4R and tachyplesins, whereas the corresponding Orn-scaffolded α- and δ-branched peptides are salt sensitive and much less potent, displaying MICs ranging from 1 to >500 μM. Structure-activity relationship studies suggested that Lys-scaffolds, but not Orn-scaffolds, can support a reverse turn to organize RLYR tetrapeptides as parallel β-strands to form an amphipathic structure with Leu-Tyr as a hydrophobic core. Together, these results provide a structural approach for designing potent and salt-insensitive dendrimeric or branched peptide antimicrobials.

Keywords: peptide dendrimers; antimicrobial peptides; polymeric scaffolds; branched peptides; isolysine scaffold; lysine reverse turns

1. Introduction

Antimicrobial peptides represent an important innate host defense that often targets microbial membranes, leading to killing mechanisms by increasing membrane permeability, promoting pore formation, or disrupting membrane integrity [1–3]. Such membrane-directed antimicrobial killing mechanisms have received increasing attention because of the emerging crisis of antibiotic tolerance and antimicrobial resistance [4]. Indeed, the World Health Organization pronounced that antimicrobial resistance is one of the biggest public health challenges of our time, as thousands of people die from antibiotic-resistant infections each year. Increasing public-health concerns have prompted the need to find effective and affordable solutions to produce peptide antimicrobials.

Antimicrobial peptides are diverse in sequences, forms, and structures [5,6]. In plants, insects, and animals, antimicrobial peptides contain proteinogenic amino acids [1,2,7–9]. In contrast, antimicrobial peptides found in bacteria and fungi often contain unusual and nonproteinogenic amino acids, polyamines, and heterocycles in their backbones with cyclic, branched, or other unusual architectures [5,10–13] that are ribosomally synthesized and post-translationally modified (RiPPs). In addition, polymeric antimicrobials containing γ-, δ-, or ε-peptide backbones are also known [14–19]. The antimicrobial ε-isolysines produced by *Streptomyces* and containing 25 to 35 lysine as ε-peptides have been widely used as food preservatives [19–21].

Peptide antimicrobials, whether they are dominantly helical or β-strand, share the propensity to be amphipathic by clustering hydrophobic and charge patches that are important for membrane-active functions [22,23]. Indeed, this structural feature has been successfully used to design synthetic cationic β- and γ-peptides, adopting amphipathic helical structures as antimicrobials [24,25]. Previously, we have designed amphipathic β-strand peptide antimicrobials [24–27]. They include constrained cyclic peptides [26–28] based on tachyplesins derived from horseshoe crabs, and protegrins from pig intestines [29,30]. Both antimicrobials belong to cystine-stabilized β-strand peptides consisting of 17 to 19 amino acids. Importantly, they display potent and broad-spectrum antimicrobial activity, often with MICs < 1 μM [29,30]. Interestingly, their antiparallel β-strands can be simplified to a BHHB tetrapeptide motif (B = basic, H = hydrophobic amino acids). To mimic tachyplesins, we have reported a cascade-type antimicrobial peptide dendrimer D4R (Figure 1A) that contains four peptides with a BHHB motif: the tetrapeptide RLYR (called R4 in this paper) tethered at the carboxyl ends of a tri-Lys scaffold [31]. An advantage to producing such peptide antimicrobials is expediency because they can be prepared by controlled polymerization that simplifies their synthesis by many fold, and in turn, lowers their production cost to make them affordable [31–33].

D4R is similar in size to tachyplesins and displays similar antimicrobial profiles and potency as tachyplesins even though their forms are different. Tachyplesins and protegrins contain two disulfide bridges that stabilize two β-strands, and which are important for maintaining their antimicrobial activity. Using end-to-end cyclization to replace the disulfide constraints with cyclic tachyplesins and protegrins also retains salt insensitivity [26,27,34]. However, removing the disulfide or cyclic constraints results in a loss of antimicrobial activity under high-salt (physiological) conditions [26,27,29,30]. Thus, it remains to be determined how the unconstrained peptide dendrimer D4R is salt insensitive and maintains its potency like the disulfide-constrained tachyplesins or their cyclic analogs. These findings suggest that the antimicrobial activity of peptide dendrimer D4R may have a structural basis. Furthermore, the salt sensitivity of antimicrobial activity provides a functional correlation in assessing the potency of antimicrobial peptide dendrimers.

Herein, we report the design and structure-activity relationships of four types of linear branched peptide antimicrobials using Orn-, iso-Orn, Lys-, or iso-Lys scaffolds to tether R4 peptides and assess their antimicrobial potency under high-salt conditions (Figure 1B–E). With molecular weights of around 2 kDa, these peptides are dendrimeric polymers of the tetrapeptide R4, which can self-assemble to form high-molecular weight polymers. For comparison, we also prepared the cascade dendrimer D4R. Our results showed that Orn- and Lys-scaffold branched peptides display distinctively different antimicrobial profiles and salt sensitivity. The Lys-based scaffold of iso-Lys ε-branched peptides and their corresponding isomeric α-branched peptides are as broadly active as D4R and tachyplesin-1 when tested against ten organisms under both low- and high-salt conditions. In contrast, the Orn-based scaffold of iso-Orn δ-branched peptides and their corresponding isomeric α-branched peptides are less active and unable to retain their antimicrobial activity against certain bacteria under high-salt conditions. These results provide a structural basis for designing Lys-based scaffolds that could organize parallel R4 peptides into β-structures to affect the antimicrobial activity of dendrimeric and branched peptides under high-salt conditions.

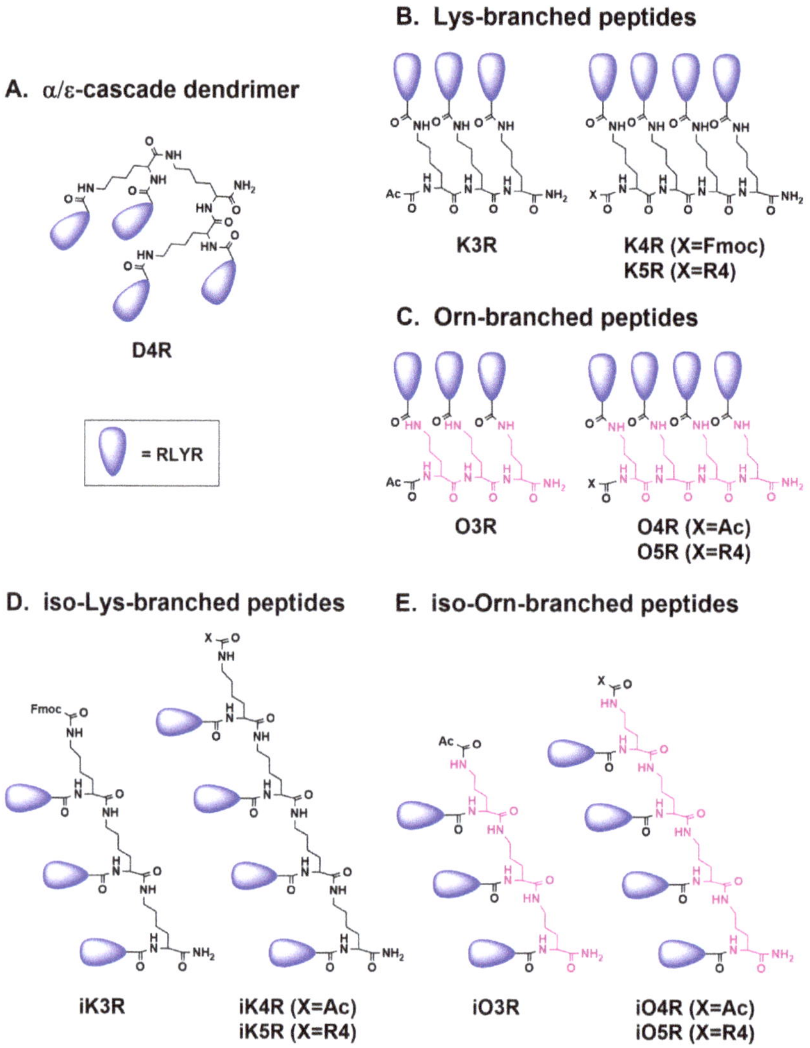

Figure 1. Schematic representation of Lys- and Orn-based scaffolds. (**A**) The Lys-scaffolded α/ε-cascade peptide dendrimer D4R contains a cascade branched tri-Lys scaffold Lys_2Lys (α/ε-peptides) tethered with 4 strands of RLYR (R4) tetrapeptides, whereas the α-dendrimers (**B**) KnR and (**C**) OnR contain the unbranched, linear tri-, or tetrapeptides of Lys or Orn as the α-peptide scaffolds, tethered with n = 3, 4, or 5 strands of RLYR tetrapeptides. (**D**) The ε-branched peptides (iso-Lys) scaffolds include ε-tripeptide in iK3R and ε-tetrapeptides in iK4R and iK5R. Similarly, (**E**) the δ-branched peptides (iso-Orn) scaffolds include δ-tripeptide in iO3R and δ-tetrapeptides in iO4R and iO5R. The unused amino groups of these di-amino scaffolds were employed for tethering the parallel peptide strands. In some cases, the R4 peptides were replaced by a hydrophobic Fmoc protecting group, which is a tricyclic aromatic moiety, or an acetyl protecting group representing a hydrophilic moiety.

2. Materials and Methods

Materials. Solvents, all HPLC grade, Fmoc amino acid derivatives, N-hydroxybenzotriazole (HOBt), N,N-diisopropylethylamine (DIEA), trifluoroacetic acid (TFA), and Fmoc-DPA and Fmoc-MBHA resin (Rink Amide MBHA resin) were obtained from Sigma Aldrich, USA.

Ten organisms obtained from the American Type Culture Collection (ATCC; Rockville, MD) were used for antimicrobial assays. Four Gram-negative bacteria included *Escherichia coli* ATCC 25922, *Pseudomonas aeruginosa* ATCC 27853, *Klebsiella oxytoca* ATCC 49131, and *Proteus vulgaris* ATCC 49132. The three Gram-positive bacteria were *Staphylococcus aureus* 29213, *Micrococcus luteus* ATCC 49,732, and *Enterococcus faecalis* ATCC 29212. The three fungi were *Candida albicans* ATCC 37092, *Candida kefyr* ATCC 37095, and *Candida tropicalis* ATCC 37097. The clinical isolates were obtained from Singapore General Hospital (SGH). They are *Enterococcus faecium, Staphylococcus aureus, Klebsiella pneumoniae, Acinetobacter baumannii, Pseudomonas aeruginosa,* and *Enterobacter* species. The strains were incubated in trypticase soy broth (TSB), which was prepared in double distilled water and autoclaved for sterilization. TSB was purchased from Becton-Dickinson (Cockeysville, MD, USA).

Peptide synthesis and purification. All peptides were assembled by solid-phase peptide synthesis using Boc- or Fmoc- chemistry [35,36]. Purification and identification were performed by reverse phase-high performance liquid chromatography (HPLC) and MS. Analytical HPLC was conducted on a Shimadzu LG-6A system with a C_{18} Vydac column (4.6 × 250 mm). A linear gradient of 10–90% buffer B ran for 30 min at 1 mL/min with detection at 225 nm. Eluent A: contained 0.04% TFA/H_2O; B: 0.04% TFA/60% CH_3CN/H_2O. Preparative RP-HPLC was performed on a Waters 600 system with a C_{18} Vydac column (20 × 250 mm). Matrix-assisted laser desorption ionization mass spectrometry (MALDI-MS) was measured on a 4800 MALDI TOF/TOF Analyzer (Applied Biosystems, Waltham, MA, USA) instrument. Samples were dissolved in 1 mL of a 1:2 mixture of H_2O-CH_3CN. Measurements were made in a linear model, with a-cyano-4-hydroxycinnamic acid (CHCA) as the matrix.

Preparation of scaffolds. (a) α/ε-dendrimeric peptide scaffold. Syntheses of the dendrimer peptide D4R scaffold required two coupling cycles using a four-molar excess of Fmoc-Lys (Fmoc) and BOP/DIEA as a coupling reagent in DMF on the Rink Amide MBHA-resin. The Fmoc group was removed using 20% piperidine/DMF. Each coupling cycle doubled the branching level of the lysyl scaffold to afford the tetrabranching Fmoc-Lys_2-Lys-MBHA-resin after two cycles. *(b) α-, δ-, and ε-branched peptide scaffolds.* All branched peptide scaffolds were prepared on a TFA-stable diphenylamine (DPA) resin using 3 to 5 equivalents of Fmoc-Lys(Boc) in DMF and BOP/DIEA as the coupling reagent. After removing the Fmoc protecting group, the coupling cycle was repeated two or three times to generate Lys_3 or Lys_4 resin supports, respectively, with the Fmoc-protecting group at the N-terminal and the Boc-protecting group at the α-, δ-, or ε-amine. Similarly, Orn_3- and Orn_4-resin supports were obtained from Fmoc-Orn(Boc), and δ- and ε-peptide resin supports from Boc-Orn(Fmoc) and Boc-Lys(Fmoc) were used as starting materials.

Dendrimer Syntheses. (a) α/ε-dendrimer. Solid-phase peptide synthesis was performed manually using Fmoc-tButyl chemistry and a single coupling protocol with BOP/DIEA in DMF. After peptide chain elongation, the resin was treated with TFA with a scavenge cocktail (TFA:Phenol:Water:EDT:Tris, 86.5:5:5:2.5:1, *v*/*v*) for 4 h. Crude peptides were precipitated in ether and purified by preparative C_{18} reverse-phase HPLC. *(b) α-, δ-, and ε-dendrimers.* R4 peptides were assembled manually by solid-phase synthesis using α-, δ-, or ε-peptide-scaffold resins through conventional Boc chemistry and DCC-HOBt coupling protocol. After completion of the peptide chain assembly, the resin was treated with a high-HF procedure. After the removal of HF, the residual solid was washed with ether. The crude peptides were dissolved in TFA, precipitated in dried ether, and purified by preparative C_{18} reverse-phase HPLC.

Antimicrobial Assays. A sensitive and reproducible two-stage radial diffusion antimicrobial assay of Lehrer et al. [37] was employed. A 1–4 × 10^6 colony-forming unit/mL of the test organism was mixed with 10 mL of molten underlay gel solution and poured into Petri dishes to form a uniform layer. The gel solution consisted of 10 mM sodium phosphate buffer, 0.03% TSB, and 0.02% Tween-20, with the addition of 100 mM NaCl in the case of high-salt conditions. Upon gel solidification, gel wells were made by a 3-mm diameter template in an evenly spaced array. A 5 µL aliquot of a serial half-log dilution of test peptides at seven concentrations was added to each well after removal of gel plugs. The dishes were incubated at 37 °C for 3 h to allow peptide diffusion into the underlay gels. Gels were overlaid with 10 mL of 1% agarose in 6% TSB (wt/vol). After additional incubation at 37 °C for 16–24 h, the diameter of the clear zone surrounding the wells (colony-free) was measured under a microscope. Antimicrobial activities were expressed in units (0.1 mm = 1 unit), and the MICs (in μM) were determined from the x-intercepts of the dose-response curves. Experiments were performed in triplicates.

3. Results

3.1. Hypothesis and Design

Apart from their disulfide bond constraints, a structural feature found in tachyplesins and protegrins is the reverse turn. In tachyplesin-1, a type-II turn stabilizes the antiparallel β-strands as a β-hairpin [38]. The ability of peptide scaffolds in peptide dendrimers to organize short parallel peptide strands such as R4 to form β-hairpin structures appeared to be worthy of consideration. Gellman and his coworkers [39,40] found that parallel peptides linked by a diamine moiety, D-Pro-DADME (1,2-diamino-1,1-dimethylethane), can fold into stable β-sheet hairpins in water (Figure 2). The Lys linker which forms an ε-peptide in the Lys-scaffold of the cascade branched dendrimer D4R has similar spacing as the D-Pro-DADME linker and could serve as its mimetic. In contrast, Nowick and Brower [41] found that ornithine (Orn) linking anti-parallel peptides can form stable β-hairpin structures through an Orn reverse turn to afford a hairpin structure (Figure 2). However, substituting a Lys-linker for an Orn-linker in the same sequence fails to form a hairpin. Thus, we hypothesized that Orn-scaffold branched peptides could not support reverse turns, but Lys-scaffold branched peptides could form a putative "Lys-turn" to stabilize a β-hairpin structure (Figure 2).

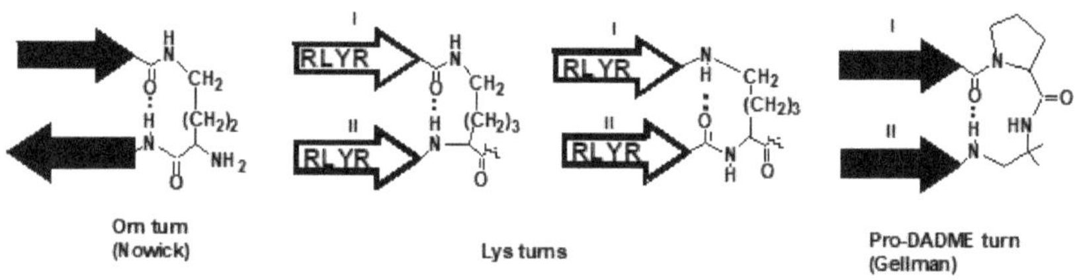

Figure 2. Orn and proposed Lys turns. In an Orn turn, Orn serves as a dipeptide mimetic linking antiparallel strands with an $i, i+3$ 10-member hydrogen bond. In the putative Lys turn, the connecting strands are parallel, which could contribute to two Lys turns with a 10-member hydrogen bonding. The D-Pro-DADME (1,2-diamino-1,1-dimethylethane) turn connecting two parallel peptide strands (I and II) has a similar spacing as the Lys turn.

To test our hypothesis that Lys-scaffolds in α- and ε-branched peptides could form reverse turns to confer structural advantages when compared to an Orn-scaffold in α- and δ-branched peptides, we prepared four series of branched peptides and their analogs containing either a Lys- or Orn-scaffold tethered with R4 peptides (RLYR). The proposed four series of branched peptides organized by either Orn(O)-, isoOrn(iO)- or Lys(K)-, or isoLys(iK)-scaffolds contain four different forms: α-, α′-, δ-, and ε-branched peptides tethered with three to five parallel strands of R4 peptides (n = 3 to 5) to give KnR, OnR, iOnR, and iKnR, respectively (Figure 1B–E). Figure 1 also shows their analogs with three to five R4 peptides for structure-activity relationship studies.

3.2. Synthesis by Controlled Polymerization

All branched peptides consisting of 15 to 24 amino acid residues were prepared by a "controlled polymerization" using a stepwise solid-phase method that is procedurally similar to conventional peptide synthesis aside from the stoichiometric adjustments of reagents and protecting group chemistries (Figure 3). In controlled polymerization, multiple peptide strands are assembled simultaneously. Thus, the number of assembling steps in their synthesis was reduced by *n* fold, where *n* is the number of peptide strands. For example, the synthesis of the 17-residue cascade dendrimer D4R with four R4 peptide strands required six assemblage cycles, which is equivalent to the synthesis of a hexapeptide. In contrast, a conventional 17-residue peptide would require 17 synthetic cycles. While the cascade dendrimer D4R was prepared entirely by Fmoc chemistry, the corresponding α-, δ-, and ε-branched peptides [2,7–10,12,13,42] were prepared by a combination of Boc and Fmoc chemistries. Their α-, δ-, and ε-peptide scaffolds were first assembled by Fmoc chemistry and then R4 peptide strands were tethered using Boc chemistry. All branched peptides were aqueous soluble. They were purified by C18 reversed-phase HPLC and their molecular weights were confirmed by mass-spectrometric analysis.

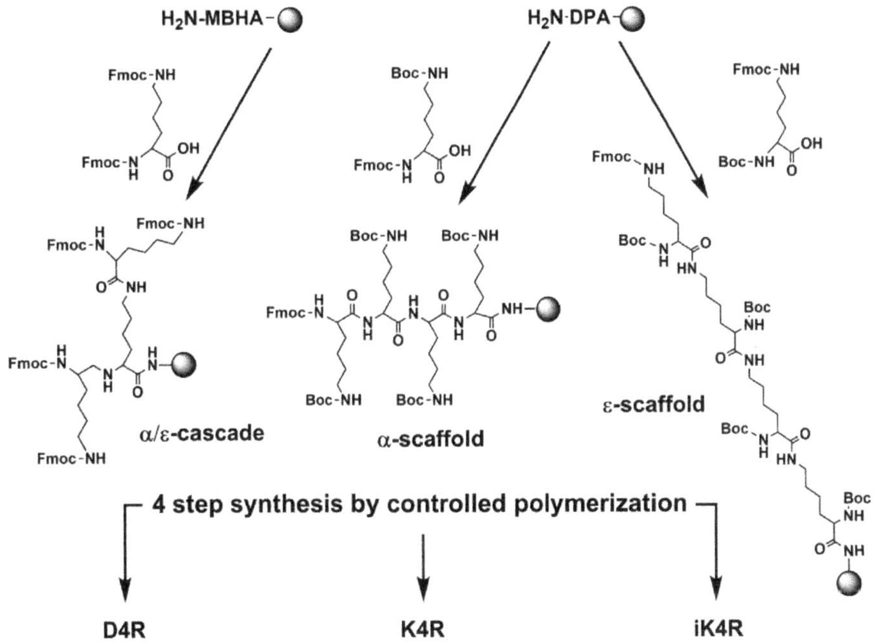

Figure 3. Controlled polymerization schemes of α/ε-cascade dendrimer D4R, and α- and ε-branched peptides K4R and iK4R. The 4-step synthesis of adding the peptide RLYR to these and other dendrimers are common steps in their synthetic schemes. The synthesis of the Orn series was similar to the Lys series, and Boc-Orn(Fmoc)-OH or Fmoc-Orn(Boc)-OH was used.

3.3. *Antimicrobial Profiles of Orn- and Lys-Scaffold Branched Peptides*

The minimum inhibitory concentrations (MICs) of four series of branched peptides and two standards, tachyplesin-1 and the dendrimer D4R, against four Gram-negative bacteria, three Gram-positive bacteria, and three fungi were determined using a two-stage radial diffusion assay in both low- (without NaCl) and high-salt (with 100 mM NaCl) conditions. Under high-salt conditions, tachyplesin-1 was broadly active, displaying MIC values of 0.4 to 1.3 μM against ten test organisms. The cascade dendrimer D4R was also broadly active and salt-insensitive (Table 1). Furthermore, its potency varied within a narrow range, displaying MICs at 0.4 to 1.0 μM under low-salt conditions and 0.6 to 1.9 μM under high-salt conditions. We also tested 31 antibiotic-resistant strains of ESKAPE bacterial pathogens (*Enterococcus faecium*, *Staphylococcus aureus*, *Klebsiella pneumoniae*, *Acinetobacter baumannii*, *Pseudomonas aeruginosa* and *Enterobacter* species). D4R displayed MICs at 0.3 to 1.6 μM under high-salt conditions (Supplementary Table S1).

Table 1. Antimicrobial activity of tachyplesin, α/ε-Lys-cascade peptide dendrimer D4R, Lys α-branched (K4R), iso-Lys ε-branched (iK4R), Orn α′-branched (O4R) and iso-Orn δ-branched (iO4R) peptides, tethered with four strands of RLYR tetrapeptides, under low (L-) and high (H-) salt conditions. Activities against multiple strains are expressed as the minimum inhibitory concentration (MIC, μM). Experiments were performed in triplicates.

	MIC (μM) *											
	Controls				**Lys-Scaffolded †**				**Orn-Scaffolded †**			
	Tachyplesin		**D4R**		**K4R**		**iK4R**		**O4R**		**iO4R**	
Organism	**L-Salt**	**H-Salt**	**L-Salt**	**H-Salt**	**L-Salt**	**H-Salt**	**L-Salt**	**H-Salt**	**L-Salt**	**H-Salt**	**L-Salt**	**H-Salt**
Gram-negative												
E. coli	0.3	0.4	0.6	0.7	0.7	0.7	0.4	0.4	0.7	**4.1**	0.9	**4.4**
P. aeruginosa	0.9	0.5	0.5	1.2	0.8	0.9	0.5	0.7	2.1	**4.8**	**4.4**	4
P. vulgaris	0.7	1	1	1.9	1.1	1	0.8	0.7	**5.6**	2.2	**4.7**	**23.2**
K. oxytoca	0.2	0.5	0.4	0.9	1.3	1.2	0.5	0.7	1.8	2.4	1.6	**6.8**
Gram-positive												
S. aureus	0.4	0.5	0.8	0.6	0.8	0.9	0.7	0.6	**7.3**	2	**6.7**	**4.8**
M. luteus	1	1.1	0.5	0.7	0.6	1	0.9	0.8	1.8	**6.9**	2.1	2
E. faecalis	0.3	0.4	0.8	1.8	0.5	1.5	0.8	1.4	**6.6**	**>500**	**5.4**	**>500**
Fungi												
C. albicans	0.7	0.9	0.8	0.8	0.7	0.8	0.8	0.9	1.4	1.6	1.3	1.4
C. kefyr	0.9	1.3	0.9	1.3	0.9	0.9	0.8	0.8	0.7	0.8	0.8	0.8
C. tropicalis	0.5	1	0.7	0.8	1.5	1.2	0.5	1.3	1.4	1.5	1.4	0.8

* When MIC is >3 μM, the number is shown in bold. † See Figure 1 for structures.

The antimicrobial profiles and potency in the two series of Lys-scaffold α- and ε-branched peptides, irrespective of the number of tethered R4 peptides which range from three to five, are similar and comparable to D4R and tachyplesin-1 (Table 1, Figure 4, and Supplementary Table S2). They exhibited a narrow range of MICs from 0.4 to 1.5 μM under low-salt conditions, and 0.4 to 1.6 μM under high-salt conditions. In contrast, the Orn-scaffold α- and δ-branched peptides exhibited a wide range of MICs from 0.5 to >500 μM (Table 1, Figure 4, and Supplementary Table S2). They were particularly salt-sensitive against certain bacteria. Against *Proteus vulgaris, Staphylococcus aureus, Micrococcus lutenus,* and *Candida albicans*, the Orn-scaffolded α-peptide dendrimer O3R displayed a 10- to 100-fold decrease in potency under high-salt conditions.

Unlike the analogous series of Lys-scaffolded α-peptide dendrimer KnR (n = 3 to 5 strands of R4), the Orn-scaffolded α-peptide dendrimers OnR (n = 3 to 5 strands of R4) improved their potency and salt insensitivity with an increasing number of tethered peptide strands, in the order of O5R > O4R >> O3R. In general, all Orn-scaffold α- or δ-branched peptides were salt sensitive, particularly against the Gram-negative bacteria *Proteus vulgaris, Escherichia coli,* and *Klebsiella oxytoca*. They completely lost activity against *Enterococcus faecalis* under high-salt conditions. Interestingly, the salt sensitivity of the Orn-scaffold branched peptides observed in bacteria was less significant in the three test fungi. Also, the increasing number of peptide strands in O3R, O4R, and O5R improved their salt sensitivity against Gram-negative bacteria *Pseudomonas aeruginoa, Proteus vulgaris,* and Gram-positive bacteria *Staphylococcus aureus*.

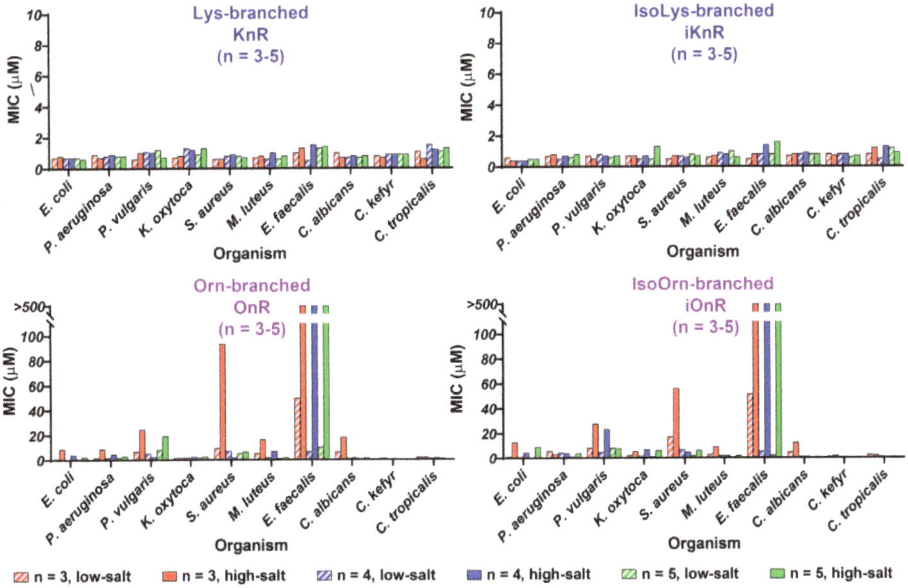

Figure 4. Comparison of antimicrobial activity of Lys-scaffolded α-branched (KnR), iso-Lys ε-branched (iKnR), Orn-scaffolded α′-branched (OnR), and iso-Orn δ-branched (iOnR) peptides, tethered with $n = 3$ (red), 4 (blue), or 5 (green) strands of RLYR tetrapeptides, under low-salt (hatched bars) and high-salt (full bars) conditions. Activities against multiple strains are expressed as the minimum inhibitory concentration (MIC, μM). Experiments were performed in triplicates.

The potencies of the δ-branched peptides improved several fold in certain bacteria when the number of R4 strands in this series increased from three in δ-branched peptide iO3R to four in iO4R. However, there was no further improvement in the δ-branched peptide iO5R with five R4 strands. Despite the increase in potency, their salt sensitivity did not improve significantly with an increase in the number of tethered R4 peptides and the corresponding number of cationic charges.

3.4. Structure-Activity Relationships

Because of the symmetry of amino acid sequences in branched or dendrimeric peptides, structure determination by NMR is difficult. Far-UV circular dichroism (CD) of D4R reveals a less intensive spectrum with a negative band at 217 nm, a crossover at 195 nm, and a positive band at about 190 nm (Figure 5), suggesting a β-structure [43,44]. This spectrum is also similar to that of poly(ε-L-Lys), which is proposed to adopt a 40% β-sheet [45]. Using the structure-promoting solvent TFE, the CD of dendrimer D4R reveals a β-turn character [43]. These observations provide partial support for their putative β-structures.

Molecular simulation of D4R with putative Lys turns suggests the importance of a hydrophobic core formed by the Leu-Tyr dipeptide of R4 tetrapeptide. To show that the BHHB motif of the R4 peptide with two consecutive bulky hydrophobic residues in the dendrimeric design is an important contributing factor that facilitates the clustering of hydrophobic and charged regions to form amphipathic structures, we prepared a cascade peptide dendrimer D4Y by replacing the BHHB motif of the R4 (RLYR) peptide with a HBBH (YRRL) motif of the Y4 peptide. Compared to D4R, D4Y was inactive under high-salt conditions against six of ten tested organisms and displayed low potency (MIC < 10 μM) in four other organisms (Figure 6).

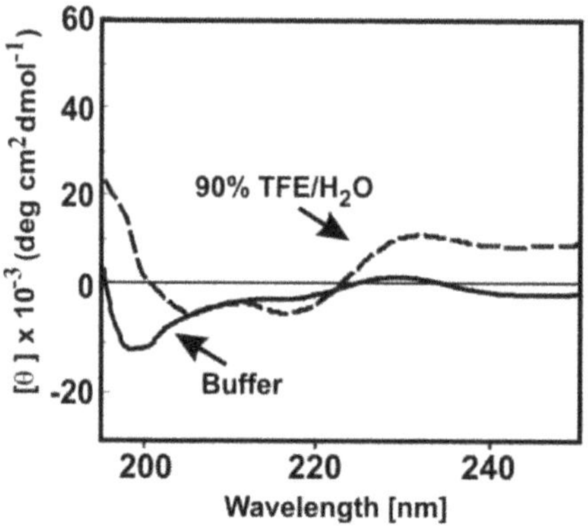

Figure 5. Far-UV circular dichroism (CD) spectra of 10 μM α/ε-Lys-cascade peptide dendrimer D4R in a phosphate buffer with 100 mM NaCl (high-salt) and in 90% TFE/H_2O.

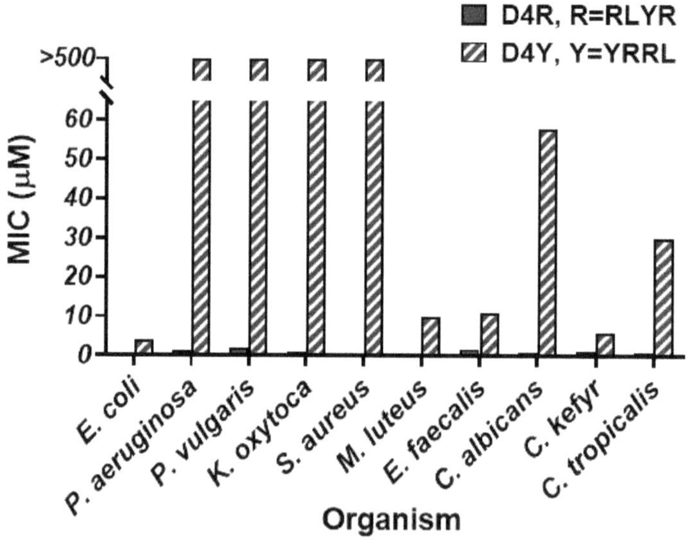

Figure 6. Comparison of antimicrobial activity of D4R (R = RLYR) with a cascade α/ε-Lys scaffold and D4Y (Y4 = YRRL) also with a cascade α/ε-Lys scaffold under high-salt conditions. Experiments were performed in radial diffusion assay with underlay gel containing 1% agarose and a 10 mM phosphate buffer with 100 mM NaCl. Activities against multiple strains are expressed as the minimum inhibitory concentration (MIC, μM). Full bars represent D4R, and hatched bars represent D4Y.

4. Discussion

In this study, we report that Lys-based scaffolds together with amphipathic structures strongly influence the antimicrobial potency and salt sensitivity of dendrimeric or branched peptides. Lys-based scaffolds tethered with three to five copies of a tetrapeptide RLYR with a BHHB motif can achieve high antimicrobial potency under high-salt physiological conditions. Such antimicrobial activity is retained when tested against the antibiotic-resistant ESKAPE bacterial pathogens, which included the vancomycin-resistant *E. faecium*, methicillin-resistant *S. aureus*, *K. pneumoniae*, *A. baumannii*, and *P. aeruginosa*, and carbapenem-resistant *Enterobacter* species, displaying MICs at about 1 μM or lower under high-salt conditions (Supplementary Table S1). In contrast, similar designs of branched peptides with Orn-based scaffolds are considerably less potent and sensitive to high-salt conditions. We showed that branched peptides are simple to prepare and could provide a cost-efficient approach to developing affordable and potent antimicrobial peptides.

Previously, we compared the antimicrobial profiles and stability of the cascade dendrimer D4R with their corresponding linear polymer $(RLYR)_4$; both are similar in size and contain the same number of cationic and hydrophobic residues. D4R and their linear poly-RLYR peptides are nontoxic to red blood cells at their effective antimicrobial IC_{50} of 0.3 to 5 µM [46]. For example, the EC_{50} for hemolysis of human red blood cells for the linear peptide $(RLYR)_4$ and D4R are 338 µM and 1510 µM, respectively. Also, consistent with the literature, branched or dendrimeric peptides are more resistant to proteolytic degradation than their linear counterparts. Finally, Figure 7 shows the membranolytic mechanism of D4R against *E. coli* and *S. aureus*, as observed using atomic force microscopy. Together, we can expect the δ- or ε-branched peptides and their respective α-branched peptides reported in this study to share similar attributes as D4R.

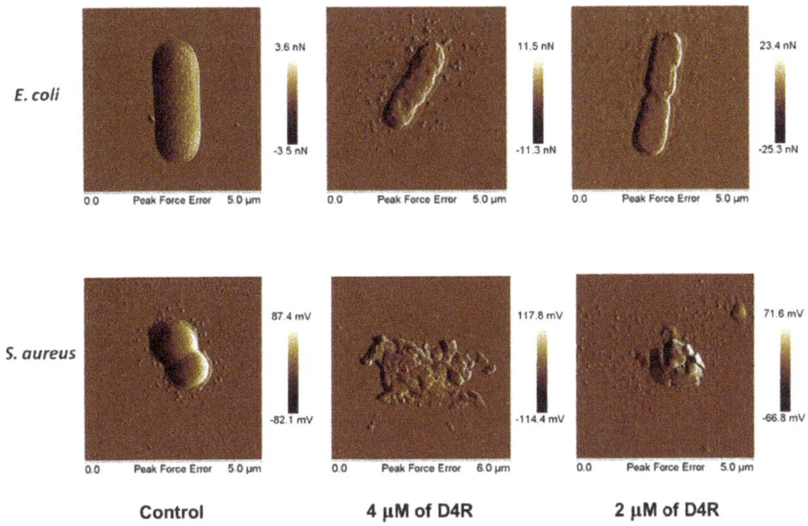

Figure 7. Atomic force microscopy images of the gram-negative bacteria *E. coli* and gram-positive bacteria *S. aureus* treated with 4 µM and 2 µM of the dendrimeric D4R, compared to the untreated controls.

From our structure-activity relationship study, we found that the antimicrobial iK3R, an ε-branched peptide with three R4 peptides, displays high potency. Indeed, the 15-residue iK3R represents one of the smallest and most potent antimicrobial peptides we have prepared so far. Increasing the number of peptide strands in the ε-branched peptide series from three to four or five, corresponding to an increase of cationic amino acids from six to eight or ten residues, respectively, did not significantly increase their potency. In contrast, all three forms of α- and δ-branched peptides formed by the Orn-scaffolds are many-fold less potent and are particularly salt-sensitive to certain bacteria, including *Escherichia coli*, *Proteus vulgaris*, *Klebsiella oxytoca*, and *Staphylococcus aureus*. Against *Enterococcus faecalis*, the δ-branched peptides are completely inactive under high-salt conditions. Unlike the ε-branched peptides with Lys-based scaffolds, the δ-branched peptides with Orn-based scaffolds are dependent on the number of R4 peptide strands. Their potency, but not their salt insensitivity, improves with the increase in the number of strands from three to five. Taken together, these results suggest that there is a rough correlation of potency, salt sensitivity, and the number of cationic amino acids with the Orn-based branched peptides, but not the Lys-based antimicrobial branched peptides. The ability of the Orn- and Lys-based scaffolds to shape the structures of these branched peptides may be a major contributor to the observed differences. The Orn scaffolds are unable to form reverse turns to organize parallel peptides as extended structures, and their antimicrobial activity and salt sensitivity improve with an increasing number of R4 strands, and correspondingly, also increase in cationic charges. In contrast, the Lys-scaffolds can form reverse turns to organize parallel peptides to β-structures, and their antimicrobial activity is largely independent of cationic interactions, hence the number of tethered R4 strands.

A key element in our design is our proposal of a putative "Lys turn" for the Lys-based scaffolds that could stabilize the structures of the ε-peptide branched peptides through a 10-member reverse turn (Figure 2). The proposed putative "Lys turn" differs from the Orn turn in two aspects: the peptide orientation, and the reverse-turn connectivity. The Orn turn is formed by anti-parallel peptide strands that are linked C-terminus-to-N-terminus to the δ-amino and the α-carboxyl groups of the Orn turn, respectively. In contrast, the Lys turn found in the dendrimers or branched peptides is formed by parallel peptide strands that are linked C-terminus-to-C-terminus to the α- and ε-amines of the Lys turn. In the Orn turn, the i, $i + 3$ hydrogen bond is contributed completed by the anti-parallel peptide strands, and the Orn, a dipeptide mimetic, occupies the $i + 1$ and $i + 2$ positions of the reverse turn. Similar hydrogen bonding for a Lys-turn involving antiparallel strands produces an 11-member ring that is not favored for a reverse turn according to the study by Nowick and Bower [41]. Because of the pseudosymmetry of the parallel strands in a hairpin turn, the i, $i + 3$ hydrogen bonding in the "Lys turn" can be contributed by CO of Arg^4 of the parallel peptide strands and the $N^{\alpha}H$ or $N^{\varepsilon}H$ amine of Lys. Thus, in parallel peptides found in dendrimers or branched peptides, the Orn turn that forms a nine-member β-hairpin turn is not favored.

Interestingly, each Lys in the α/ε and ε-scaffolds formed by Lys, either through an α- or ε-peptide linkage, can form a Lys turn. Thus, we can further argue that the trilysine scaffold of α/ε or ε-peptides may form three Lys turns through a series of β-hairpins of intramolecular hydrogen bonding network, most likely through the CO of Arg^4 to the $N^{\alpha}H$ of a Lys-scaffold. These Lys turns in the α/ε- and ε-peptide dendrimers could organize the parallel R4 peptides with a BHHB motif into β-structures such as a β-sheet or β-barrel that cluster the hydrophobic Leu-Tyr (HH residues) as the hydrophobic core and the two Arg at each end to generate bipolar amphipathic structures. Such amphipathic structures would mimic tachyplesins and protegrins, and render the Lys-scaffolded branched peptides and peptide dendrimers salt-insensitive and independent of increased numbers of cationic charges (Figure 8). Since Orn-based scaffolds are not favored to form reverse turns in the Orn-based δ-dendrimers or α- and α/δ-dendrimers, they are unable to form similar β-structures. Their inability to form a stable structure may account for the low antimicrobial activities and salt sensitivity of the δ-dendrimers.

Interestingly, β-strands favor short peptides. Evidence suggested that β-sheets in model peptides become unstable when the strands are lengthened more than seven residues, whereas α-helices become more stable as they grow longer [47]. A plausible explanation is that as the β-strand length grows, the inter-strand interactions that stabilize the β-hairpin are overtaken by the intra-strand interactions of α-helices. This behavior is consistent with analyses of protein crystal structures that show that the prevalence of β-strands decreases steadily as the strands grow longer [48,49]. For example, Eisenberg and his coworkers [50] have extensively examined β-sheets of short peptides to form β-aggregates in amyloid-forming peptides. They have found that many short β-strands of four to six amino acids in length can self-interact to form large β-sheet aggregates. The ability of the putative reverse Lys turns in our Lys-based scaffold may facilitate self-interactions of R4 peptides in the formation of β-structures. It is also consistent with our past observations of α/ε-branched cascade peptide dendrimers, such as the multiple antigen peptides [31–33], which generally exhibit α-helical or random structures because the parallel peptide strands attached to the ε-peptide scaffolds as antigens are >10 amino acid residues in length. To minimize immunogenicity, we have used short peptides using oligolysine-based scaffolds.

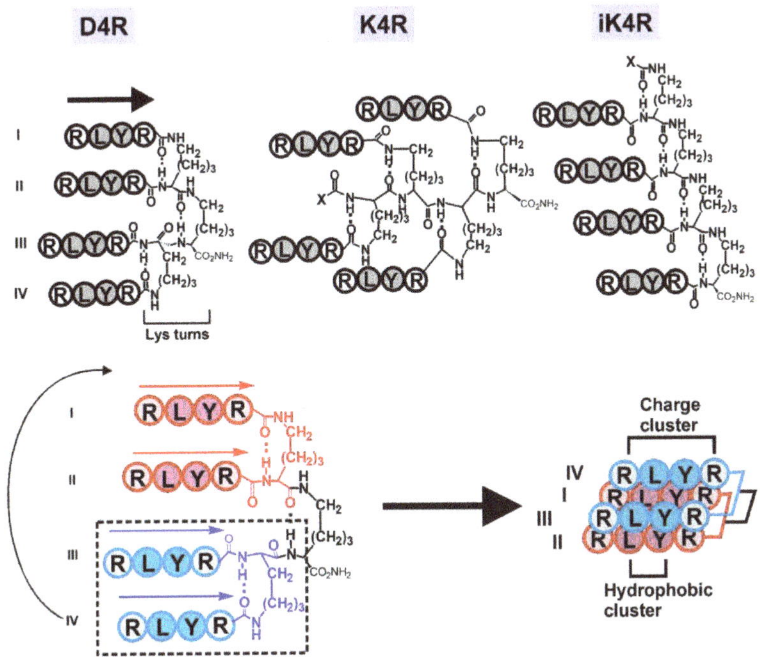

Figure 8. A schematic representation of a proposed amphipathic β-structure for α/ε-Lys-cascade peptide dendrimer D4R, α- and ε-branched peptides K4R and iK4R containing a series of putative Lys turns in their peptide scaffolds. Clustering of hydrophobic residues of the D4R peptide strands would provide an amphipathic β-structure. Roman letters I, II, III, and IV stand for the strand number. The N-to-C orientations of peptide strands are indicated by arrows.

In conclusion, peptide dendrimers and branched peptides described in this report represent a simple approach to designing and preparing antimicrobial peptides. They are small, consisting of 15 to 24 amino acid residues, and are prepared using a controlled polymerization strategy that is many fold faster than the synthesis of conventional peptides of similar sizes. The Lys-scaffolded dendrimers could support a novel "Lys-turn" to organize the tethered parallel peptides into β-structures that cluster charge and hydrophobic residues to confer the desired potency and salt-insensitivity for antimicrobial peptides.

Supplementary Materials: The following supporting information can be downloaded at: https://www.mdpi.com/article/10.3390/polym15173594/s1, Table S1: Antimicrobial activity of dendrimeric D4R against ESKAPE bacterial pathogens under high-salt conditions; Table S2: Antimicrobial activity of dendrimeric and branched peptides with Orn- and Lys-based scaffolds under low- and high-salt conditions.

Author Contributions: Conceptualization, J.P.T.; methodology, J.P.T.; formal analysis, J.T., X.Z. and J.P.T.; investigation, J.T., X.Z. and J.P.T.; resources, J.P.T.; writing—original draft preparation, J.P.T.; writing—review and editing, J.T., X.Z. and J.P.T.; funding acquisition, J.P.T. All authors have read and agreed to the published version of the manuscript.

Funding: This research was in part funded by Nanyang Technological University Internal Funding—Synzyme and Natural Products (SYNC).

Institutional Review Board Statement: Not applicable.

Data Availability Statement: The data presented in this study are available on request from the corresponding author.

Acknowledgments: The authors thank Qitao Yu, Chengwei Wu, Yi-An Lu, Jin-Long Yang, Giang K. T. Nguyen, and Mu Yuguang for their contribution to the study. We also thank Fupeng Li and Pin Jie Ho for their support.

Conflicts of Interest: The authors declare no conflict of interest.

Abbreviations

BOP, benzotrizol-1-yl-oxy-tris(dimethylamino)phosphonium-hexafluorophosphate; CHCA, a-cyano-4-hydroxycinnamic acid; D-Pro-DADME, 1,2-diamino-1,1-dimethylethane; DCC, N,N-dicyclohexylcarbodiimide; DCM, dichloromethane; DIC, N,N-diisopropylcarbodiimide; DIEA, N,N-diisopropylethylamine; DMF, dimethylformamide; DPA, diphenylamine; HOBt, N-hydroxybenzotriazole; MALDI-MS, Matrix-assisted laser desorption ionization mass spectrometry; MBHA resin, methylbenzhydrylamine resin; TFA, trifluoroacetic acid.

References

1. Ganz, T.; Lehrer, R.I. Antimicrobial peptides of vertebrates. *Curr. Opin. Immunol.* **1998**, *10*, 41–44. [CrossRef] [PubMed]
2. Zasloff, M. Antimicrobial peptides of multicellular organisms. *Nature* **2002**, *415*, 389–395. [CrossRef] [PubMed]
3. Lyu, Z.; Yang, P.; Lei, J.; Zhao, J. Biological Function of Antimicrobial Peptides on Suppressing Pathogens and Improving Host Immunity. *Antibiotics* **2023**, *12*, 1037. [CrossRef] [PubMed]
4. Moretta, A.; Scieuzo, C.; Petrone, A.M.; Salvia, R.; Manniello, M.D.; Franco, A.; Lucchetti, D.; Vassallo, A.; Vogel, H.; Sgambato, A.; et al. Antimicrobial Peptides: A New Hope in Biomedical and Pharmaceutical Fields. *Front. Cell. Infect. Microbiol.* **2021**, *11*, 668632. [CrossRef]
5. Montalbán-López, M.; Scott, T.A.; Ramesh, S.; Rahman, I.R.; Van Heel, A.J.; Viel, J.H.; Bandarian, V.; Dittmann, E.; Genilloud, O.; Goto, Y.; et al. New developments in RiPP discovery, enzymology and engineering. *Nat. Prod. Rep.* **2021**, *38*, 130–239. [CrossRef]
6. Chen, N.; Jiang, C. Antimicrobial peptides: Structure, mechanism, and modification. *Eur. J. Med. Chem.* **2023**, *255*, 115377. [CrossRef]
7. Meister, M.; Lemaitre, B.; Hoffmann, J.A. Antimicrobial peptide defense in Drosophila. *BioEssays* **1997**, *19*, 1019–1026. [CrossRef]
8. Ouellette, A.J.; Selsted, M.E. Paneth cell defensins: Endogenous peptide components of intestinal host defense. *FASEB J.* **1996**, *10*, 1280–1289. [CrossRef]
9. Nicolas, P.; Mor, A. Peptides as weapons against microorganisms in the chemical defense system of vertebrates. *Annu. Rev. Microbiol.* **1995**, *49*, 277–304. [CrossRef]
10. Danders, W.; Marahiel, M.A.; Krause, M.; Kosui, N.; Kato, T.; Izumiya, N.; Kleinkauf, H. Antibacterial action of gramicidin S and tyrocidines in relation to active transport, in vitro transcription, and spore outgrowth. *Antimicrob. Agents Chemother.* **1982**, *22*, 785–790. [CrossRef]
11. Bernhard, W.; Avrameas, S. Ultrastructural visualization of cellular carbohydrate components by means of concanavalin A. *Exp. Cell Res.* **1971**, *64*, 232–236. [CrossRef] [PubMed]
12. Vogler, K.; Studer, R.O. The chemistry of the polymyxin antibiotics. *Experientia* **1966**, *22*, 345–354. [CrossRef] [PubMed]
13. Tang, Y.Q.; Yuan, J.; Ösapay, G.; Ösapay, K.; Tran, D.; Miller, C.J.; Ouellette, A.J.; Selsted, M.E. A cyclic antimicrobial peptide produced in primate leukocytes by the ligation of two truncated α-defensins. *Science* **1999**, *286*, 498–502. [CrossRef]
14. Seebach, D.; Matthews, J.L. β-peptides: A surprise at every turn. *Chem. Commun.* **1997**, 2015–2022. [CrossRef]
15. Gellman, S.H. Foldamers: A Manifesto. *Acc. Chem. Res.* **1998**, *31*, 173–180. [CrossRef]
16. Liu, D.; DeGrado, W.F. De novo design, synthesis, and characterization of antimicrobial β-peptides. *J. Am. Chem. Soc.* **2001**, *123*, 7553–7559. [CrossRef]
17. Hintermann, T.; Gademann, K.; Jaun, B.; Seebach, D. γ-Peptides Forming More Stable Secondary Structures than α-Peptides: Synthesis and Helical NMR-Solution Structure of the γ-Hexapeptide Analog of H-(Val-Ala-Leu)2-OH. *Helv. Chim. Acta* **1998**, *81*, 983–1002. [CrossRef]
18. Karig, G.; Fuchs, A.; Büsing, A.; Brandstetter, T.; Scherer, S.; Bats, J.W.; Eschenmoser, A.; Quinkert, G. δ-Peptide analogues of pyranosyl-RNA. Part 1. Nucleo-δ-peptides derived from conformationally constrained nucleo-δ-amino acids: Preparation of monomers. *Helv. Chim. Acta* **2000**, *83*, 1049–1078. [CrossRef]
19. Shima, S.; Matsuoka, H.; Iwamoto, T.; Sakai, H. Antimicrobial action of epsilon-poly-L-lysine. *J. Antibiot.* **1984**, *37*, 1449–1455. [CrossRef]
20. Szókén, G.; Almas, M.; Krizsán, K.; Khlafulla, A.R.; Tyihák, E.; Szende, B. Structure determination and synthesis of lysine isopeptides influencing on cell proliferation. *Biopolym. Nucleic Acid Sci. Sect.* **1997**, *42*, 305–318. [CrossRef]
21. Lee, D.U.; Park, Y.J.; Yu, H.H.; Jung, S.C.; Park, J.H.; Lee, D.H.; Lee, N.K.; Paik, H.D. Antimicrobial and Antibiofilm Effect of ε-Polylysine against Salmonella Enteritidis, Listeria monocytogenes, and Escherichia coli in Tryptic Soy Broth and Chicken Juice. *Foods* **2021**, *10*, 2211. [CrossRef] [PubMed]

22. Matthyssen, T.; Li, W.; Holden, J.A.; Lenzo, J.C.; Hadjigol, S.; O'Brien-Simpson, N.M. The Potential of Modified and Multimeric Antimicrobial Peptide Materials as Superbug Killers. *Front. Chem.* **2021**, *9*, 795433. [CrossRef] [PubMed]
23. Zhang, Q.-Y.; Yan, Z.-B.; Meng, Y.-M.; Hong, X.-Y.; Shao, G.; Ma, J.-J.; Cheng, X.-R.; Liu, J.; Kang, J.; Fu, C.-Y. Antimicrobial peptides: Mechanism of action, activity and clinical potential. *Mil. Med. Res.* **2021**, *8*, 48. [CrossRef]
24. Muhle, S.A.; Tam, J.P. Design of Gram-negative selective antimicrobial peptides. *Biochemistry* **2001**, *40*, 5777–5785. [CrossRef] [PubMed]
25. Tam, J.P.; Lu, Y.A.; Yang, J.L. Design of salt-insensitive glycine-rich antimicrobial peptides with cyclic tricystine structures. *Biochemistry* **2000**, *39*, 7159–7169. [CrossRef]
26. Tam, J.P.; Lu, Y.A.; Yang, J.L. Marked increase in membranolytic selectivity of novel cyclic tachyplesins constrained with an antiparallel two-β strand cystine knot framework. *Biochem. Biophys. Res. Commun.* **2000**, *267*, 783–790. [CrossRef]
27. Tam, J.P.; Lu, Y.A.; Yang, J.L.; Chiu, K.W. An unusual structural motif of antimicrobial peptides containing end-to-end macrocycle and cystine-knot disulfides. *Proc. Natl. Acad. Sci. USA* **1999**, *96*, 8913–8918. [CrossRef]
28. Hemu, X.; Taichi, M.; Qiu, Y.; Liu, D.X.; Tam, J.P. Biomimetic synthesis of cyclic peptides using novel thioester surrogates. *Biopolymers* **2013**, *100*, 492–501. [CrossRef]
29. Kokryakov, V.N.; Harwig, S.S.L.; Panyutich, E.A.; Shevchenko, A.A.; Aleshina, G.M.; Shamova, O.V.; Korneva, H.A.; Lehrer, R.I. Protegrins: Leukocyte antimicrobial peptides that combine features of corticostatic defensins and tachyplesins. *FEBS Lett.* **1993**, *327*, 231–236. [CrossRef]
30. Nakamura, T.; Furunaka, H.; Miyata, T.; Tokunaga, F.; Muta, T.; Iwanaga, S.; Niwa, M.; Takao, T.; Shimonishi, Y. Tachyplesin, a class of antimicrobial peptide from the hemocytes of the horseshoe crab (*Tachypleus tridentatus*). Isolation and chemical structure. *J. Biol. Chem.* **1988**, *263*, 16709–16713. [CrossRef]
31. Tam, J.P. Synthetic peptide vaccine design: Synthesis and properties of a high-density multiple antigenic peptide system. *Proc. Natl. Acad. Sci. USA* **1988**, *85*, 5409–5413. [CrossRef] [PubMed]
32. Posnett, D.N.; McGrath, H.; Tam, J.P. A novel method for producing anti-peptide antibodies. Production of site-specific antibodies to the T cell antigen receptor β-chain. *J. Biol. Chem.* **1988**, *263*, 1719–1725. [CrossRef] [PubMed]
33. Tam, J.P. Recent advances in multiple antigen peptides. *J. Immunol. Methods* **1996**, *196*, 17–32. [CrossRef] [PubMed]
34. Hemu, X.; Tam, J.P. Macrocyclic Antimicrobial Peptides Engineered from ω-Conotoxin. *Curr. Pharm. Des.* **2017**, *23*, 2131–2138. [CrossRef]
35. Merrifield, R.B. Solid Phase Peptide Synthesis. I. The Synthesis of a Tetrapeptide. *J. Am. Chem. Soc.* **1963**, *85*, 2149–2154. [CrossRef]
36. Castro, B.; Dormoy, J.; Evin, G.; Selve, C. Reactifs de couplage peptidique I (1)-l'hexafluorophosphate de benzotriazolyl N-oxytrisdimethylamino phosphonium (BOP). *Tetrahedron Lett.* **1975**, *16*, 1219–1222. [CrossRef]
37. Lehrer, R.I.; Rosenman, M.; Harwig, S.S.; Jackson, R.; Eisenhauer, P. Ultrasensitive assays for endogenous antimicrobial polypeptides. *J. Immunol. Methods* **1991**, *137*, 167–173. [CrossRef]
38. Matsuzaki, K.; Yoneyama, S.; Fujii, N.; Miyajima, K.; Yamada, K.-i.; Kirino, Y.; Anzai, K. Membrane permeabilization mechanisms of a cyclic antimicrobial peptide, tachyplesin I, and its linear analog. *Biochemistry* **1997**, *36*, 9799–9806. [CrossRef]
39. Fisk, J.D.; Gellman, S.H. A parallel β-sheet model system that folds in water. *J. Am. Chem. Soc.* **2001**, *123*, 343–344. [CrossRef]
40. Woll, M.G.; Lai, J.R.; Guzei, I.A.; Taylor, S.J.C.; Smith, M.E.B.; Gellman, S.H. Parallel Sheet Secondary Structure in γ-Peptides. *J. Am. Chem. Soc.* **2001**, *123*, 11077–11078. [CrossRef]
41. Nowick, J.S.; Brower, J.O. A New Turn Structure for the Formation of β-Hairpins in Peptides. *J. Am. Chem. Soc.* **2003**, *125*, 876–877. [CrossRef] [PubMed]
42. Gross, E.; Morell, J.L. The Structure of Nisin. *J. Am. Chem. Soc.* **1971**, *93*, 4634–4635. [CrossRef] [PubMed]
43. Woody, R.W. Circular dichroism. *Methods Enzym.* **1995**, *246*, 34–71. [CrossRef]
44. Sreerama, N.; Venyaminov, S.Y.; Woody, R.W. Estimation of the number of alpha-helical and beta-strand segments in proteins using circular dichroism spectroscopy. *Protein Sci.* **1999**, *8*, 370–380. [CrossRef] [PubMed]
45. Kushwaha, D.; Mathur, K.; Balasubramanian, D. Poly (ε-L-lysine): Synthesis and conformation. *Biopolym. Orig. Res. Biomol.* **1980**, *19*, 219–229.
46. Tam, J.P.; Lu, Y.A.; Yang, J.L. Antimicrobial dendrimeric peptides. *Eur. J. Biochem.* **2002**, *269*, 923–932. [CrossRef]
47. Stanger, H.E.; Syud, F.A.; Espinosa, J.F.; Giriat, I.; Muir, T.; Gellman, S.H. Length-dependent stability and strand length limits in antiparallel β-sheet secondary structure. *Proc. Natl. Acad. Sci. USA* **2001**, *98*, 12015–12020. [CrossRef]
48. Kabsch, W.; Sander, C. Dictionary of protein secondary structure: Pattern recognition of hydrogen-bonded and geometrical features. *Biopolymers* **1983**, *22*, 2577–2637. [CrossRef]
49. Dill, K.A. Dominant forces in protein folding. *Biochemistry* **1990**, *29*, 7133–7155. [CrossRef]
50. Nelson, R.; Sawaya, M.R.; Balbirnie, M.; Madsen, A.Ø.; Riekel, C.; Grothe, R.; Eisenberg, D. Structure of the cross-β spine of amyloid-like fibrils. *Nature* **2005**, *435*, 773–778. [CrossRef]

polymers

MDPI

Article

Effect of Fe_3O_4 Nanoparticles Modified by Citric and Oleic Acids on the Physicochemical and Magnetic Properties of Hybrid Electrospun P(VDF-TrFE) Scaffolds

Vladimir Botvin [1,*], Anastasia Fetisova [2], Yulia Mukhortova [1,2], Dmitry Wagner [3], Sergey Kazantsev [4], Maria Surmeneva [1,2], Andrei Kholkin [1,5] and Roman Surmenev [1,2]

1 International Research & Development Center "Piezo- and Magnetoelectric Materials", Research School of Chemistry & Applied Biomedical Sciences, National Research Tomsk Polytechnic University, 634050 Tomsk, Russia
2 Physical Materials Science and Composite Materials Center, Research School of Chemistry & Applied Biomedical Sciences, National Research Tomsk Polytechnic University, 634050 Tomsk, Russia
3 Scientific Laboratory for Terahertz Research, National Research Tomsk State University, 634050 Tomsk, Russia
4 Institute of Strength Physics and Materials Science of Siberian Branch of Russian Academy of Sciences, 634055 Tomsk, Russia
5 School of Natural Sciences and Mathematics, Ural Federal University, 620000 Ekaterinburg, Russia
* Correspondence: v.v.botvin@gmail.com

Abstract: This study considers a fabrication of magnetoactive scaffolds based on a copolymer of vinylidene fluoride and trifluoroethylene (P(VDF-TrFE)) and 5, 10, and 15 wt.% of magnetite (Fe_3O_4) nanoparticles modified with citric (CA) and oleic (OA) acids by solution electrospinning. The synthesized Fe_3O_4-CA and Fe_3O_4-OA nanoparticles are similar in particle size and phase composition, but differ in zeta potential values and magnetic properties. Pure P(VDF-TrFE) scaffolds as well as composites with Fe_3O_4-CA and Fe_3O_4-OA nanoparticles demonstrate beads-free 1 μm fibers. According to scanning electron (SEM) and transmission electron (TEM) microscopy, fabricated P(VDF-TrFE) scaffolds filled with CA-modified Fe_3O_4 nanoparticles have a more homogeneous distribution of magnetic filler due to both the high stabilization ability of CA molecules and the affinity of Fe_3O_4-CA nanoparticles to the solvent used and P(VDF-TrFE) functional groups. The phase composition of pure and composite scaffolds includes a predominant piezoelectric β-phase, and a γ-phase, to a lesser extent. When adding Fe_3O_4-CA and Fe_3O_4-OA nanoparticles, there was no significant decrease in the degree of crystallinity of the P(VDF-TrFE), which, on the contrary, increased up to 76% in the case of composite scaffolds loaded with 15 wt.% of the magnetic fillers. Magnetic properties, mainly saturation magnetization (M_s), are in a good agreement with the content of Fe_3O_4 nanoparticles and show, among the known magnetoactive PVDF or P(VDF-TrFE) scaffolds, the highest M_s value, equal to 10.0 emu/g in the case of P(VDF-TrFE) composite with 15 wt.% of Fe_3O_4-CA nanoparticles.

Keywords: P(VDF-TrFE); magnetite; electrospinning; modification; magnetoactive scaffolds

1. Introduction

Piezo-, magnetoactive and magnetoelectric materials, due to their unique properties, are of considerable interest to material scientists [1]. Such materials are widely used in medicine [2], pharmaceuticals [3], microelectronics [4], the manufacture of modern smart devices [5], textiles [6], and other promising areas. From the point of view of practical application, magnetoelectric materials exhibiting a magnetoelectric effect occupy a special place [7]. The fabrication of such materials is of key importance for modern regenerative medicine and tissue engineering, since it allows the development of high-tech devices and treatment techniques and their implementation in medical practice, improving the quality of care provided.

Currently known magnetoelectric materials are mainly inorganic single-phase compounds ($Bi_5Ti_3FeO_{15}$, $LiFe_5O_8$, etc.) and composites based on ferromagnetic materials (metals—Co, Ni, alloys—FeCo, FeGa, Metglas, Terfenol-D, $NiFe_2O_4$, $CoFe_2O_4$, etc.) and piezoelectric ceramics ($PbZr_xTi_{(1-x)}O_3$, $BaTiO_3$, AlN, etc.) [8]. These materials, despite the high values of their magnetoelectric and piezoelectric coefficients, have a number of disadvantages in terms of biocompatibility, mechanical, and other properties, which limit their use and implementation in biomedicine. To eliminate the mentioned drawbacks, it was proposed to use composites, including a piezopolymer matrix and a nanosized magnetic filler [9]. Among polymers, poly(vinylidene fluoride) (PVDF) and a copolymer of vinylidene fluoride with trifluoroethylene (P(VDF-TrFE)) have the highest piezoelectric properties due to the presence of a piezoactive β-phase (TTTT-conformation), along with other phases with lower piezoelectric properties (γ-phase—TTTGTTTG′ conformation and δ-phase—TGTG′ conformation) [10]. PVDF and P(VDF-TrFE) can also exist at a non-piezoelectric α- (TGTG′ conformation) and ε-phase (TTTGTTTG′ conformation) [10]. The formation of certain P(VDF-TrFE) phases depends on the thermodynamic conditions and the methods of materials synthesis [10]. Therefore, the stabilization of the piezoactive β-phase in PVDF as well as in PVDF-based copolymers is a significant challenge. In the case of P(VDF-TrFE), in contrast to PVDF, the β-phase is stabilized due to a steric hindrance caused by the presence of bulky trifluoroethylene repeating units of the copolymer [11]. Magnetite (Fe_3O_4) nanoparticles can act as a magnetic filler of the magnetoelectric composites since they are quite easily synthesized, undergo the necessary surface modification, have a high saturation magnetization, and are biocompatible [12].

The method of obtaining P(VDF-TrFE)- or PVDF-based materials also affects the formation of the β-phase. The most suitable method for the fabrication of magnetoactive materials is electrospinning, because it contributes to the stabilization of the β-phase as a result of in situ poling [13] and makes it possible to obtain materials with a given morphology and physicochemical properties [14]. It is known that Fe_3O_4 nanoparticles are prone to agglomeration, which affects the size of nanoparticles and other properties, including magnetic ones [12]. To reduce the agglomeration of Fe_3O_4 nanoparticles during the synthesis and upon fabrication of PVDF or P(VDF-TrFE) composites, a surface of the magnetic filler is functionalized with various organic and high-molecular-weight modifiers, such as mono- and polyfunctional carboxylic acids [15,16], poly(N-vinylpyrrolidone) [17], xanthan gum [18], poly(dopamine) [19], and others. The distribution of magnetic nanoparticles in the electrospun PVDF or P(VDF-TrFE) fibers requires comprehensive study, since it can affect many physicochemical and operational properties of materials. However, there are a small number of the studies considering features of magnetic nanoparticles distribution in the electrospun P(VDF-TrFE) (PVDF) fibers loaded, as a rule, with unmodified Fe_3O_4 nanoparticles [20,21]. The addition of unmodified and modified Fe_3O_4 nanoparticles as well as other magnetic fillers to PVDF or P(VDF-TrFE) can also affect the degree of crystallinity of the polymer, the content of the piezoactive β-phase, and consequently the piezoelectric properties of the resulting materials [22]. Depending on the concentration and chemical nature of the surface of the magnetic fillers, both an increase and a decrease in the degree of crystallinity and the content of the β-phase can be observed. In this regard, when developing magnetoactive piezoelectric materials, the influence of both factors should be taken into account. An analysis of known studies has shown that to date there are no studies on the effect of Fe_3O_4 nanoparticles functionalized with modifiers of various natures on the physicochemical, magnetic, and other properties of P(VDF-TrFE)-based materials. To assess the effect of functional groups on the surface of Fe_3O_4 nanoparticles and the nature of their interactions with P(VDF-TrFE) functional groups, we proposed to use nanoparticles modified with hydrophobic and hydrophilic modifiers, for example, with citric and oleic acids.

Previously, we had obtained magnetoactive P(VDF-TrFE) scaffolds of various compositions filled with Fe_3O_4 nanoparticles modified with oleic acid (OA) demonstrating hydrophobic properties [23]. Considering the nature of the functional groups of the P(VDF-

TrFE), it is of considerable interest to fabricate and study composite magnetoactive scaffolds with modifiers, which, on the one hand, exhibit hydrophilic and hydrophobic properties, and, on the other hand, reduce the tendency of Fe_3O_4 nanoparticles to agglomerate. Citric acid (CA) and OA can act as hydrophilic and hydrophobic modifiers, respectively. Thus, the present research is devoted to the fabrication of electrospun magnetoactive scaffolds based on P(VDF-TrFE) and Fe_3O_4 nanoparticles functionalized with hydrophilic and hydrophobic modifiers and a comprehensive study of their structure, morphology, particle distribution, and physicochemical properties.

2. Materials and Methods

2.1. Materials

The powder of P(VDF-TrFE) (70/30) (MW = 300,000 Solvene 300/P300), $FeCl_3{\cdot}6H_2O$, and $FeSO_4{\cdot}7H_2O$ were purchased from Sigma Aldrich (Steinheim, Germany). Dimethylformamide (DMF), acetone, ammonia solution, CA, and OA were purchased from local vendors and used without further purification.

2.2. Synthesis of Magnetite Nanoparticles Modified by Citric and Oleic Acids

Synthesis of magnetite nanoparticles modified by citric (Fe_3O_4-CA) and oleic (Fe_3O_4-OA) acids was carried out according to the previously reported method (Figure 1a) [24]. Briefly, $FeCl_3{\cdot}6H_2O$ (3.32 g) and $FeSO_4{\cdot}7H_2O$ (1.22 g) were dissolved in 200 mL of deionized water under an argon atmosphere by magnetic stirrer (1300 rpm) and heated to 80 °C. Then, 20 mL of ammonia solution was added and stirring was continued for an additional 5 min. Thereafter, 1.4 mL of CA or OA solution (0.5 g/mL) was added to the reaction mixture and synthesis was carried out for another 90 min at 90 °C. At the end of the synthesis, thus-obtained Fe_3O_4-CA or Fe_3O_4-OA nanoparticles were separated by the magnet, washed with water and ethanol until neutral pH was achieved, and then used as a suspension in acetone. The concentration of synthesized modified nanoparticles in suspension was determined by the gravimetry method [23].

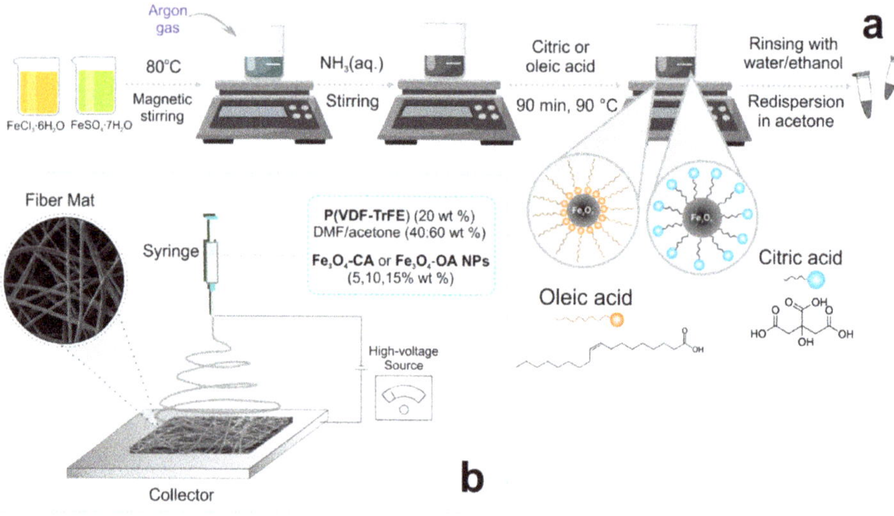

Figure 1. Scheme of the synthesis of modified Fe_3O_4 nanoparticles (**a**) and fabrication of magnetoactive P(VDF-TrFE) scaffolds by electrospinning (**b**).

2.3. Preparation of Electrospinning Solutions

The calculated amount of P(VDF-TrFE) was dissolved in a mixture of DMF: acetone (40:60 wt.%) to obtain polymer solutions with the concentration of 20 wt.% suitable for the fabrication of scaffolds with a fiber diameter of about 1 μm (Figure 1b). The solution was mixed using a shaker at room temperature overnight until the formation of a clear solution.

In the case of composite scaffolds with Fe_3O_4-CA and Fe_3O_4-OA concentrations of 5, 10, and, 15 wt.% (relative to polymer weight), the calculated volume of DMF and the portion of magnetite suspension in acetone were mixed with P(VDF-TrFE) powder in a shaker overnight.

2.4. Electrospinning Parameters

Pure P(VDF-TrFE) and composite scaffolds doped with Fe_3O_4-CA nanoparticles were fabricated by a custom-made electrospinning setup. The prepared polymer solution was loaded into a 5-mL plastic syringe with a stainless-steel needle (inner diameter 21 gauge). The flow rate of the solution was regulated by a syringe pump (Aitecs 2016 Universal). Polymer and composite fibers were deposited on the cylindrical rotating collector. All the samples were obtained using the following parameters: distance between the needle and the collector of 13 cm, applied voltage of 15.0 kV (15.5 kV in the case of composite scaffold obtainment), solution flow rate of 0.2 mL·h^{-1}, and collector rotation speed of 200 rpm.

2.5. Characterization of the Samples

Fourier transform infrared (FTIR) spectra were recorded using a Cary 630 FTIR spectrometer (Agilent Technologies, Santa Clara, CA, USA) in the wavelength range 4000–400 cm^{-1}. IR spectra of Fe_3O_4-CA and Fe_3O_4-OA nanoparticles were recorded in absorbance mode (mixing with KBr and subsequent formation of thin tablets). Spectra of pure P(VDF-TrFE) and composite scaffolds were acquired as received in an attenuated total reflectance (ATR) mode (diamond crystal).

The phase compositions of Fe_3O_4-CA, Fe_3O_4-OA nanoparticles, P(VDF-TrFE), and composite scaffolds were analyzed by X-ray diffraction (XRD) on a MiniFlex 600 diffractometer (Rigaku corporation, Tokyo, Japan) at scattering angles ranging from 10 to 70° (CuKα radiation). The crystallite size was calculated using Scherrer's equation (Equation (1)).

$$D = \frac{k\lambda}{\beta cos\theta}, \quad (1)$$

where k is the proportionality constant, λ is the X-ray wavelength (nm), β is the enlargement of the measured diffraction line at the mid-height of its maximum intensity (in radian units), and θ is the XRD peak position.

Raman spectra Fe_3O_4-CA and Fe_3O_4-OA nanoparticles were recorded on an NT-MDT system (NT-MDT Spectrum Instruments, Zelenograd, Russia) equipped with a 100× objective. Excitation was performed with a laser at wavelengths of 633 nm with a maximum power of 60 mW. To prevent heating and oxidation of magnetite, no more than 1% of the laser power was used during measurements.

The thermal properties of pure and composite scaffolds were studied on STA 449 F1 (Netzsch, Selb, Germany) in the range of 25–300 °C in an argon atmosphere at a heating rate of 10 °C/min. The crystallinity (X_c) of the polymer and composite scaffolds was evaluated using Equation (2):

$$X_c = \frac{\Delta H_f}{(1-\omega)\Delta H_f^0} \cdot 100\%, \quad (2)$$

where ω is a weight fraction of Fe_3O_4-CA and Fe_3O_4-OA in composite scaffolds, and the heat of fusion for 100% crystalline P(VDF-TrFE) with VDF:TrFE ratio of 70:30 mol.% (ΔH_f^0) equaling 45 J·g^{-1} [25].

The morphology of Fe_3O_4-CA, Fe_3O_4-OA nanoparticles and pure and composite P(VDF-TrFE) scaffolds was studied by scanning electron microscopy (SEM) with Apreo 2 SEM electron microscope (Thermo Fisher, Tokyo, Japan).

Distribution of modified Fe_3O_4 nanoparticles in the P(VDF-TrFE) fibers was studied by transmission electron microscopy (TEM) on a JEM-2100 electron microscope (JEOL,

Tokyo, Japan). To prepare samples for TEM measurements, a fragment of the scaffolds was fixed using Araldide epoxy resin, and then layers of 80–120 nm thickness were cut using ultratome «Ultrotome III» («LKB», Stockholm, Sweden).

The zeta potential of Fe_3O_4-CA and Fe_3O_4-OA nanoparticles was determined from their electrophoretic mobility in a U-shaped cell with gold electrodes with front fixation by dynamic light scattering on a ZetaSizer Nano Zs device (Malvern, Worcestershire, UK).

The magnetic properties of Fe_3O_4-CA, Fe_3O_4-OA nanoparticles and composite scaffolds were investigated at 300 K with an external pulsed magnetic field of up to 6.5 kOe using a pulsed magnetometer (Tomsk State University, Tomsk, Russia). The technical characteristics of the magnetometer and the technique for measuring the magnetic properties are described in detail in [26].

3. Results and Discussion

Synthesis conditions have a significant influence on the morphology, phase composition, and, subsequently, magnetic characteristics of Fe_3O_4 nanoparticles [12]. Due to features of a physicochemical nature, unmodified Fe_3O_4 nanoparticles are prone to agglomeration, which affects their particle size. In order to reduce the agglomeration of nanoparticles during synthesis, various groups of modifiers are used [27]. CA and OA were chosen as modifiers, since they should result in a decrease in the agglomeration of nanoparticles during their synthesis, and subsequently when obtaining magnetoactive P(VDF-TrFE) scaffolds [15,16]. SEM micrographs of Fe_3O_4-CA and Fe_3O_4-OA nanoparticles (Figure 2a,b) showed a uniform size distribution of the particles without any visible agglomerates, which confirms the influence of carboxylic acid modifiers. The average sizes of Fe_3O_4-CA and Fe_3O_4-OA nanoparticles were 17.7 and 24.7 nm, respectively (Figure 2c,d).

XRD patterns of both types of nanoparticles (Figure 2e) contained characteristic reflections of Fe_3O_4 at 2θ = 18.23°, 30.0°, 35.36°, 37.0°, 42.99°, 53.37°, 56.87°, and 62.45° corresponding to d_{hkl} crystal planes at (111), (220), (311), (222), (400), (422), (511), and (440), respectively [28]. The crystallite sizes of Fe_3O_4-CA and Fe_3O_4-OA nanoparticles calculated by Scherrer's equation (Equation (1)) are 11.6 and 13.0 nm.

IR spectroscopy makes it possible to establish not only the absorption bands of Fe_3O_4 nanoparticles, but also the absorption bands of modifier acids. Thus, the IR spectra of Fe_3O_4-CA and Fe_3O_4-OA (Figure 2f) contain, along with the absorption band related to the vibrations of the Fe–O bond of magnetite at 568 cm^{-1}, the absorption bands corresponding to the vibrations of the functional groups of CA and OA in the regions 3000–2950, 1607, and 1372 cm^{-1}, which are related to the C–H stretching vibrations, symmetric and asymmetric vibrations of the carboxylate anion of the modifier acids, respectively [29,30]. The presence of a band corresponding to the carboxylate anion confirms the binding of carboxyl groups of CA and OA onto the surface of magnetite nanoparticles via chemisorption. Raman spectra of modified Fe_3O_4 nanoparticles (Figure 2g) supplement data regarding chemical and phase composition. Both spectra contain characteristic peaks at 302, 539, and 670 cm^{-1}, related to magnetite [31]. In the case of Fe_3O_4-OA, the Raman spectrum additionally demonstrates a low-intensity shoulder at 700 cm^{-1}, which corresponds to the maghemite phase (γ-Fe_2O_3). The last could have formed both during magnetite synthesis and due to the effect of a laser beam during Raman measurements [23]. The crystal structures of Fe_3O_4 and γ-Fe_2O_3 have similar crystal lattice parameters; therefore, it is difficult to distinguish them by XRD [32]. Thus, a comprehensive study of Fe_3O_4 nanoparticles by XRD, IR, and Raman spectroscopy led to a determination of a real phase composition of magnetite nanoparticles and confirmed the presence of modifier acids on their surface.

Modification of Fe_3O_4 nanoparticles with CA and OA differed in their functionality, leading to a change in the properties of the nanoparticle surface. Such modification should affect the distribution and stability of nanoparticles in the P(VDF-TrFE) solution as well as the nature of the polymer–magnetic filler interactions. The measured values of the zeta potential for Fe_3O_4-CA and Fe_3O_4-OA turned out to be −35.3 and −2.6 mV, respectively, which is consistent with the functionality of modifier acids [33,34]. The CA molecules located on the surface of magnetite nanoparticles have two free carboxyl groups capable of ionization, defining the measured values of the zeta potential [34]. In contrast to the CA structure, the OA molecule includes a polar carboxyl group and a nonpolar long alkyl substituent, therefore acting as a surfactant and promoting the formation of micelle-like structures upon interaction with the surface of oxide nanoparticles [35]. Depending on the OA concentration, a single- or a double-layer structure can be formed during the functionalization of the Fe_3O_4 surface, affecting the physicochemical properties, including the zeta potential value [33]. At a low concentration of OA, single-layer structures are predominantly formed, as in the case of our Fe_3O_4-OA nanoparticles. This is consistent with the measured zeta potential, which is slightly negative.

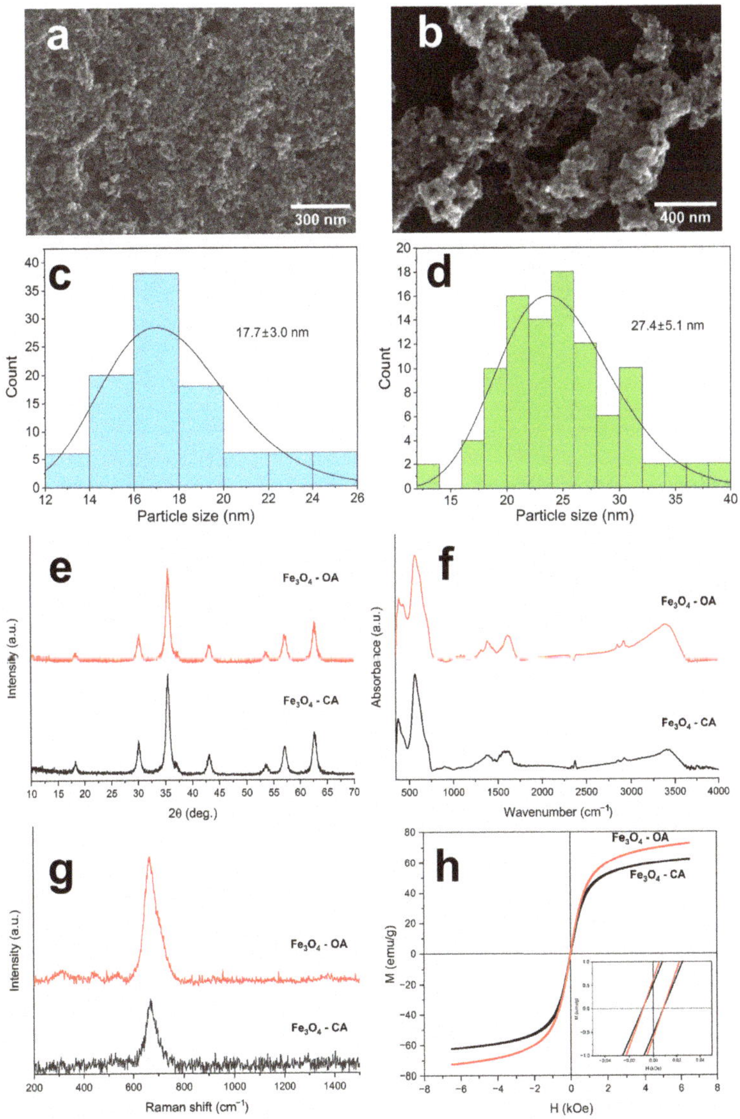

Figure 2. SEM images (**a**,**b**) and size distribution (**c**,**d**) of Fe_3O_4-CA (**a**,**c**) and Fe_3O_4-OA (**b**,**d**) nanoparticles, their XRD patterns (**e**), IR (**f**) and Raman (**g**) spectra, and magnetic hysteresis loops (**h**).

Modification of the surfaces of Fe_3O_4 nanoparticles also naturally results in a change of magnetic properties, which are important in the synthesis of the magnetoactive polymer composite scaffolds based on them. Figure 2h shows magnetic hysteresis loops of Fe_3O_4-CA and Fe_3O_4-OA nanoparticles, and their main magnetic characteristics are presented in Table 1. The saturation magnetization (M_s) of Fe_3O_4-CA and Fe_3O_4-OA nanoparticles was 61.9 and 72.6 emu/g, which is inferior to the M_s value of bulk unmodified magnetite [36]. The decrease in M_s in the case of modified Fe_3O_4 nanoparticles is caused by the small sizes of the particles. A surface layer with unordered magnetic spins is found on them, and they convert to a paramagnetic state at room temperature [37,38]. The decrease in M_s is also associated with the content in the non-magnetic layer of the modifier acid in the samples under study [39]. Higher M_s, in the case of Fe_3O_4-OA, is caused by a lower concentration of OA on the nanoparticle surface. The presence of only one carboxyl group of OA compared to the three carboxyl groups of CA, and the good complexation ability of citrates in general, leads to a more effective binding of CA molecules to the Fe_3O_4 nanoparticle surface. This additionally confirms the formation of a single OA layer on the surface of Fe_3O_4 nanoparticles. Our previous study of Fe_3O_4-OA nanoparticles demonstrates that according to thermogravimetric measurements, a weight loss equal to the amount of OA is nearly of 6% [23]. The thermogravimetric curve of Fe_3O_4-CA nanoparticles shows weight loss of about of 13% [40]. In addition, a more significant decrease in the magnetization of Fe_3O_4-CA nanoparticles can also be associated with a smaller crystallite size than that of Fe_3O_4-OA, which leads to an increase in the previously mentioned surface layer. The coercive forces (H_c) for both types of nanoparticles have close to non-zero values, which characterizes them as ferromagnets. Such materials are capable of magnetization reversal in weak magnetic fields.

Table 1. Magnetic properties of Fe_3O_4-CA and Fe_3O_4-OA nanoparticles.

Sample	M_s, emu/g	M_r [a], emu/g	H_c, Oe
Fe_3O_4-CA	61.9 ± 1.9	0.5 ± 0.1	10.0 ± 0.5
Fe_3O_4-OA	72.6 ± 3.6	0.6 ± 0.03	8.0 ± 0.4

[a] M_r—remanent magnetization.

A fabrication of magnetoactive scaffolds based on P(VDF-TrFE) and Fe_3O_4 nanoparticles by electrospinning facilitates the formation of materials with controlled properties. However, electrospinning proceeding at high voltages in a medium of various organic solvents, as a rule, results in changes in morphology, phase composition, and other physicochemical properties of all components of the scaffolds. When obtaining electrospun composite scaffolds filled with Fe_3O_4 nanoparticles, a special attention should be paid to the morphology and distribution of the magnetic filler in the P(VDF-TrFE) fibers. To obtain defects-free fibers, optimally selected parameters of the electrospinning, including the appropriate solvent, should be chosen [39]. Figure 3 shows SEM micrographs and the fiber diameter distributions of pure P(VDF-TrFE) and composite scaffolds filled with different contents of Fe_3O_4-CA and Fe_3O_4-OA nanoparticles.

The fabricated scaffolds do not contain any defects and have an average fiber diameter of about 1 μm. The known studies indicate that such a fiber diameter should have a positive effect on human cells in the biomedical application of the developed materials, since it is comparable to the fiber diameters of many tissues. [40]. The SEM micrographs of composite scaffolds, as well as the diameters of the obtained fibers, allow us to evaluate the distribution of filler nanoparticles on their surface. Regardless of the magnetic filler type, the surface of composite scaffolds contains Fe_3O_4 nanoparticles, whose amount increases with increasing concentration. When comparing the distribution of Fe_3O_4 nanoparticles modified with CA and OA on the surfaces of composite fibers, P(VDF-TrFE)/Fe_3O_4-CA scaffolds demonstrate a more uniform distribution without formation of large agglomerates. Such observations indicate that CA molecules have a greater stabilizing effect on Fe_3O_4 nanoparticles, and the modified nanoparticles themselves have a greater affinity to P(VDF-

TrFE) macromolecules. Together, CA molecules prevent an agglomeration of Fe_3O_4-CA nanoparticles and contribute to the most consistent distribution of the nanoparticles in the P(VDF-TrFE) fibers. Despite the presence of nanoparticles on the surface of P(VDF-TrFE) fibers, all of them are coated with a polymer layer. Previously, we observed such an effect for magnetoactive polymer scaffolds based on unmodified Fe_3O_4 nanoparticles [41]. When P(VDF-TrFE) solution in the mixture of DMF and acetone is stretched into a polymer jet under an electric field and deposited on a collector, dispersed Fe_3O_4 nanoparticles remain covered with a polymer layer in the fiber due to surface tension forces.

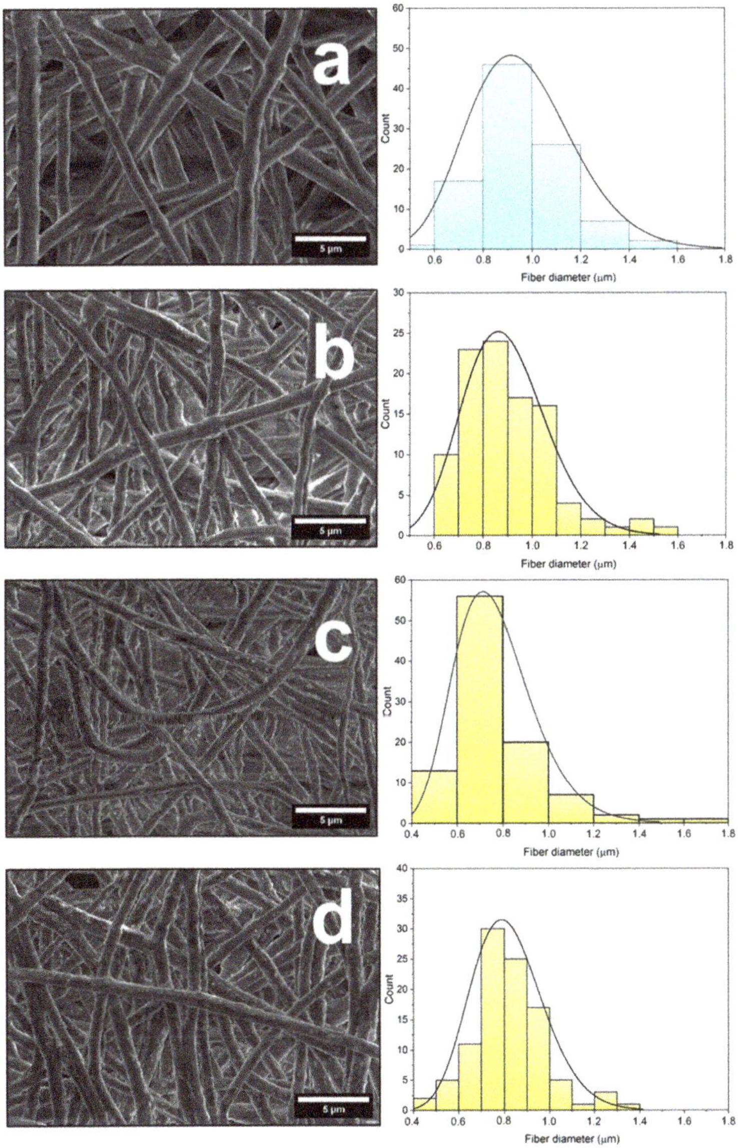

Figure 3. *Cont.*

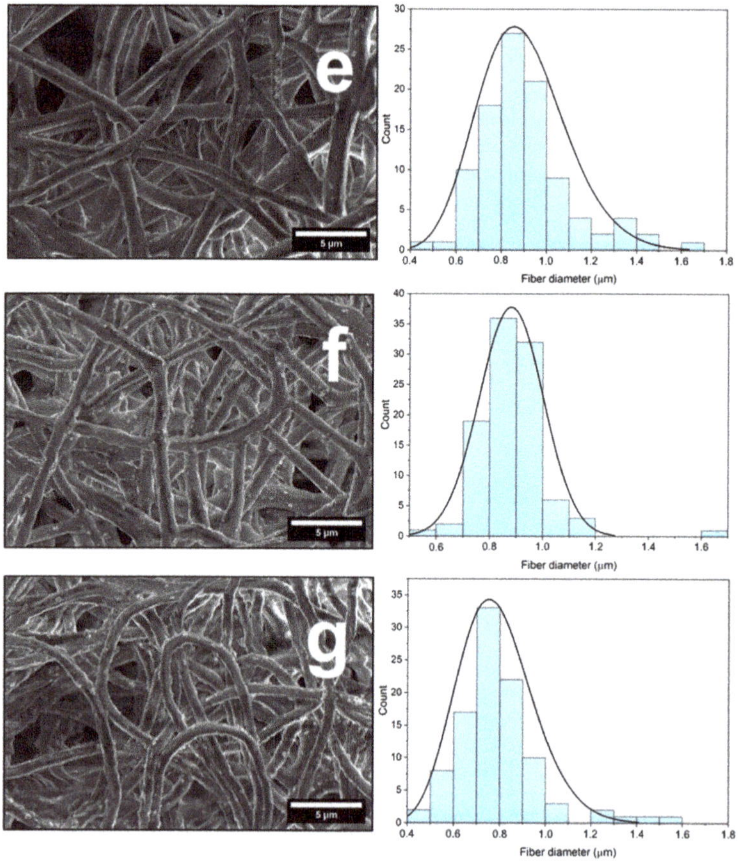

Figure 3. SEM images and fiber diameter distributions of pure P(VDF-TrFE) (**a**) and composite scaffolds filled with 5, 10, and 15 wt.% of Fe_3O_4-CA (**b–d**) and Fe_3O_4-OA (**e–g**) nanoparticles.

The properties of magnetoactive composite scaffolds are certainly affected by the distribution of magnetic filler nanoparticles, not only on the surfaces of P(VDF-TrFE) fibers, but also in the bulk. Figure 4 shows TEM micrographs of P(VDF-TrFE) composite scaffolds filled with Fe_3O_4-CA and Fe_3O_4-OA nanoparticles. The analysis of the micrographs indicates that in the case of P(VDF-TrFE)/Fe_3O_4-CA scaffolds (Figure 4a–c), a uniform distribution of the filler is observed in proportion to a given concentration, which is not accompanied by the formation of large agglomerates. The resulting small agglomerates have a morphology elongated in the direction of the flow and deposition of the polymer solution jet during electrospinning. P(VDF-TrFE)/Fe_3O_4-OA scaffolds (Figure 4d–f) are characterized by a less uniform distribution of the particles due to the nature of the particles themselves, the solvents used, and the way they were introduced into the electrospinning solution.

One of the factors affecting the uniform distribution of modified Fe_3O_4 nanoparticles, mainly with CA molecules, is the introduction of a magnetic filler into the P(VDF-TrFE) solution as a suspension. The suspension method prevents agglomeration of Fe_3O_4 nanoparticles, which is not always possible when a magnetic filler is added to a solution as a dry powder [41]. Another factor affecting the uniform distribution of Fe_3O_4 nanoparticles is the presence of modifier carboxylic acids on their surface. For example, the presence of free carboxyl groups on the surfaces of Fe_3O_4 nanoparticles modified with CA prevents their agglomeration due to an electrostatic repulsion [34]. Moreover, the polar carboxyl groups of Fe_3O_4-CA nanoparticles have a high affinity to the polar solvents (DMF, acetone) used in the polymer solution, as well as to the functional groups of P(VDF-TrFE). In the case of Fe_3O_4-OA nanoparticles, agglomeration can decrease both due to a steric hindrance caused by the bulky OA alkyl moiety [33] and hydrophobic interactions with low-polar groups of hydrophobic P(VDF-TrFE). However, the greatest stabilizing effect (decrease of agglomeration) of both factors in the fabrication of P(VDF-TrFE) composite scaffolds has been observed when Fe_3O_4 nanoparticles were modified with CA, which was confirmed by SEM and TEM results.

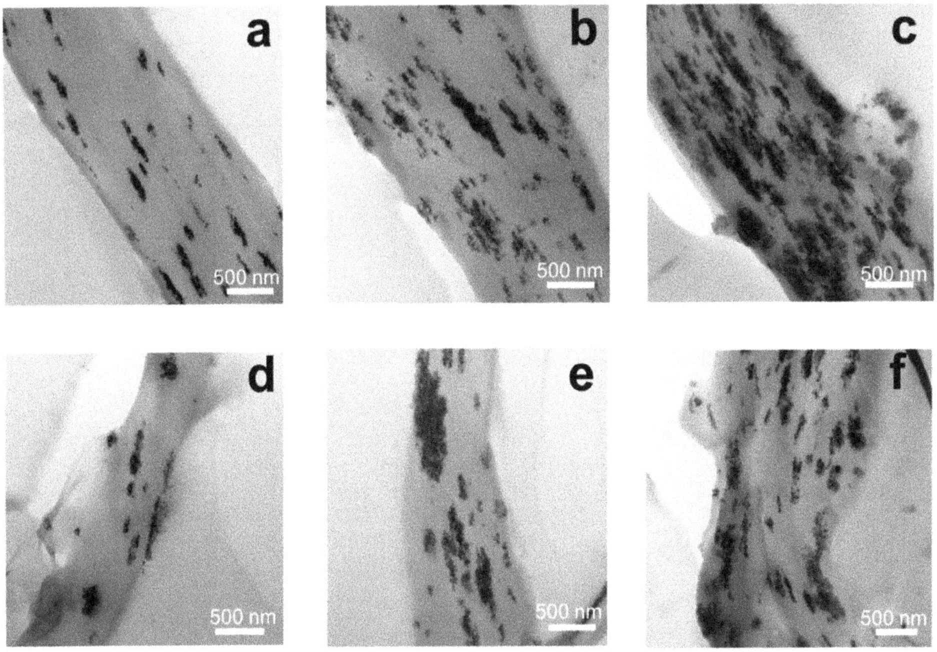

Figure 4. TEM images of P(VDF-TrFE) composite scaffolds filled with 5, 10, and 15 wt.% of Fe_3O_4-CA (**a**–**c**) and Fe_3O_4-OA (**d**–**f**) nanoparticles.

A phase composition of polymeric and inorganic components of magnetoactive scaffolds is of great importance, since they significantly affect physicochemical and, subsequently, operational properties. The phase composition of P(VDF-TrFE) correlates with a piezoresponse, whereas the phase composition of Fe_3O_4 nanoparticles is associated with magnetic properties, mainly with saturation magnetization. Figure 5 presents diffraction patterns of pure and composite P(VDF-TrFE) scaffolds filled with Fe_3O_4-CA and Fe_3O_4-OA nanoparticles of various content.

The XRD pattern (Figure 5a,i) of pure P(VDF-TrFE) scaffolding includes reflections of the β-phase at 19.8°, 35.3°, and 41.0°, related to the crystallographic planes (110)/(200), (001), and (201), respectively [23]. There is also a low-intensity shoulder at 18.3°, indicating the contribution of the (020) crystal plane of the monoclinic γ-phase, which has less pronounced piezoelectric properties [10]. At the same time, the characteristic reflections of the P(VDF-TrFE) α-phase at 17.6°, 19.9°, and 26.6° are absent in the XRD pattern [10]. XRD patterns of composite scaffolds (Figure 5a, ii–vii), along with reflections of the β- and γ-phases of P(VDF-TrFE), demonstrate reflections of the Fe_3O_4 phase with an intensity proportional to the concentration of the magnetic filler. The absence of side reflections of other iron oxides and oxyhydroxide phases indicates that electrospinning does not affect the phase composition of the magnetic filler of the magnetoactive scaffolds.

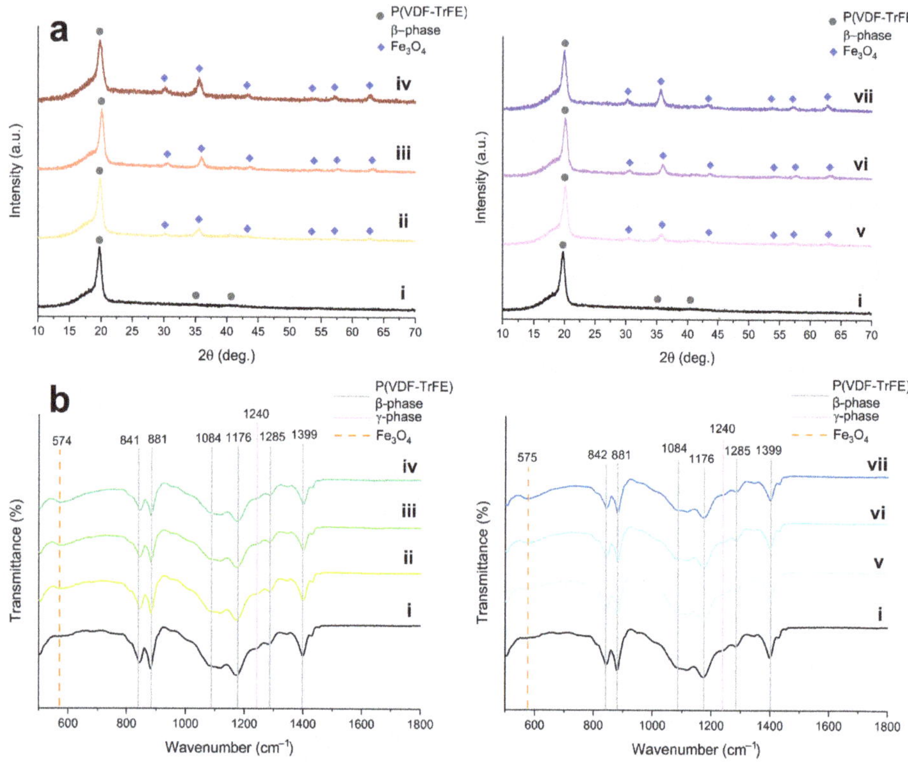

Figure 5. XRD patterns (**a**) and IR spectra (**b**) of pure P(VDF-TrFE) (i) and composite scaffolds filled with 5, 10, 15 wt.% of Fe_3O_4-CA (ii–iv) and Fe_3O_4-OA (v–vii) nanoparticles.

For a more detailed study of the phase composition and conformation of pure P(VDF-TrFE) and composite scaffolds filled with Fe_3O_4-CA and Fe_3O_4-OA nanoparticles, the fabricated scaffolds were studied by IR spectroscopy. This method allows analysis of the macromolecules' conformation, determining a qualitative composition and, for some systems, the quantitative content of the P(VDF-TrFE) phases [42]. IR spectra of pure P(VDF-TrFE) scaffolds and composites with Fe_3O_4-CA and Fe_3O_4-OA nanoparticles are presented in Figure 5b.

The IR spectra in the range from 500 to 1800 cm^{-1} contain characteristic absorption bands at 842, 881, 1084, 1176, 1285, and 1399 cm^{-1}, related to the crystalline β-phase of P(VDF-TrFE), mainly to vinylidene fluoride repeating units [43,44]. The band at 1399 cm^{-1} indicates bending vibrations of CH_2 groups and stretching vibrations of C–C groups of P(VDF-TrFE), the bands at 1283 and 1176 cm^{-1} correspond to symmetric and asymmetric stretching vibrations of CF_2 groups, and the bands at 880 cm^{-1} and 841 cm^{-1} refer to rocking bending vibrations of CH_2 groups and asymmetric and symmetric stretching vibrations of CF_2 groups, respectively [44]. The band at 1116 cm^{-1} refers to the characteristic vibrations of the trifluoroethylene units of the copolymer, along with a low-intensity band at 1341 cm^{-1}, which belongs to the bending vibrations of C–H in the CHF group [45]. A detailed analysis of IR spectra demonstrates an absence of characteristic absorption bands of the P(VDF-TrFE) α-phase at 762, 976, and 1207 cm^{-1} [45]. This observation proves the role of the trifluoroethylene repeating units of the copolymer in the stabilization of the β-phase.

The IR spectra exhibit, additionally, a low-intensity band at 1240 cm^{-1}, which belongs to a weaker piezoelectric γ-phase of the P(VDF-TrFE) [46]. Another characteristic band of the γ-phase at 840 cm^{-1} coincides with one of the bands of the β-phase, which makes it less informative. The formation of the γ-phase during the electrospinning of PVDF and its derivatives was also noted elsewhere [47–49]. Such a phase transition occurs due to the influence of the electric field [47] and the nature of solvents used, as well as their ratio [50], on the crystallization of P(VDF-TrFE) during electrospinning. Mentioned factors lead to a partial transformation of the TTTT conformation of the β-phase to the TTTGTTTG′ of the γ-phase conformation. A reliable quantitative calculation of the phase composition in P(VDF-TrFE) scaffolds in comparison with PVDF is difficult, because, on the one hand, unlike the α-phase, most of the bands of the β- and γ-phases are superimposed due to the similarity of conformation of their repeating units [51]. On the other hand, one of the characteristic bands of the γ-phase at 1240 cm^{-1} is present in the spectrum as a low-intensive shoulder, which complicates the baseline correction and subsequent calculation [10]. The spectra of composite scaffolds also contain a band observed at 574 cm^{-1}, which refers to the Fe–O stretching vibrations of Fe_3O_4 nanoparticles [40]. The intensity of this band naturally increases with an increase in the concentration of the magnetic filler in the scaffolds.

It is known that the incorporation of nanoparticles into polymer scaffolds results in a change in thermal properties and crystallization behavior. The latter is important in assessing the degree of crystallinity of the polymer, which is consistent with the content of the β-phase and the piezoelectric properties of P(VDF-TrFE). To study the effect of modified Fe_3O_4 nanoparticles on the thermal properties and degree of crystallinity, fabricated scaffolds were studied by DSC. Figure 6 demonstrates the DSC curves of pure and composite P(VDF-TrFE) scaffolds, and Table 2 includes the main thermal characteristics and the degree of crystallinity.

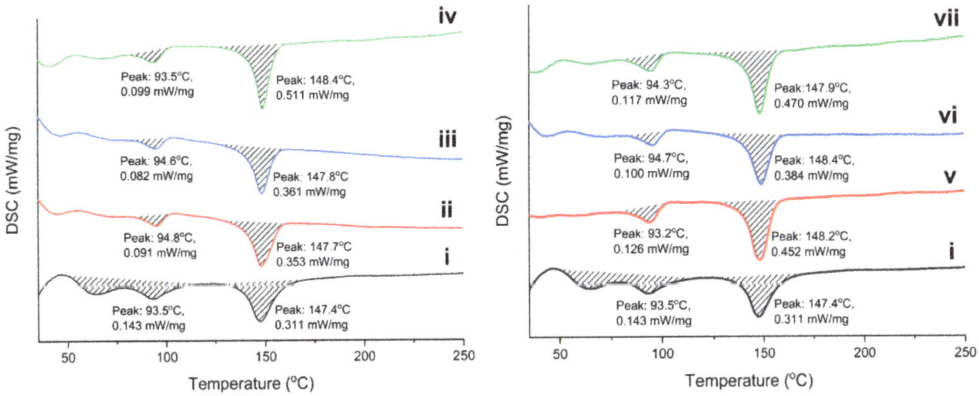

Figure 6. DSC curves of pure P(VDF-TrFE) (i) and composite scaffolds filled with 5, 10, 15 wt.% of Fe_3O_4-CA (ii–iv) and Fe_3O_4-OA (v–vii) nanoparticles.

Table 2. Thermal properties of pure P(VDF-TrFE) and composite scaffolds filled with Fe_3O_4-CA and Fe_3O_4-OA nanoparticles.

Sample	T_m, °C	ΔH_m, J/g	X_c,%
P(VDF-TrFE)	147.4	27.30	60.7
P(VDF-TrFE)/Fe_3O_4-CA 5%	147.7	25.58	59.8
P(VDF-TrFE)/Fe_3O_4-CA 10%	147.8	22.80	56.3
P(VDF-TrFE)/Fe_3O_4-CA 15%	148.4	29.04	75.9
P(VDF-TrFE)/Fe_3O_4-OA 5%	148.2	32.00	74.8
P(VDF-TrFE)/Fe_3O_4-OA 10%	148.4	26.68	65.9
P(VDF-TrFE)/Fe_3O_4-OA 15%	147.9	29.42	76.9

All DSC curves contain an endo peak in the region of 93.1–94.8 °C, which refers to the P(VDF-TrFE) phase transition from the ferroelectric to the paraelectric state (Curie transition). Such a temperature of transition is characteristic of a copolymer with a ratio of VDF:TrFE units of 70:30 mol.% [52]. The melting temperature, which is characterized by the second endo peak on the DSC curves, is in the region of 147.4–148.4 °C. A slight increase in the melting temperature in the case of composite scaffolds is due to the influence of the magnetic filler on the melting of P(VDF-TrFE) crystallites. The degree of crystallinity of the fabricate scaffolds has an unusual behavior. In a comparison of both types of nanoparticles, it can be noticed that the degree of crystallinity varies nonmonotonically, especially in the case of composite scaffolds filled with Fe_3O_4-OA nanoparticles. Such regularity is in agreement with a less uniform distribution of Fe_3O_4-OA nanoparticles in the fibers, which was confirmed by SEM and TEM. With an increase in the concentration of the magnetic filler, no significant decrease in the degree of crystallinity is observed. Moreover, it is important to note that for composite scaffolds containing 15 wt.% of both types of Fe_3O_4 nanoparticles, the highest value of the degree of crystallinity of about 76% is observed. This result is a consequence of the interactions of CA and OA located on the surface of Fe_3O_4 nanoparticles with functional groups of P(VDF-TrFE) macromolecules promoting their crystallization. An increase in the degree of crystallinity of magnetoactive composite scaffolds can have a positive effect on the piezoelectric properties of materials, which depend both on the total crystallinity and the content of the β-phase.

A chemical nature of the modifier, the phase composition and distribution of the magnetic filler in the polymer fibers of composite scaffolds, has a significant effect on their magnetic properties. Figure 7 shows the magnetic hysteresis loops of P(VDF-TrFE) composite scaffolds with different contents of Fe_3O_4-CA and Fe_3O_4-OA nanoparticles, and Table 3 lists their main magnetic characteristics.

With an increase in the concentrations of both types of Fe_3O_4 nanoparticles in the composition of the scaffolds, a regular increase in M_s is observed. Thus, for P(VDF-TrFE) scaffolds with 5, 10, and 15 wt.% of Fe_3O_4-CA nanoparticles, the M_s values were 4.4, 7.9, and 10.0 emu/g, respectively. For scaffolds with similar concentrations of Fe_3O_4-OA nanoparticles, M_s was 4.2, 6.8, 8.8 emu/g. All samples of composite scaffolds have a non-zero H_c, which characterizes them as ferromagnetic materials. The increase in the H_c of all composite scaffolds is associated with a low content of magnetite particles. The particles cease to interact magnetostatically with each other; as a result, magnetite chain aggregates disrupt in the composite [53]. The value of H_c facilitates a distribution of magnetic nanoparticles in the polymer fibers. For P(VDF-TrFE) scaffolds with 5 and 10 wt.% Fe_3O_4-OA nanoparticles, large H_c values can be observed, which indicates a less uniform distribution of filler particles and the formation of agglomerates. This is in agreement with the results of SEM, TEM, and DSC. The lower affinity of P(VDF-TrFE), solvents, and Fe_3O_4-OA nanoparticles can also explain a decrease in M_s of composite scaffolds as being due to the lower concentration of the filler than theoretically calculated values. It is important to note that the achieved M_s for P(VDF-TrFE) scaffolds with Fe_3O_4-CA nanoparticles, to the best of our knowledge, has the highest value among known PVDF or P(VDF-TrFE) composite scaffolds, considering the introduction of a high concentration of magnetic filler and its uniform distribution in the polymer fibers.

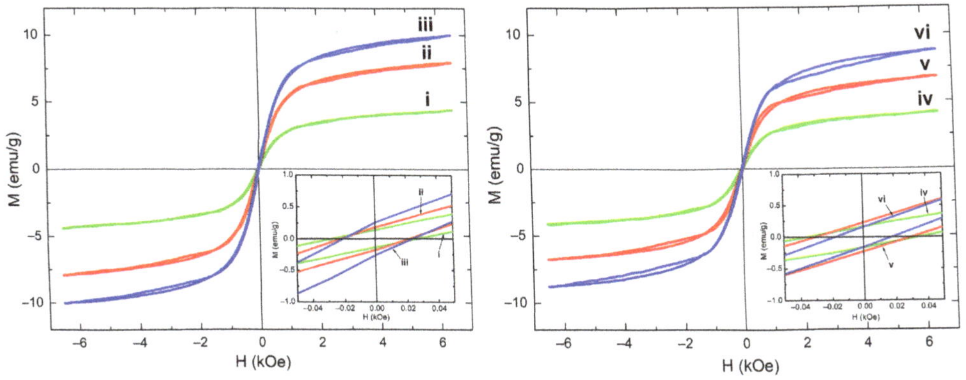

Figure 7. Hysteresis loops of magnetoactive P(VDF-TrFE) scaffolds filled with 5, 10, 15 wt.% of Fe_3O_4-CA (i–iii) and Fe_3O_4-OA (iv–vi) nanoparticles.

Table 3. Magnetic properties of composite P(VDF-TrFE) scaffolds filled with different contents of Fe_3O_4-CA and Fe_3O_4-OA nanoparticles.

Sample	M_s, emu/g	M_r [a], emu/g	H_c, Oe
P(VDF-TrFE)/Fe_3O_4-CA (5 wt.%)	4.4 ± 0.2	0.14	28 ± 1
P(VDF-TrFE)/Fe_3O_4-CA (10 wt.%)	7.9 ± 0.4	0.18	22 ± 1
P(VDF-TrFE)/Fe_3O_4-CA (15 wt.%)	10.0 ± 0.5	0.26	24 ± 1
P(VDF-TrFE)/Fe_3O_4-OA (5 wt.%)	4.2 ± 0.2	0.16	35 ± 2
P(VDF-TrFE)/Fe_3O_4-OA (10 wt.%)	6.8 ± 0.3	0.23	30 ± 2
P(VDF-TrFE)/Fe_3O_4-OA (15 wt.%)	8.8 ± 0.4	0.16	17 ± 1

[a] M_r—remanent magnetization.

Taking into account the results of studying the morphology, phase composition, and physicochemical properties of P(VDF-TrFE) scaffolds with Fe_3O_4 nanoparticles modified with CA and OA, the possible types of polymer–magnetic filler interactions leading to a high degree of affinity are shown in Figure 8.

The most probable type of interaction is intermolecular hydrogen bonding of highly electronegative fluorine atoms of P(VDF-TrFE) and hydrogen atoms of carboxylic acid modifiers [54]. The presence of polarized carboxyl groups on the surface of modified Fe_3O_4 nanoparticles allows them to enter into dipolar interactions with dipoles of P(VDF-TrFE) macromolecules. A similar interaction mechanism was proposed for PVDF composites with organic molecules and polymers containing polarized functional groups [55,56]. In the case of Fe_3O_4-OA nanoparticles, as already noted, hydrophobic interactions of the long alkyl OA substituent with low-polarity P(VDF-TrFE) moieties are possible [57]. It should be noted that such types of interaction can have a lesser effect on P(VDF-TrFE) macromolecules, since the β-phase in them is stabilized by bulky trifluoroethylene repeating units, but, at the same time, they can significantly affect the $\alpha \rightarrow \beta$ phase transitions in PVDF macromolecules, resulting in an improvement of the piezoelectric properties of piezoactive polymer.

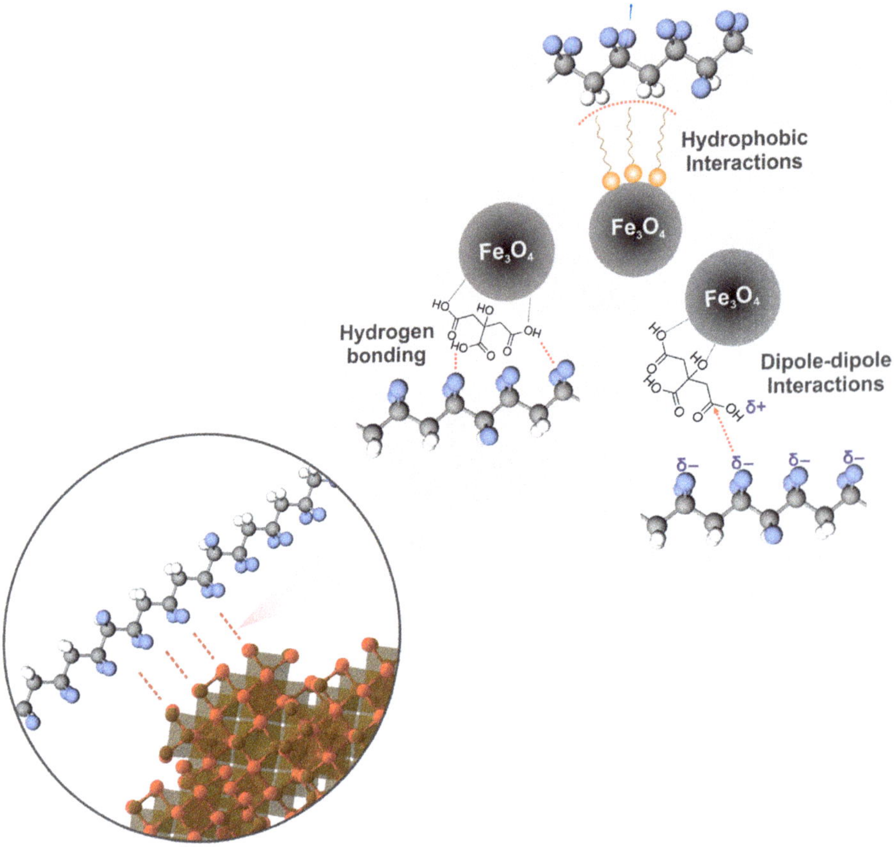

Figure 8. Proposed scheme of interactions of P(VDF-TrFE) and Fe_3O_4 nanoparticles modified with CA and OA.

4. Conclusions

Magnetoactive P(VDF-TrFE) composite scaffolds filled with 5, 10, and 15 wt.% of Fe_3O_4 nanoparticles modified with CA and OA were fabricated by electrospinning. The phase compositions of the modified Fe_3O_4 nanoparticles were represented by a pure magnetite phase, but they naturally differed in surface functionality and magnetic properties due to the differences in the properties of the modifier carboxylic acids used and their concentrations. Pure P(VDF-TrFE) and composite scaffolds demonstrate a defect-free morphology and a fiber diameter of about 1 μm. The surface of the P(VDF-TrFE) fibers contains Fe_3O_4 nanoparticles, which have a more uniform distribution in the case of Fe_3O_4-CA nanoparticles due to the stabilizing effect of the free polar carboxyl groups of CA and their good affinity to the polar solvents used as well as the P(VDF-TrFE) functional groups. In the case of Fe_3O_4-OA nanoparticles, stabilization, to a lesser extent, can occur due to a steric hindrance of the bulky alkyl substituent OA and its hydrophobic interactions with low-polarity P(VDF-TrFE) moieties.

The XRD patterns of P(VDF-TrFE) composite scaffolds with Fe_3O_4 nanoparticles contain only a magnetite phase, demonstrating that selected electrospinning conditions do not affect the phase composition of both types of Fe_3O_4 nanoparticles. The intensity of Fe_3O_4 reflections increases proportionally with an increase in the magnetic filler concentration. In this case, the P(VDF-TrFE) composite scaffolds contain a predominantly piezoactive β-phase, along with a small amount of γ-phase, which exhibits weaker piezoelectric properties. The formation of the latter is a consequence of the influence of the electric field, the nature and composition of the solvents used on the behavior, and the rate of P(VDF-TrFE) crystallization. When a magnetic filler was added, no significant decrease in the degree of crystallinity is observed, and for composite scaffolds containing 15 wt.% of both types of Fe_3O_4 nanoparticles, the highest degree of crystallinity of about 76% is observed. This result is a consequence of the interaction of CA and OA located on the surfaces of Fe_3O_4 nanoparticles with functional groups of P(VDF-TrFE) macromolecules, which promotes their crystallization and, by increasing the total content of the β-phase, improves the piezoelectric properties of magnetoactive scaffolds. The saturation magnetization naturally increases in proportion to the content of the magnetic filler. The highest saturation magnetization value of 10.0 emu/g was obtained for P(VDF-TrFE) scaffolds with 15 wt.% of Fe_3O_4-CA nanoparticles. All fabricated composite P(VDF-TrFE) scaffolds exhibit a non-zero coercive force, characterizing their ferromagnetic behavior. An increase in the coercive force in the case of P(VDF-TrFE)/Fe_3O_4-OA scaffolds indicates a less uniform distribution of the filler particles and the formation of agglomerates. Based on the obtained experimental data, a scheme of possible interactions between modified Fe_3O_4 nanoparticles and P(VDF-TrFE) macromolecules via hydrogen bonding and dipolar and hydrophobic interactions is proposed. The developed magnetoactive composite scaffolds based on P(VDF-TrFE) and modified Fe_3O_4 nanoparticles with high saturation magnetization and uniform distribution of magnetic filler should be considered as promising materials for biomedical applications. Moreover, the suggested modified Fe_3O_4 nanoparticles, suspension method of their introduction, and electrospinning parameters can be used for the improvement of β-phase content and piezoresponce, not only of P(VDF-TRFE), but also of PVDF.

Author Contributions: Conceptualization, V.B.; methodology, M.S., A.K. and R.S.; validation, Y.M.; investigation, A.F., D.W. and S.K.; writing—original draft preparation, V.B. and A.F.; writing—review and editing, V.B. and D.W.; supervision, A.K. and R.S.; funding acquisition, V.B. All authors have read and agreed to the published version of the manuscript.

Funding: This research was funded by the Russian Science Foundation, grant number 22-73-00228, https://rscf.ru/project/22-73-00228/.

Data Availability Statement: The data presented in this study are available on request from the corresponding author.

Acknowledgments: The authors kindly thank T. Botvina, T. Bugrova, Andrey A. Miller, and I. Mishin for their assistance in the study of the samples by IR spectroscopy, XRD, DSC, TEM, and SEM.

Conflicts of Interest: The authors declare no conflict of interest.

References

1. Narita, F.; Fox, M. A review on piezoelectric, magnetostrictive, and magnetoelectric materials and device technologies for energy harvesting applications. *Adv. Eng. Mater.* **2018**, *20*, 1700743. [CrossRef]
2. Kapat, K.; Shubhra, Q.T.; Zhou, M.; Leeuwenburgh, S. Piezoelectric nano-biomaterials for biomedicine and tissue regeneration. *Adv. Funct. Mater.* **2020**, *30*, 1909045. [CrossRef]
3. Thévenot, J.; Oliveira, H.; Sandre, O.; Lecommandoux, S. Magnetic responsive polymer composite materials. *Chem. Soc. Rev.* **2013**, *42*, 7099–7116. [CrossRef] [PubMed]
4. Liang, X.; Chen, H.; Sun, N.X. Magnetoelectric materials and devices. *APL Mater.* **2021**, *9*, 041114. [CrossRef]
5. Bahl, S.; Nagar, H.; Singh, I.; Sehgal, S. Smart materials types, properties and applications: A review. *Mater. Today Proc.* **2020**, *28*, 1302–1306. [CrossRef]
6. Mokhtari, F.; Spinks, G.M.; Fay, C.; Cheng, Z.; Raad, R.; Xi, J.; Foroughi, J. Wearable electronic textiles from nanostructured piezoelectric fibers. *Adv. Mater. Technol.* **2020**, *5*, 1900900. [CrossRef]
7. Kopyl, S.; Surmenev, R.; Surmeneva, M.; Fetisov, Y.; Kholkin, A. Magnetoelectric effect: Principles and applications in biology and medicine—A review. *Mater. Today Bio.* **2021**, *12*, 100149. [CrossRef]
8. Pereira, N.; Lima, A.C.; Lanceros-Mendez, S.; Martins, P. Magnetoelectrics: Three centuries of research heading towards the 4.0 industrial revolution. *Materials* **2020**, *13*, 4033. [CrossRef]
9. Palneedi, H.; Annapureddy, V.; Priya, S.; Ryu, J. Status and perspectives of multiferroic magnetoelectric composite materials and applications. *Actuators* **2016**, *5*, 9. [CrossRef]

10. Martins, P.; Lopes, A.; Lanceros-Mendez, S. Electroactive phases of poly(vinylidene fluoride): Determination, processing and applications. *Prog. Polym. Sci.* **2014**, *39*, 683–706. [CrossRef]
11. Wang, S.; Li, Q. Design, synthesis and processing of PVDF-based dielectric polymers. *IET Nanodielectrics* **2018**, *1*, 80–91. [CrossRef]
12. Niculescu, A.-G.; Chircov, C.; Grumezescu, A.M. Magnetite nanoparticles: Synthesis methods–A comparative review. *Methods* **2022**, *199*, 16–27. [CrossRef] [PubMed]
13. Wu, C.M.; Chou, M.H. Polymorphism, piezoelectricity and sound absorption of electrospun PVDF membranes with and without carbon nanotubes. *Compos. Sci. Technol.* **2016**, *127*, 127–133. [CrossRef]
14. Haider, A.; Haider, S.; Kang, I.-K. A comprehensive review summarizing the effect of electrospinning parameters and potential applications of nanofibers in biomedical and biotechnology. *Arab. J. Chem.* **2018**, *11*, 1165–1188. [CrossRef]
15. Adebayo, L.L.; Soleimani, H.; Yahya, N.; Abbas, Z.; Ridwan, A.T.; Wahaab, F.A. Investigation of the broadband microwave absorption of citric acid coated Fe_3O_4/PVDF composite using finite element method. *Appl. Sci.* **2019**, *9*, 3877. [CrossRef]
16. Wang, T.; Li, W.; Luo, L.; Zhu, Y. Ultrahigh dielectric constant composites based on the oleic acid modified ferroferric oxide nanoparticles and polyvinylidene fluoride. *Appl. Phys. Lett.* **2013**, *102*, 092904. [CrossRef]
17. Lakshmi, N.; Tambe, P.; Panda, B. Surface modified iron oxide (Fe_3O_4) nanosheets reinforced PVDF nanocomposites: Influence on morphology, thermal and magnetic properties. *Plast. Rubber Compos.* **2022**, *51*, 205–216. [CrossRef]
18. Koyuncu, I.; Gul, B.Y.; Esmaeili, M.S.; Pekgenc, E.; Teber, O.O.; Tuncay, G.; Karimi, H.; Parvaz, S.; Maleki, A.; Vatanpour, V. Modification of PVDF membranes by incorporation Fe_3O_4@ Xanthan gum to improve anti-fouling, anti-bacterial, and separation performance. *J. Environ. Chem. Eng.* **2022**, *10*, 107784. [CrossRef]
19. Wu, J.; Wang, N.; Zhao, Y.; Jiang, L. Simple synthesis of smart magnetically driven fibrous films for remote controllable oil removal. *Nanoscale* **2015**, *7*, 2625–2632. [CrossRef]
20. Venugopal, A.; Cespedes, O.; Russell, S. Controlling dielectric and magnetic properties of PVdF/magnetite nanocomposite fibre webs. *Int. J. Polym. Sci.* **2014**, *2014*, 102946. [CrossRef]
21. Liu, F.; Li, M.; Li, F.; Weng, K.; Qi, K.; Liu, C.; Ni, Q.; Tao, X.; Zhang, J.; Shao, W. Preparation and properties of PVDF/Fe_3O_4 nanofibers with magnetic and electret effects and their application in air filtration. *Macromol. Mater. Eng.* **2020**, *305*, 1900856. [CrossRef]
22. Wu, L.; Jin, Z.; Liu, Y.; Ning, H.; Liu, X.; Hu, N. Recent advances in the preparation of PVDF-based piezoelectric materials. *Nanotechnol. Rev.* **2022**, *11*, 1386–1407. [CrossRef]
23. Botvin, V.V.; Surmeneva, M.A.; Mukhortova, Y.R.; Belyakova, E.O.; Wagner, D.V.; Chelobanov, B.P.; Laktionov, P.P.; Sukhinina, E.V.; Pershina, A.G.; Kholkin, A.L. Magnetoactive electrospun hybrid scaffolds based on poly(vinylidene fluoride-co-trifluoroethylene) and magnetite particles with varied sizes. *Polym. Eng. Sci.* **2022**, *62*, 1593–1607. [CrossRef]
24. Mukhortova, Y.R.; Pryadko, A.S.; Chernozem, R.V.; Pariy, I.O.; Akoulina, E.A.; Demianova, I.V.; Zharkova, I.I.; Ivanov, Y.F.; Wagner, D.V.; Bonartsev, A.P. Fabrication and characterization of a magnetic biocomposite of magnetite nanoparticles and reduced graphene oxide for biomedical applications. *Nano-Struct. Nano-Objects* **2022**, *29*, 100843. [CrossRef]
25. Kuang, X.; Gao, Q.; Zhu, H. Effect of calcination temperature of TiO_2 on the crystallinity and the permittivity of PVDF-TrFE/TiO_2 composites. *J. Appl. Polym. Sci.* **2013**, *129*, 296–300. [CrossRef]
26. Wagner, D.V.; Kareva, K.V.; Zhuravlev, V.A.; Dotsenko, O.A.; Minin, R.V. Investigation of $BaFe_{12}O_{19}$ Hexaferrites Manufactured by Various Synthesis Methods Using a Developed Pulsed Magnetometer. *Inventions* **2023**, *8*, 26. [CrossRef]
27. Favela-Camacho, S.E.; Samaniego-Benítez, E.J.; Godínez-García, A.; Avilés-Arellano, L.M.; Pérez-Robles, J.F. How to decrease the agglomeration of magnetite nanoparticles and increase their stability using surface properties. *Colloids Surf. A Physicochem. Eng. Asp.* **2019**, *574*, 29–35. [CrossRef]
28. Chernova, E.; Botvin, V.; Galstenkova, M.; Mukhortova, Y.; Wagner, D.; Gerasimov, E.; Surmeneva, M.; Kholkin, A.; Surmenev, R. A Comprehensive Study of Synthesis and Analysis of Anisotropic Iron Oxide and Oxyhydroxide Nanoparticles. *Nanomaterials* **2022**, *12*, 4321. [CrossRef]
29. Ye, L.; Zhou, L.; Lu, Y. Direct Continuous Synthesis of Oleic Acid-Modified Fe_3O_4 Nanoparticles in a Microflow System. *Ind. Eng. Chem. Res.* **2022**, *61*, 4320–4328. [CrossRef]
30. Dheyab, M.A.; Aziz, A.A.; Jameel, M.S.; Noqta, O.A.; Khaniabadi, P.M.; Mehrdel, B. Simple rapid stabilization method through citric acid modification for magnetite nanoparticles. *Sci. Rep.* **2020**, *10*, 10793. [CrossRef]
31. Shebanova, O.N.; Lazor, P. Raman spectroscopic study of magnetite ($FeFe_2O_4$): A new assignment for the vibrational spectrum. *J. Solid State Chem.* **2003**, *174*, 424–430. [CrossRef]
32. Kim, W.; Suh, C.-Y.; Cho, S.-W.; Roh, K.-M.; Kwon, H.; Song, K.; Shon, I.-J. A new method for the identification and quantification of magnetite–maghemite mixture using conventional X-ray diffraction technique. *Talanta* **2012**, *94*, 348–352. [CrossRef] [PubMed]
33. Soares, P.I.; Laia, C.A.; Carvalho, A.; Pereira, L.C.; Coutinho, J.T.; Ferreira, I.M.; Novo, C.M.; Borges, J.P. Iron oxide nanoparticles stabilized with a bilayer of oleic acid for magnetic hyperthermia and MRI applications. *Appl. Surf. Sci.* **2016**, *383*, 240–247. [CrossRef]
34. Liu, J.; Dai, C.; Hu, Y. Aqueous aggregation behavior of citric acid coated magnetite nanoparticles: Effects of pH, cations, anions, and humic acid. *Environ. Res.* **2018**, *161*, 49–60. [CrossRef] [PubMed]
35. Baharuddin, A.A.; Ang, B.C.; Hussein, N.A.A.; Andriyana, A.; Wong, Y.H. Mechanisms of highly stabilized ex-situ oleic acid-modified iron oxide nanoparticles functionalized with 4-pentynoic acid. *Mater. Chem. Phys.* **2018**, *203*, 212–222. [CrossRef]
36. Smit, J.; Wijn, H.P.J. *Ferrites*; Philips Technical Library: Eindhoven, The Netherlands, 1959; Volume 278.

37. Arefi, M.; Kazemi Miraki, M.; Mostafalu, R.; Satari, M.; Heydari, A. Citric acid stabilized on the surface of magnetic nanoparticles as an efficient and recyclable catalyst for transamidation of carboxamides, phthalimide, urea and thiourea with amines under neat conditions. *J. Iran. Chem. Soc.* **2019**, *16*, 393–400. [CrossRef]
38. Kodama, R. Magnetic nanoparticles. *J. Magn. Magn. Mater.* **1999**, *200*, 359–372. [CrossRef]
39. Hatamie, S.; Parseh, B.; Ahadian, M.M.; Naghdabadi, F.; Saber, R.; Soleimani, M. Heat transfer of PEGylated cobalt ferrite nanofluids for magnetic fluid hyperthermia therapy: In vitro cellular study. *J. Magn. Magn. Mater.* **2018**, *462*, 185–194. [CrossRef]
40. Nigam, S.; Barick, K.; Bahadur, D. Development of citrate-stabilized Fe_3O_4 nanoparticles: Conjugation and release of doxorubicin for therapeutic applications. *J. Magn. Magn. Mater.* **2011**, *323*, 237–243. [CrossRef]
41. Pryadko, A.A.S.; Mukhortova, Y.R.; Chernozem, R.V.; Pariy, I.; Alipkina, S.I.; Zharkova, I.I.; Dudun, A.A.; Zhuikov, V.A.; Moisenovich, A.M.; Bonartseva, G.A. Electrospun Magnetic Composite Poly-3-hydroxybutyrate/Magnetite Scaffolds for Biomedical Applications: Composition, Structure, Magnetic Properties, and Biological Performance. *ACS Appl. Bio Mater.* **2022**, *5*, 3999–4019. [CrossRef]
42. Cai, X.; Lei, T.; Sun, D.; Lin, L. A critical analysis of the α, β and γ phases in poly(vinylidene fluoride) using FTIR. *RSC Adv.* **2017**, *7*, 15382–15389. [CrossRef]
43. Simoes, R.; Rodriguez-Perez, M.; De Saja, J.; Constantino, C. Tailoring the structural properties of PVDF and P (VDF-TrFE) by using natural polymers as additives. *Polym. Eng. Sci.* **2009**, *49*, 2150–2157. [CrossRef]
44. Mayeen, A.; Kala, M.; Jayalakshmy, M.; Thomas, S.; Philip, J.; Rouxel, D.; Bhowmik, R.; Kalarikkal, N. Flexible and self-standing nickel ferrite–PVDF-TrFE cast films: Promising candidates for high-end magnetoelectric applications. *Dalton Trans.* **2019**, *48*, 16961–16973. [CrossRef] [PubMed]
45. Arrigoni, A.; Brambilla, L.; Bertarelli, C.; Serra, G.; Tommasini, M.; Castiglioni, C. P(VDF-TrFE) nanofibers: Structure of the ferroelectric and paraelectric phases through IR and Raman spectroscopies. *RSC Adv.* **2020**, *10*, 37779–37796. [CrossRef] [PubMed]
46. Jia, N.; He, Q.; Sun, J.; Xia, G.; Song, R. Crystallization behavior and electroactive properties of PVDF, P(VDF-TrFE) and their blend films. *Polym. Test.* **2017**, *57*, 302–306. [CrossRef]
47. Yu, L.; Cebe, P. Crystal polymorphism in electrospun composite nanofibers of poly(vinylidene fluoride) with nanoclay. *Polymer* **2009**, *50*, 2133–2141. [CrossRef]
48. Zhong, G.; Zhang, L.; Su, R.; Wang, K.; Fong, H.; Zhu, L. Understanding polymorphism formation in electrospun fibers of immiscible poly(vinylidene fluoride) blends. *Polymer* **2011**, *52*, 2228–2237. [CrossRef]
49. Zheng, J.; He, A.; Li, J.; Han, C.C. Polymorphism control of poly(vinylidene fluoride) through electrospinning. *Macromol. Rapid Commun.* **2007**, *28*, 2159–2162. [CrossRef]
50. Lei, T.; Cai, X.; Wang, X.; Yu, L.; Hu, X.; Zheng, G.; Lv, W.; Wang, L.; Wu, D.; Sun, D. Spectroscopic evidence for a high fraction of ferroelectric phase induced in electrospun polyvinylidene fluoride fibers. *RSC Adv.* **2013**, *3*, 24952–24958. [CrossRef]
51. Zhou, Z.; Li, J.; Xia, W.; Zhu, X.; Sun, T.; Cao, C.; Zhang, L. Enhanced piezoelectric and acoustic performances of poly(vinylidene fluoride-trifluoroethylene) films for hydroacoustic applications. *Phys. Chem. Chem. Phys.* **2020**, *22*, 5711–5722. [CrossRef]
52. Gregorio, R., Jr.; Botta, M.M. Effect of crystallization temperature on the phase transitions of P(VDF/TrFE) copolymers. *J. Polym. Sci. Part B Polym. Phys.* **1998**, *36*, 403–414. [CrossRef]
53. Pryadko, A.S.; Mukhortova, Y.R.; Chernozem, R.V.; Shlapakova, L.E.; Wagner, D.V.; Romanyuk, K.; Gerasimov, E.Y.; Kholkin, A.; Surmenev, R.A.; Surmeneva, M.A. Comprehensive Study on the Reinforcement of Electrospun PHB Scaffolds with Composite Magnetic Fe_3O_4–rGO Fillers: Structure, Physico-Mechanical Properties, and Piezoelectric Response. *ACS Omega* **2022**, *7*, 41392–41411. [CrossRef] [PubMed]
54. Zhang, X.; Xia, W.; Liu, J.; Zhao, M.; Li, M.; Xing, J. PVDF-based and its Copolymer-Based Piezoelectric Composites: Preparation Methods and Applications. *J. Electron. Mater.* **2022**, *51*, 5528–5549. [CrossRef]
55. Chen, J.; Xiong, X.; Zhang, Q.; Shui, L.; Shen, S.; Yang, H.; Zhu, Z.; Zhang, F. P(VDF-TrFE)/PMMA blended films with enhanced electrowetting responses and superior energy storage performance. *Polymers* **2019**, *11*, 526. [CrossRef] [PubMed]
56. Liang, C.-L.; Mai, Z.-H.; Xie, Q.; Bao, R.-Y.; Yang, W.; Xie, B.-H.; Yang, M.-B. Induced formation of dominating polar phases of poly(vinylidene fluoride): Positive ion–CF_2 dipole or negative ion–CH_2 dipole interaction. *J. Phys. Chem. B* **2014**, *118*, 9104–9111. [CrossRef]
57. Zheng, L.; Tang, M.; Wang, Y.; Hou, D.; Li, X.; Wang, J. A novel Cu-BTC@ PVA/PVDF Janus membrane with underwater-oleophobic/hydrophobic asymmetric wettability for anti-fouling membrane distillation. *Sep. Purif. Technol.* **2022**, *299*, 121807. [CrossRef]

polymers

MDPI

Systematic Review

Hydrogel-Based Biomaterial as a Scaffold for Gingival Regeneration: A Systematic Review of In Vitro Studies

Dimas Ilham Hutomo [1], Lisa Amir [2,*], Dewi Fatma Suniarti [2], Endang Winiati Bachtiar [2] and Yuniarti Soeroso [1]

1 Department of Periodontology, Faculty of Dentistry, Universitas Indonesia, Jakarta 10430, Indonesia; dimas.hutomo@ui.ac.id (D.I.H.); yuniarti@ui.ac.id (Y.S.)
2 Department of Oral Biology, Faculty of Dentistry, Universitas Indonesia, Jakarta 10430, Indonesia; dewi.fatma@ui.ac.id (D.F.S.); endang04@ui.ac.id (E.W.B.)
* Correspondence: lisa.amir@ui.ac.id

Abstract: Background: Hydrogel is considered a promising scaffold biomaterial for gingival regeneration. In vitro experiments were carried out to test new potential biomaterials for future clinical practice. The systematic review of such in vitro studies could synthesize evidence of the characteristics of the developing biomaterials. This systematic review aimed to identify and synthesize in vitro studies that assessed the hydrogel scaffold for gingival regeneration. Methods: Data on experimental studies on the physical and biological properties of hydrogel were synthesized. A systematic review of the PubMed, Embase, ScienceDirect, and Scopus databases was conducted according to the Preferred Reporting System for Systematic Reviews and Meta-Analyses (PRISMA) 2020 statement guidelines. In total, 12 original articles on the physical and biological properties of hydrogels for gingival regeneration, published in the last 10 years, were identified. Results: One study only performed physical property analyses, two studies only performed biological property analyses, and nine studies performed both physical and biological property analyses. The incorporation of various natural polymers such as collagen, chitosan, and hyaluronic acids improved the biomaterial characteristics. The use of synthetic polymers faced some drawbacks in their physical and biological properties. Peptides, such as growth factors and arginine–glycine–aspartic acid (RGD), can be used to enhance cell adhesion and migration. Based on the available primary studies, all studies successfully present the potential of hydrogel characteristics in vitro and highlight the essential biomaterial properties for future periodontal regenerative treatment.

Keywords: hydrogel; gingival tissue regeneration; in vitro studies; systematic review

1. Introduction

Gingival recession, thin gingival phenotype, and lack of keratinized tissue around natural teeth and dental implants are the most common gingival tissue problems which require soft tissue reconstruction treatment. To date, a coronally advanced flap in combination with a connective tissue graft is still considered as the gold standard treatment that results in a high success rate, high esthetic outcome, and long-term soft tissue stability [1,2]. However, this technique has some disadvantages, as it requires additional surgical area to obtain the donor, resulting in a longer surgical procedure, increased patient morbidity, prolonged intra- and post-operative bleeding, palatal sensory dysfunction, and infection [3–5]. The lack of available tissue for autografts has instigated researchers across the globe to find alternatives to autografts. Research in soft tissue regeneration conducted in the last two decades has addressed this widespread issue by developing and integrating highly biocompatible yet sensitive materials to obtain alternative biomaterial scaffolds that can substitute connective tissue grafts for soft tissue regeneration [6,7].

Scaffolds are defined as biomaterials with a three-dimensional solid porous structure that plays a role in the promotion of interactions between cells and biomaterials, adhesion of cells, and the deposition of ECM; allows suitable transport of gasses and nutrients that permits the survival, differentiation, and proliferation of cells; and triggers a minimal toxicity or inflammation degree in vivo [8,9]. Biophysical cues such as scaffold physical properties, degradation, and architectural morphology and biochemical cues for exploiting natural molecules and spatiotemporal delivery of biomolecules are important factors to consider in designing biomaterial scaffolds for tissue regeneration [10,11]. A scaffold should maintain its shape during application, be easy to handle, and not cause damage to the tissue [12].

Hydrogels are defined as an insoluble three-dimensional network of polymer matrices made from macromers, hydrophilic homopolymers, and copolymers [13]. Different types of hydrogels have been tested for their potential use in tissue engineering. The most used hydrogels are made of synthetic monomers such as poly-ethylene glycol (PEG), poly-vinyl alcohol (PVA), and poly-2-hydroxyethyl methacrylate (PHEMA) and natural monomers such as agarose, alginate, hyaluronic acid, fibrin, and collagen [14]. The first use of hydrogels in the biomedical field consisting of crosslinking 2-hydroxyethyl methacrylate was introduced by Wictherle and Lim in 1960 [15]. Hydrogels are hydrophilic polymer networks that are capable of imbibing large amounts of water to mimic the extracellular matrix (ECM) [16]. They act as a temporary matrix when they are employed in order to secure the proliferation of cells, the deposition of the ECM, and tissue ingrowth until the regeneration of the newly desired tissue is achieved [17].

In recent years, hydrogels received attention in tissue engineering since they have a unique composition and structure resembling the natural ECM, serving as a desirable scaffold for cellular migration, proliferation, differentiation, neovascularization, and biomineralization [13,18]. The other advantages of hydrogels are their ability to absorb exudate from the wound surface and their ability to promote fibroblast proliferation, cell migration, and keratinization [15]. Furthermore, the shape adaptability characteristics of hydrogels for minimally invasive procedures make these biomolecules attractive as a scaffold biomaterial for tissue engineering purposes, including for oral tissue regeneration [18–20]. Studies have demonstrated the potential of hydrogels for the regeneration of dentin and dental pulp and tooth-supporting structures of periodontal tissues. The goal is to reconstruct the architecture and function of dental tissues and periodontal tissues including gingiva, periodontal ligament, cementum, and alveolar bone [13] (Figure 1).

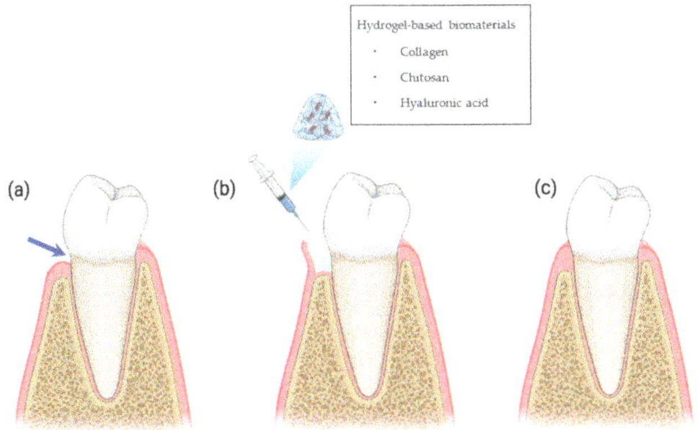

Figure 1. Schematic illustration of hydrogels for soft tissue regeneration. (**a**) Gingival recession indicated by blue arrow; (**b**) root coverage procedure followed by application of hydrogel-based scaffold; (**c**) healing of gingival tissue by the regeneration of the epithelial and the underlying connective tissue.

Gingival tissues are the part of the oral epithelium that covers the alveolar bone of mandibular and maxillary bone. Numerous reports on the physical and biological properties of hydrogels demonstrated the increasing interest in this polymer matrix for its potential use in gingival tissue regeneration. Before its therapeutic use in human tissue, in vitro experiments were an important approach to understand the physical properties and biological properties of the biomaterials. The present systematic review aimed to identify and synthesize in vitro studies on hydrogels as biomaterial scaffolds for gingival regeneration based on their physical and biological properties. Systematic reviews of in vitro studies present an opportunity to synthesize evidence from numerous studies that address the same topic as new approaches for clinical translation.

2. Materials and Methods

The review protocol of this study was registered to the Open Science Framework Database (Registration DOI https://doi.org/10.17605/OSF.IO/ZP9DT), accessed on 30 April 2023.

2.1. Search Strategy

The review process followed the Preferred Reporting System for Systematic Reviews and Meta-Analyses (PRISMA) 2020 statement guidelines to review the physical and biological properties of hydrogels for gingival tissue regeneration (Figure 2). The research question was formulated using the population, intervention, comparison, and outcome (PICO) approach. Full-text papers written in English from PubMed, Embase, ScienceDirect, and Scopus data and a manual search of publications in the last 10 years were used (February 2012–February 2022). The search strategies, including the combination of keywords and the number of articles retrieved, are described in Table 1.

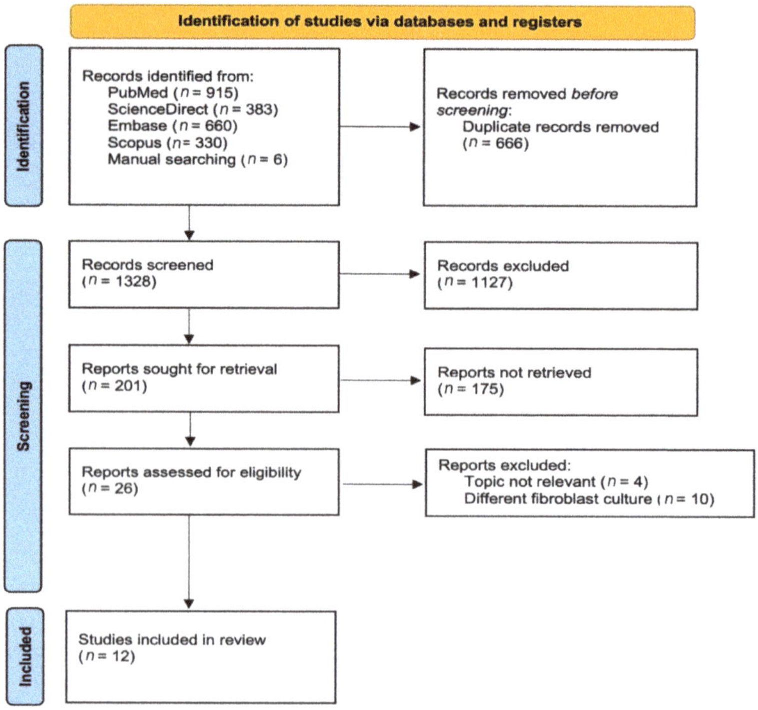

Figure 2. PRISMA flowchart for the screening protocol.

Table 1. Databases and research method.

Database	Search Strategy
PubMed	(hydrogel) AND (injectable)) OR (gingival [MeSH Terms])) OR (periodontal [MeSH Terms])) AND (regeneration [MeSH Terms])) AND (natural polymer)) OR (synthetic polymer)) AND (in vitro [MeSH Terms])
Science Direct	hydrogel OR injectable AND natural polymer OR synthetic polymer AND gingival regeneration OR periodontal regeneration AND in vitro
Embase and Scopus	(hydrogel:ti OR injectable:ti) AND ('natural polymers' OR ('natural'/exp OR natural) AND ('polymers'/exp OR polymers)) OR 'hyaluronic acid'/exp OR 'hyaluronic acid' OR (hyaluronic AND ('acid'/exp OR acid)) OR collagen OR 'chitosan'/exp OR chitosan OR 'gelatin'/exp OR gelatin OR 'cellulose'/exp OR cellulose OR 'hyaluronan'/exp OR hyaluronan OR 'agarose'/exp OR agarose OR 'natural scaffold' OR (('natural'/exp OR natural) AND ('scaffold'/exp OR scaffold)) OR 'synthetic polymer'/exp OR 'synthetic polymer' OR (synthetic AND (polymer'/exp OR polymer)) OR 'polyester'/exp OR polyester OR 'poly aci' OR (poly AND lactic AND aciD) OR 'poly glycolic acid'/exp OR 'poly glycolic acid' OR 'plga'/exp OR plga OR 'pga'/exp OR pga OR 'peg'/exp OR peg OR (poly AND ('ethylene glycol'/exp OR 'ethylene glycol' OR (('ethylene'/exp OR ethylene) AND ('glycol'/exp OR alcohol)) I) OR (poly AND ('vinyl alcohol'/exp OR 'vinyl alcohol' OR (('vinyl'/exp OR vinyl) AND ('alcohol'/exp OR alcohol)) AND 'gingival regeneration' OR 'gingival fibroblasts' OR (gingival AND ('regeneration'/exp OF regeneration))

2.2. Eligibility Criteria of the Articles

Two reviewers (D.I.H and L.R.A) independently searched the titles and abstracts from electronic databases. Full reports were retrieved and examined independently in duplicate, and any disagreement was resolved with a third reviewer (D.F.S) if necessary to obtain relevant data for data analyses. The articles were selected for full-text reading and were assessed by individual reviews independently to obtain relevant data for qualitative review.

2.3. Inclusion and Exclusion Criteria

Full-text articles were considered eligible if they satisfied the following criteria: (1) in vitro experimental setup on primary or cell line of gingival fibroblast culture, (2) studies that analyzed the physical and/or biological properties of hydrogel. Meanwhile, as the current systematic review focused on the physical and biological properties of hydrogels, the exclusion criteria were as follows: (1) experiments other than on primary or cell line of gingival fibroblast culture, (2) clinical or randomized control trials or case reports or case studies, (3) narrative reviews or meta-analyses, (3) no abstract presented, and (4) duplicate publications.

2.4. Quality Assessment

To analyze the quality of the included studies, we used the clinical experimental checklist for non-randomized experimental studies by the Joanna Briggs Institute. The checklists were modified based on the research questions and PICOS structure [21]. The quality of each study was assessed using the nine questions presented in Table 2.

Table 2. Quality assessment.

Authors	Joanna Briggs Institute Items									Raw Score and %	Risk
	X	Q2	Q3	Q4	Q5	Q6	Q7	Q8	Q9		
Cheung [22]	1	1	1	1	1	1	1	1	U	89%	Low
Choi [23]	1	1	1	1	1	1	1	1	U	89%	Low
Colangelo [24]	1	1	U	1	1	1	1	1	U	78%	Low
Cozens [25]	1	1	0	1	1	1	1	1	U	78%	Low
Kang [26]	1	1	1	0	1	1	1	1	U	78%	Low
Laird [27]	1	1	1	1	1	1	1	1	U	89%	Low
Miranda [28]	1	1	1	1	1	1	1	1	U	89%	Low
Moatary [29]	1	1	1	1	1	1	1	1	U	89%	Low
Montalbano [30]	1	1	1	1	1	1	1	1	U	89%	Low
Rosdiani [31]	1	1	0	1	1	1	1	1	U	78%	Low
Tabatabaei [32]	1	1	1	1	1	1	1	1	U	89%	Low
Zhou [33]	1	1	0	0	1	1	1	1	U	67%	Moderate

Q1: Were experimental conditions identical across study groups? Q2: Were all measured outcomes reported? Q3: Was there a control group? Q4: Was the rationale for hydrogel mixture concentration used explained? Q5: Were there multiple measurements of the outcome both before and after the intervention/exposure? Q6: Was follow-up complete and, if not, were differences between groups in terms of their follow-up adequately described and analyzed? Q7: Were the outcomes of participants included in any comparisons measured in the same way? Q8: Were the quantitative data mentioned? Q9: Were the experiments performed in replicates?

3. Results

3.1. Selection of Studies

The electronic and manual searches found 1328 articles of potential relevance after the removal of duplicates. Title and abstract screening were performed to meet the inclusion criteria. In total, 39 studies were fully analyzed to precisely check for the exclusion criteria, of which 27 articles were excluded. Finally, 12 articles were identified and included in this review. From the 12 included articles, 1 study performed a physical property analysis, 2 studies performed a biological property analysis, and 9 studies performed both physical and biological property analyses.

3.2. Characteristics of Studies

The selection of biomaterial combinations is an important consideration for the formulation of hydrogel-based scaffolds. Table 3 presents the 12 articles included in this review, which used 13 different biomaterials. The most common biomaterial used was collagen (five articles), followed by chitosan (three articles) and hyaluronic acid, PEG, and fibrin (two articles).

Table 3. Characteristics of the included studies.

No.	Author	Biomaterial	Crosslinking Method	Fabrication
1	Cheung [22]	DVO, PEG	Chemical	D-PHI flat films were generated using a divinyl oligomer (DVO); PEG was used as porogen
2	Choi [23]	Collagen, growth factor (TGF-b1), nanoparticles (gold, TaO, dextran, or ferritin)	Chemical	Collagen type I from rat tail tendons was loaded with either TGF-b1 or other nanoparticles such as gold, TaO, dextran, or ferritin
3	Colangelo [24]	PN, HyA	Chemical	PN extraction from salmon trout gonads mixed with HyA

Table 3. *Cont.*

No.	Author	Biomaterial	Crosslinking Method	Fabrication
4	Cozens [25]	PAA, Tyr, Cys, BA, BP	Chemical	Coupling amine to PAA using 4-(4,6-dimethoxy-1,3,5-triazin-2-yl)-4-methylmorpholinium chloride (DMTMM) and then crosslinking with Tyr/Cys/BA/BP
5	Kang [26]	KRT, FIB	Chemical	Keratin from human hair mixed with fibrinogen from human plasma
6	Laird [27]	PEG, gyrase B (with/without RGD motifs), coumermycin, novobiocin	Chemical	PEG added to mixture of Novobiocin (antibiotic) or Coumermycin (dimeric form of novobiocin) and GyrB with/without RGD motifs (protein expressed by *Eschericia coli*)
7	Miranda [28]	Chitosan, HyA	Chemical	Medium-molecular-weight chitosan was succinylated, and HyA oxides were mixed
8	Moatary [29]	Nano diopside, b-chitin, chitosan	Chemical	Nano diopside ceramic was prepared by a modified sol–gel method then added to sulfate derivative of chitin and chitosan
9	Montalbano [30]	Collagen, alginate, fibrin	Chemical	Collagen type I from calf skin, low-viscosity alginate from brown algae, fibrinogen from bovine plasma
10	Tabatabaei [32]	Collagen, GelMA	Chemical	Cell-embedded collagen and collagen-embedded gelatin methacryloyl (GelMA)
11	Rosdiani [31]	Collagen, chitosan, glycerol	Chemical	Collagen and chitosan powders from tissue banks were dissolved into the solvents, and then glycerol was added
12	Zhou [33]	PVA, collagen	Chemical	Tilapia collagen type I mixed with PVA

PN = polynucleotides, HyA = hyaluronic acid, Cys = cystine, Tyr = tyramine, BA = bromine moieties, BP = boronic acid moieties, KRT = keratin, CS = chitosan, FIB = fibrin/fibrinogen, Col = collagen, TGF = transforming growth factor, TaO = tantalum oxide, PEG = polyethylene glycol, DVO = divinyl oligomer.

As biomaterial scaffolds, including hydrogel, are used for in situ tissue regeneration, the scaffold should have the ideal physical and biological properties before its application to human tissue. Of the 12 articles, we categorized the selected articles into two categories: (i) 9 articles that studied the physical properties of hydrogel and (ii) 10 articles that studied the biological properties of hydrogel in vitro.

3.3. *Result of Studies Included*

3.3.1. Physical Property Analyses

Hydrogels are hydrated systems that consist of crosslinked hydrophilic units forming compact and stable polymer networks [34]. The fabrication of the hydrogels mostly involves physical or chemical crosslinking methods to obtain stable polymer networks [35]. All studies in this review performed a chemical crosslinking process to synthesize the hydrogels. The most popular chemical crosslinking techniques are complementing the group's chemical reaction, high-energy radiation, free radical polymerization, and enzyme-mediated crosslinking [36]. The chemical crosslinking method produces hydrogels with higher mechanical properties and stability compared to physical crosslinking [37]. Increasing crosslinking density improved the stiffness, toughness, and degradation time due to a relatively more stable structure and more elasticity [38].

Of the nine studies that performed a physical analysis of the hydrogel properties, five studies analyzed the characteristics and swelling ratio. Scaffolds for tissue engineering initially fill a space in tissue and serve as a temporary matrix for newly regenerated tissue. Physical properties such as gel formation, mechanical characteristics, morphology, and biodegradation behavior of the biomaterials are important to the success of the scaffold [39]. Table 4 presents the biophysical characteristics of hydrogels designed for gingival tissue regeneration. Pores are important for cell proliferation and for nutrient and gas transfer while maintaining good mechanical properties of the scaffold [40]. Fibroblasts showed optimal cell proliferation in scaffolds with a pore size range of 50–160 μm and 86% porosity [41,42]. Hydrogels composed of combinations of PEG DVO, keratin fibrin, collagen alginate fibrin, collagen chitosan glycerol, and PVA collagen showed suitable pore sizes for fibroblast ingrowth [22,26,30,31,33].

Table 4. Physical properties of hydrogels.

No.	Authors	Type of Scaffold	Physical Property Analyses	Main Findings
1	Cheung et al. [22]	- PEG + DVO	- Characterization	- Size of pores: 30–250 μm.
2	Choi et al. [23]	- Collagen + TGFb-1 - Collagen + gold - Collagen + TaO - Collagen + dextran - Collagen + ferritin	- Characterization	- Hydrogel loaded with gold or TaO had smallest pore size, while hydrogel loaded with dextran showed largest pore size.
3	Cozens et al. [25]	- PAA + Cys - PAA + Tyr - PAA + BP - PAA + BA	- Morphology - Rheology - Hydrogel bonding	- PAA-BP and PAA-Cys showed the highest adhesion strength and energy density for porcine keratinized gingiva. - PAA-BP produced high moduli and elasticity, resulting in strong adhesion to tissues. - PAA-Tyr and PAA-BA displayed weak mechanical properties.
4	Kang et al. [26]	- Keratin + Fibrinogen (in different concentration)	- Morphology - Rheology - Injectable performances - Swelling capacity - Biodegradation	- The porosity and viscosity of keratin–fibrinogen hydrogel were improved by increasing the molar ratio of KRT to FIB so the hydrogel could be injected through needle extrusion. - The molar ratio of KRT:FIB = 3:1 was the most suitable for the biomaterial scaffold with a highly porous structure (10–100 mm). - The 3:1 mol ratio of KRT:FIB had a swelling ratio almost 6 times higher than that of FIB.

Table 4. Physical properties of hydrogels.

No.	Authors	Type of Scaffold	Physical Property Analyses	Main Findings
5	Miranda et al. [28]	- Chitosan - HyA - Chitosan + HyA (in different concentration)	- Characterization - Swelling capacity	- HyA and CS-HA hydrogels showed irregular structures with higher pore sizes compared to CS hydrogels. - HyA and CS scaffolds showed higher swelling ratios.
6	Moatary et al. [29]	- Chitin hydrogel +/chitosan/nano diopside	- Characterization - Porosity - Water uptake - Mechanical - Swelling capacity - Degradation	- Chitin/chitosan/nano diopside had homogeneous porous structures, suitable for scaffolds. - They were to retain water, so the internal surface area was increased. - Increased mechanical stability. - Addition of nano diopside controlled the swelling rate and reduced the biodegradation rate.
7	Montalbano et al. [30]	- Collagen + Alginate + Fibrin	- Characterization - Gelation time - Swelling - Degradation - Rheology - Degradation	- Pore size: 40–120 µm. - Increased collagen concentration and less interconnected pores were observed, along with more stiffness and a decrease in the swelling ratio. - Gelation time happened at temperature of 37. - Completely dissolved in 3 h; degradation rate faster in low collagen concentration.
8	Rosdiani et al. [31]	- Collagen-Chitosan-Glycerol (in different concentration)	- Morphology - Mechanical strength - Swelling capacity	- The pore size obtained ranges between 102.4 and 143.5 µm, which is suitable for periodontal application. - Greater collagen concentration showed better mechanical strength but the lowest swelling capacity.
9	Zhou et al. [33]	- PVA + Col (in different concentration)	- Morphology - Swelling capacity	- PVA/Col (90:10) blended hydrogel exhibited a homogeneous dense surface layer, and small pores were observed at high magnifications; the addition of Col increased the porosity and pore size of the hydrogel and the swelling rate.

3.3.2. Biological Property Analyses

Figure 1 shows the 12 studies included in this systematic review. Overall, 10 studies produced hydrogel formulations using natural polymers (Col, CS, and HyA) as the biomaterials (Table 5). Collagen originates from the tissues of humans or any other species, has special physicochemical properties and architectural features, and is the most widely used natural polymer in biomedical fields [23,30–33,43]. It has the ability to promote cell adhesion, chemotaxis, homeostasis, physical degradation, and low toxicity, as well as induce the stimulation of fibroblasts' DNA [42,43]. The characteristics of collagen were improved in combination with other polymers such as chitosan, gelatin, alginate, and PVA;

growth factors; or bioparticles or by modifying its functional chain in the crosslinking process [30,31,33,44].

Table 5. Biological properties of hydrogels.

No.	Author	Type of Scaffold	HGF Cell Culture	Biological Property Analyses	Main Findings
1	Cheung et al. [22]	- PEG + DVO	ATCC	- Cytotoxicity - Cell proliferation - DNA mass quantification - Metabolic activity - Histology - Cell morphology - Protein measurement	- Experiments were performed in static and dynamic cell cultures. - HGF population remained viable and increased over 14 days of culture. - Metabolic activity increased significantly in the fourth week. - More cells were found in the inner regions of the scaffold in the dynamic culture, particularly on days 14 and 28. - Accumulation of Col I in the dynamic culture was greater than that in the static culture.
2	Choi et al. [23]	- Collagen + TGFb-1 - Collagen + gold - Collagen + TaO - Collagen + dextran - Collagen + ferritin	Primary HGF	- Cytotoxicity - Cell proliferation	- TGF-β1, in facilitating the recovery and proliferation of human gingival cells, was assessed when TGF-β1 was incorporated into and excreted from physiologically degraded collagen hydrogels. - TGF-β1 and TGF β1-carrying hydrogels showed larger amounts and metabolic activity of live cells. - Cells became attached and grew well due to the surface of collagen hydrogels.
3	Colangelo et al. [24]	- PN - HyA - PN + HyA	ATCC	- Cytotoxicity	- PN increased the assay signal in a visible and significant way; PN + HA failed to further increase cell growth as compared to PN alone.
4	Kang et al. [26]	- Keratin + Fibrinogen (in different concentration)	Primary HGF from ScienCell	- Cytotoxicity	- KFH exhibited much higher cell viabilities and supported cell proliferation. - KFH was more resistant to high-temperature and acidic conditions than FIB-H, protecting the integrity of its structure.
5	Laird et al. [27]	- PEG + gyrase B (with/without RGD motifs) + coumermycin, - PEG + gyrase B (with/without RGD motifs) + novobiocin	Primary HGF	- Cytotoxicity - Cell morphology	- Cells grown on hydrogels with RGD motifs showed normal morphology; meanwhile, cells exhibited unusual morphology indicating cell apoptosis on hydrogels without RGD motifs. - Hydrogels containing novobiocin exhibited a reduction in cell number.

Table 5. *Cont.*

No.	Author	Type of Scaffold	HGF Cell Culture	Biological Property Analyses	Main Findings
6	Miranda et al. [28]	- Chitosan - HyA - Chitosan + HyA (in different concentration)	NIH3T3	- Cytotoxicity - Cell morphology	- An increase in cell viability was observed after the cells were seeded on the scaffold. - Migration rates of the cells were higher when they were seeded on the scaffold.
7	Moatary et al. [29]	- Chitin hydrogel +/chitosan/nano diopside	ESK-1	- Cytotoxicity - Cell attachment	- Chitin/chitosan/nano diopside composite scaffolds were cytocompatible and non-toxic to fibroblast cells. - Nanosurfaces had larger surface areas, which allowed the attachment of more cells.
8	Montalbano et al. [30]	- Collagen + Alginate + Fibrin	L929	- Cytotoxicity - DNA quantification - Metabolic activity - Cell morphology	- Cells were still viable with intact cell membranes. - No significant change in cell morphology was observed with changing collagen concentration. - Cells were proliferating; no significant differences in cell number and proliferation rate were observed.
9	Tabatabaei et al. [32]	- Gelatin methacryloyl (GelMA) - Collagen	Primary HGF	- Cytotoxicity - Morphology	- Cells were more likely to survive in collagen hydrogels. - GelMA hydrogels maintain cell initial shape and size without noticeable contraction.
10	Zhou et al. [33]	- PVA + Col (in different concentration)	HGFs #2620	- Cytotoxicity - Cell morphology	- Pure PVA showed lowest cell viability and cell adhesion; meanwhile, the highest was in PVA/Col (50:50). - On PVA/Col (100:0, 90:10, 70:30) blended hydrogels, HGFs appeared as round cells without pseudopods. HGFs with pseudopods were found in the scaffolds of the PVA/Col (50:50).

All studies included in this review performed a cytotoxicity assay, as toxicity screening is mandatory in the development of new biomaterials for pharmaceutical safety and highlights a crucial step in formulating scaffolds [45]. However, the lack of bioactive components is still a major challenge for tissue engineering as the cells cannot proliferate, differentiate, or migrate. The chemical modification allows the incorporation of natural scaffolds or bioactive molecules to increase its biological capability to ensure cell biology performance similar to the native environment [46].

4. Discussion

The trend of the development of biomaterial scaffolds in the periodontology field has increased in the last decade. Tissue engineering scaffolds provide a substitute for autografts for gingival tissue regeneration. Scaffolds interact with human tissue through their physical and biological properties by altering tissue microenvironments via modulating the host immune system and controlling the kinetics and healing capabilities of endogenous cells [47]. This systematic review has evidenced the biophysical and biological properties of various hydrogel formulations in vitro.

Based on origin, hydrogels can be classified as natural, synthetic, or a combination of both. Natural polymers are derived from biological materials present in nature that are extracted through physical or chemical methods. Instead, synthetic hydrogels are produced from synthetic polymers such as plastics, elastomers, and synthetic fibers. Both natural and synthetic hydrogels are developed through physical or chemical crosslinking, and they can be combined to improve their physical characteristics [44,48,49]. Natural polymers show good biological properties since they have physical and biological characteristics recognized in human tissue, while synthetic polymers have greater control of the characteristics and properties for tissue regeneration [13,50–52]. The structure of synthetic polymers can be modified to improve their biological performance. Hence, to improve hydrogel properties, it is applicable to combine natural polymers with synthetic polymers [53]. Pure PVA is a super-hydrophilic polymer that has poor cell adhesion properties due to its low affinity to protein. To overcome this limitation, physical modifications, such as extracellular matrix coating and air plasma treatment, have been shown to increase hydrogel surface roughness and have a positive effect on the cytoskeleton arrangement to promote cell attachment [54].

Ideally, scaffolds should have mechanical properties consistent with the desired tissue formed that are strong enough to facilitate good handling during implantation [55]. Due to their high water content, porosity, and mechanical tunability, hydrogels are particularly attractive as extracellular matrices to stimulate the regeneration of soft tissue [56]. The elastic characteristic of hydrogels resembles that of rubber, and existing theories of rubber elasticity are in line with hydrogel deformation during swelling [57]. The increase in pore size permits cells to avail the maximum internal surface of the scaffolds and facilitates cell infiltration into the scaffolds [29]. Six studies have demonstrated a high swelling rate of hydrogels [26,28–31,33]. The high swelling rate exhibits the water-absorbing capability of hydrogels making a moist environment for tissue engineering [58]. The swelling rate of hydrogels can be regulated by the mixture concentration. Montalbano and Rosdiani demonstrated low collagen concentration in the hydrogel mixture resulting in a decrease in swelling ratio [30,31]. Collagen has many hydrophilic groups to absorb water from its surroundings [31]. The swelling ratio is a parameter that can be used to examine the increase in a hydrogel's weight due to water absorption, which could alter the cell growth on the hydrogel scaffold. A swelling ratio that is too fast may cause hydrogel network collapse before the desired new tissue has formed, whereas a swelling ratio that is too slow may inhibit the growth of cells inside the scaffold, thus preventing tissue formation within the scaffold [59].

Hydrogels must have appropriate mechanical and biological properties consistent with the desired tissue formed. They must support tissue growth and proliferation without causing toxic reactions in the cells. The structure of the hydrogels must be compatible with human cells and tissue without causing toxicity or altered immune reactions in the host after it degrades [60,61]. Previously, it was reported that most hydrogel mixtures have good biocompatibility due to the elastic and soft nature of the hydrogels that minimizes their irritability toward cells [36]. The existence of synthetic polymers in hydrogels has shown lower cell viability [32,33]. Synthetic polymers were reported to be less biocompatible compared to natural polymers, as they have no bioactive capacity [62]. The biofunctionalization of synthetic polymer structures was reported to increase scaffold–cell or scaffold–tissue interactions. Synthetic polymers can also be mixed with natural polymers or growth factors [63].

The extracellular matrix is a bioactive scaffold that contains many natural polymers that have the capacity to promote various types of tissue-specific cues, giving biological cues to the microenvironment surrounding the scaffold and ultimately promoting tissue regeneration [64]. Growth factors (GFs) are blood-derived peptide factors that direct inflammatory cells to migrate onto the wound, attract fibroblasts, and stimulate cell proliferation [65]. The addition of GFs as bioactive molecules into a hydrogel scaffold is a strategy to enhance the healing process [66]. For example, the incorporation of TGFβ-1 into a collagen hydrogel was shown to enhance fibroblast proliferation [23]. Gingival fibroblasts differentiate to myofibroblasts in response to the presence of TGFβ-1 in a stiff collagen substrate [67]. Myofibroblasts are a contractile phenotype of fibroblasts due to the upregulation of α-smooth muscle actin and matrix metalloproteinase [22]. They exhibit reinforced adhesion to the ECM and increased ECM remodeling [68]. Even though GFs are useful in designing a scaffold, several drawbacks of growth factor application in tissue regeneration have been reported, such as their short half-life after being delivered in vivo, poor stability, and systemic toxicity due to over-release [69–71]. Studies have shown hydrogels' property of absorbing a large number of tissue exudates at the wound site to maintain a moist environment. This condition is favorable for growth factor release and subsequently promotes cell proliferation and differentiation to obtain rapid wound healing.

Gelation kinetics play a role in hydrogel delivery. Mechanisms of gel formation dictate how cells and molecules are integrated into the scaffold and how that scaffold is delivered [39]. Injectable hydrogels can be formed as a free-flowing solution that can be transformed into a semi-solid form under certain circumstances [72]. Thermo-responsive hydrogels are supramolecular hydrogels that undergo a gelation process via hydrophobic interaction. Hydrogels can undergo sol–gel phase transition because they consist of amphiphilic polymers with hydrophilic and hydrophobic components [73]. Hydrogels can undergo a gelation process from room temperature to body temperature (range of 25–37 °C), which is beneficial for their application in tissue engineering [58]. Collagen/alginate/fibrin hydrogels showed sol-phase transition at a temperature of 37 °C [30]. Slow gel formation results in poor adaptation and network formation, altering the cell encapsulation as the therapeutics may flow away from the targeted site. Furthermore, fast gel formation makes it difficult to inject in the targeted site [74,75].

The biodegradable property of biomaterials resembles an extracellular matrix and promotes tissue regeneration [34]. The biodegradation of hydrogels occurs based on several mechanisms, such as hydrolysis, photolysis, separation, or combinations thereof [76]. Scaffold degradation is mainly a chemical process, but it can be determined by physical stimuli to influence cell function and behavior. Degradation of the scaffold is followed by a decrease in its stiffness and is controlled by different polymer concentrations [30,77]. Longer hydrogel degradation occurred when the collagen concentration was reduced due to the presence of collagenase in the tissue host [30]. Ideally, the scaffold degradation rate is parallel to the rate of new tissue regeneration.

Fibroblasts have been regularly used to examine scaffold biocompatibility for soft tissue engineering and are the most abundant resident cells in the periodontium [22,78,79]. Approximately 2.10^8 fibroblasts are estimated per 1 cm^3 of connective tissue, equal to 5% volume. The goal of designing a biomaterial scaffold is to provide a temporary framework for the migration and attachment of fibroblasts from areas adjacent to the wound [65,80]. In soft tissue regeneration, the first step is gingival fibroblast migration to the scaffold and closing the dehiscence wound [81]. A recent study demonstrated that the addition of HyA to chitosan hydrogels and HyA showed higher fibroblast migration into the scaffold compared to chitosan alone [28]. The *CD44* receptors in fibroblast surfaces bind to HyA [82,83]. The ability to promote cell adhesion is also an important hydrogel property. The tripeptide arginine–glycine–aspartic acid (RGD) motif is a well-known general recognition motif acting via cell surface integrin receptions and is present in various extracellular matrix proteins [84]. The RGD peptide is commonly used to improve cell adhesion properties [84,85]. Cells seeded on hydrogels containing RGD showed good cell morphology and a greater number of cells compared to the non-RGD hydrogels [27]. The incorporation of the RGD peptide is useful for promoting the cell adhesion of synthetic materials [38]. RGD binds to several surface receptors due to the presence of integrin in the cell membranes. To date, 24 integrins have been identified to bind with ECM proteins containing RGD peptide sequences [86].

A mixture of gelatin and methacryloyl showed a decrease in cell viability because the high-density scaffold network reduced its porosity and, therefore, the transport of cell nutrients [32,87]. A similar result was found using PVA as the combination [33]. PVA has very low interactions with protein, resulting in less cell deposition on the scaffold surface. The residual charge of this polymer caused a disruption in the cell membrane [88]. Furthermore, any initiators, crosslinkers, or other additives used in hydrogels must not cause cellular toxicity. The polymerization process of hydrogels should not cause increased temperatures that could result in thermal necrosis on the tissue host [38]. PVA/collagen (50:50) hydrogels induced the differentiation of fibroblasts to form pseudopods [33]. Pseudopodia are essential for cell movement since they determine the trajectory, direction, and speed of cell migration [89]. In contrast, GelMA hydrogels failed to reveal fibroblast contraction after hydrogel implantation; this may alleviate the proliferation capacity of fibroblasts [32]. The migration and proliferation of fibroblasts at the defect sites lead to collagen deposition [90]. Collagen is a major protein secreted by fibroblasts in gingival ECM that serves as a structural scaffold in tissues [91,92]. The ultimate goal in tissue regeneration is to restore the integrity of the ECM structure and collagen networks under which the physiological regeneration occurs [93].

Although the majority of hydrogels are currently still under investigation, in vitro studies as well as numerous in vivo studies reported promising results that may lead to clinical studies. In vivo biocompatibility of hydrogels was observed, confirming the non-toxic nature of the biomaterials toward normal tissues [94]. Future research development of hydrogels as biomaterial scaffolds for gingival regeneration is promising. For example, the development of 3D printing technologies in tissue engineering has been increasing over the last decade [95,96]. This technology is able to produce an individualized scaffold to mimic tissue behavior with precise geometry matching the defects [97]. Peng et al. successfully constructed a 3D printing scaffold consisting of acellular dermal matrix/gelatin-sodium alginate (ADM/A/G) with living gingival fibroblasts which was able to promote cell proliferation and increase the expression of specific tissue biomarkers in vitro [98].

Strength and Limitations

This article presents a systematic review of the biophysical and biological properties of various hydrogels for gingival tissue regeneration. Although there are many reviews that have focused on hydrogel use in tissue engineering, we still found a lack of articles that have focused on hydrogel use in gingival tissue regeneration. One limitation of this study is that few previous studies have analyzed hydrogels as gingival scaffolds, and some authors did not reveal the uniform analysis performed in their studies. Further research is needed to develop hydrogel scaffolds with tunable mechanical properties that induce the regenerative process through the stimulation of cell function, the matching of the degradation rate and the physiological remodeling process, and functionalization to enhance cell interaction. Hence, extensive in vivo and clinical trial studies should be further investigated for the application of hydrogel scaffolds in gingival tissue engineering, which is also one of the main directions of biomaterial research and development in the future.

5. Conclusions

The trend of substitute biomaterial research in the periodontology field, especially for gingival regeneration, is increasing, as it potentially reduces the use of autografts and thereby reduces patient morbidity. Hydrogels are potential biomaterials to be used as a substitute for autografts. In this systematic review, the authors summarized research on hydrogels as scaffold biomaterials in vitro. Natural hydrogels are still the most popular polymers in designing hydrogels. Further research is necessary to determine which composition suits clinical uses in daily practice following the parameters that are discussed in this systematic review.

Author Contributions: Conceptualization, D.I.H., D.F.S. and L.A.; methodology, D.I.H., D.F.S. and L.A.; formal analysis, D.I.H., D.F.S., E.W.B., Y.S. and L.A.; data curation, D.I.H. and L.A.; writing—original draft preparation, D.I.H.; writing—review and editing, D.F.S., E.W.B., Y.S. and L.A.; supervision, D.F.S., E.W.B., Y.S. and L.A.; funding acquisition, D.I.H. and L.A. All authors have read and agreed to the published version of the manuscript.

Funding: Faculty of Dentistry Universitas Indonesia supported the publication funding.

Institutional Review Board Statement: Ethical review and approval were waived for this systematic review study.

Informed Consent Statement: Not applicable.

Data Availability Statement: Not applicable.

Conflicts of Interest: The authors declare no conflict of interest.

References

1. Zangrando, M.S.; Valle, L.A.; Stuani, V.D.; Costa, M.R.; Shibuy, S.K.; Damante, C.A. Clinical Outcomes of Root Cov-erage with Subepithelial Connective Tissue Graft According to Site Specific Factors—Longitudinal Retrospective Clinical Study. *J. Int. Acad. Periodontol.* **2019**, *21*, 159–167. [PubMed]
2. Adam, K.; Staufenbiel, I.; Geurtsen, W.; Günay, H. Root coverage using a connective tissue graft with epithelial striation in combination with enamel matrix derivatives—A long-term retrospective clinical interventional study. *BMC Oral Health* **2019**, *19*, 148. [CrossRef] [PubMed]
3. Wang, H.-L.; Romanos, G.E.; Geurs, N.C.; Sullivan, A.; Suárez-López Del Amo, F.; Eber, R.M. Comparison of Two Differently Processed Acellular Dermal Matrix Products for Root Coverage Procedures: A Prospective, Randomized Multicenter Study. *J. Periodontol.* **2014**, *85*, 1693–1701. [CrossRef]
4. Tavelli, L.; Barootchi, S.; Di Gianfilippo, R.; Modarressi, M.; Cairo, F.; Rasperini, G.; Wang, H.-L. Acellular Dermal Matrix and Coronally Advanced Flap or Tunnel Technique in the Treatment of Multiple Adjacent Gingival Recessions. A 12-Year Follow-up from a Randomized Clinical Trial. *J. Clin. Periodontol.* **2019**, *46*, 937–948. [CrossRef] [PubMed]
5. Tavelli, L.; McGuire, M.K.; Zucchelli, G.; Rasperini, G.; Feinberg, S.E.; Wang, H.; Giannobile, W.V. Extracellular matrix-based scaffolding technologies for periodontal and peri-implant soft tissue regeneration. *J. Periodontol.* **2019**, *91*, 17–25. [CrossRef] [PubMed]
6. Mantha, S.; Pillai, S.; Khayambashi, P.; Upadhyay, A.; Zhang, Y.; Tao, O.; Pham, H.M.; Tran, S.D. Smart Hydrogels in Tissue Engineering and Regenerative Medicine. *Materials* **2019**, *12*, 3323. [CrossRef]
7. Xu, Y.; Yuan, S.; Han, J.; Lin, H.; Zhang, X. Design and fabrication of a chitosan hydrogel with gradient structures via a step-by-step cross-linking process. *Carbohydr. Polym.* **2017**, *176*, 195–202. [CrossRef]
8. Dhandayuthapani, B.; Yoshida, Y.; Maekawa, T.; Kumar, D.S. Polymeric Scaffolds in Tissue Engineering Application: A Re-view. *Int. J. Polym. Sci.* **2011**, *2011*, 290602. [CrossRef]
9. Chaudhary, S.; Chakraborty, E. Hydrogel based tissue engineering and its future applications in personalized disease modeling and regenerative therapy. *Beni-Suef Univ. J. Basic Appl. Sci.* **2022**, *11*, 1–15. [CrossRef]
10. Carnes, M.E.; Pins, G.D. Skeletal Muscle Tissue Engineering: Biomaterials-Based Strategies for the Treatment of Volumetric Muscle Loss. *Bioengineering* **2020**, *7*, 85. [CrossRef]
11. Du, Y.; Yu, M.; Lu, W.; Kong, J. Three-dimensional (3D), macroporous, elastic, and biodegradable nanocomposite scaffold for in situ bone regeneration: Toward structural, biophysical, and biochemical cues integration. *Compos. Part B Eng.* **2021**, *225*, 109270. [CrossRef]
12. Mao, Z.; Fan, B.; Wang, X.; Huang, X.; Guan, J.; Sun, Z.; Xu, B.; Yang, M.; Chen, Z.; Jiang, D.; et al. A Systematic Review of Tissue Engineering Scaffold in Tendon Bone Healing in vivo. *Front. Bioeng. Biotechnol.* **2021**, *9*, 621483. [CrossRef] [PubMed]
13. Ayala-Ham, A.; López-Gutierrez, J.; Bermúdez, M.; Aguilar-Medina, M.; Sarmiento-Sánchez, J.I.; López-Camarillo, C.; Sanchez-Schmitz, G.; Ramos-Payan, R. Hydrogel-Based Scaffolds in Oral Tissue Engineering. *Front. Mater.* **2021**, *8*, 708945. [CrossRef]
14. Slaughter, B.V.; Khurshid, S.S.; Fisher, O.Z.; Khademhosseini, A.; Peppas, N.A. Hydrogels in Regenerative Medicine. *Adv. Mater.* **2009**, *21*, 3307–3329. [CrossRef]
15. Zhang, L.; Yin, H.; Lei, X.; Lau, J.N.Y.; Yuan, M.; Wang, X.; Zhang, F.; Zhou, F.; Qi, S.; Shu, B.; et al. A Systematic Review and Meta-Analysis of Clinical Effectiveness and Safety of Hydrogel Dressings in the Management of Skin Wounds. *Front. Bioeng. Biotechnol.* **2019**, *7*, 342. [CrossRef]
16. Yacob, N.; Hashim, K. Morphological effect on swelling behaviour of hydrogel. *AIP Conf. Proc.* **2014**, *1584*, 153–159. [CrossRef]
17. Toledano, M.; Toledano-Osorio, M.; Carrasco-Carmona, Á.; Vallecillo, C.; Lynch, C.D.; Osorio, M.T.; Osorio, R. State of the Art on Biomaterials for Soft Tissue Augmentation in the Oral Cavity. Part I: Natural Polymers-Based Biomaterials. *Polymers* **2020**, *12*, 1850. [CrossRef]
18. El-Sherbiny, I.M.; Yacoub, M.H. Hydrogel scaffolds for tissue engineering: Progress and challenges. *Glob. Cardiol. Sci. Pract.* **2013**, *2013*, 316–342. [CrossRef]

19. Sun, Y.; Nan, D.; Jin, H.; Qu, X. Recent Advances of Injectable Hydrogels for Drug Delivery and Tissue Engineering Applica-tions. *Polym. Test* **2020**, *81*, 106283. [CrossRef]
20. Ying, H.; Zhou, J.; Wang, M.; Su, D.; Ma, Q.; Lv, G.; Chen, J. In situ formed collagen-hyaluronic acid hydrogel as biomimetic dressing for promoting spontaneous wound healing. *Mater. Sci. Eng. C* **2019**, *101*, 487–498. [CrossRef]
21. Tufanaru, C.; Munn, Z.; Aromataris, E.; Campbell, J.; Hopp, L. Systematic Reviews of Effectiveness. In *Joanna Briggs Institute Reviewer's Manual*; The Joanna Briggs Institute: Adelaide, Australia, 2017; pp. 1–7.
22. Cheung, J.W.; Rose, E.E.; Santerre, J.P. Perfused culture of gingival fibroblasts in a degradable/polar/hydrophobic/ionic polyurethane (D-PHI) scaffold leads to enhanced proliferation and metabolic activity. *Acta Biomater.* **2013**, *9*, 6867–6875. [CrossRef]
23. Choi, J.; Jeong, Y.; Gilad, A.; Kim, T.; Oh, M.H.; Hyeon, T.; Park, H.; Lee, K.H. Engineered collagen hydrogels for the sustained release of biomolecules and imaging agents: Promoting the growth of human gingival cells. *Int. J. Nanomed.* **2014**, *9*, 5189–5201. [CrossRef] [PubMed]
24. Colangelo, M.T.; Belletti, S.; Govoni, P.; Guizzardi, S.; Galli, C. A Biomimetic Polynucleotides–Hyaluronic Acid Hydrogel Promotes Wound Healing in a Primary Gingival Fibroblast Model. *Appl. Sci.* **2021**, *11*, 4405. [CrossRef]
25. Cozens, E.J.; Roohpour, N.; Gautrot, J.E. Comparative Adhesion of Chemically and Physically Crosslinked Poly(Acrylic Acid)-Based Hydrogels to Soft Tissues. *Eur. Polym. J.* **2021**, *146*, 110250. [CrossRef]
26. Kang, H.J.; Ko, N.; Oh, S.J.; An, S.Y.; Hwang, Y.S.; Kim, S.Y. Injectable Human Hair Keratin–Fibrinogen Hydrogels for En-gineering 3d Microenvironments to Accelerate Oral Tissue Regeneration. *Int. J. Mol. Sci.* **2021**, *22*, 14538–14547. [CrossRef]
27. Laird, D.; El-Baba, M.D.; Hamri, G.C.-E.; Eberwein, P.; Nelson, K.; Tomakidi, P.; Steinberg, T. In vitro and in vivo biocompatibility evaluation of a novobiocin stimulus-responsive poly(ethylene glycol)-based hydrogel designed for soft tissue regeneration. *J. Bioact. Compat. Polym.* **2015**, *30*, 319–339. [CrossRef]
28. Miranda, D.G.; Malmonge, S.M.; Campos, D.M.; Attik, N.G.; Grosgogeat, B.; Gritsch, K. A chitosan-hyaluronic acid hydrogel scaffold for periodontal tissue engineering. *J. Biomed. Mater. Res. Part B Appl. Biomater.* **2016**, *104*, 1691–1702. [CrossRef]
29. Moatary, A.; Teimouri, A.; Bagherzadeh, M.; Chermahini, A.N.; Razavizadeh, R. Design and fabrication of novel chitin hydro-gel/chitosan/nano diopside composite scaffolds for tissue engineering. *Ceram. Int.* **2017**, *43*, 1657–1668. [CrossRef]
30. Montalbano, G.; Toumpaniari, S.; Popov, A.; Duan, P.; Chen, J.; Dalgarno, K.; Scott, W.E.; Ferreira, A.M. Synthesis of Bioin-spired Collagen/Alginate/Fibrin Based Hydrogels for Soft Tissue Engineering. *Mater. Sci. Eng.* **2018**, *91*, 236–246. [CrossRef]
31. Rosdiani, A.F.; Widiyanti, P.; Rudyarjo, D.I. Synthesis and Characterization Biocomposite Collagen-Chitosan- Glycerol as Scaffold for Gingival Recession Therapy. *J. Int. Dent. Med. Res.* **2017**, *10*, 118–122.
32. Tabatabaei, F.; Moharamzadeh, K.; Tayebi, L. Fibroblast encapsulation in gelatin methacryloyl (GelMA) versus collagen hydrogel as substrates for oral mucosa tissue engineering. *J. Oral Biol. Craniofacial Res.* **2020**, *10*, 573–577. [CrossRef]
33. Zhou, T.; Zheng, K.; Sui, B.; Boccaccini, A.R.; Sun, J. In Vitro Evaluation of Poly (Vinyl Alcohol)/Collagen Blended Hydrogels for Regulating Human Periodontal Ligament Fi Broblasts and Gingival Fi Broblasts. *Int. J. Biol. Macromol.* **2020**, *163*, 1938–1946. [CrossRef]
34. Radulescu, D.-M.; Neacsu, I.A.; Grumezescu, A.-M.; Andronescu, E. New Insights of Scaffolds Based on Hydrogels in Tissue Engineering. *Polymers* **2022**, *14*, 799. [CrossRef] [PubMed]
35. Hu, W.; Wang, Z.; Xiao, Y.; Zhang, S.; Wang, J. Advances in crosslinking strategies of biomedical hydrogels. *Biomater. Sci.* **2018**, *7*, 843–855. [CrossRef] [PubMed]
36. Akhtar, M.F.; Hanif, M.; Ranjha, N.M. Methods of Synthesis of Hydrogels … A Review. *Saudi Pharm. J.* **2016**, *24*, 554–559. [CrossRef] [PubMed]
37. Xue, X.; Hu, Y.; Wang, S.; Chen, X.; Jiang, Y.; Su, J. Fabrication of physical and chemical crosslinked hydrogels for bone tissue engineering. *Bioact. Mater.* **2022**, *12*, 327–339. [CrossRef] [PubMed]
38. Chang, B.; Ahuja, N.; Ma, C.; Liu, X. Injectable Scaffolds: Preparation and Application in Dental and Craniofacial Regenera-tion. *Mater. Sci. Eng. R* **2017**, *111*, 1–26. [CrossRef]
39. Drury, J.L.; Mooney, D.J. Hydrogels for tissue engineering: Scaffold design variables and applications. *Biomaterials* **2003**, *24*, 4337–4351. [CrossRef]
40. Mandal, B.B.; Kundu, S.C. Cell proliferation and migration in silk fibroin 3D scaffolds. *Biomaterials* **2009**, *30*, 2956–2965. [CrossRef]
41. Choi, D.J.; Park, S.J.; Gu, B.K.; Kim, Y.-J.; Chung, S.; Kim, C.-H. Effect of the pore size in a 3D bioprinted gelatin scaffold on fibroblast proliferation. *J. Ind. Eng. Chem.* **2018**, *67*, 388–395. [CrossRef]
42. Bružauskaitė, I.; Bironaitė, D.; Bagdonas, E.; Bernotienė, E. Scaffolds and cells for tissue regeneration: Different scaffold pore sizes—Different cell effects. *Cytotechnology* **2016**, *68*, 355–369. [CrossRef]
43. Xue, X.; Hu, Y.; Deng, Y.; Su, J. Recent Advances in Design of Functional Biocompatible Hydrogels for Bone Tissue Engineering. *Adv. Funct. Mater.* **2021**, *31*, 2009432. [CrossRef]
44. Zhang, Y.; Huang, Y. Rational Design of Smart Hydrogels for Biomedical Applications. *Front. Chem.* **2021**, *8*, 615665. [CrossRef]
45. Bácskay, I.; Nemes, D.; Fenyvesi, F.; Váradi, J.; Vasvári, G.; Fehér, P.; Vecsernyés, M.; Ujhelyi, Z. Role of Cytotoxicity Experi-ments in Pharmaceutical Development. In *Cytotoxicity*; InTech Open: Apulia, Italy, 2018.
46. Wu, D.T.; Munguia-Lopez, J.G.; Cho, Y.W.; Ma, X.; Song, V.; Zhu, Z.; Tran, S.D. Polymeric Scaffolds for Dental, Oral, and Craniofacial Regenerative Medicine. *Molecules* **2021**, *26*, 7043. [CrossRef] [PubMed]
47. Gaharwar, A.K.; Singh, I.; Khademhosseini, A. Engineered biomaterials for in situ tissue regeneration. *Nat. Rev. Mater.* **2020**, *5*, 686–705. [CrossRef]

48. Bashir, S.; Hina, M.; Iqbal, J.; Rajpar, A.H.; Mujtaba, M.A.; Alghamdi, N.A.; Wageh, S.; Ramesh, K.; Ramesh, S. Fundamental Concepts of Hydrogels: Synthesis, Properties, and Their Applications. *Polymers* **2020**, *12*, 2702. [CrossRef] [PubMed]
49. Mancha Sánchez, E.; Gómez-Blanco, J.C.; López Nieto, E.; Casado, J.G.; Macías-García, A.; Díaz Díez, M.A.; Carrasco-Amador, J.P.; Torrejón Martín, D.; Sánchez-Margallo, F.M.; Pagador, J.B. Hydrogels for Bioprinting: A Systematic Review of Hydrogels Synthesis, Bioprinting Parameters, and Bioprinted Structures Behavior. *Front Bioeng. Biotechnol.* **2020**, *8*, 776. [CrossRef] [PubMed]
50. Bustamante-Torres, M.; Romero-Fierro, D.; Arcentales-Vera, B.; Palomino, K.; Magaña, H.; Bucio, E. Hydrogels Classification According to the Physical or Chemical Interactions and as Stimuli-Sensitive Materials. *Gels* **2021**, *7*, 182. [CrossRef]
51. Peña, B.; Laughter, M.; Jett, S.; Rowland, T.J.; Taylor, M.R.G.; Mestroni, L.; Park, D. Injectable Hydrogels for Cardiac Tissue Engineering. *Macromol. Biosci.* **2018**, *18*, e1800079. [CrossRef]
52. Jiang, Y.; Wang, Y.; Li, Q.; Yuheng, J.; Chu, W. Natural Polymer-based Stimuli-responsive Hydrogels. *Curr. Med. Chem.* **2020**, *27*, 2631–2657. [CrossRef]
53. Sánchez-Cid, P.; Jiménez-Rosado, M.; Romero, A.; Pérez-Puyana, V. Novel Trends in Hydrogel Development for Biomedical Applications: A Review. *Polymers* **2022**, *14*, 3023. [CrossRef] [PubMed]
54. Firoozi, M.; Entezam, M.; Masaeli, E.; Ejeian, F.; Nasr-Esfahani, M.H. Physical modification approaches to enhance cell supporting potential of poly (vinyl alcohol)-based hydrogels. *J. Appl. Polym. Sci.* **2021**, *139*, 51485. [CrossRef]
55. O'Brien, F.J. Biomaterials & scaffolds for tissue engineering. *Mater. Today* **2011**, *14*, 88–95. [CrossRef]
56. Rosales, A.M.; Anseth, K.S. The design of reversible hydrogels to capture extracellular matrix dynamics. *Nat. Rev. Mater.* **2016**, *1*, 1–15. [CrossRef]
57. Islam, R.; Tanveer, S.; Chen, C.-C. Modeling swelling behavior of hydrogels in aqueous organic solvents. *Chem. Eng. Sci.* **2021**, *242*, 116744. [CrossRef]
58. Wang, M.; Li, J.; Li, W.; Du, Z.; Qin, S. Preparation and Characterization of Novel Poly (Vinyl Alcohol)/Collagen Double-Network Hydrogels. *Int. J. Biol. Macromol.* **2018**, *118*, 41–48. [CrossRef] [PubMed]
59. Park, H.; Guo, X.; Temenoff, J.S.; Tabata, Y.; Caplan, A.I.; Kasper, F.K.; Mikos, A.G. Effect of Swelling Ratio of Injectable Hy-drogel Composites on Chondrogenic Differentiation of Encapsulated Rabbit Marrow Mesenchymal Stem Cells in Vitro. *Biol. Biomacromol.* **2009**, *10*, 541–546. [CrossRef]
60. Fraczyk, J.; Wasko, J.; Walczak, M.; Kaminski, Z.J.; Puchowicz, D.; Kaminska, I.; Bogun, M.; Kolasa, M.; Stodolak-Zych, E.; Scislowska-Czarnecka, A.; et al. Conjugates of Copper Alginate with Arginine-Glycine-Aspartic Acid (RGD) for Potential Use in Regenerative Medicine. *Materials* **2020**, *13*, 337. [CrossRef]
61. Chamkouri, H. A Review of Hydrogels, Their Properties and Applications in Medicine. *Am. J. Biomed. Sci. Res.* **2021**, *11*, 485–493. [CrossRef]
62. Mathew, A.P.; Uthaman, S.; Cho, K.-H.; Cho, C.-S.; Park, I.-K. Injectable hydrogels for delivering biotherapeutic molecules. *Int. J. Biol. Macromol.* **2018**, *110*, 17–29. [CrossRef]
63. Naahidi, S.; Jafari, M.; Logan, M.; Wang, Y.; Yuan, Y.; Bae, H.; Dixon, B.; Chen, P. Biocompatibility of hydrogel-based scaffolds for tissue engineering applications. *Biotechnol. Adv.* **2017**, *35*, 530–544. [CrossRef] [PubMed]
64. Londono, R.; Badylak, S.F. Biomaterials from Decellularized Tissues. In *Biomaterials from Nature for Advanced Devices and Therapies*; Wiley Blackwell: Hoboken, NJ, USA, 2016; pp. 190–210. ISBN 9781119126218.
65. Yamaguchi-Ueda, K.; Akazawa, Y.; Kawarabayashi, K.; Sugimoto, A.; Nakagawa, H.; Miyazaki, A.; Kurogoushi, R.; Iwata, K.; Kitamura, T.; Yamada, A.; et al. Combination of Ions Promotes Cell Migration via Extracellular Signal-Regulated Kinase 1/2 Signaling Pathway in Human Gingival Fibroblasts. *Mol. Med. Rep.* **2019**, *19*, 5039–5045. [CrossRef] [PubMed]
66. Wang, T.; Yi, W.; Zhang, Y.; Wu, H.; Fan, H.; Zhao, J.; Wang, S. Sodium alginate hydrogel containing platelet-rich plasma for wound healing. *Colloids Surf. B* **2023**, *222*, 113096. [CrossRef] [PubMed]
67. Bolívar-Monsalve, E.J.; Alvarez, M.M.; Hosseini, S.; Espinosa-Hernandez, M.A.; Ceballos-González, C.F.; Sanchez-Dominguez, M.; Shin, S.R.; Cecen, B.; Hassan, S.; Di Maio, E.; et al. Engineering bioactive synthetic polymers for biomedical applications: A review with emphasis on tissue engineering and controlled release. *Mater. Adv.* **2021**, *2*, 4447–4478. [CrossRef]
68. Arora, R.; Narula, S.C.; Sharma, R.K.; Tewari, S. Evaluation of Supracrestal Gingival Tissue After Surgical Crown Lengthening: A 6-Month Clinical Study. *J. Periodontol.* **2013**, *84*, 934–940. [CrossRef]
69. Smith, P.C.; Martínez, C.; Martínez, J.; McCulloch, C.A. Role of Fibroblast Populations in Periodontal Wound Healing and Tissue Remodeling. *Front. Physiol.* **2019**, *10*, 270. [CrossRef]
70. Kaigler, D.; Cirelli, J.A.; Giannobile, W.V. Growth factor delivery for oral and periodontal tissue engineering. *Expert Opin. Drug Deliv.* **2006**, *3*, 647–662. [CrossRef]
71. Mitchell, A.C.; Briquez, P.S.; Hubbell, J.A.; Cochran, J.R. Engineering growth factors for regenerative medicine applications. *Acta Biomater.* **2016**, *30*, 1–12. [CrossRef]
72. Ouyang, Y.; Zhao, J.; Wang, S. Multifunctional Hydrogels Based on Chitosan, Hyaluronic Acid and Other Biological Macromolecules for the Treatment of Inflammatory Bowel Disease: A Review. *Int. J. Biol. Macromol.* **2023**, *227*, 505–523. [CrossRef]
73. Jiao, W.; Li, X.; Shan, J.; Wang, X. Study of Several Alginate-Based Hydrogels for In Vitro 3D Cell Cultures. *Gels* **2022**, *8*, 147. [CrossRef]
74. Zhang, K.; Xue, K.; Loh, X.J. Thermo-Responsive Hydrogels: From Recent Progress To. *Gels* **2021**, *7*, 1–17. [CrossRef] [PubMed]

75. Mavlyanova, R.; Yang, R.; Tao, T.; Aquib, M.; Kesse, S.; Maviah, M.B.J.; Boakye-Yiadom, K.O.; Farooq, M.A.; Wang, B. In-jectable Hydrogels for Targeted Delivering of Therapeutic Molecules for Tissue Engineering and Disease Treatment. *Polym. Adv.Technol.* **2020**, *31*, 192–203. [CrossRef]
76. Wang, H.; Heilshorn, S.C. Adaptable Hydrogel Networks with Reversible Linkages for Tissue Engineering. *Adv. Mater.* **2015**, *27*, 3717–3736. [CrossRef]
77. Unangolla, J.M.; Jayasuriya, A.C. Hydrogel-Based 3D Bioprinting: A Comprehensive Review on Cell- Laden Hydrogels, Bioink Formulations, and Future Perspectives. *Appl. Mater. Today* **2020**, *176*, 139–148.
78. Li, X.; Sun, Q.; Li, Q.; Kawazoe, N.; Chen, G. Functional Hydrogels with Tunable Structures and Properties for Tissue Engi-neering Applications. *Front Chem.* **2018**, *6*, 1–20. [CrossRef] [PubMed]
79. Chung, T.-W.; Wang, S.-S.; Wang, Y.-Z.; Hsieh, C.-H.; Fu, E. Enhancing growth and proliferation of human gingival fibroblasts on chitosan grafted poly (ε-caprolactone) films is influenced by nano-roughness chitosan surfaces. *J. Mater. Sci. Mater. Med.* **2008**, *20*, 397–404. [CrossRef] [PubMed]
80. Schreier, T.; Degen, E.; Baschong, W. Fibroblast migration and proliferation during in vitro wound healing. *Res. Exp. Med.* **1993**, *193*, 195–205. [CrossRef]
81. Sterczała, B.; Grzech-Leśniak, K.; Michel, O.; Trzeciakowski, W.; Dominiak, M.; Jurczyszyn, K. Assessment of Human Gingival Fibroblast Proliferation after Laser Stimulation In Vitro Using Different Laser Types and Wavelengths (1064, *980*, 635, 450, and 405 nm)—Preliminary Report. *J. Pers. Med.* **2021**, *11*, 98. [CrossRef]
82. Amberg, R.; Elad, A.; Rothamel, D.; Fienitz, T.; Szakacs, G.; Heilmann, S.; Witte, F. Design of a migration assay for human gingival fibroblasts on biodegradable magnesium surfaces. *Acta Biomater.* **2018**, *79*, 158–167. [CrossRef]
83. Misra, S.; Hascall, V.C.; Markwald, R.R.; Ghatak, S. Interactions between Hyaluronan and Its Receptors (CD44, RHAMM) Regulate the Activities of Inflammation and Cancer. *Front. Immunol.* **2015**, *6*, 201. [CrossRef]
84. Marinho, A.; Nunes, C.; Reis, S. Hyaluronic Acid: A Key Ingredient in the Therapy of Inflammation. *Biomolecules* **2021**, *11*, 1518. [CrossRef] [PubMed]
85. Amir, L.R.; Soeroso, Y.M.; Fatma, D.; Sunarto, H.; Sulijaya, B.; Idrus, E.; Rahdewati, H.; Tjokrovonco, A.M.; Izumi, K.; Abbas, B.; et al. Periodontal Ligament Cell Sheets and RGD-Modified Chitosan Improved Regeneration in the Horizontal Periodontal Defect Model. *Eur. J. Dent.* **2020**, *14*, 306–314. [CrossRef] [PubMed]
86. Hernández, A.M.P.; Pohlit, H.; Sjögren, F.; Shi, L.; Ossipov, D.; Antfolk, M.; Tenje, M. A simplified approach to control cell adherence on biologically derived in vitro cell culture scaffolds by direct UV-mediated RGD linkage. *J. Mater. Sci. Mater. Med.* **2020**, *31*, 1–10. [CrossRef] [PubMed]
87. Oliveira, É.R.; Nie, L.; Podstawczyk, D.; Allahbakhsh, A.; Ratnayake, J.; Brasil, D.L.; Shavandi, A. Advances in Growth Factor Delivery for Bone Tissue Engineering. *Int. J. Mol. Sci.* **2021**, *22*, 903. [CrossRef]
88. Billiet, T.; Gevaert, E.; De Schryver, T.; Cornelissen, M.; Dubruel, P. The 3D printing of gelatin methacrylamide cell-laden tissue-engineered constructs with high cell viability. *Biomaterials* **2013**, *35*, 49–62. [CrossRef]
89. Pathan, S.G.; Fitzgerald, L.M.; Ali, S.M.; Damrauer, S.M.; Bide, M.J.; Nelson, D.W.; Ferran, C.; Phaneuf, T.M.; Phaneuf, M.D. Cytotoxicity associated with electrospun polyvinyl alcohol. *J. Biomed. Mater. Res. Part B* **2015**, *103*, 1652–1662. [CrossRef]
90. Allena, R. Cell Migration with Multiple Pseudopodia: Temporal and Spatial Sensing Models. *Cell Migr. Mult. Pseudopodia* **2013**, *75*, 288–316. [CrossRef]
91. Asparuhova, M.B.; Kiryak, D.; Eliezer, M.; Mihov, D.; Sculean, A. Activity of two hyaluronan preparations on primary human oral fibroblasts. *J. Periodontal Res.* **2018**, *54*, 33–45. [CrossRef]
92. De Souza, V.Z.; Manfro, R.; Joly, J.C.; Elias, C.N.; Peruzzo, D.C.; Napimoga, M.H.; Martinez, E.F. Viability and Collagen Se-cretion by Fibroblasts on Titanium Surfaces with Different Acid-Etching Protocols. *Int. J. Implant. Dent.* **2019**, *5*, 1–6. [CrossRef]
93. Rujirachotiwat, A.; Suttamanatwong, S. Curcumin Promotes Collagen Type I, Keratinocyte Growth Factor-1, and Epidermal Growth Factor Receptor Expressions in the In Vitro Wound Healing Model of Human Gingival Fibroblasts. *Eur. J. Dent.* **2020**, *15*, 063–070. [CrossRef]
94. Cen, L.; Liu, W.; Cui, L.; Zhang, W.; Cao, Y. Collagen Tissue Engineering: Development of Novel Biomaterials and Applications. *Pediatr. Res.* **2008**, *63*, 492–496. [CrossRef] [PubMed]
95. Li, H.; Tan, C.; Li, L. Review of 3D printable hydrogels and constructs. *Mater. Des.* **2018**, *159*, 20–38. [CrossRef]
96. Donderwinkel, I.; van Hest, J.C.M.; Cameron, N.R. Bio-inks for 3D bioprinting: Recent advances and future prospects. *Polym. Chem.* **2017**, *8*, 4451–4471. [CrossRef]
97. Nesic, D.; Durual, S.; Marger, L.; Mekki, M.; Sailer, I.; Scherrer, S.S. Could 3D Printing Be the Future for Oral Soft Tissue Re-generation? *Bioprinting* **2020**, *20*, e00100. [CrossRef]
98. Liu, P.; Li, Q.; Yang, Q.; Zhang, S.; Lin, C.; Zhang, G.; Tang, Z. Three-dimensional cell printing of gingival fibroblast/acellular dermal matrix/gelatin–sodium alginate scaffolds and their biocompatibility evaluation in vitro. *RSC Adv.* **2020**, *10*, 15926–15935. [CrossRef]

polymers

MDPI

Article

The Reliability of PCL/Anti-VEGF Electrospun Scaffolds to Support Limbal Stem Cells for Corneal Repair

Emilija Zdraveva [1,*], Tamara Dolenec [2], Mirna Tominac Trcin [3], Emi Govorčin Bajsić [4], Tamara Holjevac Grgurić [5], Antoneta Tomljenović [1], Iva Dekaris [6], Josip Jelić [1] and Budimir Mijovic [1,*]

1 Faculty of Textile Technology, University of Zagreb, 10000 Zagreb, Croatia; antoneta.tomljenovic@ttf.unizg.hr (A.T.); josip.jelic@ttf.unizg.hr (J.J.)
2 Department of Transfusion and Regenerative Medicine, Sestre Milosrdnice University Hospital Center, 10000 Zagreb, Croatia; tamara.dolenec@kbcsm.hr
3 The Institute of Immunology, 10000 Zagreb, Croatia; mirna.tomtrcin@gmail.com
4 Faculty of Chemical Engineering and Technology, University of Zagreb, 10000 Zagreb, Croatia; egovor@fkit.hr
5 School of Medicine, Catholic University of Croatia, 10000 Zagreb, Croatia; tamara.holjevac.grguric@unicath.hr
6 Faculty of Medicine, University of Rijeka, Bilić Vision Polyclinic, 10000 Zagreb, Croatia; iva.dekaris@gmail.com
* Correspondence: emilija.zdraveva@ttf.unizg.hr (E.Z.); budimir.mijovic@ttf.unizg.hr (B.M.)

Abstract: Since only few reported studies propose anti-vascular endothelial growth factor (anti-VEGF) delivery through electrospun scaffolds, this study greatly contributes to the potential prevention of patient's vision loss, as it explores electrospun polycaprolactone (PCL) coated with anti-VEGF for the blockage of abnormal cornea vascularization. In terms of physicochemical properties, the biological component increased the PCL scaffold fiber diameter (by ~24%) and pore area (by ~82%), while ut slightly reduced its total porosity as the anti-VEGF solution filled the voids of the microfibrous structure. The addition of the anti-VEGF increased the scaffold stiffness almost three-fold at both strains of 5 and 10%, as well as its biodegradation rate (~36% after 60 days) with a sustained release profile after Day 4 of phosphate buffered saline incubation. In terms of scaffold application function, the PCL/Anti-VEGF scaffold proved to be more favorable for the adhesion of cultured limbal stem cells (LSCs); this was confirmed by the SEM images, where the cells showed flat and elongated conformations. Further support of the LSC growth and proliferation was confirmed by the identified p63 and CK3 markers after cell staining. These results demonstrate the advantageous effect of the surface-adsorbed anti-VEGF to stop vision loss and help damaged corneal tissue repair.

Keywords: electrospinning; PCL; anti VEGF; scaffolds; LSCs; physicochemical performance; immunocytochemistry

1. Introduction

Corneal neovascularization (NV) is a pathological condition that causes excessive blood vessel ingrowth from the limbal zone to the stroma, with time resulting in the reduction of the eye visual acuity. Diseases associated with this condition include: trauma (i.e., chemical burns), inflammation (i.e., corneal graft rejection, atopic conjunctivitis), infection (fungal, bacterial, viral), degenerative conditions, congenital disease (i.e., limbal stem cell deficiency, LSCD), and so forth [1,2]. Vascular endothelial growth factor (VEGF) is a protein that promotes new blood vessel growth (angiogenesis) in case of i.e., eye inflammation after injury [3]. The anti-vascular endothelial growth factor (anti-VEGF), on the other hand, is a medical agent used to block the function of the VEGF [4], which is why it is of great importance in controlling corneal NV. The anti-VEGF is usually injected directly into the vitreous (back) cavity of the eye (so called intravitreal drug delivery). Unfortunately, this type of drug administration is the cause of many serious complications, including infectious endophthalmitis, intraocular inflammation, rhegmatogenous retinal detachment, intraocular pressure elevation, ocular hemorrhage, or rare ocular, and systemic side effects [5]. Some of the limitations in the anti-VEGF topical administration include the physical structure of the cornea, the vitreous, blood vessels, and the tear film, all slowing down the drug delivery to the retina [5,6]. Due to the aforementioned issues, the application of this biological component requires special attention in terms of the design of a target sustained drug-delivery system. As reported, the anti-VEGF drug-delivery system needs to fulfill the following requirements: biocompatibility, sustained and target delivery over one month, transparency, adequate concentration, and bioactivity [7]. An innovative approach in the design of this type of drug-delivery systems in recent years is the technique of electrospinning [8], where the system is a nanofibrous material. The technique of electrospinning uses an electrostatic force to stretch a viscoelastic polymer solution for the fabrication of nanofibers (or microfibers). In the field of nanomedical applications, the design of drug-delivery systems garners special attention since this technique successfully encapsulates medical or biological species in the nano-fibrous/micro-fibrous materials functionalized for fast, sustained, or tunable drug release. At the same time, these structures sufficiently support specific cell-type cultures and can help in the regeneration of damaged tissues by promoting their adhesion, migration, and proliferation [9]. Recently reported studies concerned with the fabrication of electrospun fibrous scaffolds mainly incorporate VEGF or (in a few studies) anti-VEGF compounds, generally for tissue-engineering applications. Electrospun polycaprolactone (PCL) was plasma treated to immobilize VEGF by covalent bonding. Human umbilical vein endothelial cells (HUVECs) were cultured on the scaffolds, and it was observed that the biological activity of the immobilized VEGF induced higher cell numbers after 9 days of culture due to its sustained action [10]. VEGF was also loaded into core/shell electrospun dextran (DEX)/poly(lactide-co-glycolide) (PLGA) fibrous scaffolds for vascular tissue regeneration. The scaffolds showed sustained release over a period of 28 days, while the VEGF promoted HUVECs proliferation, spreading cell–membrane interactions [11]. Co-electrospinning was also conducted for the fabrication of PCL/gelatin scaffolds. Scaffold-surface immobilization with VEGF was further carried after gelatin functionalization with heparin. Significantly improved endothelial cell proliferation was observed over the sustained release of the VEGF for 25 days. Excellent vascularization after 4 weeks of subcutaneous implantation of the scaffolds in rats was confirmed by almost full cellularization of the same [12]. Electrospun polyurethane (PU) scaffolds were loaded with polyethylene glycol (PEG)-coated cerium oxide nanoparticles (CNPs) and VEGF to be used as small-diameter artificial blood vessels for endothelialization in vitro. The scaffolds were cultured with endothelial progenitor cells (EPCs). The scaffolds were designed to obtain a synergistic effect of both the VEGF and the CNPs, i.e., the effect on angiogenesis and anti-EPCs apoptosis effect, respectively. The release of the CNPs and VEGF was sustained over 30 days. The scaffolds inhibited H_2O_2-induced EPCs apoptosis, while the CNP–PEG enhanced VEGF activity to support mobilization and differentiation of the EPCs [13]. Scaffolds with orthogonally aligned poly(lactic-co-glycolic acid) (PLGA) nanofibers and controlled dual release of VEGF and platelet-derived growth factor (PDGF) were fabricated by emulsion electrospinning. The scaffolds successfully directed and stimulated the function of HUVECs and human aortic smooth muscle cells to help in early-stage vascularization and smooth muscle maturation [14]. VEGF is also known for its role in bone osteoblast growth and mineralization [15]. Three-dimensional-printed hydroxyapatite/calcium sulfate/PCL scaffolds were loaded with VEGF to be utilized in bone-tissue engineering. In-vitro results showed enhanced human mesenchymal stem cells (hMSCs) and HUVEC proliferation, while in-vivo studies resulted in rabbit's femur bone regeneration after 8 weeks of implantation [16]. The frequent invasive dosing of anti-VEGF in case of age-related macular degeneration treatment was reported to be replaced by electrospun core/shell bevacizumab/PCL system with a constant rate of release of the medicine over 2 months [8]. Anti-bone morphogenic protein 2 (anti-BMP-2) and anti-VEGF were immobilized on the surface (separate areas) of UV-ozone irradiation-activated electrospun PCL with a parallel pattern design. Human bone marrow-derived mesenchymal stem cells were positive on osteogenic and angiogenic markers in areas where the BMP-2 and VEGF were bound, respectively. A significantly increased angiogenic response was obtained in an in-vivo chick chorioallantoic membrane (CAM) assay. Thus, the scaffolds have confirmed their role in guided bone regeneration, i.e., for internal bone fixtures or flat bone healing [17]. Other techniques used in the encapsulation of both VEGF and anti-VEGF compounds (except electrospinning and 3D printing) included the gas-foaming process, employed for a simultaneous stable release of the VEGF and a dose-dependent release of the anti-VEGF agent [18]. Freeze-drying [19] and hydrogelling [20], as well as colloidal crystal templating [21], were reported for the fabrication of hydroxyapatite-collagen scaffold loaded with VEGF, collagen/gold nanoparticles scaffold incorporating anti-VEGF, and 3D porous polydimethylsiloxane (PDMS) scaffold for a 30-month delivery of anti-VEGF, respectively. In this study, electrospun PCL was incorporated with an anti-VEGF agent via a post-processing wetting treatment. The PCL/anti-VEGF scaffolds were evaluated for their physicochemical properties and final role in the support of limbal stem cells (LSCs) to be used in corneal tissue engineering. To the best of our knowledge, the majority of reported studies explore electrospun scaffolds for the delivery of the aforementioned VEGF agent. Thus, the current study greatly contributes to the earlier discussed topic in both effects of the anti-VEGF agent on the scaffold properties and cultured cells.

2. Materials and Methods

The polymer used as the matrix in this study was polycaprolactone (PCL) with Mn = 80,000 (Sigma–Aldrich, Burlington, MA, USA). The solvents used were glacial acetic acid and acetone (Ru–Ve). Anti-vascular endothelial growth factor (anti-VEGF) solution (25 mg/mL bevacizumab, Avastin, Roche) was kindly supplied by Bilić Vision Polyclinic, Zagreb, Croatia. Eighteen percent of PCL solution concentration was prepared by dissolving the polymer in the glacial acetic acid and acetone mixture, with the volume ratio of 8:2, respectively, under constant stirring for more than 24 h.

Single electrospun PCL was fabricated with the electrospinning device NT–ESS–300, NTSEE Co. Ltd., Gwangju, Republic of Korea, on a ribbed 3D-printed collector (Form 2 3D printer FormLabs, Somerville, MA, USA) treated with a graphite coating, Figure 1A. The dimensions of the collector ribs were the following: height of 1.2/0.6 mm, width of 0.3 mm, and rib distance of 0.8 mm. The processing conditions were electrical voltage of 14–20 kV, needle tip to collector distance of 18 cm, and volume flow rate of 1 mL/h. The environmental conditions were temperature of 20 °C and humidity of 40%. The electrospun PCL scaffolds, Figure 1B, were soaked in the 25 mg/mL anti-VEGF solution post-processing for 2 h and dried for 24 h at room temperature afterwards.

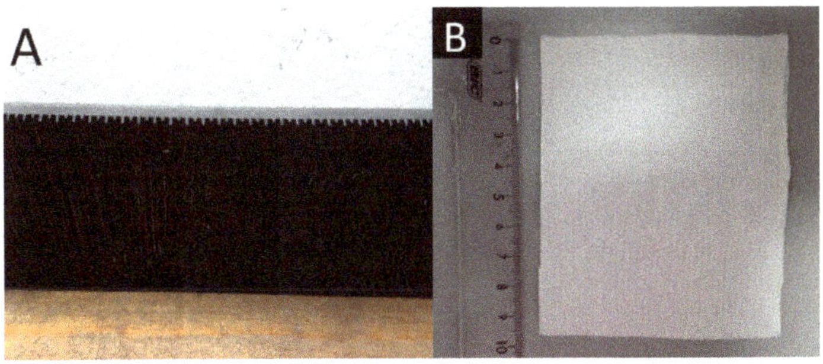

Figure 1. Pattern electrospinning; (**A**) 3D-printed collector with ribbed geometry, (**B**) Electrospun PCL scaffold.

2.1. Evaluation of the Scaffolds' Physicochemical Performance

Attenuated total reflectance–Fourier transform infrared (ATR–FTIR) spectroscopy Perkin Elmer Spectrum One was used to confirm the presence of the anti-VEGF on the surface of the electrospun scaffold. The measurements were carried out at room temperature in the wavelength range of 4000 to 650 cm^{-1}.

The morphology of the PCL and PCL/anti-VEGF electrospun scaffolds was analyzed by scanning electron microscopy Jeol, JSM 7000F (Tokyo, Japan) and Tescan Vega TS 5136 MM (Brno, Czech Republic). The scaffold's fiber diameter and pore area were determined based on the SEM images by measuring 100 randomly selected fibers or pores, respectively, using the ImageJ–NIH software. The porosity was calculated according to the equation provided elsewhere [22] based on scaffolds' thickness, weight, area, polymer, and blend densities. The thickness of the scaffolds was measured by a Digi Micrometer, Mitutoyo, Kawasaki, Japan.

The water contact angle was measured based on the images (Dino Capture 2.0 microscopy) of a water droplet (volume of 0.5 mL) placed on the surface of the scaffolds by using the Low Bond–Axisymmetric Drop Shape Analysis (LB–ADSA) tool in the ImageJ software [23]. The wettability was evaluated at 0 and 3–5 s after drop placement.

The scaffolds were tested under tensile load with Tensile Strength Tester Tensolab 3000, Mesdan S.p.A., Puegnago Del Garda, Italy. The load cell was 100 N, the gauge length was 20 mm, and the rate of extension was 20 mm/min. The dimensions of the samples were 10 × 90 mm.

Dynamic mechanical analysis was conducted on DMA 983, TA Instruments with the following testing conditions: frequency of 1 Hz, amplitude of 0.2 mm, heating rate of 3 °C/min, and a temperature range between −100 °C and 100 °C. The cooling was conducted under liquid nitrogen.

The release profile of the biological component, the anti-VEGF, was determined by monitoring the behavior of the electrospun 18% PCL/anti-VEGF scaffold (dimensions of 20 × 20 mm) incubated in 10 mL of PBS (phosphate buffered saline) solution at 37 °C (with continuous stirring at 80 rpm) for a period of 7 days. A portion (2–3 mL) of the solution was withdrawn every 24 h to determine the absorbance by the UV–Vis transmission spectrophotometer Cary 50, Varian. The solutions were returned to the bulk to maintain the same volume during the testing period. The concentrations of the released biological components were calculated based on a standard curve obtained from known concentrations (from 0.13 to 0.5 mg/mL) of the anti-VEGF in the PBS.

The biodegradability of the PCL and the PCL/anti-VEGF electrospun scaffolds was examined through incubation in PBS solution for 7 to 60 days at a temperature of 37 °C. The weight loss of the scaffolds was calculated based on the initial dry mass and the dry mass of the samples after incubation [24]. It is known that the degradation rate of the PCL is very slow, and it usually takes from at least 6 months up to 4 years [25]. Therefore, the degradability of the scaffolds during the period envisaged as the therapy period (up to 14 days) can be attributed mostly to the release of the bioactive component. All the aforementioned tests were carried out in triplicates.

2.2. Human LSCs Isolation, Culture and Immunocytochemistry

Detailed procedure of the human LSC cultivation and immunocytochemical analysis was described in our previous study [26]. Briefly, the first step is the preparation of the 3T3 (mouse cells, ATCC-CCL-92, Swiss albino) feeder layer in a growth medium containing Dulbecco's Modified Eagle Medium (DMEM) (Gibco, Invitrogen, Waltham, MA, USA), 10% of heat inactivated Fetal Bovine Serum (FBS) (Gibco, Invitrogen, Waltham, MA, USA), antibiotic-antimycotic (ABAM), and 1% L-glutamine. Further proliferation of the trypsinized 3T3s was inhibited with the FBS medium and γ-rays. The second step involves the isolation of the LSCs from surgical remains of human cadaveric cornea with permission of the Ethics Committee of the University Eye Hospital Svjetlost (Zagreb, Croatia). Previously cut (15 mm in diameter) and disinfected (under UV-light) scaffolds were seeded with the 3T3s and LSCs on top in 24 well plates with 1.5×10^4 cells/cm^2 per well with the ratio of 1:1. The LSCs were grown until 70–100% of confluence with the change of the keratinocyte growth medium every third day.

Indirect immunocytochemistry (ICC) first involved fixation of the cells with 4% paraformaldehyde (Sigma–Aldrich, Burlington, MA, USA), permeabilization with 0.5% triton X-100 (Sigma–Aldrich, Burlington, MA, USA), and antibody incubation. The primary and secondary antibodies used were the mouse monoclonal antibody [4A4] IgG2a to human p63 protein (Abcam, Cambridge, UK), and mouse monoclonal antibody [AE5] IgG1 to human cytokeratin 3/CK-3 (Abcam, Cambridge, UK), as well as goat anti-mouse IgG Alexa Fluor 488 (Invitrogen, Thermo Fisher Scientific, Waltham, MA, USA) diluted in phalloidin-tetramethylrhodamine B isothiocyanate TRITC (Sigma–Aldrich, St. Louis, MO, USA) and goat anti-mouse IgG Alexa Fluor 568 (Invitrogen Thermo Fisher Scientific, Waltham, MA, USA), respectively. Cell nuclei were marked with 4′,6-diamidino-2-phenylindole (DAPI) (Sigma–Aldrich, Burlington, MA, USA) and imaged under confocal microscopy Leica, TCS SP2 AOBS (Leica Microsystems CMS GmbH, Mannheim, Germany) at the Ruder Bošković Institute in Zagreb, Croatia.

3. Results

Figure 2 shows the FTIR transmission spectra of the electrospun single PCL and anti-VEGF-coated PCL scaffolds. The analysis was conducted to confirm the anti-VEGF coating on the surface of the electrospun PCL scaffold. The characteristic functional groups for the single PCL (black line) are at the wavelengths of 2944 cm^{-1} and 2865 cm^{-1}, which corresponds to the –CH$_2$– group. Further, the amorphous phase of the PCL is presented by the carbonyl group C=O at the wavelength of 1720 cm^{-1}, while the band at 1292 cm^{-1} corresponds to the C–C and C–O bonds in the crystalline phase of the PCL. The absorption bands at 1238 cm^{-1} and 1163 cm^{-1} correspond to the asymmetric and symmetric C–O–C bond stretching, respectively [27]. The anti-VEGF coating (red line) showed many vibrational frequencies, but the most relevant to the protein molecule are at the wavelength of 1644 cm^{-1} related to Amide I, originated from C=O stretching, and at the wavelength of 1545 cm^{-1} related to Amide II band [28].

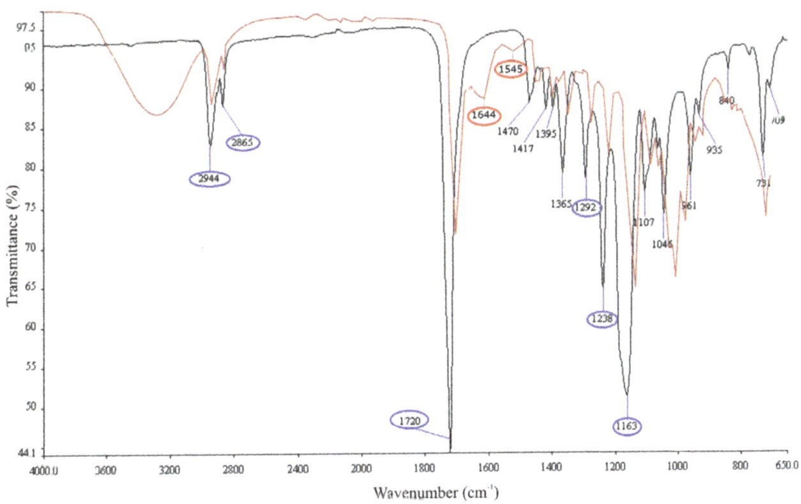

Figure 2. FTIR transmission spectra of the electrospun PCL (black line) and PCL/Anti-VEGF (red line) scaffolds.

Figure 3 shows the SEM images of the electrospun single PCL; Figure 3A,A1 and anti-VEGF coated PCL scaffolds; Figure 3B,B1. Generally, the fibers showed a homogeneous appearance with smooth surfaces, random deformations (beads), and a few thicker fibers on the scaffold surfaces. The anti-VEGF coating filled the pores in between the fibers, while some of the fibers observed showed a swollen appearance due to anti-VEGF solution adsorption, Figure 3B1.

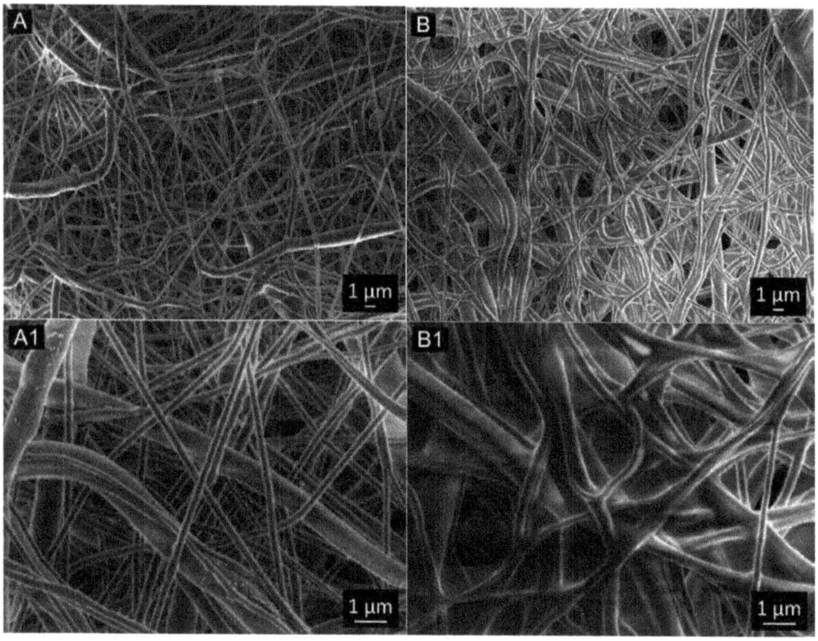

Figure 3. SEM photomicrographs of the electrospun scaffolds: (**A**,**A1**) PCL and (**B**,**B1**) PCL/Anti-VEGF.

Figure 4 shows the distributions (histograms) of the electrospun scaffolds' fiber diameter, Figure 4A, and pore area, Figure 4B. The ranges of diameters and areas were almost fully overlapping for the electrospun scaffolds. The mean fiber diameters of the electrospun PCL and PCL/anti-VEGF scaffolds were 0.342 ± 0.252 μm and 0.424 ± 0.260 μm, respectively, while the mean pore areas were 1.006 ± 0.812 μm^2 and 1.829 ± 1.170 μm, respectively.

ANOVA one-way and Tukey tests revealed that the mean values of the fiber diameter and pore areas are significantly different at the level of 0.05 between the PCL and the PCL/anti-VEGF electrospun scaffolds. The total porosity of the electrospun single PCL was calculated to be almost 85% (84.60 ± 9.93), while that of the scaffolds with the added biological component were calculated to be about 75% (75.43 ± 9.13). Statistically, there was no significant difference between the two means of electrospun scaffolds.

Figure 5 shows the wetting ability on the surface of the electrospun scaffolds. The hydrophobic nature of the electrospun single PCL scaffold is confirmed by the large water contact angle measured (122.18 ± 1.80°), which does not change even after a few seconds, Figure 5A. On the other hand, the addition of the anti-VEGF results in high scaffolds' hydrophilicity or immediate adsorption of the drop of water on the scaffolds' surface, Figure 5B.

Figure 6 shows the stress–strain curves, and, Figure 6A, those of the electrospun scaffolds after tensile testing. The calculated Young moduli at the strain of 5 and 10% are shown in Figure 6B. The aforementioned strain values were determined as the elastic region limit values.

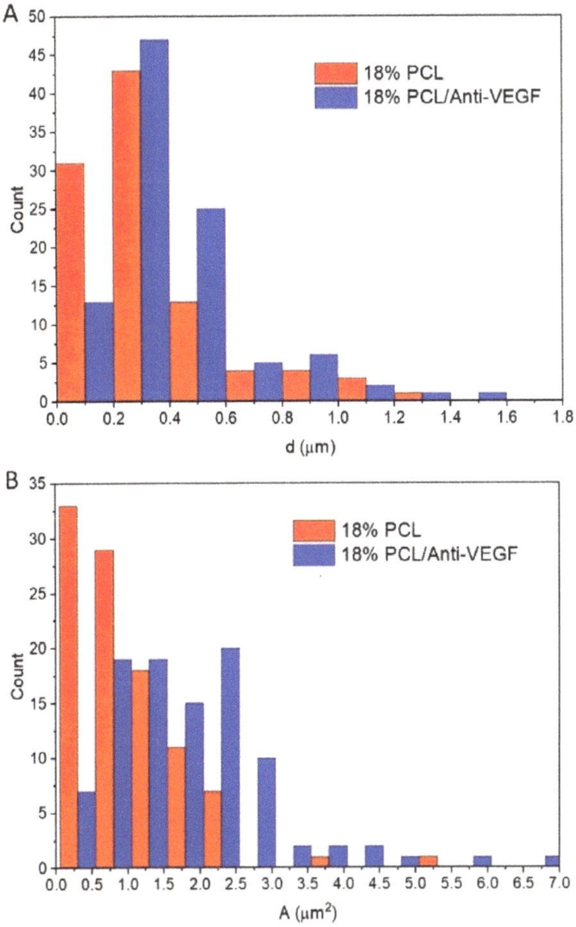

Figure 4. (**A**) Fiber diameter and (**B**) pore-area distribution of the electrospun PCL and PCL/anti-VEGF scaffolds.

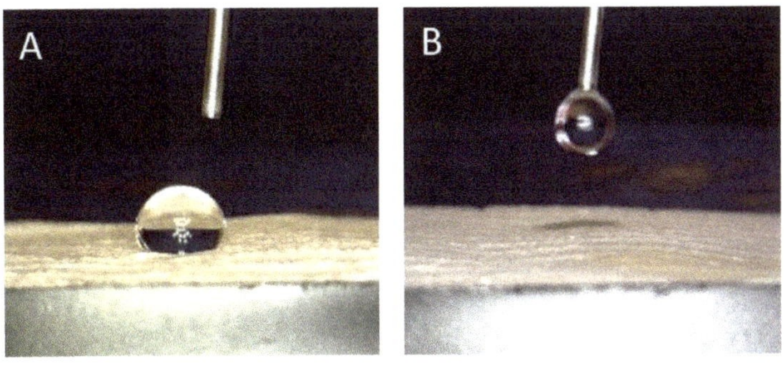

Figure 5. Water contact angle measurement on the surface of the electrospun scaffolds: (**A**) PCL and (**B**) PCL/anti-VEGF.

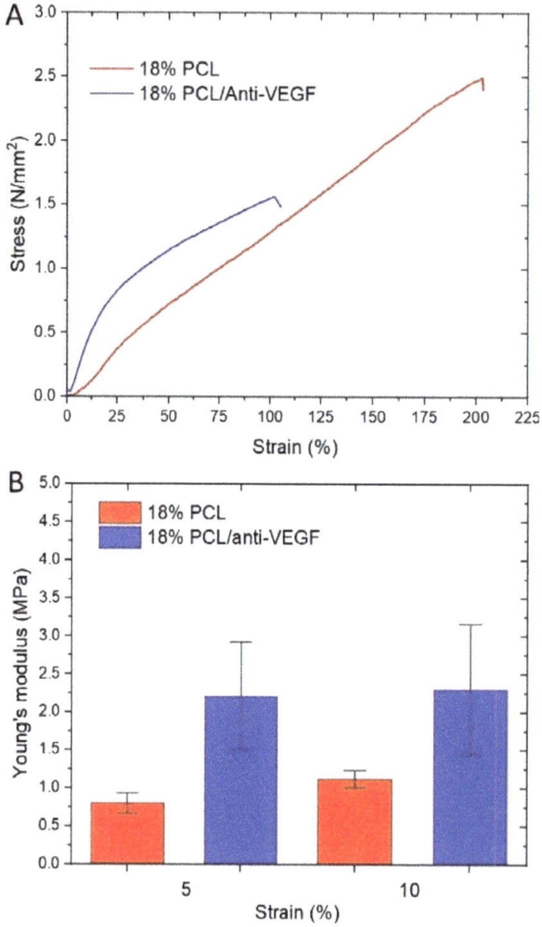

Figure 6. (**A**) Tensile stress–strain curves and (**B**) Young moduli of the electrospun scaffolds.

Table 1 shows the maximum force, elongation at break, and tensile strength of the electrospun scaffolds.

Table 1. Maximum force, elongation at break and tensile strength of the electrospun scaffolds.

Electrospun Scaffolds	F_{max} (N)	ε (%)	σ (N/mm^2)
18% PCL	3.84 ± 0.30	181.85 ± 21.65	2.31 ± 0.18
18% PCL/anti-VEGF	5.43 ± 0.88	99.00 ± 18.19	1.41 ± 0.14

The addition of the anti-VEGF has evidently reduced the stretching ability of the PCL scaffolds from 181.85 ± 21.65 to 99.00 ± 18.19%, as well as the maximum tensile strength from 2.31 ± 0.18 to 1.41 ± 0.14 N/mm^2. At both strain values (5 and 10%), the electrospun PCL/anti-VEGF scaffold shows a significantly higher elastic modulus (from 0.80 ± 0.27 to 2.21 ± 1.41 MPa, and from 1.12 ± 0.23 to 2.73 ± 1.39 MPa), although after the strain of 10%, the scaffold behaves plastically, which is not the case with the single PCL. The results indicate that the addition of the biological component directly affects the elasticity of the scaffolds.

In terms of the anti-VEGF effect, the mechanical behavior of the electrospun scaffolds was further analyzed to determine the viscoelastic properties of the electrospun scaffolds, including storage modulus, E′, and loss tangent tanδ as a function of temperature, as shown in Figure 7A,B, respectively.

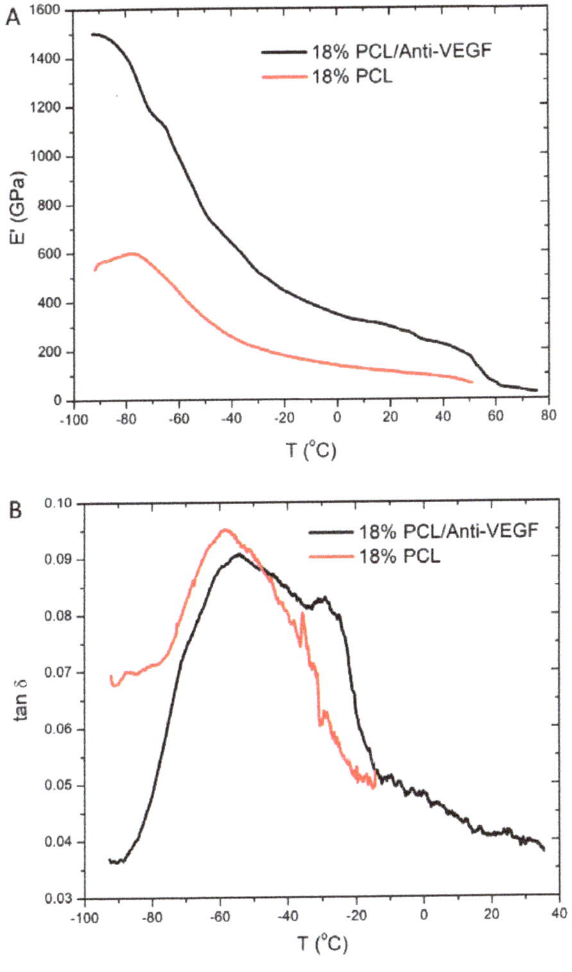

Figure 7. DMA analysis: (**A**) Storage modulus as a function of temperature and (**B**) Damping as a function of temperature of the electrospun scaffolds.

The storage modulus curve is typical for the semi-crystalline PCL polymer. A smaller drop in the storage modulus can be observed below the glass transition temperature (T_g) of −60 °C. While above it, a sharp decrease in the storage modulus is recorded, which corresponds to the amorphous phase of the polymer. Up to a temperature of 40 °C, the drop in the storage modulus is due to the progressive melting of the polymer. While above a temperature of 50–60 °C, plastic deformations occur due to the complete melting of the crystalline phase [29]. A significant increase in the storage modulus is observed with the addition of the anti-VEGF, which indicates an improvement in the mechanical behavior of the scaffold due to the strengthening effect of the biological component. The relaxation maxima at temperatures below zero corresponds to the glass transition temperature of the amorphous phase of the PCL. The addition of the anti-VEGF results in a shift of the T_g to higher temperatures (−55 °C), compared to the single electrospun PCL scaffold (−60 °C). The results indicate the placement of the particles of the anti-VEGF in the amorphous phase of the PCL and the reduction of its mobility.

The release profile of the anti-VEGF in the PBS solution (Figure 8) is determined based on the standard curve concentrations of the biological component (from 0.13 to 0.5 mg/mL) in the PBS. The release profile includes two phases: 1. Immediate release of the biological component of about 19%, with a maximum of 30% of the anti-VEGF (from Day 1 to Day 5), and 2. Continuous release of the remaining amount of the anti-VEGF, which means a relatively constant concentration of the biological component. The described profile and calculated concentrations correspond to a period of 7 days, after which it was not possible to determine the maximum absorbance peak. In the period from 12 to 14 days, a slightly smaller peak appeared (not given here), i.e., the absorbance was read at another wavelength of 760 nm, which can be assumed to correspond to the degradation signal of the biological component. The largest amount of the anti-VEGF component was released in the first 24 h due to its placement on the surface of the electrospun PCL scaffold, which is the result of the post-processing treatment.

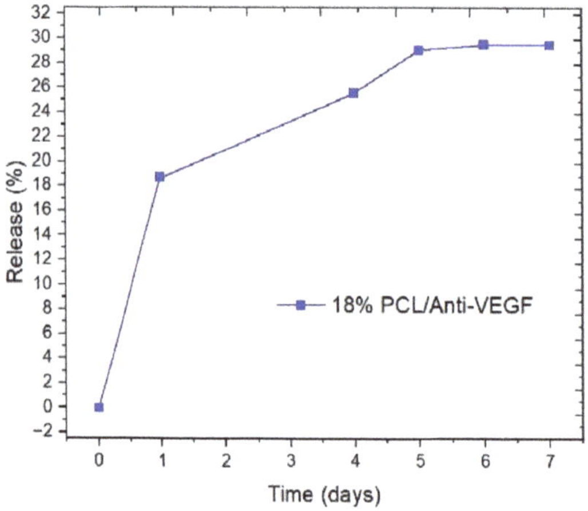

Figure 8. Anti-VEGF release profile from the electrospun PCL/anti-VEGF scaffold.

The release profile is based on the diffusion of the encapsulated anti-VEGF component, which slows down over time. The slow release is the result of the hydrophobic PCL matrix, taking more time to absorb the PBS solution during incubation.

Table 2 shows the percentage of weight loss for the electrospun scaffolds incubated in the PBS solution for the period of 7 to 60 days. The weight loss of the electrospun PCL/anti-VEGF showed a gradual increase with an initial percentage loss of 14% to almost 36% after 60 days of incubation. This behavior of the scaffolds favors the healing process in corneal therapy, i.e., it enables blockage of the vascularization of the visual eye field during the regeneration of the eye tissue. A negative percentage of the weight loss was observed in cases of the electrospun single PCL scaffold, which indicates that no degradation of the scaffold occurs after the total time of incubation. Certainly, the addition of the active biological component improves the degradability of the scaffold, which is in favor of eye therapy.

Figure 9 shows representative SEM photomicrographs of the electrospun PCL (Figure 9A,A1), and PCL/anti-VEGF scaffolds after LSC cultivation (Figure 9B,B1). The scaffolds showed partial or full coverage of their surface by the adhered LSCs, with a more favorable cell adherence in case of the hydrophilic scaffold surface of the PCL/anti-VEGF system. The shapes of the LSCs were more clearly observed in case of the PCL/anti-VEGF scaffolds, where most of the cells showed flat and elongated conformations (Figure 9B). The cells were mostly joint and filling the voids in between the fibers.

Table 2. Biodegradation of the electrospun scaffolds after incubation in PBS solution.

Electrospun Scaffolds	7 Days of Incubation	60 Days of Incubation
	Weight Loss (%)	
18% PCL	negative	negative
18% PCL/Anti-VEGF	13.49 ± 5.69	35.89 ± 1.69

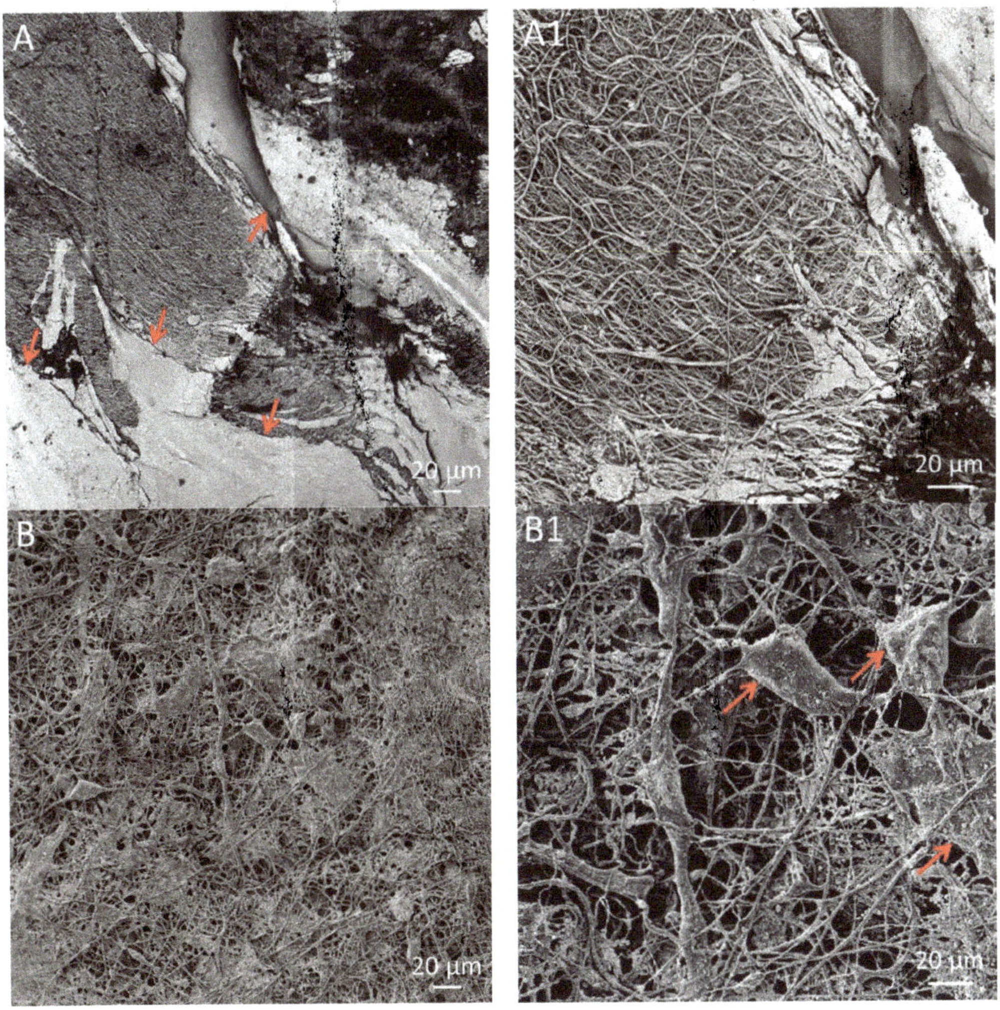

Figure 9. SEM photomicrographs of the adhered LSCs on the electrospun scaffolds: (**A**,**A1**) PCL and (**B**,**B1**) PCL/anti-VEGF.

Immunofluorescence analysis of the LSCs cultured on the PCL and PCL/anti-VEGF scaffolds was conducted to evaluate the presence of the limbal stem cell p63 marker, a nuclear transcription factor, and cytokeratin 3 (CK3), which is a corneal epithelial differentiation marker. Figure 10 shows the confocal images of the LSCs on the electrospun

PCL (Figure 10A,B), and PCL/anti-VEGF scaffolds after immunofluorescence staining (Figure 10C,D).

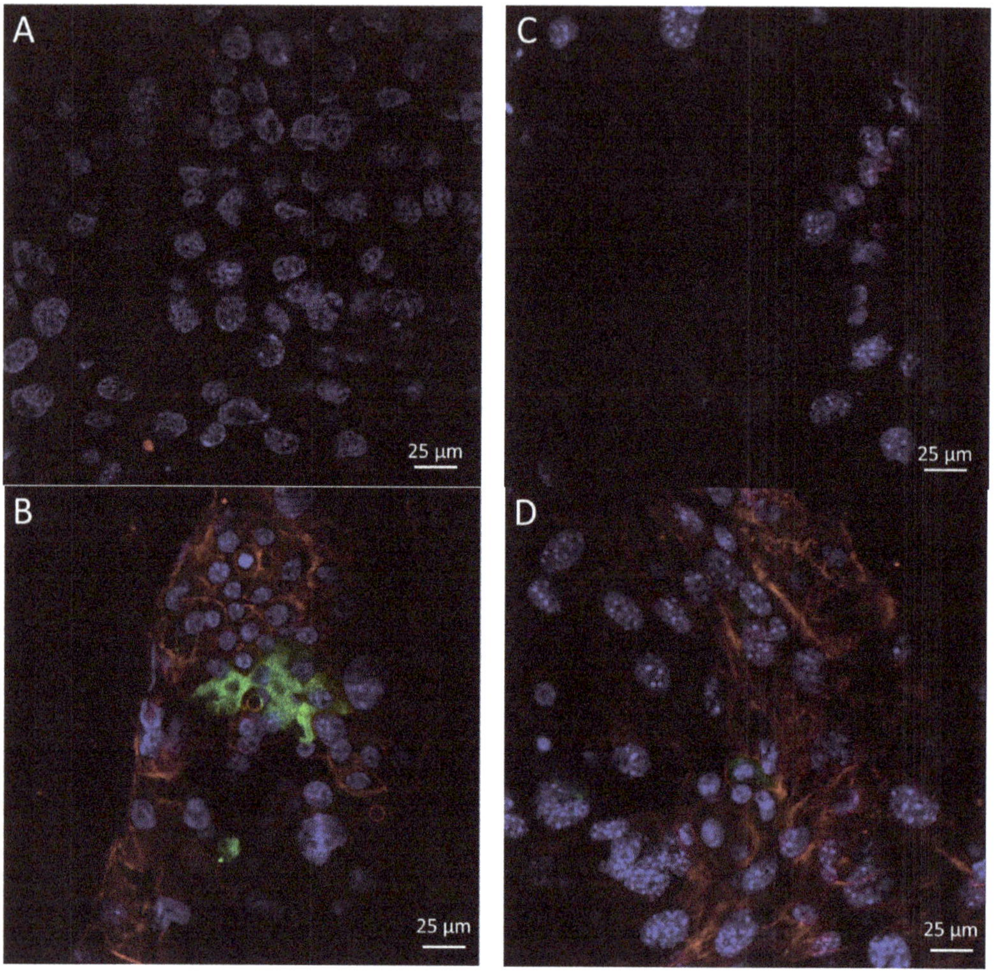

Figure 10. Immunofluorescence analysis of the limbal stem cells cultured on: (**A**,**B**) PCL and (**C**,**D**) PCL/anti-VEGF scaffolds. Cells positive on cornea marker CK3 have cytoplasm colored red (**A**,**C**) and cells positive on stem cell marker p63 show nuclei stained with turquoise. Cytoskeleton is stained red with phalloidin TRITC (**B**,**D**). All nuclei are counterstained with blue stain 4′,6-diamidino-2-phenylindole (DAPI).

The LSCs were identified as positive on cornea marker CK3 when their cytoplasm was colored red, while their nuclei was colored turquoise when the cells were positive on stem cell marker p63. The cytoskeleton of the LSCs was red and all the nuclei were counterstained with DAPI staining. Generally, the confocal images confirmed the observation of the SEM analysis, i.e., both electrospun PCL and PCL/anti-VEGF scaffolds showed support of the LSC adhesion, growth, and proliferation, since both limbal stem cells and differentiated corneal cells were detected on the scaffold surfaces, while the scaffold with the added biological component was expectedly more favorable.

4. Discussion

The human cornea is the outermost transparent part of the eye, with a multi-scale structure due to its different layers and fiber pattern organization. The most characteristic architecture includes the one of the Bowman's layer with its randomly oriented collagen fibrils, while collagen fibrils organized into parallel lamellae and lamellae above one another in the center represent the corneal stroma layer [30–32]. The major contribution to the corneal mechanical strength comes from the stroma and the difference in the degree of fibrils alignment and the diameter. Its strength varies with depth as well [33]. The LSCs are situated in the limbus, which is the transition zone between the clear cornea and the opaque sclera (the white of the eye) [34]. The structure of the electrospun PCL/anti-VEGF scaffolds tends to simulate the aforementioned corneal architecture that consists of parallel densely packed fibrous bundles, as well as of randomly loose packed fibers in between. The scaffolds' structure is a result of the 3D-printed collector ribbed geometry (Figure 1A), as mentioned in earlier sections. This pattern was generated to facilitate cell adhesion and growth. In terms of chemical composition, the addition of the anti-VEGF agent is vital for corneal tissue regeneration, as it prevents vascularization and closure of the eye's visual field. The evaluation of physicochemical properties of the scaffolds included morphology, porosity, surface wettability, static tensile and dynamic mechanical behavior, biological component release activity, and scaffold degradation ability. Visually, the SEM photomicrographs displayed a swollen appearance (Figure 3B,B1), with the addition of the anti-VEGF agents. This is in accordance with the mean values of the measured fiber diameter, which increased by ~24%. A similar effect was reported in a study of electrospun PLGA fibers treated with plasma, chitosan, and heparin, as well as mineral (hydroxyapatite) coating afterwards, to result in the diameter increase of 200 nm, or up to 1 μm [35]. A significant increase of ~82% in the pore area was also observed after the anti-VEGF coating. This was expected, as the increase in fiber diameter results in increased pore size, as reported in the case of electrospun PCL–hydroxyapatite and PCL–hydroxyapatite/b-tricalcium phosphate with a two-fold increase in fiber diameter and average pore-size area compared to PCL [36]. The increase in pore size usually results in the increase of the scaffolds' total porosity [37], but an opposite effect was observed in this study, although the difference was not significant statistically (reduction of the porosity after coating for 10.8%). This can be explained through the application of the anti-VEGF solution, which filled the voids in the microporous structure of the scaffolds in depth. The study also reports on the sensitivity of this scaffold's property to very small changes in the process conditions; thus, the porosity may vary even across the surface of the same electrospun sample [38,39]. Although the overall porosity of the PCL/anti-VEGF scaffold is slightly reduced, it is still high enough (almost 80%) to fulfill the requirements of a tissue-engineering scaffold. High porosity enables the effective release or exchange of bioactive components (i.e., proteins, genes, cells) and nutrients/waste, respectively. At the same time, very high porosity may compromise scaffolds' mechanical integrity; thus, an optimal value of the same is preferred [40]. Optimal cell adhesion is also possible in case of moderate scaffold hydrophilicity [41], although the addition of the anti-VEGF reduced the water contact angle to zero in this study, whereas other studies reported on the improvement of cell infiltration, proliferation, and differentiation in cases of scaffold high water retention [42]. In terms of scaffold tensile properties, the anti-VEGF coating has reduced the tensile strength of electrospun PCL by ~39%, Table 1, but the single PCL scaffold had a lower strength initially. Generally, lower porosity will increase the scaffolds' tensile strength [43], but an opposite effect in the current study is in accordance with our previous research in the case of electr-spun PCL/5 wt% Cefuroxime scaffolds [26]. On the other hand, an increase (for almost three-fold at both strains of 5 and 10%) in the scaffold's stiffness (Young's modulus, Figure 6B) is observed with the addition of the biological component, which can be explained by the increase in the scaffold's fiber mean diameter, as reported elsewhere [44,45].

The functionality of the PCL/anti-VEGF scaffold is also evaluated through its release profile (Figure 8). The release pace of the scaffold needs to be in accordance with the patient's therapy. The initial burst release of the anti-VEGF is explained by the post-processing treatment; i.e., the coating that was mainly present on the fibers' surfaces [46], was, to a lesser extent, adsorbed inside the fibrous structure. The adsorption rate is affected mainly by the polymer's high hydrophobicity. After Day 4 of PBS incubation, the PCL/anti-VEGF scaffold showed a sustained release of the bioactive component, which is preferred in retinal disease therapies. It is reported that the physical adsorption of a drug or bioactive component negatively affects their sustained release [47] but, at the same time, maintains their activity or prevents their destabilization and denaturation during the electrospinning process [48]. These results are comparable to a similar study where core-shell electrospun anti-VEGF (Avastin solution)/PCL scaffolds showed biphasic release profiles, beginning with a burst phase over 24 h (total release of ~25% at Day 3), and a further constant rate of release over a period of 19 days (total release of ~60%) [8]. Another study reported on the release of the anti-VEGF compound from core/shell PVA(Anti-VEGF)/PCL (gelatin) scaffolds, showing a smaller released amount and a constant release over a 6-day period, compared to blended electrospun scaffolds [49]. Scaffold biodegradability is another important property among the essential ones required in the process of tissue repair. As expected, the anti-VEGF coating introduced the scaffold with gradual weight loss ability from Day 7 to Day 60, with an almost three-fold increase of its biodegradation (Table 2). The single electrospun PCL scaffold has not shown any weight-loss ability, even after 60 days of PBS incubation. This is explained by the low permeability of the scaffold due to its highly hydrophobic nature. The degradation process of an implanted scaffold begins when the material starts interacting with the tissue fluids, thus adsorbing water and changing its structure to promote the hydrolysis process. The material's hydrophilicity is not influential in the case of enzymatic degradation [50,51]. The result of the single electrospun PCL is in accordance with another reported study, where the electrospun PCL showed insignificant weight loss of 1.44% after 90 days of PBS incubation. An opposite effect of the same scaffold was observed in the case of PBS/lipase medium. Thus, the enzymatic degradation of the PCL was 97.11% after 90 days of incubation [24]. As reported, the degradation rate of PCL electrospun scaffolds can be tailored through its combination with other biomaterials (e.g., PCL/gelatin plus cellulose nanocrystals) [52], or through the design of different configurations, i.e., core/shell fibers, layered structures, and so forth.

Human LSCs were successfully cultivated on both electrospun PCL and PCL/anti-VEGF scaffolds, as first confirmed by the SEM (Figure 9) and the confocal images (Figure 10). LSC coverage was more evident in case of the biologically active scaffold, as shown in Figure 10B,B1, where the cells can be identified in more detailed conformations. The identification of the cultured LSCs in terms of stem cell presence (cells positive on p63 marker) or cell corneal epithelial differentiation (cells positive on CK3 marker) is essential as the success of a transplant in corneal tissue repair is reported to be in case of more than 3% of detected p63-bright cells [53]. The confocal-identified p63- and CK3-positive LSCs are in accordance with the SEM observations, whereby the PCL/Anti-VEGF scaffolds showed to be more favorable for the adhesion of the cultured cells, which means the same can support further growth and proliferation of the LSCs. The results confirmed the advantageous effect of the biological component on cell life, which is a dominant factor in damaged tissue regeneration.

The limitation of the current work concerns the statistical analysis of the anti-VEGF release profile, the evaluation of its bioactivity maintenance, and cell proliferation analysis restricted by the availability of human donor LSCs that would further confirm the biocompatibility of the PCL/Anti-VEGF scaffolds. Thus, future research will explore the anti-VEGF function sustainability, in terms of its in-vitro release and long-term bioactivity, as well as the quantitative analysis of viable LSCs adhered on the scaffolds, including both stem cells positive on the p63 marker or differentiated corneal epithelial cells positive on the CK3 marker.

5. Conclusions

To prevent uncontrollable ocular vascularization, VEGF blocking agents are incorporated into engineered biomaterials to support their function and sustained target release. This study considered the design of electrospun scaffolds based on PCL treated with anti-VEGF through simple physical adsorption. The aim was to evaluate the PCL/anti-VEGF scaffold reliability in terms of its physicochemical function to support LSCs for corneal repair. The addition of the biological component influenced PCL scaffold morphology as it resulted in fiber diameter and pore area increases by ~24% and ~82%, respectively, while it slightly reduced its total porosity due to the closure of microfibrous structure voids. The anti-VEGF had a positive effect on the PCL wetting ability and its tensile property, as it fully reduced its hydrophobicity and increased its Young modulus by almost three-fold at both strains of 5 and 10%. The gradual weight loss of the scaffold was significantly beneficial as the biodegradation rate completely increased to ~36% after 60 days of PBS incubation, which was in accordance with its sustained agent delivery (total of 30%) after Day 4 of PBS incubation. The results further confirmed PCL/anti-VEGF greater biocompatibility compared to a single PCL, which showed to be more favorable to LSC adhesion, growth, and differentiation as observed by the SEM, and p63 (stem cells presence) and CK3 (cell corneal epithelial differentiation) marker identification after cells staining. The results will contribute to the design of electrospun scaffolds for sustained anti-VEGF delivery to inhibit angiogenesis in the process of LSC ocular tissue engineering. To overcome the limitations of the current study, future research will focus on the evaluation of the biological compound sustained bioactivity, its release pace statistics, and cell proliferation quantitative analysis through the identification of p63 and CK3 positive stem cells or differentiated corneal epithelial cells, respectively.

Author Contributions: Conceptualization, E.Z., M.T.T., I.D. and B.M.; methodology, E.Z., T.D., M.T.T., E.G.B., T.H.G., A.T., I.D. and B.M.; software, E.Z.; validation, E.Z., T.D., M.T.T., E.G.B., T.H.G., A.T., I.D., J.J. and B.M.; formal analysis, E.Z., M.T.T., E.G.B. and B.M.; investigation, E.Z., T.D., M.T.T., E.G.B., T.H.G., A.T., I.D., J.J. and B.M.; resources, E.Z., T.D., M.T.T., I.D. and B.M.; data curation, E.Z., M.T.T. and B.M.; writing—original draft preparation, E.Z.; writing—review and editing, E.Z., T.D., M.T.T., E.G.B., T.H.G., A.T., I.D., J.J. and B.M.; visualization, E.Z.; supervision, E.Z., M.T.T., A.T., I.D. and B.M.; project administration, E.Z. and B.M.; funding acquisition, E.Z. and B.M. All authors have read and agreed to the published version of the manuscript.

Funding: This research was funded by CROATIAN SCIENCE FOUNDATION under the project IP-2016-06-6878, Custom-Tailored Fibrous Scaffold Prototype for Tissue Cells Culture via Combined Electrospinning, COMBOELECTROSPUN, PI. Budimir Mijovic. The APC was funded by the University of Zagreb Faculty of Textile Technology under the short-term financial support for research for 2023, TP20-23 PI: Emilija Zdraveva.

Informed Consent Statement: Informed consent was obtained from all subjects involved in the study in accordance with Ethical Approval of Sestre Milosrdnice University Hospital Center (EP-15333/14), Declaration of Helsinki (Doh) and International Conference on Harmonisation of technical requirements for registration of pharmaceuticals for human use (ICH GCP).

Data Availability Statement: Not applicable.

Conflicts of Interest: The authors declare no conflict of interest.

References

1. Labelle, P. The Eye. In *Pathologic Basis of Veterinary Disease*, 6th ed.; Zachary, J.F., Ed.; Elsevier, Inc.: St. Louis, MO, USA, 2017.
2. Feizi, S. Corneal Angiogenesis: Etiologies, Complications and Management. In *Physiologic and Pathologic Angiogenesis-Signaling Mechanisms and Targeted Therapy*; Simionescu, D., Simionescu, A., Eds.; IntechOpen: Rijeka, Croatia, 2017; pp. 57–76.
3. Xia, B.; Lv, Y. Dual-delivery of VEGF and NGF by emulsion electrospun nanofibrous scaffold for peripheral nerve regeneration. *Mater. Sci. Eng. C* **2018**, *82*, 253–264. [CrossRef] [PubMed]
4. Song, D.; Liu, P.; Shang, K.; Ma, Y. Application and mechanism of anti-VEGF drugs in age-related macular degeneration. *Front. Bioeng. Biotechnol.* **2023**, *10*, 943915. [CrossRef] [PubMed]

5. Ghasemi Falavarjani, K.; Nguyen, Q. Adverse events and complications associated with intravitreal injection of anti-VEGF agents: A review of literature. *Eye* **2013**, *27*, 787–794. [CrossRef] [PubMed]
6. Edwards, A.; Prausnitz, M.R. Predicted permeability of the cornea to topical drugs. *Pharm. Res.* **2001**, *18*, 1497–1508. [CrossRef] [PubMed]
7. Seah, I.; Zhao, X.; Lin, Q.; Liu, Z.; Su, S.Z.Z.; Yuen, Y.S.; Hunziker, W.; Lingam, G.; Loh, X.J.; Su, X. Use of biomaterials for sustained delivery of anti-VEGF to treat retinal diseases. *Eye* **2020**, *34*, 1341–1356. [CrossRef] [PubMed]
8. Angkawinitwong, U.; Awwad, S.; Khaw, P.T.; Brocchini, S.; Williams, G.R. Electrospun formulations of bevacizumab for sustained release in the eye. *Acta Biomater.* **2017**, *64*, 126–136. [CrossRef]
9. Zdraveva, E.; Mijovic, B. *Frontier Electrospun Fibers for Nanomedical Applications*; IntechOpen: Rijeka, Croatia, 2023.
10. Guex, A.; Hegemann, D.; Giraud, M.-N.; Tevaearai, H.; Popa, A.; Rossi, R.; Fortunato, G. Covalent immobilisation of VEGF on plasma-coated electrospun scaffolds for tissue engineering applications. *Colloids Surf. B Biointerfaces* **2014**, *123*, 724–733. [CrossRef]
11. Jia, X.; Zhao, C.; Li, P.; Zhang, H.; Huang, Y.; Li, H.; Fan, J.; Feng, W.; Yuan, X.; Fan, Y. Sustained release of VEGF by coaxial electrospun dextran/PLGA fibrous membranes in vascular tissue engineering. *J. Biomater. Sci. Polym. Ed.* **2011**, *22*, 1811–1827. [CrossRef]
12. Wang, K.; Chen, X.; Pan, Y.; Cui, Y.; Zhou, X.; Kong, D.; Zhao, Q. Enhanced vascularization in hybrid PCL/gelatin fibrous scaffolds with sustained release of VEGF. *BioMed Res. Int.* **2015**, *2015*, 865076. [CrossRef]
13. Dai, W.-W.; Guo, H.-F.; Qian, D.-H.; Qin, Z.-X.; Lei, Y.; Hou, X.-Y.; Wen, C. Improving endothelialization by the combined application of polyethylene glycol coated cerium oxide nanoparticles and VEGF in electrospun polyurethane scaffolds. *J. Mater. Chem. B* **2017**, *5*, 1053–1061. [CrossRef]
14. Zhou, Y.; Zhao, Q.; Wang, M. Dual release of VEGF and PDGF from emulsion electrospun bilayer scaffolds consisting of orthogonally aligned nanofibers for gastrointestinal tract regeneration. *MRS Commun.* **2019**, *9*, 1098–1104. [CrossRef]
15. Casarrubios, L.; Gómez-Cerezo, N.; Sánchez-Salcedo, S.; Feito, M.; Serrano, M.; Saiz-Pardo, M.; Ortega, L.; De Pablo, D.; Díaz-Güemes, I.; Fernández-Tomé, B. Silicon substituted hydroxyapatite/VEGF scaffolds stimulate bone regeneration in osteoporotic sheep. *Acta Biomater.* **2020**, *101*, 544–553. [CrossRef] [PubMed]
16. Chen, C.-Y.; Chen, C.-C.; Wang, C.-Y.; Lee, A.K.-X.; Yeh, C.-L.; Lin, C.-P. Assessment of the release of vascular endothelial growth factor from 3D-printed poly-ε-caprolactone/hydroxyapatite/calcium sulfate scaffold with enhanced osteogenic capacity. *Polymers* **2020**, *12*, 1455. [CrossRef] [PubMed]
17. Casanova, M.R.; Oliveira, C.; Fernandes, E.M.; Reis, R.L.; Silva, T.H.; Martins, A.; Neves, N.M. Spatial immobilization of endogenous growth factors to control vascularization in bone tissue engineering. *Biomater. Sci.* **2020**, *8*, 2577–2589. [CrossRef] [PubMed]
18. Yuen, W.W.; Du, N.R.; Chan, C.H.; Silva, E.A.; Mooney, D.J. Mimicking nature by codelivery of stimulant and inhibitor to create temporally stable and spatially restricted angiogenic zones. *Proc. Natl. Acad. Sci. USA* **2010**, *107*, 17933–17938. [CrossRef]
19. Li, B.; Wang, H.; Zhou, G.; Zhang, J.; Su, X.; Huang, Z.; Li, Q.; Wu, Z.; Qiu, G. VEGF-loaded biomimetic scaffolds: A promising approach to improve angiogenesis and osteogenesis in an ischemic environment. *RSC Adv.* **2017**, *7*, 4253–4259. [CrossRef]
20. Buznyk, O.; Azharuddin, M.; Islam, M.M.; Fagerholm, P.; Pasyechnikova, N.; Patra, H.K. Collagen-based scaffolds with infused anti-VEGF release system as potential cornea substitute for high-risk keratoplasty: A preliminary in vitro evaluation. *Heliyon* **2020**, *6*, e05105. [CrossRef]
21. Zhou, C.; Singh, A.; Qian, G.; Wolkow, N.; Dohlman, C.H.; Vavvas, D.G.; Chodosh, J.; Paschalis, E.I. Microporous drug delivery system for sustained anti-VEGF delivery to the eye. *Transl. Vis. Sci. Technol.* **2020**, *9*, 5. [CrossRef]
22. Balan, K.K.; Sivanesan, V.; Moorthy, N.; Budhhan, D.; Jeyaseelan, S.; Sundaramoorthy, S. Effect of thickness of mat and testing parameters on tensile strength variability of electrospun nanofibrous mat. *Mater. Today Proc.* **2016**, *3*, 1320–1329. [CrossRef]
23. Stalder, A.F.; Melchior, T.; Müller, M.; Sage, D.; Blu, T.; Unser, M. Low-bond axisymmetric drop shape analysis for surface tension and contact angle measurements of sessile drops. *Colloids Surf. A Physicochem. Eng. Asp.* **2010**, *364*, 72–81. [CrossRef]
24. Dias, J.R.; Sousa, A.; Augusto, A.; Bártolo, P.J.; Granja, P.L. Electrospun polycaprolactone (PCL) degradation: An in vitro and in vivo study. *Polymers* **2022**, *14*, 3397. [CrossRef] [PubMed]
25. Waterkotte, T.; He, X.; Wanasathop, A.; Li, S.K.; Park, Y.C. Long-Term Antibody Release Polycaprolactone Capsule and the Release Kinetics in Natural and Accelerated Degradation. *ACS Biomater. Sci. Eng.* **2022**, *8*, 4428–4438. [CrossRef] [PubMed]
26. Tominac Trcin, M.; Zdraveva, E.; Dolenec, T.; Vrgoč Zimić, I.; Bujić Mihica, M.; Batarilo, I.; Dekaris, I.; Blažević, V.; Slivac, I.; Holjevac Grgurić, T. Poly (ε-caprolactone) titanium dioxide and cefuroxime antimicrobial scaffolds for cultivation of human limbal stem cells. *Polymers* **2020**, *12*, 1758. [CrossRef]
27. Elzein, T.; Nasser-Eddine, M.; Delaite, C.; Bistac, S.; Dumas, P. FTIR study of polycaprolactone chain organization at interfaces. *J. Colloid Interface Sci.* **2004**, *273*, 381–387. [CrossRef] [PubMed]
28. Ferreira, N.N.; Caetano, B.L.; Boni, F.I.; Sousa, F.; Magnani, M.; Sarmento, B.; Cury, B.S.F.; Gremião, M.P.D. Alginate-based delivery systems for bevacizumab local therapy: In vitro structural features and release properties. *J. Pharm. Sci.* **2019**, *108*, 1559–1568. [CrossRef]
29. Zoppe, J.O.; Peresin, M.S.; Habibi, Y.; Venditti, R.A.; Rojas, O.J. Reinforcing poly (ε-caprolactone) nanofibers with cellulose nanocrystals. *ACS Appl. Mater. Interfaces* **2009**, *1*, 1996–2004. [CrossRef]
30. DelMonte, D.W.; Kim, T. Anatomy and physiology of the cornea. *J. Cataract. Refract. Surg.* **2011**, *37*, 588–598. [CrossRef]

31. Bergmanson, J.P.; Horne, J.; Doughty, M.J.; Garcia, M.; Gondo, M. Assessment of the number of lamellae in the central region of the normal human corneal stroma at the resolution of the transmission electron microscope. *Eye Contact Lens* **2005**, *31*, 281–287. [CrossRef]
32. Li, H.F.; Petroll, W.M.; Møller-Pedersen, T.; Maurer, J.K.; Cavanagh, H.D.; Jester, J.V. Epithelial and corneal thickness measurements by in vivo confocal microscopy through focusing (CMTF). *Curr. Eye Res.* **1997**, *16*, 214–221. [CrossRef] [PubMed]
33. Blackburn, B.J.; Jenkins, M.W.; Rollins, A.M.; Dupps, W.J. A review of structural and biomechanical changes in the cornea in aging, disease, and photochemical crosslinking. *Front. Bioeng. Biotechnol.* **2019**, *7*, 66. [CrossRef]
34. Spaniol, K.; Witt, J.; Mertsch, S.; Borrelli, M.; Geerling, G.; Schrader, S. Generation and characterisation of decellularised human corneal limbus. *Graefe's Arch. Clin. Exp. Ophthalmol.* **2018**, *256*, 547–557. [CrossRef] [PubMed]
35. Liu, W.; Yeh, Y.-C.; Lipner, J.; Xie, J.; Sung, H.-W.; Thomopoulos, S.; Xia, Y. Enhancing the stiffness of electrospun nanofiber scaffolds with a controlled surface coating and mineralization. *Langmuir* **2011**, *27*, 9088–9093. [CrossRef] [PubMed]
36. Nelson, M.T.; Keith, J.P.; Li, B.-B.; Stocum, D.L.; Li, J. Electrospun composite polycaprolactone scaffolds for optimized tissue regeneration. *Proc. Inst. Mech. Eng. Part N J. Nanoeng. Nanosyst.* **2012**, *226*, 111–121. [CrossRef]
37. Chen, Y.-P.; Liu, H.-Y.; Liu, Y.-W.; Lee, T.-Y.; Liu, S.-J. Determination of electrospinning parameters' strength in Poly (D, L)-lactide-co-glycolide micro/nanofiber diameter tailoring. *J. Nanomater.* **2019**, *2019*, 2626085. [CrossRef]
38. Liu, Y.; Fan, F.; Dull, T.; Chaparro, F.; Franz, C.; Abdalbaqi, A.; McElroy, C.; Lannutti, J. Physical characterization of electrospun polycaprolactone via laser micrometry: Porosity and condition-dependent jet instabilities. *Polymer* **2020**, *211*, 123044. [CrossRef]
39. Liu, Y.-X.; Chaparro, F.J.; Tian, Z.; Jia, Y.; Gosser, J.; Gaumer, J.; Ross, L.; Tafreshi, H.; Lannutti, J.J. Visualization of porosity and pore size gradients in electrospun scaffolds using laser metrology. *PLoS ONE* **2023**, *18*, e0282903. [CrossRef] [PubMed]
40. Loh, Q.L.; Choong, C. Three-dimensional scaffolds for tissue engineering applications: Role of porosity and pore size. *Tissue Eng. Part B Rev.* **2013**, *19*, 485–502. [CrossRef]
41. Raftery, R.M.; Woods, B.; Marques, A.L.; Moreira-Silva, J.; Silva, T.H.; Cryan, S.-A.; Reis, R.L.; O'Brien, F.J. Multifunctional biomaterials from the sea: Assessing the effects of chitosan incorporation into collagen scaffolds on mechanical and biological functionality. *Acta Biomater.* **2016**, *43*, 160–169. [CrossRef] [PubMed]
42. Bacakova, L.; Filova, E.; Parizek, M.; Ruml, T.; Svorcik, V. Modulation of cell adhesion, proliferation and differentiation on materials designed for body implants. *Biotechnol. Adv.* **2011**, *29*, 739–767. [CrossRef]
43. Zhu, X.; Cui, W.; Li, X.; Jin, Y. Electrospun fibrous mats with high porosity as potential scaffolds for skin tissue engineering. *Biomacromolecules* **2008**, *9*, 1795–1801. [CrossRef]
44. Bölgen, N.; Menceloğlu, Y.Z.; Acatay, K.; Vargel, I.; Pişkin, E. In vitro and in vivo degradation of non-woven materials made of poly (ε-caprolactone) nanofibers prepared by electrospinning under different conditions. *J. Biomater. Sci. Polym. Ed.* **2005**, *16*, 1537–1555. [CrossRef]
45. Dalgic, A.D.; Atila, D.; Karatas, A.; Tezcaner, A.; Keskin, D. Diatom shell incorporated PHBV/PCL-pullulan co-electrospun scaffold for bone tissue engineering. *Mater. Sci. Eng. C* **2019**, *100*, 735–746. [CrossRef] [PubMed]
46. Zeng, J.; Yang, L.; Liang, Q.; Zhang, X.; Guan, H.; Xu, X.; Chen, X.; Jing, X. Influence of the drug compatibility with polymer solution on the release kinetics of electrospun fiber formulation. *J. Control. Release* **2005**, *105*, 43–51. [CrossRef] [PubMed]
47. Meinel, A.J.; Germershaus, O.; Luhmann, T.; Merkle, H.P.; Meinel, L. Electrospun matrices for localized drug delivery: Current technologies and selected biomedical applications. *Eur. J. Pharm. Biopharm.* **2012**, *81*, 1–13. [CrossRef]
48. Liu, Z.; Ramakrishna, S.; Liu, X. Electrospinning and emerging healthcare and medicine possibilities. *APL Bioeng.* **2020**, *4*, 030901. [CrossRef] [PubMed]
49. de Souza, S.O.L.; Guerra, M.C.A.; Heneine, L.G.D.; de Oliveira, C.R.; Cunha Junior, A.d.S.; Fialho, S.L.; Oréfice, R.L. Biodegradable core-shell electrospun nanofibers containing bevacizumab to treat age-related macular degeneration. *J. Mater. Sci. Mater. Med.* **2018**, *29*, 173. [CrossRef] [PubMed]
50. Azevedo, H.S.; Gama, F.M.; Reis, R.L. In vitro assessment of the enzymatic degradation of several starch based biomaterials. *Biomacromolecules* **2003**, *4*, 1703–1712. [CrossRef]
51. Castilla-Cortázar, I.; Más-Estellés, J.; Meseguer-Dueñas, J.M.; Ivirico, J.E.; Marí, B.; Vidaurre, A. Hydrolytic and enzymatic degradation of a poly (ε-caprolactone) network. *Polym. Degrad. Stab.* **2012**, *97*, 1241–1248. [CrossRef]
52. Hivechi, A.; Bahrami, S.H.; Siegel, R.A.; Milan, P.B.; Amoupour, M. In vitro and in vivo studies of biaxially electrospun poly (caprolactone)/gelatin nanofibers, reinforced with cellulose nanocrystals, for wound healing applications. *Cellulose* **2020**, *27*, 5179–5196. [CrossRef]
53. Rama, P.; Matuska, S.; Paganoni, G.; Spinelli, A.; De Luca, M.; Pellegrini, G. Limbal stem-cell therapy and long-term corneal regeneration. *N. Engl. J. Med.* **2010**, *363*, 147–155. [CrossRef]

polymers

MDPI

Article

In Vitro Evaluation of Biphasic Calcium Phosphate Scaffolds Derived from Cuttlefish Bone Coated with Poly(ester urea) for Bone Tissue Regeneration

Patrícia Pereira [1,2,†], Ana S. Neto [3,†], Ana S. Rodrigues [4], Inês Barros [4,5,6], Catarina Miranda [4,5,6], João Ramalho-Santos [4,7], Luís Pereira de Almeida [4,5,8,9], José M. F. Ferreira [3], Jorge F. J. Coelho [1,2] and Ana C. Fonseca [2,*]

[1] IPN, Instituto Pedro Nunes, Associação para a Inovação e Desenvolvimento em Ciência e Tecnologia, Rua Pedro Nunes, 3030-199 Coimbra, Portugal
[2] CEMMPRE, Department of Chemical Engineering, University of Coimbra, Rua Sílvio Lima-Pólo II, 3030-790 Coimbra, Portugal
[3] Department of Materials and Ceramic Engineering/CICECO–Aveiro Institute of Materials, University of Aveiro, 3810-193 Aveiro, Portugal; jmf@ua.pt (J.M.F.F.)
[4] CNC-Center for Neuroscience and Cell Biology, University of Coimbra, 3004-504 Coimbra, Portugal
[5] CIBB-Center for Innovative Biomedicine and Biotechnology, University of Coimbra, 3004-504 Coimbra, Portugal
[6] III-Institute for Interdisciplinary Research, University of Coimbra, 3030-789 Coimbra, Portugal
[7] DCV-Department of Life Sciences, University of Coimbra, 3000-456 Coimbra, Portugal
[8] Faculty of Pharmacy, University of Coimbra, 3000-548 Coimbra, Portugal
[9] Viravector-Viral Vector for Gene Transfer Core Facility, University of Coimbra, 3004-504 Coimbra, Portugal
* Correspondence: anafs@eq.uc.pt
† These authors contributed equally to this work.

Abstract: This study investigates the osteogenic differentiation of umbilical-cord-derived human mesenchymal stromal cells (hUC-MSCs) on biphasic calcium phosphate (BCP) scaffolds derived from cuttlefish bone doped with metal ions and coated with polymers. First, the *in vitro* cytocompatibility of the undoped and ion-doped (Sr^{2+}, Mg^{2+} and/or Zn^{2+}) BCP scaffolds was evaluated for 72 h using Live/Dead staining and viability assays. From these tests, the most promising composition was found to be the BCP scaffold doped with strontium (Sr^{2+}), magnesium (Mg^{2+}) and zinc (Zn^{2+}) (BCP-6Sr2Mg2Zn). Then, samples from the BCP-6Sr2Mg2Zn were coated with poly(ε-caprolactone) (PCL) or poly(ester urea) (PEU). The results showed that hUC-MSCs can differentiate into osteoblasts, and hUC-MSCs seeded on the PEU-coated scaffolds proliferated well, adhered to the scaffold surfaces, and enhanced their differentiation capabilities without negative effects on cell proliferation under *in vitro* conditions. Overall, these results suggest that PEU-coated scaffolds are an alternative to PCL for use in bone regeneration, providing a suitable environment to maximally induce osteogenesis.

Keywords: cuttlefish bone; biphasic calcium phosphate; polymer coatings; *in vitro* cell culture; osteogenic differentiation

1. Introduction

Bone defects and bone destruction caused by disease (osteoporosis, bacterial infections, osteoarthritis and tumor) or accidental factors (car accidents and trauma) are increasingly common and have a huge impact on a patient's quality of life [1]. Because of the limited availability of biological bone substitutes, several tissue engineering strategies have been widely considered in the reconstruction of vascularized bone tissue and in the treatment of bone defects, namely, the combination of cells, biological molecules and/or (bio)materials [2]. Regarding the materials, they must be biocompatible and provide a suitable environment to accommodate a sufficient number of cells at the injury site and, therefore, promote cell adhesion, proliferation and differentiation without adverse effects on the host tissue [3]. It should preferably be biodegradable at a similar rate to that of bone formation, and the resulting by-products should be non-toxic (biosafety) [4,5].

Among the various biomaterials used as bone graft substitutes, calcium phosphate (CaP)-based biomaterials have received special attention in recent years due to their chemical similarity to the inorganic matrix of natural bone tissue [6]. CaP-based biomaterials represent a large family of materials formed by the reaction of calcium and phosphate ions. The most common CaP-based biomaterials are hydroxyapatite [HAp, $Ca_5(PO_4)_3OH$)], β-tricalcium phosphate [β-TCP, $Ca_3(PO_4)_2$] and their mixtures, which are called biphasic calcium phosphates (BCPs) [7]. HAp offers biocompatibility, chemical stability in body fluids, and microporosity but is poorly resorbable and fragile, whereas β-TCP is biodegradable and allows a faster release of constituent ions [2,8,9]. Compared with HAp or β-TCP, BCP-based biomaterials have better biocompatibility, superior mechanical properties, higher biological activity, tunable degradation and resorbable properties that provide a physiological environment to support nutrient transport and bone tissue ingrowth [10–14].

It is also known that co-doped CaP-based biomaterials can mimic the chemical composition, functionality and properties of natural bone and can therefore clearly favor the process of osteogenesis and rapid healing [15,16]. The ideal doping content depends on the type of ion used. Several reports can be found in the literature: (strontium (Sr^{2+}) [17,18], magnesium (Mg^{2+}) [13,19,20], cobalt (Co^{2+}) [21,22], copper (Cu^{2+}) [22,23], chromium (Cr^{3+}) [22], lithium (Li^{+}) [24], silicon (Si^{4+}) [25,26], cerium (Ce^{3+}) [25], nickel (Ni^{2+}) [27] and selenium (Se^{4+}) [28]). Each has different amounts of doped ions. Sr^{2+} and Zn^{2+} ions are known to promote bone formation by enhancing the osteoclastic resorption process and increasing pre-osteoblastic cell proliferation. Mg^{2+} ions play a role in the activity of osteoblasts and stimulate bone growth [15]. When the amount of doped metal ions is adequate, their release triggers osteogenesis and angiogenesis and enables effective regeneration of bone tissue [16]. On the other hand, excessive accumulation of dopant ions may trigger cytotoxicity and the inhibition of biological activity [29]. Additionally, few studies have investigated the effect of co-doping and its biological relevance; for example, Sr^{2+}/Fe^{3+} co-doped BCPs [30], Sr^{2+}/Mg^{2+} co-doped BCPs [31] and Ce^{3+}/Si^{4+} co-doped BCPs [32]. Basu and et al. showed that BCPs co-doped with Sr^{2+}/Fe^{3+} support osteoblast proliferation, while the single-doped BCP (at a dopant content of 10 mol% or higher) exhibited a significant reduction in cell viability [30]. Similar results have been reported for Sr^{2+}/Mg^{2+} co-doped BCPs [31] in osteoblasts. Priyadarshini and Vijayalakshmi also proved that dual-doped HAp (Ce^{3+}/Si-HAP@5% in a concentration up to 800 μg/mL) has excellent bone-like apatite layer formation on its surfaces and good metabolic activity on the MG-63 692 osteoblast-like cell [32].

Ferreira's research group was the first to produce HAp from cuttlefish bone, which has a unique porous structure and the ability of the aragonite mineral to be hydrothermally converted to CaPs [11,33]. However, HAp scaffolds derived from cuttlefish bone have poor mechanical properties, especially high brittleness and poor load-bearing capacity. To overcome these low mechanical properties, the scaffolds can be coated with polymers such as hyaluronic acid/gelatin [34], poly(ε-caprolactone) (PCL)/poly(lactide-*co*-glycolide) [35] and PCL [36,37] to achieve the desired functional and physicochemical properties of CaPs. In our recent publications [38–40], we reported the preparation and physicochemical characterization of multifunctional BCP scaffolds coated with PCL or poly(ester urea) (PEU). The results showed a significant improvement of the compressive strength of the scaffold, especially with the PEU coating [39]. The conducted studies also showed that the scaffold coated with PEU exhibited an excellent cytocompatibility with hUC-MSCs, promoting their adhesion and proliferation on its surface [38].

Considering the excellent previous results, in this work, we investigate the effect of doping polymer-coated BCP scaffolds with three ions, namely, Sr^{2+}, Mg^{2+} and Zn^{2+} (PCL-coated BCP-6Sr2Mg2Zn and PEU-coated BCP-6Sr2Mg2Zn) to promote the differentiation of human mesenchymal stromal cells from umbilical cord matrix (hUC-MSCs) into an osteogenic lineage. Although the effects of these ions on osteoblast functionality are known, their combined effect has never been investigated. To identify the osteogenic responses of the doped scaffolds, the behavior of hUC-MSCs with coated doped scaffolds was examined using the MTS assay, alkaline phosphatase (ALP) staining and Alizarin Red S (ARS) staining.

2. Materials and Methods

2.1. Preparation of BCP Scaffolds Derived from Cuttlefish Bone

The scaffolds used in this work were prepared and characterized as described elsewhere [25–27]. Briefly, cuttlefish bones were carefully cut into small square-shaped pieces (~10 mm × 10 mm × 2 mm) and then subjected to hydrothermal transformation into BCP scaffolds [40]. Undoped (BCP) and doped compositions containing strontium (BCP-6Sr), strontium and magnesium (BCP-6Sr2Mg), strontium and zinc (BCP-6Sr2Zn), or strontium, magnesium, and zinc (BCP-6Sr2Mg2Zn) were prepared. Briefly, di-ammonium hydrogen phosphate $(NH_4)_2HPO_4$ was used as a starting chemical precursor for phosphorous, while strontium nitrate ($Sr(NO_3)_2$), magnesium nitrate hexahydrate ($Mg(NO_3)_2.6H_2O$) and zinc nitrate hexahydrate ($Zn(NO_3)_2.6H_2O$) were used as the starting precursors for strontium, magnesium and zinc. Cuttlefish bones (CBs) obtained from *Sepia officinalis* were used as a source of calcium. The CB pieces with a known amount of $CaCO_3$ were mixed with the required amount and concentration of the phosphorous precursor solution to obtain the undoped scaffold (BCP). For the doped scaffolds, four different doped compositions with a fixed Sr^{2+} content (6 mol%, BCP-6Sr), alone or in various combinations with other doping ions (2 mol% Mg^{2+}, BCP-6Sr2Mg; 2 mol% Zn^{2+}, BCP-6Sr2Zn; and 2 mol% Mg^{2+} plus 2 mol% Zn^{2+}, BCP-6Sr2Mg2Zn), were prepared. These solutions were placed into a poly(tetrafluorethylene) (PTFE) stainless steel autoclave together with the CB pieces, followed by an HT at 200 °C for 24 h. Subsequently, the scaffolds obtained were washed with distilled water and dried in an oven at 40 °C. The heat treatment occurred at 700 °C using a heating rate of 0.5 °C min^{-1} for 1 h, followed by sintering at 1200 °C for 2 h at a heating rate of 2 °C min^{-1}.

Posteriorly, the BCP-6Sr2Mg2Zn scaffolds were coated with two polymers: PCL (a well-known commercial polymer, CAPA™6800, Mn = 80.000 g mol^{-1}) and PEU (a polymer synthesized in the laboratory, Mn = 63.000 g mol^{-1}). The polymers were dissolved in dichloromethane (Sigma-Aldrich, Darmstadt, Germany) at a concentration of 5% (*w*/*v*). To improve the dissolution of PEU, the addition of 0.9% (*v*/*v*) dimethyl sulfoxide as a co-solvent was required. The impregnation of the scaffolds with the polymer solutions occurred under vacuum conditions. After removing the excess solutions by capillarity, the samples were dried in a vacuum oven to ensure total removal of the solvent.

2.2. Determination of the Porosity and Mechanical Properties of the Scaffolds

Porosity was determined with density measurements using the buoyancy method (Archimedes' method). For this purpose, the scaffolds were immersed in distilled water, according to the European Standard EN 993-1. From the data related to the composition of the mineralogical phase compositions and, if necessary, the percentage of adsorbed polymer in each sample, it was possible to calculate the theoretical density of the scaffold and, consequently, its porosity [39]. Compression tests were performed using a universal testing machine (AG-IS10kN, Shimadzu, Kyoto, Japan) by applying a maximum force of 200 N perpendicular to the lamella of cubic scaffolds, with a side length of about 3 mm, at a constant crosshead speed of 0.5 mm min^{-1} under dry ambient conditions.

2.3. In Vitro Degradation Assay

An *in vitro* degradation profile of BCP-6Sr2Mg2Zn, BCP-6Sr2Mg2Zn-PEU and BCP-6Sr2Mg2Zn-PCL scaffolds (~10 mm × 10 mm) with an average thickness of 2 mm was evaluated in α-MEM medium without ribonucleosides and deoxyribonucleosides (GIBCO™ Invitrogen Corporation, Carlsbad, CA, USA) supplemented with 10% fetal bovine serum (Cytiva HyClone™ Fetal Bovine Serum (FBS) U.S. Origin, Fisher Scientific, Loughborough, UK), 1% Penicillin/Streptomycin and 1% Amphotericin B (Gibco) at pH = 7.4, according to the method previously used by Jeong and co-workers [41]. Briefly, the dried scaffolds were immersed in the culture medium at a 1:10 ratio of scaffold weight (g) to solution volume (mL) in a constant temperature incubator shaker (37 °C, 100 rpm). The samples were removed from the medium at predetermined time intervals, i.e., after 1, 7 and 14 days. The medium was renewed every 3 days. The samples were washed gently with deionized water to eliminate non-adherent particles, dehydrated with absolute ethanol and dried in an oven (40 °C) to constant weight. The weight loss (%) at each time point was calculated using Equation (1) [41],

$$\text{Weight loss } (\%) = \frac{\text{Wo} - \text{Wt}}{\text{Wo}} \times 100 \tag{1}$$

where Wo is the initial weight before the *in vitro* degradation test, and Wt is the dry weight at a predetermined time.

2.4. Isolation and Culture of Human Mesenchymal Stromal Cells from Umbilical Cord Matrix (hUC-MSCs)

Human umbilical cords from healthy donors were kindly donated after birth, with parental consent, by Crioestaminal Saúde e Tecnologia, Biocant Park, Portugal. The umbilical cords were stored in sterile 50 mL tubes at room temperature between 12 and 48 h before tissue processing. The samples were cut into small pieces of approximately 5 cm. The pieces were washed with sterile phosphate-buffered saline (PBS) solution to remove the blood. The umbilical veins were also washed to remove blood and blood clots. To avoid contamination by endothelial cells, the umbilical veins and arteries were also removed. The specimens were then dried in tissue culture plates to promote adhesion of the fragment to the polystyrene surface, and, after adhesion, hUC-MSC proliferation medium [α-MEM (1x) with L-glutamine and without ribonucleosides and deoxyribonucleosides (GIBCO™ Invitrogen Corporation, Carlsbad, CA, USA) supplemented with 10% FBS (Cytiva HyClone™ Fetal Bovine Serum (FBS) U.S. Origin, Fisher Scientific, Loughborough, UK), 1% Penicillin/Streptomycin and 1% Amphotericin B (GIBCO™ Invitrogen Corporation, Carlsbad, CA, USA)] was added to the cell culture plate. Samples were cultured at 37 °C, 5% CO_2 and 95% humidity for 10 days until hUC-MSCs migrated from the umbilical cord matrix and defined colonies formed. Finally, the fragments of the umbilical cord matrix were removed, and the cells were passaged. hUC-MSCs at passages 2–4 were used in the following experiments.

2.5. Cytocompatibility Assessments

The scaffolds were sterilized with UV irradiation light for 15 min and pre-wetted in culture medium for 2 h before their use in cell experiments. After sterilization, the scaffolds were transferred to the culture plate and cells were seeded dropwise at a cell density of 1.5×10^5 cells/well. To promote cell adhesion, the samples were incubated for 30 min before the addition of culture medium (α-MEM with L-glutamine and without ribonucleosides and deoxyribonucleosides supplemented with 10% FBS, 1% Penicillin/Streptomycin and 1% Amphotericin B). The seeded scaffolds were cultured in an incubator at 37 °C and a humidified atmosphere (5% CO_2 and 95% air) for a period of 1–14 days. The medium was changed every 2–3 days.

2.5.1. Cell Viability and Proliferation

Cells (1.5×10^5 cells/well) seeded on scaffolds were cultured for 24, 48 and 72 h. At the end of each period, the scaffolds were washed with PBS solution and then a mixture of Hoechst 33342 and propidium iodide in PBS solution was added to each well and incubated for an additional 15 min at 37 °C. The Live/Dead™ Cell Imaging Kit (Invitrogen™) was used to monitor live and dead cells seeded into the well plates in the presence of scaffolds.

The kit contains solutions of Hoechst 33342 and propidium iodide that stain both live and dead cells. The stained cells were then observed using an inverted fluorescence microscope. Cell proliferation and the viability of seeded scaffolds were also assessed using the MTT assay (Sigma-Aldrich, Darmstadt, Germany) or the MTS assay (CellTiter 96® AQueous One Solution Cell Proliferation colorimetric assay, Promega Corporation, Madison, WI, USA), according to the manufacturers' instructions. After each time point, an MTT solution (0.5 mg/mL in PBS) or an MTS reagent (10% v/v) was added to each well, and plates were incubated for 4 h in the dark at 37 °C, in a humidified atmosphere. For the MTT assay, after incubation, the MTT reagent was removed from the wells and an acidified isopropanol solution (0.04 N HCl in isopropanol) was added to dissolve the formazan crystals. Then, 100 µL of each well was transferred to a 96-well plate, and the absorbance was measured at 570 nm. Using the MTS assay, 100 µL of culture medium containing an MTS reagent was transferred to a 96-well plate, and absorbance was measured at 490 nm. A negative control (untreated cells), i.e., cells cultured without being exposed to the scaffolds, was performed. Cell viability was calculated as the percentage of viable cells relative to the untreated control cells, which were assumed to be 100% viable. Three replicate samples were tested for each condition.

2.5.2. Cell Attachment

Cell adhesion and the morphology of cells adhering to the scaffolds were studied using scanning electron microscopy (SEM). After 7 and 14 days of culture, the scaffolds seeded with the human MSCs were carefully washed with PBS and fixed in 4% paraformaldehyde solution for 1 h at 4 °C. After fixation, the samples were washed with PBS and distilled water. Samples were then dehydrated in ascending ethanol solutions (30, 50, 70 and 90%) for 15 min each and, finally, dehydrated in absolute ethanol for 30 min. The dried cell-seeded scaffolds were sputter-coated with a gold layer before visualization using SEM. SEM images were acquired at various magnifications and at an accelerating voltage of 10 kV using a high-resolution field emission scanning electron microscope with Energy-dispersive X-ray spectroscopy (EDS) and Wavelength-dispersive X-ray spectroscopy (WDS) (STEM ZEISS, Merlin, Oberkochen, Germany).

2.6. *Osteogenic Differentiation Assessment*

For osteogenic differentiation, the cells were seeded on the scaffolds (1×10^5 cell/well in 24-well plate), and, after 2 days, the culture medium was replaced by an osteogenic differentiation medium. The composition of the differentiation medium was as follows: StemPro® Osteocyte/Chondrocyte Differentiation Basal Medium, StemPro® Osteogenesis Supplement, 1% gentamicin solution, 10 nM dexamethasone, 10 mM β-glycerophosphate, 50 µM ascorbate-2-phosphate (GIBCO™ Invitrogen Corporation, Carlsbad, CA, USA). The medium was replaced with a fresh portion every 3 days for 14 days.

2.6.1. Alizarin Red S Staining

Alizarin Red S was used to assess matrix mineralization on days 7 and 14 after cell seeding. The differentiation culture medium was removed, and the scaffolds seeded with cells were washed with PBS 1x. Then, samples were incubated in cold 4% paraformaldehyde solution for 30 min. Cells were rinsed again with PBS 1x and stained with 40 mM Alizarin Red S (Sigma-Aldrich, Darmstadt, Germany) at pH 4.2 for 10 min at 37 °C. To remove nonspecific staining (i.e., not associated with calcium mineral deposits), the samples were washed out several times with distilled water. The Alizarin Red S staining of hUC-MSCs without differentiation medium was also performed to compare with cells differentiated in differentiation medium. Stained samples were imaged with a stereomicroscope (Stemi 508 Stereo Microscope, Zeiss, Germany) for qualitative analysis and examined for orange-red-colored deposits.

2.6.2. Alkaline Phosphatase Activity

Alkaline phosphatase (ALP) activity was measured using the hydrolysis of *p*-nitrophenyl phosphate (*p*NPP) to *p*-nitrophenol. hUC-MSC-seeded scaffolds were washed with PBS and incubated with 20 µL of *p*NPP 0.67 M, 960 µL reaction buffer (1 M Diethanolamine and 0.5 mM of $MgCl_2$, pH 9.8, alkaline phosphatase, Diethanolamine Detection Kit, Sigma-Aldrich) and 20 µL of the test sample for 30 min at 37 °C in the dark. For enzyme activity control, 20 µL of diluted alkaline buffer solution was added. The reaction was quenched with 400 µL of 0.5 M NaOH, and the absorbance of 150 µL of the resulting solution was measured at 405 nm.

2.7. Statistical Analysis

Experiments were performed in triplicate with two replicates in each experiment. Quantitative data are reported as mean ± standard deviation (SD). Two-way analysis of variance (ANOVA) and Student's *t*-test were used for the statistical study. When more than two groups were compared (multiple comparisons), two-way analysis ANOVA with Tukey comparison was used. The results were considered statistically different if the *p* value was less than 0.05. Significance is indicated in the graphs. All statistical analyses were performed using the Prism software package (PRISM 5.0; GraphPad Software, San Diego, CA, USA, 2007).

3. Results

Scaffolds obtained from marine skeletons have been explored over the last few years as a promising material for the development of bone tissue. In our previous works [38–40], BCP scaffolds were obtained using hydrothermal transformation of cuttlefish bones, and Sr^{2+}, Mg^{2+} and/or Zn^{2+} ions were incorporated by partial replacement of calcium or by coating bioactive glass derived from sol-gel. The mechanical properties of the hydrothermally transformed samples were improved by applying PCL or PEU coatings. Table 1 presents the main physical-mechanical properties of the scaffolds under study in this work. More details can be found elsewhere [39].

Table 1. Physical-mechanical properties of the scaffolds under study in this work [39].

Scaffold	Porosity (%)	Mechanical Properties	
		Compressive Strength (MPa)	Young's Modulus (MPa)
BCP	92.73 ± 0.27	0.21 ± 0.02	0.33 ± 0.03
BCP-6Sr	92.56 ± 0.35	0.20 ± 0.02	0.27 ± 0.02
BCP-6Sr2Mg	92.85 ± 0.50	0.40 ± 0.02	0.51 ± 0.09
BCP-6Sr2Zn	92.85 ± 0.54	0.20 ± 0.01	0.21 ± 0.01
BCP-6Sr2Mg2Zn	92.76 ± 0.26	0.36 ± 0.02	0.52 ± 0.05
PCL-coated BCP-6Sr2Mg2Zn	89.27 ± 0.08	0.46 ± 0.04	0.66 ± 0.05
PEU-coated BCP-6Sr2Mg2Zn	91.28 ± 0.24	1.05 ± 0.12	1.54 ± 0.18

It is important to highlight that the polymer coatings did not compromise the porosity of the scaffolds. In this work, the potential use of coated doped scaffolds (Figure 1) [38–40] in promoting the differentiation of hUC-MSCs into an osteoblastic lineage was evaluated.

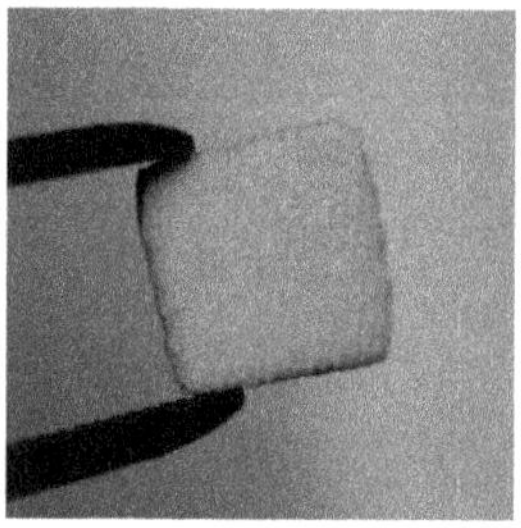

Figure 1. Representative image of the coated doped scaffolds (top view).

3.1. Human MSCs Seeded onto BCP-6Sr2Mg2Zn Scaffolds Coated with PCL and PEU Show High Cell Viability Rates

The first set of experiments investigated whether doping elements introduced by partial replacement of calcium would have a positive effect on the viability and proliferation of hUC-MSCs in direct contact with (un)doped scaffolds after 24, 48 and 72 h of incubation. The Live/Dead cell assay was used to directly monitor the proportion of viable and dead cells upon contact with the undoped and doped BCP scaffolds. The fluorescence microscopy images suggest that hUC-MSCs remained viable over 72 h of culture for all examined scaffolds (Figure 2a). Viable cells (stained with Hoechst 33342) were present in large numbers as detected by the high bright blue fluorescence in the experiments. Very few dead cells (stained with propidium iodide), which are red in color, were observed in the images (Figure 2a). Taken together, both tests show that the undoped and doped scaffolds are minimally cytotoxic and do not negatively affect cell viability. The results from the viability tests also indicate that the cells seeded on the BCP-6Sr2Mg2Zn scaffolds exhibited higher cell viability compared to the neat BCP scaffolds after 72 h of incubation (**** $p < 0.0001$) (Figure 2b). Furthermore, the cells' proliferation levels on the BCP-6Sr2Mg2Zn samples were significantly greater than those obtained at 24 h (* $p < 0.05$). There were no statistically significant differences between any of the samples based on Student *t*-test results (Figure 2b). All doped scaffolds are non-toxic and cytocompatible (cell viability > 5%).

Considering that the BCP-6Sr2Mg2Zn scaffolds exhibited the highest cell viability, it was decided to proceed with additional *in vitro* cytotoxicity tests on the scaffolds coated with PCL and PEU. To this end, hUC-MSCs were seeded onto the surface of PCL-coated BCP-6Sr2Mg2Zn and PEU-coated BCP-6Sr2Mg2Zn scaffolds, and cell proliferation studies were performed at different time points over a 14-day period. For comparison purposes, the BCP-6Sr2Mg2Zn scaffold was also subjected to the same test. As shown in Figure 2c, the uncoated samples (BCP-6Sr2Mg2Zn) showed a tendency toward increased cell viability over time, but the PCL- and PEU-coated scaffolds resulted in superior cell viability at all time points, indicating that the polymer coatings did not negatively affect cell behavior over time. One-way ANOVA and Tukey's multiple comparison tests revealed a statistically significant increase in the viability of cells grown on BCP-6Sr2Mg2Zn scaffolds coated with PCL and PEU compared to cells on BCP-6Sr2Mg2Zn scaffolds (**** $p < 0.0001$) (Figure 2c). While 73.08 ± 4.41% and 78.10 ± 6.74% of seeded cells were attached in PCL-coated BCP-6Sr2Mg2Zn and PEU-coated BCP-6Sr2Mg2Zn, respectively, after 1 day, only 46.52 ± 6.03% of cells were attached in BCP-6Sr2Mg2Zn (Figure 2c). On the seventh day of incubation, statistically significant increases were observed between PCL-coated BCP-6Sr2Mg2Zn (79.86 ± 7.42%) and PEU-coated BCP-6Sr2Mg2Zn (86.32 ± 6.56%) scaffolds compared to their uncoated counterparts (57.20 ± 7.02%) (Figure 2c). Finally, on the 14th day of incubation, there was, again, a significant difference in supporting cell growth on the scaffolds between the BCP-6Sr2Mg2Zn scaffolds (76.73 ± 9.36%) and the BCP-6Sr2Mg2Zn scaffolds coated with PCL (89.75 ± 5.01%) and PEU (94.31 ± 5.86%). Briefly, the metabolic activity of hUC-MSCs on BCP-6Sr2Mg2Zn scaffolds coated with PCL and PEU was higher than BCP-6Sr2Mg2Zn samples after 1, 7 and 14 days of culture. So, the viability results showed that the PCL and PEU coating process seems to, in fact, improve cell viability.

Together, this indicates that the scaffolds are cytocompatible and able to support the viability of hUC-MSCs after their attachment.

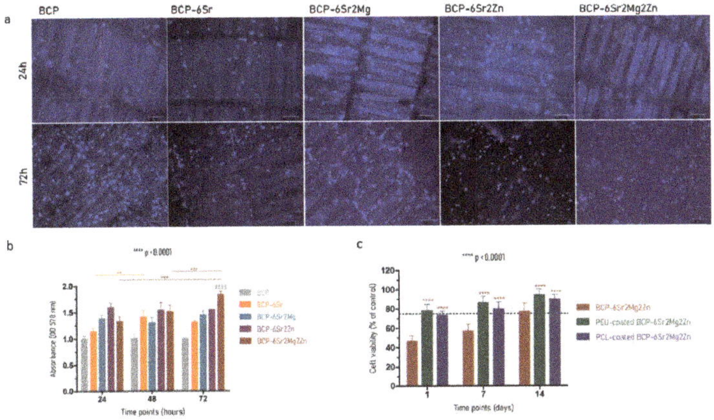

Figure 2. (**a**) Live/Dead fluorescence images after 24 and 72 h of cell culture. The nuclei of healthy cells were stained with Hoechst 33342 (blue), and dead cells were stained with propidium iodide (red). Scale bar represents 100 µm. (**b**) Viability results from undoped and doped scaffolds. Data represent mean ± SD, n = 3. The * indicates significant difference (*** $p < 0.001$ and ** $p < 0.05$) in the same sample at different times of culture and the $ denotes significant difference between the samples on the same culture day ($$$$ $p < 0.0001$). (**c**) Viability results from the doped (un)coated scaffolds. Data represent mean ± SD, n = 5 (asterisks denote significant differences relative to the control group BCP-6Sr2Mg2Zn, **** $p < 0.0001$).

3.2. *hUC-MSCs Show Normal Morphology and Adhesion Properties When Growing onto BCP-6Sr2Mg2Zn Scaffolds Coated with PCL and PEU Scaffolds*

Cell growth was also monitored with morphological/adhesion studies using inverted and electron microscopes. The process of cell adhesion involves four stages: attachment, filopodial growth, cytoplasmic web and flattening of the cell mass. It depends on several factors such as surface topography, surface chemistry, surface energy, and hydrophilicity, as well as the chemical composition of the sample [42]. Although some morphological heterogeneity was observed in the adherent fraction during the initial culture period, morphological homogeneity was gradually achieved over time. According to the inverted microscope images (Figure 3), hUC-MSCs seeded on the surface of BCP-6Sr2Mg2Zn scaffolds coated with PCL and PEU showed a similar morphology to control cells (hUC-MSCs seeded on plates). After 14 days in culture (Figure 3), the cells were viable, exhibiting a normal, spread and fusiform morphology characteristic of this cell type (fibroblast-like), and no spontaneous differentiation was observed. It can also be seen that hUC-MSCs tend to grow toward the scaffolds (red arrows point to scaffold, i.e., the black shadow seen in the figures is the scaffold). In turn, the cell density in the BCP-6Sr2Mg2Zn scaffolds was lower, and the morphology showed an unhealthy appearance (spheroidal shape) compared to the control group (hUC-MSCs seeded on plates).

The adhesion, spreading and morphology of hUC-MSC cells seeded on the prepared scaffolds were also studied using SEM. Figure 4a,b show SEM images of hUC-MSCs grown on coated scaffolds after 7 and 14 days of cell culture, respectively. After 7 days of culture, hUC-MSCs proliferated well, exhibited a flat appearance and formed confluent monolayers, indicating that our scaffolds supported uniform cell attachment and growth. In the high-magnification images in Figure 4a referring to the PEU-coated BCP-6Sr2Mg2Zn scaffold, it can be seen that the hUC-MSCs exhibit a partially spread morphology with some filipodial protrusions, indicating that the cells are starting to interact with the substrate. In turn, the surface of the PCL-coated BCP-6Sr2Mg2Zn scaffold is covered by a layer of cells, making it difficult to draw conclusions about the morphology of the cells. After 14 days of culture, the surface of the BCP-6Sr2Mg2Zn scaffolds coated with PCL and PEU was almost completely covered with dense layers of cells and the extracellular matrix secreted by them (Figure 4b), suggesting better spreading of the cells. The cell density on the BCP-6Sr2Mg2Zn scaffolds was lower and the formation of a cell layer is not visible. Only single cells exist on the scaffold surface.

In Figure 4b, the cells cultured for 14 days grew in colonies on the surface and showed a greater number of cells with the formation of more clusters. The surface of the coated doped scaffolds was almost completely covered with dense layers of hUC-MSCs, which formed smooth, mesh-like structures (the entire surface was covered), in contrast to the rough surface of the BCP-6Sr2Mg2Zn scaffold. hUC-MSCs showed adhesion to the scaffold itself by forming single-layered sheets. In some cases, cell sheets branched through open pores to reach the opposite edges of the pore walls and spread further over a larger part of the scaffold surface (Figure 4b). In addition, cells were observed bridging scaffold filaments, suggesting that coated scaffolds are capable of supporting cell growth and proliferation. The observed results suggest that the coated scaffolds promote intercellular contact and spatial arrangement of cells. Finally, the performed analyses suggest good cell viability, proliferation and distribution, pointing out a good cytocompatibility profile of the PEU-coated BCP-6Sr2Mg2Zn scaffold in contact with hUC-MSCs, providing a suitable environment for cell growth.

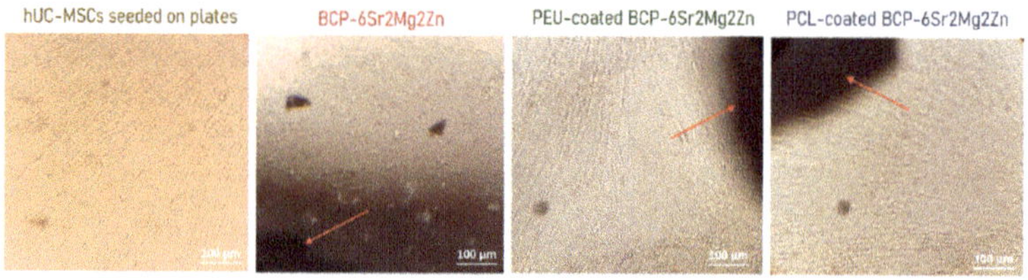

Figure 3. Inverted microscopic images (scale bar, 100 µm) of hUC-MSCs after 14 days of culture: control group (regular cell culture plate), BCP-6Sr2Mg2Zn, PCL-coated BCP-6Sr2Mg2Zn and PEU-coated BCP-6Sr2Mg2Zn scaffolds. The black specks in the figures are the shadows of the scaffolds, so the red arrows indicate the presence of scaffolds on the culture plates.

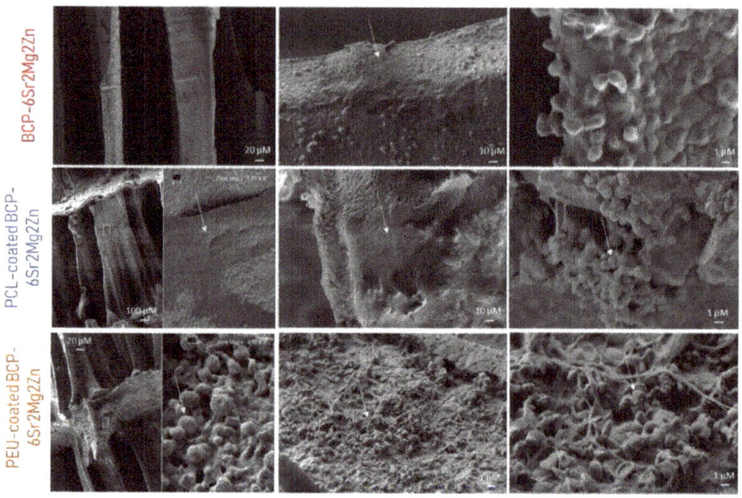

(**a**)

Figure 4. *Cont.*

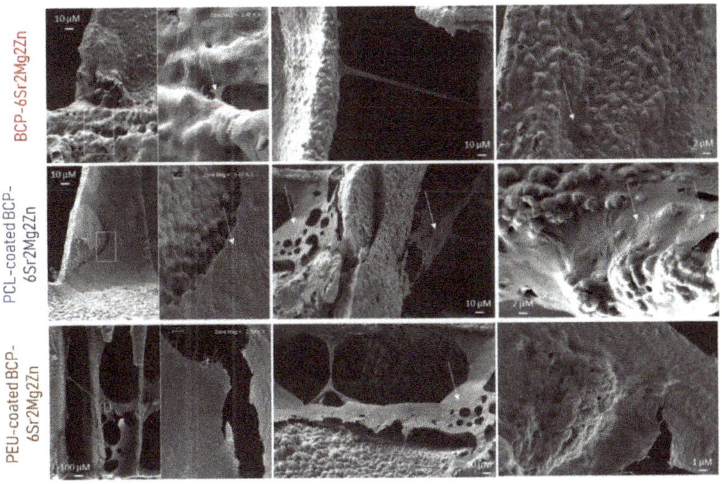

(b)

Figure 4. (a) Representative scanning electron micrographs of cell attachment to BCP-6Sr2Mg2Zn and BCP-6Sr2Mg2Zn scaffolds coated with PCL and PEU after 7 days of culture. (b) Representative scanning electron micrographs of cell attachment to BCP-6Sr2Mg2Zn and BCP-6Sr2Mg2Zn scaffolds coated with PCL and PEU after 14 days of culture. Attachment of hUC-MSCs to the surface of scaffolds is highlighted with white arrows.

3.3. *In Vitro Degradation*

The *in vitro* degradability of the uncoated and coated doped scaffolds was investigated in serum-enriched culture medium (α-MEM with L-glutamine and without ribonucleosides and deoxyribonucleosides supplemented with 10% FBS, 1% Penicillin/Streptomycin and 1% Amphotericin B). Figure 5 shows the degradation curves of the BCP-6Sr2Mg2Zn, PCL-coated BCP-6Sr2Mg2Zn and PEU-coated BCP-6Sr2Mg2Zn scaffolds after 1, 3, 7 and 14 days. As shown in Figure 5, the weight loss of the PEU-coated BCP-6Sr2Mg2Zn samples was measured to be 11.25 ± 0.35% after 24 h of incubation in serum-containing medium, indicating a slightly faster degradation. On the other hand, under the same conditions, the weight loss was 0.66 ± 0.40% and 1.39 ± 0.58% for BCP-6Sr2Mg2Zn and PCL-coated BCP-6Sr2Mg2Zn scaffolds, respectively. After 3 days of incubation, the *in vitro* degradation rate of BCP-6Sr2Mg2Zn scaffolds increased, and the weight loss was 2.33 ± 1.27%. In turn, the coated scaffolds maintained their degradation percentage at 3 days of incubation: 10.98 ± 0.19% for PEU-coated BCP-6Sr2Mg2Zn scaffolds and 0.15 ± 0.14% for PCL-coated BCP-6Sr2Mg2Zn scaffolds. Compared to other scaffolds, PEU-coated BCP-6Sr2Mg2Zn scaffolds continued exhibiting a higher weight loss of 11.82 ± 0.74% after 7 days of incubation: 0.57 ± 0.14% for PCL-coated BCP-6Sr2Mg2Zn scaffolds and 1.75 ± 0.24% for BCP-6Sr2Mg2Zn scaffolds. After 14 days of incubation, the average percentage of degradation for PEU-coated BCP-6Sr2Mg2Zn scaffolds was 9.77 ± 0.48% (Figure 5). The weight loss was 2.96 ± 0.75% and 0,24 ± 0.01% for BCP-6Sr2Mg2Zn and PCL-coated BCP-6Sr2Mg2Zn scaffolds, respectively.

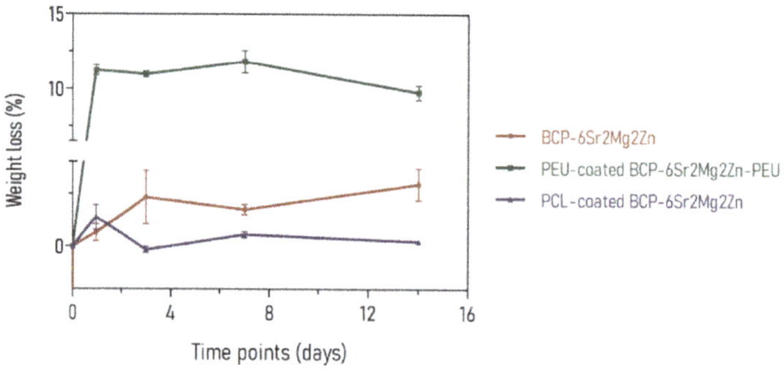

Figure 5. *In vitro* degradation curves of BCP-6Sr2Mg2Zn, PEU-coated BCP-6Sr2Mg2Zn and PCL-coated BCP-6Sr2Mg2Zn porous scaffolds at 1, 3, 7 and 14 days in serum-containing medium.

3.4. hUC-MSCs Growing onto BCP-6Sr2Mg2Zn Scaffolds Coated with PCL and PEU Scaffolds Can Differentiate into Osteoblasts

To assess the effect of scaffold composition on the ability of hUC-MSCs to promote matrix mineralization, Alizarin Red S staining assay was used [43,44]. This staining can reveal the presence of calcium deposits, which is indicative of the differentiation of hUC-MSCs into osteoblasts (osteogenesis) [44]. When the cells reached 90% confluence (~3 days) and exhibited a fusiform shape, the basal medium was removed and osteogenic differentiation was initiated by adding osteogenic medium. Optical microscopy images of the mineralized matrix deposits on differentiating cultures of hUC-MSCs in direct contact with the scaffolds are shown in Figure 6a. Staining with Alizarin Red S showed that hUC-MSCs on scaffolds began to form calcified nodules after 7 days of cultivation in osteogenic medium. This positive result is indicated by bright red extracellular calcium deposits. In contrast, cells in the control medium (undifferentiated) did not form calcium deposits, even after 7 days (Figure 6a). After 14 days of osteogenic differentiation, cells seeded on the coated scaffolds showed a more intense Alizarin Red S staining (significantly darker) than in the BCP-6Sr2Mg2Zn scaffolds. The results indicate that the hUC-MSCs on these scaffolds produced a high number of larger calcified nodules (calcium content) that promote osteoblastic activity. hUC-MSCs in direct contact with coated scaffolds synthesized bone mineral nodules, as demonstrated with Alizarin Red staining, indicating that exposure to the coating promoted/increased the ability of hUC-MSCs to differentiate along the osteogenic lineage. The results suggest that the mineralization was mediated by the hUC-MSCs and was not the result of nonspecific calcium precipitation.

In addition to the calcium deposits stained with the Alizarin Red assay, the potential of the scaffolds for the differentiation process was confirmed by the activity of alkaline phosphatase (ALP) at 7 and 14 days (Figure 6b). ALP plays an important role in the bone mineralization process, and its activity is a phenotypic marker for osteoblastic cells [45,46]. As shown in Figure 6b, the intracellular ALP activity of hUC-MSCs grown on coated scaffolds increased with the increasing cultivation time between Days 7 and 14, indicative of the process of osteogenic differentiation. No statistically significant difference was observed between groups in the first week after induction, but, after 14 days, hUC-MSCs seeded on coated scaffolds showed significantly higher ALP activity compared with BCP-6Sr2Mg2Zn. On Day 14, ALP activity continued to increase in coated scaffold cultures and was significantly higher than in uncoated scaffold cultures (BCP-6Sr2Mg2Zn differentiated, 0.575 ± 0.082; PCL-coated BCP-6Sr2Mg2Zn differentiated, 1.113 ± 0.343; PEU-coated BCP-6Sr2Mg2Zn differentiated, 0.860 ± 0.141; **** $p < 0.0001$) (Figure 6b). The data revealed that cells in contact with PCL-coated BCP-6Sr2Mg2Zn and PEU-coated BCP-6Sr2Mg2Zn exhibited 1.9- and 1.5-fold higher ALP activity, respectively, compared to the BCP-6Sr2Mg2Zn scaffold, corroborating with the previous results of the biomineralization assays.

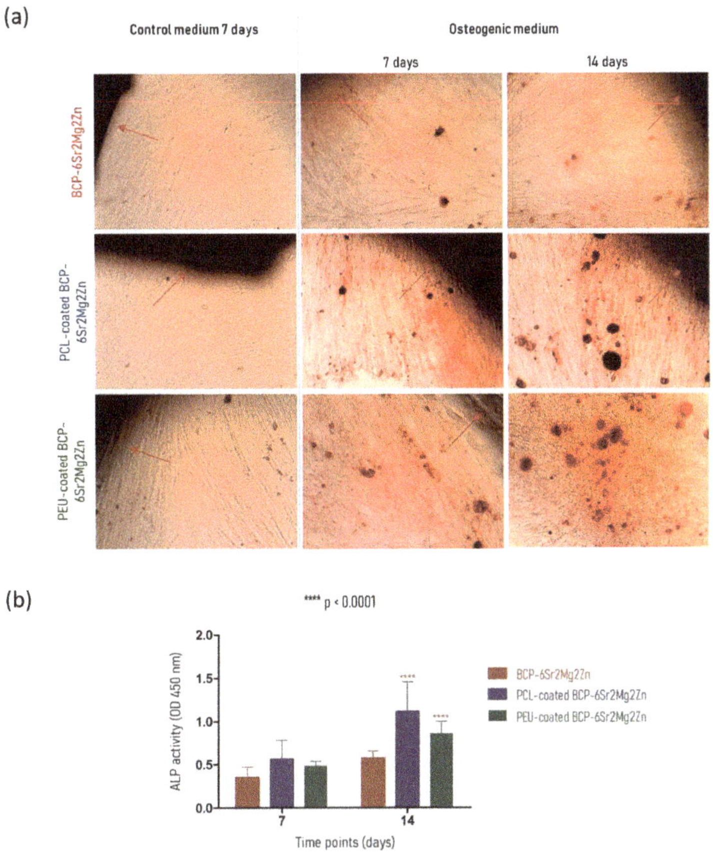

Figure 6. Induced differentiation of hUC-MSCs. (**a**) Osteogenic differentiation of hUC-MSCs with Alizarin Red to assess mineralization. hUC-MSCs were cultured in normal growth medium for 7 days followed by osteogenic medium during 7 and 14 days. Red arrows point to the scaffold. Dark red staining denotes mineralization. Original magnification: ×100. (**b**) Alkaline phosphatase (ALP) activity of hUC-MSCs that were cultured on the BCP-6Sr2Mg2Zn, PCL-coated BCP-6Sr2Mg2Zn and PEU-coated BCP-6Sr2Mg2Zn scaffolds after 7 and 14 days in osteogenic culture medium (asterisks denote significant difference between the groups at $p < 0.0001$).

4. Discussion

In our previous work, BCP scaffolds obtained using hydrothermal transformation of cuttlefish bone were fabricated, doped with different ions (Sr^{2+}, Mg^{2+} and Zn^{2+}) and coated with PCL or PEU [39]. The results demonstrated that coated doped scaffolds exhibited many desirable characteristics, including high porosity, improved swelling profile and better mechanical performance than BCP scaffolds [39]. In the present study, our investigation focused on the potential use of coated doped scaffolds to promote the differentiation of hUC-MSCs into an osteoblastic lineage. To confirm cytocompatibility, the adhesion and viability of hUC-MSCs on the surface of the scaffolds were evaluated. Live/Dead staining showed that hUC-MSCs seeded on the surface of the doped scaffolds survived well after 3 days of culture. As shown in Figure 2a, most of the hUC-MSCs cultured on the scaffolds were alive (shown in blue), and only a few cells were dead (red fluorescence). It is important to emphasize that the presence of doping elements (Sr^{2+}, Mg^{2+} and Zn^{2+}) resulted in an increase in cellular metabolic activity (cytocompatibility). The viability results showed no cytotoxic effect of the scaffolds on the hUC-MSCs over a period of

3 days, and a time-dependent proliferation pattern was observed (Figure 2b). Many studies using biological activity assays have shown that BCP scaffolds doped with metals are not toxic to MC3T3-E1 cells [25,47], human umbilical vein endothelial cells [48], bone mesenchymal stem cells [49], rabbit-adipose-derived mesenchymal stem cells [50], human osteoblast-like MG-63 cells [51] and human mesenchymal stem cells [13]. Among the tested compositions, the BCP-6Sr2Mg2Zn scaffolds proved to be the most promising. Next, the impact of PCL or PEU coatings on the interaction of BCP-6Sr2Mg2Zn scaffolds with hUC-MSCs was also investigated through similar *in vitro* studies (Figure 2c). The uncoated samples (BCP-6Sr2Mg2Zn) showed a tendency toward increased cell viability over time, but the PCL- and PEU-coated scaffolds resulted in greater cell proliferation during the 14-day incubation (Figure 2c). These results suggest that the polymers used and the coating process itself do not affect the viability of hUC-MSCs as the viability of coated scaffolds after 14 days in culture was slightly above ~90%, comparable to cells exposed to uncoated scaffolds. The percentage of cell viability was above 80%, indicating that the scaffolds are not toxic to hUC-MSCs (according to ISO 10993-5). Similar results were reported with MG-63 cells in which the PCL-coated cuttlefish-bone-derived HAp scaffold improved cell proliferation, viability and adherence [52–54]. In turn, some research groups have shown the potential of PEUs as non-toxic and bioresorbable materials on MC3T3-E1 pre-osteoblast cells seeded in PEU-blended HAp scaffolds [55] and bone marrow-derived stromal cells seeded in PEU scaffolds [56]. The morphological and proliferative characteristics of the hUC-MSCs seeded on the surface of the coated doped scaffolds were similar to those observed when cultured in tissue culture polystyrene plates (Figure 3). Within the first week, hUC-MSCs showed a well-spread morphology along the coated scaffold pore walls and were able to bridge gaps between pores (Figure 4a). After 14 days of culture, and corroborating the results from the MTS assay, SEM images showed that the cells grew on the surface and in the pores of the coated scaffolds (Figure 4b). The cells on the PEU-coated scaffold appeared to be denser compared to those on the PCL-coated scaffold, resulting in partial or even complete closure of some pores. The SEM microscopy of the cell-seeded PEU-coated scaffolds also showed that most of the outer macropores of the PEU-coated scaffold were occluded by a continuous layer of cells. The cells extended around the pore wall and formed an extracellular matrix (Figure 4b). In our opinion, these data are very promising as they show that the PEU-coated scaffolds are not toxic and do not inhibit the metabolic activity or proliferation of hUC-MSCs after 14 days of culture. PEU-coated scaffolds presented a higher number of attached cells than PCL-coated scaffolds after 14 days of seeding, probably due to slight differences in the hydrophilicity of the two substrates and their high porosity (91.28 $\pm$ 0.24% vs. 89.27 $\pm$ 0.08%), which increases the surface area (Table 1) [39]. A crucial parameter to consider in tissue engineering is the degradation profile of the scaffolds, as this property influences the structural integrity, stability and mechanical performance of the scaffold but also cellular processes including cell proliferation and tissue growth (osteogenesis) [57]. Ideally, the scaffold should degrade at a rate compatible with new bone formation. The results showed that the weight loss of the PCL-coated BCP-6Sr2Mg2Zn scaffolds was almost negligible or did not change at all time points (Figure 5). These data are similar to other studies that indicate that PCL has a long degradation time, sometimes remaining *in vivo* years after implantation [58]. Teoh and co-workers (2009) prepared PCL and PCL-based composite scaffolds, and the results up to 6 months indicated a maximum mass loss of only about 1% and 7% for the PCL scaffolds and PCL-composite scaffolds in vivo, respectively [59]. To overcome this issue, PCL has been blended with natural polymers (e.g., collagen [60,61] or silk [62]) with high susceptibility to hydrolysis and/or enzymolysis in order to improve the biodegradation rate of the scaffolds. On the other hand, the PEU-coated BCP-6Sr2Mg2Zn scaffolds lost ~12% of their original weight over a 14-day period, demonstrating their suitability as temporary scaffolds during the bone healing process (Figure 3) [63]. The higher mass loss of PEU-coated BCP-6Sr2Mg2Zn scaffolds can be attributed to their high susceptibility to hydrolysis and/or enzymolysis [56–58] and their high porosity (91.28 $\pm$ 0.24% vs.

89.27 ± 0.08%), which enable the degradation medium to penetrate more easily into the scaffold (Table 1) [39]. The next study focused on the differentiation of hUC-MSCs seeded on the scaffolds into osteoblasts. hUC-MSCs synthesized bone mineral nodules in the coated scaffolds, as shown with Alizarin Red staining, indicating that the coating process did not affect the ability of hUC-MSCs to differentiate along the osteogenic lineage (Figure 6a). Cellular differentiation was also assessed by monitoring ALP enzyme activity. ALP activity is generally considered an early-stage marker for the osteoblastic phenotype and an important indicator of differentiation and mineralization [54]. A positive effect of PCL and PEU coating on proliferation, differentiation and extracellular matrix deposition was observed compared to the uncoated scaffold. Consistent with the results on cell proliferation, the activity of ALP in osteogenic media significantly increased in cultures on coated scaffolds, peaking at day 14 of culture (Figure 6b). Based on these results, the coated scaffolds prepared in this study show cytocompatibility and provide suitable conditions for the attachment and growth of hUC-MSCs towards a mature and differentiated state, signifying their promising potential to be candidates for regenerative medicine in the treatment of bone defects. A positive effect of PEU coating on adhesion and proliferation of hUC-MSCs was observed, i.e., hUC-MSCs adhered, proliferated well and remained viable after 14 days of culture. The PEU-coated scaffold was shown to have a continuous 'blanket' of cells across its surface, indicating that it is a good substrate for hUC-MSC proliferation. In short, our study suggests that PEU can be used as an alternative to PCL in bone regeneration, as it provides a suitable local microenvironment for maximal induction of osteogenesis. However, more studies are needed to realize these promises and further validate PEU's effectiveness in bone regeneration. We plan to carry out research with the PEU-coated BCP-6Sr2Mg2Zn scaffold to repair small-sized bone defects in animal models, to fully evaluate the biological performance of these scaffolds and to validate their future clinical application. Furthermore, the PEU-coated BCP-6Sr2Mg2Zn scaffold can be further functionalized with several growth factors such as TGF-β and VEGF with sequential release.

5. Conclusions

In the present study, we demonstrated that the coated, doped BCP scaffolds obtained from the hydrothermal transformation of cuttlefish bones are not cytotoxic and provide adequate support for the adhesion and proliferation of hUC-MSCs. First, we demonstrated that the presence of Sr^{2+}, Mg^{2+} and Zn^{2+}, even at low levels, promoted better cell proliferation. The polymer-coated scaffolds led to superior cell viability rates after 7- and 14-day incubation compared to uncoated scaffolds. SEM observations showed that the hUC-MSCs seeded on the PEU-coated scaffolds proliferated well, adhered to the scaffold surfaces and spread over the scaffold surface. The PEU-coated scaffold enhanced cell differentiation ability without negatively affecting cell proliferation under *in vitro* conditions.

In short, our study suggests that PEU can be used as an alternative to PCL in bone regeneration because it provides a suitable local microenvironment for maximal induction of osteogenesis. It would be of interest to include growth factors that control cell proliferation and osteogenic differentiation in the formulation. In addition, it will be important to conduct further studies with the PEU-coated BCP-6Sr2Mg2Zn scaffold to repair small bone defects in animal models, fully evaluate the biological performance of these scaffolds and validate their future clinical application.

Author Contributions: Conceptualization, P.P., A.S.N., C.M., A.S.R. and I.B.; methodology, P.P., A.S.N., C.M., A.S.R. and I.B.; validation, P.P. and C.M.; formal analysis, P.P. and C.M.; investigation, P.P. and A.S.N.; resources, J.M.F.F., J.F.J.C. and A.C.F.; writing—original draft preparation, P.P., A.SN. and C.M.; writing—review and editing, J.R.-S., L.P.d.A., J.M.F.F., J.F.J.C. and A.C.F.; supervision, J.M.F.F., J.F.J.C. and A.C.F.; project administration, J.M.F.F., J.F.J.C. and A.C.F. All authors have read and agreed to the published version of the manuscript.

Funding: This work was developed within the scope of the project CICECO-Aveiro Institute of Materials, UIDB/50011/2020, UIDP/50011/2020 & LA/P/0006/2020 financed by national funds through the FCT/MCTES (PIDDAC).

Institutional Review Board Statement: The use of Human umbilical cord-mesenchymal stem cells (UC-MSCs) involving a partnership with Crioestaminal Saúde e Tecnologia, Biocant Park, Portugal was approved by the Faculty of Medicine of the University of Coimbra's Ethics Committee and followed appropriate informed consent of donors and data protection guidelines at the University of Coimbra (process approval #CE-131/2020).

Data Availability Statement: Data available on request from the corresponding author.

Acknowledgments: Inês Barros thanks the funding support from the Fundação para a Ciênia e Tecnologia (Grant: SFRH/BD/148877/2019).

Conflicts of Interest: The authors declare no conflict of interest.

References

1. Valtanen, R.S.; Yang, Y.P.; Gurtner, G.C.; Maloney, W.J.; Lowenberg, D.W. Synthetic and Bone tissue engineering graft substitutes: What is the future? *Injury* **2021**, *52*, S72–S77. [CrossRef] [PubMed]
2. Wang, W.; Yeung, K.W.K. Bone Grafts and Biomaterials Substitutes for Bone Defect Repair: A Review. *Bioact. Mater.* **2017**, *2*, 224–247. [CrossRef] [PubMed]
3. Yousefi, A.M.; James, P.F.; Akbarzadeh, R.; Subramanian, A.; Flavin, C.; Oudadesse, H. Prospect of stem cells in bone tissue engineering: A review. *Stem Cells Int.* **2016**, *2016*, 6180487. [CrossRef] [PubMed]
4. Ghassemi, T.; Shahroodi, A.; Ebrahimzadeh, M.H.; Mousavian, A.; Movaffagh, J.; Moradi, A. Current Concepts in Scaffolding for Bone Tissue Engineering. *Arch. Bone Jt. Surg.* **2018**, *6*, 90–99.
5. Lesage, C.; Lafont, M.; Guihard, P.; Weiss, P.; Guicheux, J.; Delplace, V. Material-Assisted Strategies for Osteochondral Defect Repair. *Adv. Sci.* **2022**, *9*, e2200050. [CrossRef]
6. Bouler, J.M.; Pilet, P.; Gauthier, O.; Verron, E. Biphasic calcium phosphate ceramics for bone reconstruction: A review of biological response. *Acta Biomater.* **2017**, *53*, 1–12. [CrossRef]
7. Kasten, P.; Luginbühl, R.; van Griensven, M.; Barkhausen, T.; Krettek, C.; Bohner, M.; Bosch, U. Comparison of human bone marrow stromal cells seeded on calcium-deficient hydroxyapatite, β-tricalcium phosphate and demineralized bone matrix. *Biomaterials* **2003**, *24*, 2593–2603. [CrossRef]
8. Sohn, H.S.; Oh, J.K. Review of Bone Graft and Bone Substitutes with an Emphasis on Fracture Surgeries. *Biomater. Res.* **2019**, *23*, 9. [CrossRef]
9. Ebrahimi, M.; Botelho, M.G.; Dorozhkin, S.V. Biphasic calcium phosphates bioceramics (HA/TCP): Concept, physicochemical properties and the impact of standardization of study protocols in biomaterials research. *Mater. Sci. Eng. C* **2017**, *71*, 1293–1312. [CrossRef]
10. Watson-Levings, R.S.; Palmer, G.D.; Levings, P.P.; Dacanay, E.A.; Evans, C.H.; Ghivizzani, S.C. Gene therapy in orthopaedics: Progress and challenges in pre-Clinical development and translation. *Front. Bioeng. Biotechnol.* **2022**, *10*, 901317. [CrossRef]
11. Rocha, J.H.; Lemos, A.F.; Agathopoulos, S.; Kannan, S.; Valério, P.; Ferreira, J.M. Hydrothermal growth of hydroxyapatite scaffolds from aragonitic cuttlefish bones. *J. Biomed. Mater. Res. Part A* **2006**, *77*, 160–168. [CrossRef] [PubMed]
12. Samavedi, S.; Whittington, A.R.; Goldstein, A.S. Calcium phosphate ceramic in bone tissue engineering: A review of properties and their influence on cell behavior. *Acta Biomater.* **2013**, *9*, 8037–8045. [CrossRef] [PubMed]
13. Ballouze, R.; Marahat, M.H.; Mohamad, S.; Saidin, N.A.; Kasim, S.R.; Ooi, J.P. Biocompatible magnesium-doped Biphasic Calcium Phosphate for Bone Regeneration. *J. Biomed. Mater. Res. Part B Appl. Biomater.* **2021**, *109*, 1426–1435. [CrossRef] [PubMed]
14. Cao, H.; Kuboyama, N. A Biodegradable Porous Composite Scaffold of PGA/β-TCP for Bone Tissue Engineering. *Bone* **2010**, *46*, 386–395. [CrossRef] [PubMed]
15. Tite, T.; Popa, A.C.; Balescu, L.M.; Bogdan, I.M.; Pasuk, I.; Ferreira, J.M.F.; Stan, G.E. Cationic substitutions in hydroxyapatite: Current status of the derived biofunctional effects and their in vitro interrogation methods. *Materials* **2018**, *11*, 2081. [CrossRef] [PubMed]
16. Bose, S.; Fielding, G.; Tarafder, S.; Bandyopadhyay, A. Understanding of dopant-induced osteogenesis and angiogenesis in calcium phosphate ceramics. *Trends Biotechnol.* **2013**, *31*, 594–605. [CrossRef]
17. Zhu, H.; Guo, D.; Qi, W.; Xu, K. Development of Sr-incorporated biphasic calcium phosphate bone cement. *Biomed. Mater.* **2017**, *12*, 015016. [CrossRef]
18. Chen, Y.; Liu, Z.; Jiang, T.; Zou, X.; Lei, L.; Yan, W.; Yang, J.; Li, B. Strontium-substituted biphasic calcium phosphate microspheres promoted degradation performance and enhanced bone regeneration. *J. Biomed. Mater. Res. Part A* **2020**, *108*, 895–905. [CrossRef]
19. Marques, C.F.; Olhero, S.; Abrantes, J.C.C.; Marote, A.; Ferreira, S.; Viera, S.I.; Ferreira, J.M.F. Biocompatability and antimicrobial activity of biphasic calcium phosphate powder doped with metal ions for regenerative medicine. *Ceram Int.* **2017**, *43*, 15719–15728. [CrossRef]

20. Kim, T.W.; Lee, H.S.; Kim, D.H.; Jin, H.H.; Hwang, K.H.; Lee, J.K.; Park, H.C.; Yoon, S.Y. In situ synthesis of magnesium-substituted biphasic calcium phosphate and in vitro biodegradation. *Mater. Res. Bull.* **2012**, *47*, 2506–2512. [CrossRef]
21. Kulanthaivel, S.; Roy, B.; Agarwal, T.; Giri, S.; Pramanik, K.; Pal, K.; Ray, S.S.; Maiti, T.K.; Banerjee, I. Cobalt doped proangiogenic hydroxyapatite for bone tissue engineering application. *Mater. Sci. Eng. C* **2016**, *58*, 648–658. [CrossRef] [PubMed]
22. Bernhardt, A.; Schamel, M.; Gbureck, U.; Gelinsky, M. Osteoclastic differentiation and resorption is modulated by bioactive metal ions Co^{2+}, Cu^{2+} and Cr^{3+} incorporated into calcium phosphate bone cements. *PLoS ONE* **2017**, *12*, e0182109. [CrossRef] [PubMed]
23. Jacobs, A.; Renaudin, G.; Charbonnel, N.; Nedelec, J.M.; Forestier, C.; Descamps, S. Copper-Doped Biphasic Calcium Phosphate Powders: Dopant Release, Cytotoxicity and Antibacterial Properties. *Materials* **2021**, *14*, 2393. [CrossRef]
24. Wang, Y.; Yang, X.; Gu, Z.; Qin, H.; Li, L.; Liu, J.; Yu, X. In vitro study on the degradation of lithium-doped hydroxyapatite for bone tissue engineering scaffold. *Mater. Sci. Eng. C* **2016**, *66*, 185–192. [CrossRef] [PubMed]
25. Kim, H.W.; Kim, Y.J. Effect of silicon or cerium doping on the anti-inflammatory activity of biphasic calcium phosphate scaffolds for bone regeneration. *Prog. Biomater.* **2022**, *11*, 421–430. [CrossRef] [PubMed]
26. Cui, L.; Xiang, S.; Chen, D.; Fu, R.; Zhang, X.; Chen, J.; Wang, X. A novel tissue-engineered bone graft composed of silicon-substituted calcium phosphate, autogenous fine particulate bone powder and BMSCs promotes posterolateral spinal fusion in rabbits. *J. Orthop. Transl.* **2021**, *26*, 151–161. [CrossRef]
27. Baradaran, S.; Moghaddam, E.; Nasiri-Tabrizi, B.; Basirun, W.J.; Mehrali, M.; Sookhakian, M.; Hamdi, M.; Alias, Y. Characterization of nickel-doped biphasic calcium phosphate/graphene nanoplatelet composites for biomedical application. *Mater. Sci. Eng. C Mater. Biol. Appl.* **2015**, *49*, 656–668. [CrossRef]
28. Nie, L.; Hou, M.; Wang, T.; Sun, M.; Hou, R. Nanostructured selenium-doped biphasic calcium phosphate with in situ incorporation of silver for antibacterial applications. *Sci. Rep.* **2020**, *10*, 13738. [CrossRef]
29. Wätjen, W.; Haase, H.; Biagioli, M.; Beyersmann, D. Induction of apoptosis in mammalian cells by cadmium and zinc. *Environ. Health Perspect.* **2002**, *110*, 865–867.
30. Basu, S.; Ghosh, A.; Barui, A.; Basu, B. (Fe/Sr) Codoped Biphasic Calcium Phosphate with Tailored Osteoblast Cell Functionality. *ACS Biomater. Sci. Eng.* **2018**, *4*, 857–871. [CrossRef]
31. Besleaga, C.; Nan, B.; Popa, A.C.; Balescu, L.M.; Nedelcu, L.; Neto, A.S.; Pasuk, I.; Leonat, L.; Popescu-Pelin, G.; Ferreira, J.M.F.; et al. Sr and Mg Doped Bi-Phasic Calcium Phosphate Macroporous Bone Graft Substitutes Fabricated by Robocasting: A Structural and Cytocompatibility Assessment. *J. Funct. Biomater.* **2022**, *13*, 123. [CrossRef] [PubMed]
32. Priyadarshini, B.; Vijayalakshmi, U. Development of cerium and silicon co-doped hydroxyapatite nanopowder and its in vitro biological studies for bone regeneration applications. *Adv. Powder Technol.* **2018**, *29*, 2792–2803. [CrossRef]
33. Rocha, J.H.; Lemos, A.F.; Agathopoulos, S.; Valério, P.; Kannan, S.; Oktar, F.N.; Ferreira, J.M. Scaffolds for bone restoration from cuttlefish. *Bone* **2005**, *37*, 850–857. [CrossRef] [PubMed]
34. Park, S.S.; Park, M.; Lee, B.T. Autologous stromal vascular fraction-loaded hyaluronic acid/gelatin-biphasic calcium phosphate scaffold for bone tissue regeneration. *Mater. Sci. Eng. C Mater. Biol. Appl.* **2022**, *132*, 112533. [CrossRef]
35. Thi Hiep, N.; Chan Khon, H.; Dai Hai, N.; Byong-Taek, L.; Van Toi, V.; Thanh Hung, L. Biocompatibility of PCL/PLGA-BCP porous scaffold for bone tissue engineering applications. *J. Biomater. Sci. Polym. Ed.* **2017**, *28*, 864–878. [CrossRef]
36. Ressler, A.; Bauer, L.; Prebeg, T.; Ledinski, M.; Hussainova, I.; Urlić, I.; Ivanković, M.; Ivanković, H. PCL/Si-Doped Multi-Phase Calcium Phosphate Scaffolds Derived from Cuttlefish Bone. *Materials* **2022**, *15*, 3348. [CrossRef]
37. Thuaksuban, N.; Monmaturapoj, N.; Luntheng, T. Effects of polycaprolactone-biphasic calcium phosphate scaffolds on enhancing growth and differentiation of osteoblasts. *Biomed. Mater. Eng.* **2018**, *29*, 159–176. [CrossRef]
38. Neto, A.S.; Pereira, P.; Fonseca, A.C.; Dias, C.; Almeida, M.C.; Barros, I.; Miranda, C.O.; de Almeida, L.P.; Morais, P.V.; Coelho, J.F.J.; et al. Highly Porous Composite scaffolds endowed with antibacterial activity for multifunctional grafts in bone repair. *Polymers* **2021**, *13*, 4378. [CrossRef]
39. Neto, A.S.; Fonseca, A.C.; Abrantes, J.C.C.; Coelho, J.F.J.; Ferreira, J.M.F. Surface functionalization of cuttlefish bone-derived biphasic calcium phosphate scaffolds with polymeric coatings. *Mater. Sci. Eng. C* **2019**, *105*, 110014. [CrossRef]
40. Neto, A.S.; Brazete, D.; Ferreira, J.M.F. Cuttlefish Bone-Derived Biphasic Calcium Phosphate Scaffolds Coated with Sol-Gel Derived Bioactive Glass. *Materials* **2019**, *12*, 2711. [CrossRef]
41. Jeong, S.I.; Jeon, O.; Krebs, M.D.; Hill, M.C.; Alsberg, E. Biodegradable photocrosslinked alginate nanofibre scaffolds with tuneable physical properties, cell adhesivity and growth factor release. *Eur. Cells Mater.* **2012**, *24*, 331–343. [CrossRef]
42. Hashemi Doulabi, A.; Mirzadeh, H.; Imani, M.; Samadi, N. Chitosan/polyethylene glycol fumarate blend film: Physical and antibacterial properties. *Carbohydr. Polym.* **2013**, *92*, 48–56. [CrossRef]
43. Saravanan, S.; Vimalraj, S.; Vairamani, M.; Selvamurugan, N. Role of mesoporous wollastonite (calcium silicate) in mesenchymal stem cell proliferation and osteoblast differentiation: A cellular and molecular study. *J. Biomed. Nanotechnol.* **2015**, *11*, 1124–1138. [CrossRef] [PubMed]
44. Sainitya, R.; Sriram, M.; Kalyanaraman, V.; Dhivya, S.; Saravanan, S.; Vairamani, M.; Sastry, T.P.; Selvamurugan, N. Scaffolds containing chitosan/carboxymethyl cellulose/mesoporous wollastonite for bone tissue engineering. *Int. J. Biol. Macromol.* **2015**, *80*, 481–488. [CrossRef] [PubMed]
45. Xing, Q.; Qian, Z.; Kannan, B.; Tahtinen, M.; Zhao, F. Osteogenic differentiation evaluation of an engineered extracellular matrix based tissue sheet for potential periosteum replacement. *ACS Appl. Mater. Interfaces* **2015**, *7*, 23239–23247. [CrossRef] [PubMed]

46. Leena, R.S.; Vairamani, M.; Selvamurugan, N. Alginate/Gelatin scaffolds incorporated with Silibinin-loaded Chitosan nanoparticles for bone formation in vitro. *Colloids Surf. B Biointerfaces* **2017**, *158*, 308–318. [CrossRef] [PubMed]
47. Pina, S.; Vieira, S.I.; Rego, P.; Torres, P.M.; da Cruz e Silva, O.A.; da Cruz e Silva, E.F.; Ferreira, J.M. Biological responses of brushite-forming Zn-and ZnSr-substituted β-Tricalcium phosphate bone cements. *Eur. Cells Mater.* **2010**, *20*, 162–177. [CrossRef]
48. Lu, T.; Ma, N.; He, F.; Liang, Y.; Ye, J. Enhanced osteogenesis and angiogenesis of biphasic calcium phosphate scaffold by synergistic effect of silk fibroin coating and zinc doping. *J. Biomater. Appl.* **2022**, *37*, 1007–1017. [CrossRef]
49. Xiao, D.; Yang, F.; Zhao, Q.; Chen, S.; Shi, F.; Xiang, X.; Deng, L.; Sun, X.; Weng, J.; Feng, G. Fabrication of a Cu/Zn co-incorporated calcium phosphate scaffold-derived GDF-5 sustained release system with enhanced angiogenesis and osteogenesis properties. *RSC Adv.* **2018**, *8*, 29526–29534. [CrossRef]
50. Sundar, R.; Joseph, J.; Babu, S.; Varma, H.; John, A.; Abraham, A. 3D-bulk to nanoforms of modified hydroxyapatite: Characterization and osteogenic potency in an in vitro 3D bone model system. *J. Biomed. Mater. Res. B Appl. Biomater.* **2022**, *110*, 1151–1164. [CrossRef]
51. Bracci, B.; Torricelli, P.; Panzavolta, S.; Boanini, E.; Giardino, R.; Bigi, A. Effect of Mg($^{2+}$), Sr($^{2+}$), and Mn($^{2+}$) on the chemico-physical and in vitro biological properties of calcium phosphate biomimetic coatings. *J. Inorg. Biochem.* **2009**, *103*, 1666–1674. [CrossRef] [PubMed]
52. Kim, B.S.; Kang, H.J.; Lee, J. Improvement of the compressive strength of a cuttlefish bone-derived porous hydroxyapatite scaffold via polycaprolactone coating. *J. Biomed. Mater. Res. Part B* **2013**, *101*, 1302–1309. [CrossRef] [PubMed]
53. Kim, B.S.; Yang, S.S.; Lee, J. A polycaprolactone/cuttlefish bone-derived hydroxyapatite composite porous scaffold for bone tissue engineering. *J. Biomed. Mater. Res. Part B* **2014**, *102*, 943–951. [CrossRef] [PubMed]
54. Milovac, D.; Gamboa-Martínez, T.C.; Ivankovic, M.; Gallego Ferrer, G.; Ivankovic, H. PCL-coated hydroxyapatite scaffold derived from cuttlefish bone: In vitro cell culture studies. *Mater. Sci. Eng. C* **2014**, *42*, 264–272. [CrossRef]
55. Yu, J.; Xu, Y.; Li, S.; Seifert, G.V.; Becker, M.L. Three-dimensional printing of nanohydroxyapatite/poly (ester urea) composite scaffolds with enhanced bioactivity. *Biomacromolecules* **2017**, *18*, 4171–4183. [CrossRef]
56. Camarero-Espinosa, S.; Tomasina, C.; Calore, A.; Moroni, L. Additive manufactured, highly resilient, elastic, and biodegradable poly(ester)urethane scaffolds with chondroinductive properties for cartilage tissue engineering. *Mater. Today Bio* **2020**, *6*, 100051. [CrossRef]
57. Lu, L.; Peter, S.J.; Lyman, M.D.; Lai, H.L.; Leite, S.M.; Tamada, J.A.; Uyama, S.; Vacanti, J.P.; Langer, R.; Mikos, A.G. In vitro and in vivo degradation of porous poly (Dl-lactic-co-glycolic acid) foams. *Biomaterials* **2000**, *21*, 1837–1845. [CrossRef]
58. Dreger, N.Z.; Fan, Z.; Zander, Z.K.; Tantisuwanno, C.; Parsell, T.; Søndergaard, C.S.; Haines, M.C.; Waggoner, M.; Hiles, M.; Premanandan, C.; et al. Amino acid-based poly (ester urea) copolymer films for hernia-repair applications. *Biomaterials* **2018**, *182*, 44–57. [CrossRef]
59. Lam, C.X.; Hutmacher, D.W.; Schantz, J.T.; Woodruff, M.A.; Teoh, S.H. Evaluation of polycaprolactone scaffold degradation for 6 months in vitro and in vivo. *J. Biomed. Mater. Res. A* **2009**, *90*, 906–919. [CrossRef]
60. Kim, J.H.; Linh, N.T.; Min, Y.K.; Lee, B.T. Surface modification of porous polycaprolactone/biphasic calcium phosphate scaffolds for bone regeneration in rat calvaria defect. *J. Biomater. Appl.* **2014**, *29*, 624–635. [CrossRef]
61. Hwang, K.S.; Choi, J.W.; Kim, J.H.; Chung, H.Y.; Jin, S.; Shim, J.H.; Yun, W.S.; Jeong, C.M.; Huh, J.B. Comparative efficacies of collagen-based 3D printed PCL/PLGA/β-TCP composite block bone grafts and biphasic calcium phosphate bone substitute for bone regeneration. *Materials* **2017**, *10*, 421. [CrossRef] [PubMed]
62. Roohani-Esfahani, S.I.; Lu, Z.F.; Li, J.J.; Ellis-Behnke, R.; Kaplan, D.L.; Zreiqat, H. Effect of self-assembled nanofibrous silk/polycaprolactone layer on the osteoconductivity and mechanical properties of biphasic calcium phosphate scaffolds. *Acta Biomater.* **2012**, *8*, 302–312. [CrossRef] [PubMed]
63. Boga, J.C.; Miguel, S.P.; de Melo-Diogo, D.; Mendonça, A.G.; Louro, R.O.; Correia, I.J. In vitro characterization of 3D printed scaffolds aimed at bone tissue regeneration. *Colloids Surf. B Biointerfaces* **2018**, *165*, 207–218. [CrossRef] [PubMed]

polymers

MDPI

Article

Gelatin/$Na_2Ti_3O_7$ Nanocomposite Scaffolds: Mechanical Properties and Characterization for Tissue Engineering Applications

Rittichai Sangkatip [1], Kaona Jongwuttanaruk [2] and Wipoo Sriseubsai [1,*]

1 Department of Industrial Engineering, School of Engineering, King Mongkut's Institute of Technology Ladkrabang, Bangkok 10520, Thailand
2 Department of Industrial Engineering, Faculty of Engineering, Rajamangala University of Technology Thanyaburi, Pathum Thani 12110, Thailand
* Correspondence: wipoo.sr@kmitl.ac.th

Abstract: Materials and manufacturing technologies are necessary for tissue engineering and developing temporary artificial extracellular matrices. In this study, scaffolds were fabricated from freshly synthesized titanate ($Na_2Ti_3O_7$) and its precursor titanium dioxide and their properties were investigated. The scaffolds with improved properties were then mixed with gelatin to form a scaffold material using the freeze-drying technique. To determine the optimal composition for the compression test of the nanocomposite scaffold, a mixture design with three factors of gelatin, titanate, and deionized water was used. Then, the scaffold microstructures were examined by scanning electron microscopy (SEM) to determine the porosity of the nanocomposite scaffolds. The scaffolds were fabricated as a nanocomposite and determined their compressive modulus values. The results showed that the porosity of the gelatin/$Na_2Ti_3O_7$ nanocomposite scaffolds ranged from 67% to 85%. When the mixing ratio was 100:0, the degree of swelling was 22.98%. The highest swelling ratio of 85.43% was obtained when the freeze-drying technique was applied to the mixture of gelatin and $Na_2Ti_3O_7$ with a mixing ratio of 80:20. The specimens formed (gelatin:titanate = 80:20) exhibited a compressive modulus of 30.57 kPa. The sample with a composition of 15.10% gelatin, 2% $Na_2Ti_3O_7$, and 82.9% DI water, processed by the mixture design technique, showed the highest yield of 30.57 kPa in the compression test.

Keywords: scaffold; freeze-drying technique; characterization; mixture design technique

1. Introduction

The scaffold is a structure required for cell growth. When living cells are in the scaffold, they multiply. Collagen, gelatin, silk, and synthetic PLA/PCL materials might be used for biomaterial scaffold development. In most cases, the biomaterial scaffold is used in experiments with various methods. [1] Currently, research is being conducted to develop biomaterials for medical applications. However, while developing material properties, the improvement of the mechanical properties of materials should be considered. The mechanical properties of materials can respond positively to real applications, especially the mechanical properties of gelatin. The mechanical properties of gelatin must be improved before it is used. [2] Gelatin consists of protein molecules composed of amino acid chains. Gelatin can be obtained from bones and animal skins. When gelatin is melted at 32 °C, it is in a viscous liquid state. Then, it becomes a gel within a few minutes. Gelatin is mainly used as an ingredient in cosmetics, medicines, foods, photographic films, and coated tablets. However, gelatin is too expensive for commercial use. Therefore, the properties of gelatin were investigated in several studies. The improvement of the properties of gelatin is usually performed by mixing it with other materials to reduce the material cost [2–7]. The use of titanium oxide (TiO_2) in the culture of muscle cells (C2C12 myoblasts), fibroblasts, and keratinocytes has been extensively studied. These cultured cells showed good adhesive properties and could be efficiently used in tissue engineering [8,9]. Nano titanium dioxide is a biocompatible material with high tensile strength and nontoxic effects [10–12]. It can improve the antibacterial properties of the scaffold material and promote the wound-healing performance of the material [13–15]. Scaffolds have also been prepared by salt leaching [16–18], solvent casting [19], phase separation [20], gas foaming [21], freeze drying, electrospinning, and fiber bonding [22,23]. The gas foaming technique offers several advantages in the fabrication of scaffolds, such as no need for aggressive organic solvents during the fabrication process, and the pore size and fiber diameter can be adjusted. However, this technique also has some disadvantages, such as the absence of pore interconnections, which in turn depend on the expansion mechanism of the gas during scaffold formation. The electrospinning processes offer simple advantages such as cost efficiency, wide pore size distribution, high porosity, and large surface areas. Therefore, electrospinning processes are used to produce samples with small pores and inhomogeneous pore distribution. However, the process leads to the formation of scaffolds with a low volume. Highly porous samples with interconnected pores can be produced using the fiber-bonding technique. However, a major limitation of this technique is that the prepared samples are toxic to cells, and this toxicity cannot be eliminated [24]. Additionally, several devices, a large room, and numerous protective systems are required to perform the gas foaming, electrospinning, and fiber-bonding techniques. In comparison, the freeze-drying technique is a simple and inexpensive choice because only solvents are required in this method [25]. A mixture of titanium dioxide (TiO_2) and polymers forms a framework that is more powerful due to an improvement in mechanical properties. A mixture of titanium dioxide (TiO_2) and polymers results in better performance compared to the use of a normal polymer regarding size and texture [26–28]. Several studies have demonstrated the effectiveness of titanium dioxide (TiO_2) in killing bacteria [29–35]. Some studies have shown that the use of nanoparticles along with a polymer improves the properties of the scaffold. The use of nanoparticles in a proportion of 5% mixed with 25–40% microcomposites can improve the properties [36–39]. The physical properties of titanium dioxide, such as structure, particle size, specific surface area, etc., directly affect the reaction mechanism. These physical properties depend on the synthesis method and conditions. In this study, the hygrothermal method was used to synthesize the material. In this method, the shape of the compounds can be controlled. Compounds with various shapes have been reported, including those with nanostructures such as nanotubes [40–43], nanoribbons [44–47], and nanowires [48,49]. The hydrothermal synthesis method increases chemical homogeneity without the need for further combustion or comminution. Additionally, the hydrothermal method allows the synthesis of compounds with a layered structure, especially a layered structure in which ions move between layers [40,49]. The hydrothermal synthesis method uses titanium dioxide (TiO_2), and thus, the synthesis process must be performed under highly alkaline conditions and at high temperatures. However, such alkaline conditions and high temperatures can accelerate the process of structural changes in the compound. For example, one study reported that the layered structure of titanium dioxide (TiO_2) changes to $Na_2Ti_3O_7$ [50].

Some studies [51,52] reported the preparation of a gelatin/$Na_2Ti_3O_7$ nanocomposite scaffold by mixing titanate ($Na_2Ti_3O_7$) and gelatin. A salt leaching technique was used to synthesize titanate ($Na_2Ti_3O_7$) from titanium dioxide (TiO_2). The prepared gelatin/$Na_2Ti_3O_7$ nanocomposite scaffold showed the best results concerning tensile strength and pore size. In this study, a mixture-design method was used to fabricate the scaffold instead of the freeze-drying technique. Then, the porosity, compression modulus, biodegradability, and swelling of the prepared samples were investigated.

2. Materials and Methods

2.1. Materials

Two main materials were used in the experiments. The first was gelatin which comprised porcine skin, 180 G Bloom, Type B from Fluke Analytical, and 2,2,2-trifluoroethanol (TFE) (purity 99.0%). Powdered gelatin was obtained from Sigma-Aldrich Corporation, Bangkok, Thailand.

The other material was titanate ($Na_2Ti_3O_7$) synthesized from titanium dioxide and sodium hydroxide as precursors. A solution of sodium hydroxide (20 mL) was mixed with 0.5 g of titanium dioxide and then stirred with a magnetic stirrer (mixing time: 90 min). The mixture was then placed in an alkaline hydrothermal reactor at 200 °C for 24 h. The slurry was then purified with deionized water until the pH reached 7 and titanate ($Na_2Ti_3O_7$) was formed. Then, the titanate was dried in an oven at 80 °C for 24 h to obtain titanate (v). The titanate had a specific surface area (BET) of 10 m^2/g, an average primary particle size of 197 nm, and a catalytic activity equivalent to that of 100% anatase.

2.2. Mixture Design

The mixture design experiment involves analyzing the result as a function of the percentage of components [53]. The objective is to explore the possible results by estimating the parameters of each component to be prepared to achieve the best value or the value desired by the experimenter. In the present study, the constrained mixture design principle or the vertex model was used. The experimental design was based on three parameters: gelatin, $Na_2Ti_3O_7$, and DI water. Minitab 21 software was used for statistical analyses. The software was customized to display the three factors as A, B, and C, where A was gelatin, B was $Na_2Ti_3O_7$, and C was DI water. Gelatin and collagen were used for fabricating the scaffold for the fibroblast cell culture. The properties of the scaffolds obtained from type A and type B gelatin were compared to the scaffolds obtained from collagen, which is used widely as a skin substitute. Titanate ribbon ($Na_2Ti_3O_7$) was synthesized to study the compressive modulus. Table 1 presents the ratio in which the three substances (A, B, and C) were mixed.

Table 1. The raw materials used for synthesizing the gelatin/$Na_2Ti_3O_7$ nanocomposite scaffolds.

Factors	Factor Levels	
	Low Level	High Level
A: Gelatin (g)	0	20
B: $Na_2Ti_3O_7$ (g)	0	2
C: DI water (mL)	0	87.8

2.3. Preparation of the Gelatin/$Na_2Ti_3O_7$ Nanocomposite Scaffold Samples Using the Freeze-Drying Technique

According to the principles of mixture design, the three raw materials (gelatin, titanate, and DI water) were mixed in a definite ratio and stirred for 30 min at 50 °C. The mixtures were then poured into the wells of a five-well culture plate, with each well containing equal volumes, followed by freezing treatment overnight at −30 °C. After vacuum freeze-drying at −50 °C, all the samples were stored at −20 °C until further use [10]. The mass ratios and the designations of the prepared samples are listed in Table 2. Some of the mixtures are shown in Figure 1.

2.4. Compressive Test

Pressure resistance is one of the key properties of scaffolds. Therefore, the impact of the prepared gelatin/$Na_2Ti_3O_7$ nanocomposite scaffolds was investigated by employing the universal testing machine (UTM, Instron No. 5566, Norwood, MA, USA) using the 10 kN load cells (Figure 2). The flexible cellular materials slab and the bonded and molded urethane foams were evaluated using the respective standard ASTM D 3574 methods [18].

The testing was conducted in a dry state at 25 °C at a speed maintained at 0.5 mm/min. The load was applied until the samples were compressed to approximately 100% of their original height. The stiffness of the scaffolds was assessed at the stress-point region. The yield strength was considered the stress-strain kPa yield point. The dimensions of the prepared samples were diameter, 1 cm and thickness, 1 cm. Each sample was tested five times. Subsequently, the average and standard deviations were determined for all samples.

Table 2. The experimental design and results.

Run Order	Factors			Compressive Test (kPa.)
	A: Gelatin	B: $Na_2Ti_3O_7$	C: DI Water	
01	0.200	0.000	0.800	27.407
02	0.200	0.002	0.798	26.764
03	0.122	0.000	0.878	24.678
04	0.120	0.002	0.878	25.240
05	0.161	0.001	0.839	30.306
06	0.180	0.001	0.819	29.861
07	0.180	0.002	0.818	28.396
08	0.141	0.001	0.858	28.511
09	0.140	0.002	0.858	27.186
10	0.200	0.000	0.800	26.687
11	0.200	0.002	0.798	26.222
12	0.122	0.000	0.878	24.038
13	0.120	0.002	0.878	25.370
14	0.161	0.001	0.839	30.586
15	0.180	0.001	0.819	29.621
16	0.180	0.002	0.818	28.866
17	0.141	0.001	0.858	28.337
18	0.140	0.002	0.858	26.974
19	0.200	0.000	0.800	25.647
20	0.200	0.002	0.798	26.182
21	0.122	0.000	0.878	25.170
22	0.120	0.002	0.878	24.970
23	0.161	0.001	0.839	30.546
24	0.180	0.001	0.819	29.941
25	0.180	0.002	0.818	28.659
26	0.141	0.001	0.858	29.871
27	0.140	0.002	0.858	27.953

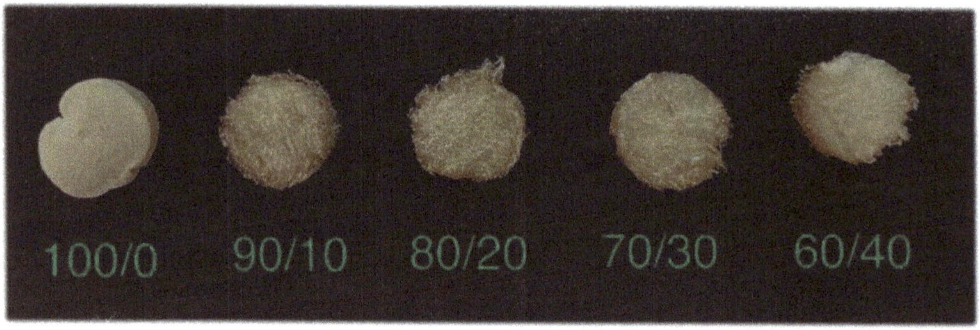

Figure 1. The gelatin/$Na_2Ti_3O_7$ nanocomposite scaffold samples prepared using the freeze-drying technique.

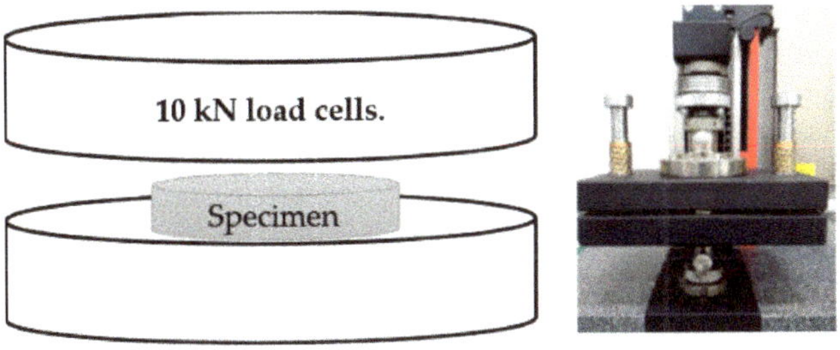

Figure 2. The testing of the prepared gelatin/$Na_2Ti_3O_7$ nanocomposite scaffold samples using the universal testing machine.

2.5. Biodegradation of the Gelatin/$Na_2Ti_3O_7$ Nanocomposite Scaffold Samples

The bioavailability of the prepared gelatin/$Na_2Ti_3O_7$ nanocomposite scaffold samples was determined through the in vitro weight biodegradation analysis of the samples. The sample fragments were first incubated in the PBS solution (pH 7.4) containing 104 U/mL of lysozyme in an oven at 37 °C for 54 h. Afterwards, the samples were incubated in an enzyme-free buffer, followed by curing and then thoroughly washing the samples with pure water to remove the enzymes. The samples were then dried and weighed several times, i.e., the degree of enzymatic biodegradation was determined after 0.5, 1, 1.5, 24, 48, and 54 h [16,17] using the following equation:

$$Weight\ remain(\%) = 100 - \left[\frac{\left(W_0 - W_f\right)}{W_0} \times 100\right] \quad (1)$$

where:

W_0 denotes the initial weight of the nanocomposite scaffold;
W_f denotes the final weight of the nanocomposite scaffold.

2.6. Swelling Test

The swelling test of scaffolds is an important tool to investigate the diffusion of water or liquid into the scaffold. In the present study, the swelling test was performed to evaluate the percentage difference between the dry weight and wet weight of the prepared biomedical scaffold. The dried nanocomposite scaffold was weighed and then soaked in a PBS buffer solution (pH 7.4) at 37 °C for 3 h [20,21]. Afterwards, both sides of the scaffold were wiped using low-lint paper for 10 s on each side, followed by weighing the scaffold immediately. The dry weight and wet weight values were then utilized to calculate the swelling ratio using the following formula:

$$\text{Swelling ratio} = \frac{W_{so} - W_o}{W_0} \quad (2)$$

where:

W_{so} denotes the weight of the scaffold absorbing the water content;
W_0 denotes the initial weight of the nanocomposite scaffold.

2.7. Statistical Analysis

The quantitative results were evaluated for data normality using Explore software and subjected to Tukey's A test. The results were presented as the mean standard error (SE). In the case of significant result values obtained ($p = 0.05$), post hoc tests were conducted. Post hoc tests are an integral part of ANOVA. When ANOVA is performed to compare the means of a minimum of three groups, obtaining statistically significant results indicates that all the compared group means are not equal. However, the ANOVA results do not identify which particular differences between the pairs of means are significant. In this context, using post hoc tests enables exploring the differences among multiple groups while controlling the experiment-wise error rate. Tukey's method is a single-step multiple-comparison procedure and statistical test, which may be used to identify the means that are significantly different from each other. Therefore, Tukey's multiple-comparison test was also conducted for analysis in the present study. All descriptive and inferential statistical analyses were conducted using MINITAB Statistical Software (v. 21).

3. Results and Discussion

3.1. Experimental Design and Results

A simplex axial design experiment was conducted for the mixture design of gelatin (A), titanate ($Na_2Ti_3O_7$) (B), and DI water (C) using the conditions stated above. In total, 27 experimental trials were performed, with three replicates in each trial condition. The results of the compressive test are presented in Table 2.

3.2. Residual Plots for Response

The quality of the result data was evaluated using the (1) normal distribution test, (2) the data independence test, and (3) the test of variance stability (Figure 3). As depicted in Figure 3, the data distribution did not have a specific pattern, indicating data independence, i.e., the data were collected randomly. In addition, the plot in which a straight line indicated the normality of the data obtained in the experimental work demonstrated that the experimental data had a stable variance as per the design. Therefore, it was concluded that the obtained data had all three properties.

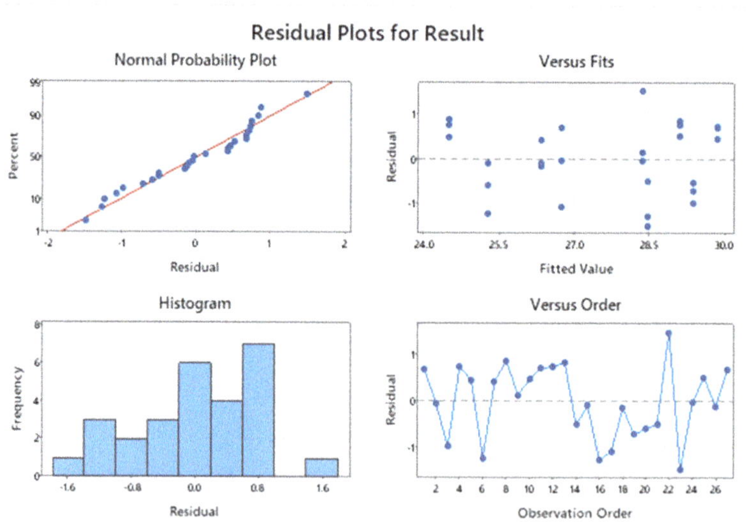

Figure 3. The residual plots for response average (kPa).

3.3. Analysis of Variance (ANOVA)

The analysis of variance, or ANOVA, is a linear modeling method for evaluating the relationship data [16]. In the present study, an ANOVA was performed to determine the compressive modulus. The results were considered statistically significant at $p < 0.05$ (Table 3). The ANOVA results revealed that the interaction term between the amount of gelatin used and the amount of titanate ($Na_2Ti_3O_7$) used was statistically significant ($p = 0.000$). The amounts of gelatin and DI water were also significantly related ($p < 0.05$). The coefficient of determination (R2) indicates the percentage variation in the dependent

variables that is explained by the independent variables in a regression analysis. The results for the regression analysis conducted in the present study for the response surface methodology results presented in Table 3 revealed a high coefficient of determination (R2 = 84.55%). This observation indicated that the independent variables [the amounts of gelatin, titanate ($Na_2Ti_3O_7$), and DI water] could explain 82.54% of the variation in the independent variables [the amounts of gelatin, titanate ($Na_2Ti_3O_7$), and DI water]. Therefore, the model was used to obtain a prediction equation to determine an accurate and appropriate response value.

Table 3. Summary of the results obtained in the analysis of variance (ANOVA).

Source	DF	Seq SS	Adj SS	Adj MS	F-Value	*p*-Value
Regression	3	87.933	87.933	29.3110	41.96	0.000
Linear	2	10.864	75.028	37.5141	53.70	0.000
Quadratic	1	77.069	77.069	77.0688	110.32	0.000
A*C	1	77.069	77.069	77.0688	110.32	0.000
Residual Error	23	16.067	16.067	0.6986		
Lack of Fit	5	11.408	11.408	2.2817	8.82	0.000
Pure Error	18	4.659	4.659	0.2588		
Total	26	104.000				
S		**R-sq**		**R-sq(adj)**	**PRESS**	**R-sq(pred)**
0.835807		84.55%		82.54%	21.6100	79.22%

3.4. Results of the Response Surface Methodology Analysis

The response surface methodology analysis is an effective method for optimizing the process conditions. It may also be used for determining the influence of various factors. As depicted in Figure 4, the response surface revealed for the compressive test of the nanocomposite scaffold-forming process involving gelatin, titanate ($Na_2Ti_3O_7$), and DI water increased with the increase in the amounts of gelatin and $Na_2Ti_3O_7$.

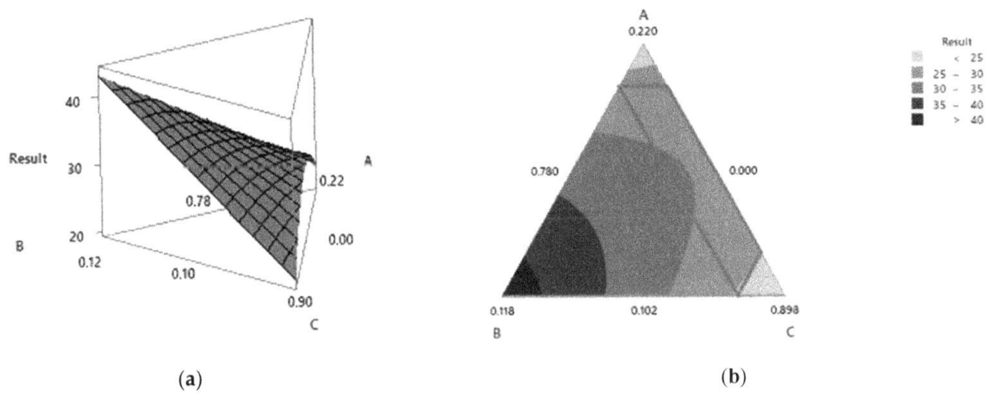

Figure 4. The mixture surface plot (**a**) and the mixture contour plot (**b**).

3.5. Optimization

A response optimizer function was adopted to determine the most appropriate value of the factors and thereby efficiently obtain the tensile strength. In addition, another function was adopted to identify the most appropriate parameter among the factors and measure composite desirability (D). The value of composite desirability ranged from 0 to 1. When D was equal to 1, the result was favorable for the overall response. The results revealed that 15.10% gelatin, 2% titanate ($Na_2Ti_3O_7$), and 82.90% DI water, or gelatin/$Na_2Ti_3O_7$ in the ratio of 80/20, yielded the highest tensile strength of 30.57 kPa and the highest desirability of 1 (Figure 5).

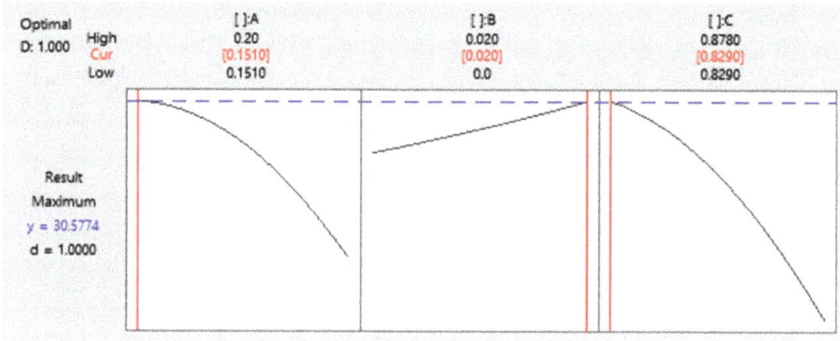

Figure 5. Optimization chart.

The samples with gelatin and $Na_2Ti_3O_7$ mixed in ratios of 100/0, 90/10, 80/20, 70/30, and 60/40 were investigated using the degradation test and the swelling test and also by analyzing the surface morphology and the pore size of the samples.

3.6. *Biodegradation of the Gelatin/$Na_2Ti_3O_7$ Nanocomposite Scaffold Samples Results*

Figure 6 depicts a comparison of the pure gelatin nanocomposite scaffolds and the gelatin/$Na_2Ti_3O_7$ nanocomposite scaffolds based on the results of the biodegradation analysis [18,19]. It was observed that the nanocomposite scaffolds prepared from pure gelatin (gelatin/$Na_2Ti_3O_7$ ratio—100/0) degraded in just one hour, while the nanocomposite scaffolds prepared from the 90/10 gelatin/$Na_2Ti_3O_7$ mixture required 24 h for degradation. Moreover, the 60/40 and 70/30 gelatin/$Na_2Ti_3O_7$ mixtures degraded in 48 h, while the 80:20 gelatin/$Na_2Ti_3O_7$ mixture degraded in 54 h.

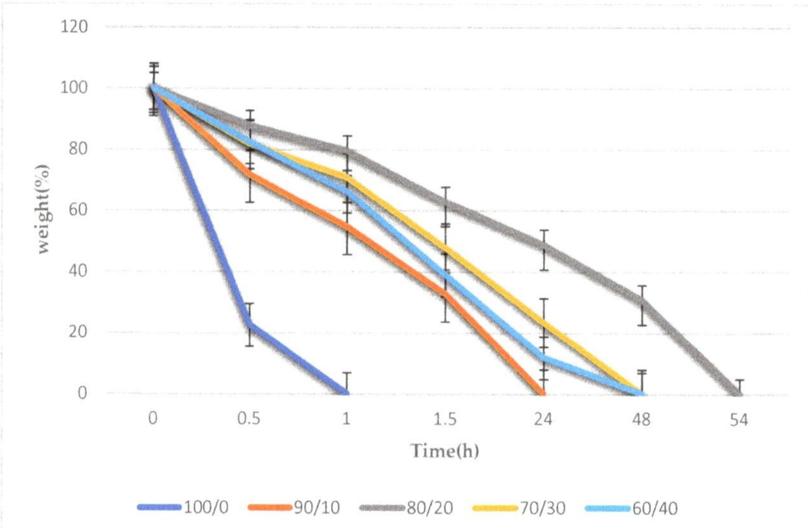

Figure 6. Results of the biodegradation of the gelatin/$Na_2Ti_3O_7$ nanocomposite scaffolds. (n = 20) at $p < 0.05$.

3.7. Swelling Test Results

Figure 7 depicts the nanocomposite scaffolds obtained using the freeze-drying technique. The swelling ratio of the scaffold with a gelatin/$Na_2Ti_3O_7$ mixture ratio of 100:0 was 22.98%. The highest swelling ratio of 85.43% was obtained for the scaffold with a gelatin/$Na_2Ti_3O_7$ mixture ratio of 80:20. This corresponded to the porosity value of the mixing ratio of 80:20, i.e., when the workpiece had high porosity, there was more space for water absorption, and thus, the rate of swelling increased.

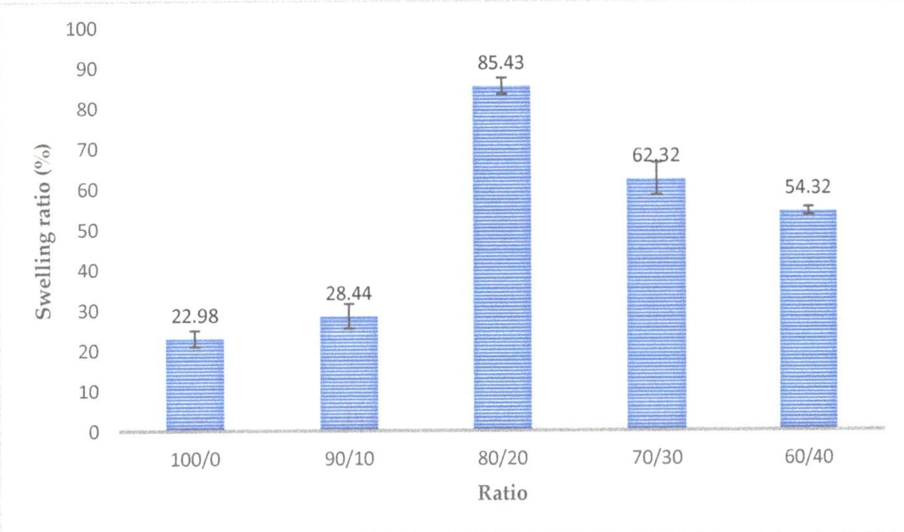

Figure 7. The swelling ratios of the different gelatin/$Na_2Ti_3O_7$ nanocomposite scaffolds prepared using the freeze-drying technique. (n = 20) at $p < 0.05$.

3.8. Porosity Percent and Pore Size of the Gelatin/$Na_2Ti_3O_7$ Nanocomposite Scaffolds Prepared Using the Freeze-Drying Technique

The porosity of the gelatin/$Na_2Ti_3O_7$ nanocomposite scaffolds was determined using the following equation [25]:

$$\textbf{Porosity}(\%) = \frac{V_{pore}}{V_{compact}} \times \mathbf{100} \quad (3)$$

where V_{Pore} denotes the volume of the pores in the cellular scaffold material and $V_{compact}$ denotes the volume of the nonporous workpiece.

The porous nature of the character scaffolds prepared using the freeze-drying technique depends on the size of the pore networks that are interconnected. In the present study, the scaffolds presented a homogeneous porous morphology. The porosity of the gelatin/$Na_2Ti_3O_7$ nanocomposite scaffold was subject to the conditions of the freeze-drying technique. The porosity was calculated using Equation (3). The result revealed that the porosity of the samples ranged from 67% to 85% (Figure 6). The porosity of the gelatin/$Na_2Ti_3O_7$ nanocomposite scaffolds was >85%, which is conducive to cell growth [16–18]. The microstructures of the pure gelatin and gelatin/$Na_2Ti_3O_7$ nanocomposite scaffolds are depicted in Figure 8a–e. The pure gelatin scaffolds were observed to be spongy, with interconnected pore networks of indeterminate sizes distributed throughout the scaffold. The gelatin/$Na_2Ti_3O_7$ scaffolds had a porous material, with evenly-distributed pores of sizes in the approximate range of 250–425 μm, which is the optimal size for the growth of bone cells [54]. However, this variation in the pore size of the prepared scaffolds could affect the movement of toxins for removal from the cells and the transmission of nutrients through the scaffold material. Despite the afore-stated limitation of the prepared scaffold samples, the fact that all prepared samples exhibited a certain degree of porosity indicated that these scaffolds would influence cell growth upon application [16–18].

The porosity percentage and the pore diameter (pore size) of the gelatin/$Na_2Ti_3O_7$ nanocomposite scaffolds prepared in the present study were determined, and the result values are listed in Table 4. As evident, the pore size decreased when excessive or pure $Na_2Ti_3O_7$ was added to the material composition [51,52]. In addition, the diameters of the pores (pore size) in the different gelatin/$Na_2Ti_3O_7$ scaffolds were not extremely small or large.

Table 4. The porosity percentage and pore size of the scaffolds (significance level—$p < 0.05$).

Gelatin/$Na_2Ti_3O_7$	Pore Size (μm)	Porosity (%)
100/0	124.23 ± 2.65	67.96 ± 2.77
90/10	195.32 ± 2.52	83.43 ± 0.82
80/20	278.71 ± 2.52	86.42 ± 1.08
70/30	257.79 ± 2.08	85.44 ± 1.20
60/40	234.56 ± 3.51	84.63 ± 2.30

3.9. *Surface Morphology of the Prepared Nanocomposite Scaffolds*

The scanning electron microscope (SEM) [1,11] was invented by M. Von Andenne in 1957. In SEM analysis, the sample preparation step is crucial for obtaining the representative material and ensuring that the results truly refer to the structure. However, the samples are not required to be of the same size. In the present study, only the surfaces of the sample scaffold structures were examined using SEM. Therefore, the number of electrons passing through the scaffold to the transmission electron microscope was not determined. Scanning electron microscopy may, therefore, be performed for analyzing the surfaces of cells and skin samples. A high-energy electron beam was aimed at the surface of the scaffold sample. When the electron beam strikes the outer surface of the sample at higher angles of incidence, the scanning electron microscope produces secondary electrons. The signals from these secondary electrons are then recorded and converted to electronic signals. The electronic signals are then projected onto a television which produces a 3D image of the sample scaffold structures. A scanning electron microscope has a lower vision capability and a lower magnification compared to a transmission electron microscope. Therefore, a scanning electron microscope is suitable for the investigation, analysis, morphological structure study, and detailed examination of the sample scaffold surfaces, such as the surface of a cell, a skin sample, the external surface of a tissue, the external surface of a cell, a metal cross-section, and the cross-section of a nonmetal material, among others.

The morphology of the nanocomposite scaffold was examined using a scanning electron microscope (SEM; JSM-5610LV, JEOL) at a voltage of 20 kV. The sample surface was first coated with gold [55]. The sample was then examined under SEM, and the pore size was measured at least 20 times. The calculated average-diameter values are listed in Table 4. The images of the pores in the prepared nanocomposite scaffold samples were obtained after the dehydrothermal treatment (DHT) of the samples at 140 °C for 48 h (Figure 6). The diameters of the pores of the gelatin/$Na_2Ti_3O_7$ nanocomposite scaffold were observed to have changed. The fabricated scaffold exhibited varied pore sizes (Figure 6). When a greater amount of $Na_2Ti_3O_7$ was added to the scaffold composition, the sizes of the pores were reduced. Figure 8c depicts the high variation in the pore sizes. As depicted in Figure 8a–f, the diameters of the pores in the prepared gelatin/$Na_2Ti_3O_7$ nanocomposite scaffolds ranged from 120 to 240 μm. However, the SEM cross-section of the 80/20 sample shown in Figure 8f also indicates uniform porosity characteristics. This indicates that when titanate (Na2Ti3O7) is added, samples can have more porosity. There is a network of interconnections between the pores compared to the 100/0 sample.

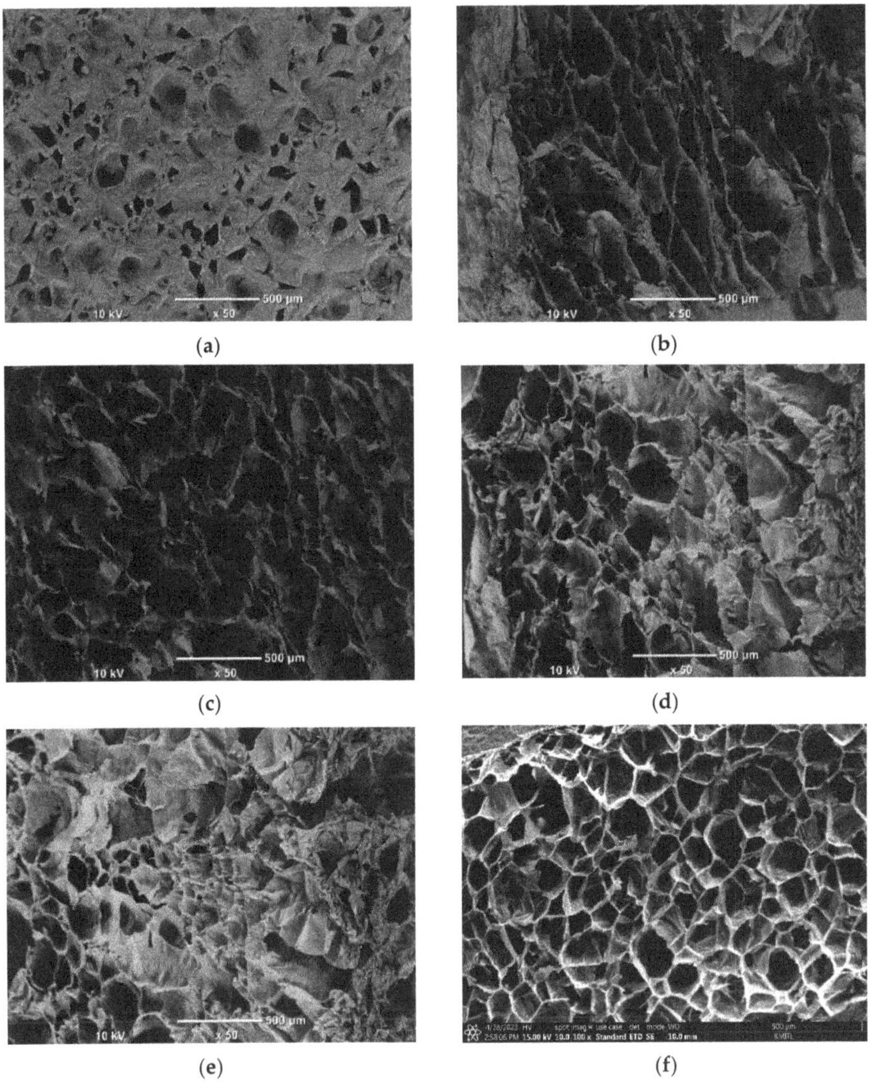

Figure 8. The SEM images of the scaffolds prepared using the freeze-drying technique, with the gelatin/$Na_2Ti_3O_7$ mixture ratio of (**a**) 100/0, (**b**) 90/10, (**c**) 80/20, (**d**) 70/30, (**e**) 60/40 and cross-sectional (**f**) 80/20.

3.10. Characterization of the Gelatin/$Na_2Ti_3O_7$ Nanocomposite Scaffolds

The FTIR spectra were obtained by analyzing the structure and molecular composition of the sample gelatin/$Na_2Ti_3O_7$ nanocomposite scaffold using a Fourier transform infrared spectrometer (Brand: PerkinElmer Scientific Model: Spectrum Two FT-IR Spectrometer (Universal ATR Sampling Accessory)). By measuring the absorbance of substances in the infrared range (Figure 9), we found that in the wavelength range of 3300–3500 cm^{-1}, aliphatic C–H stretching signals were found in the range of about 3100–2700 cm^{-1}. Additionally, in the wavelength range of 1620–1680 cm^{-1}, C = bonds were found. O stretching of amide groups was observed in the wavelength range of 1640–1590 cm^{-1}; N–H bending signals were observed in the wavelength range of 1365–1455 cm^{-1}; and C–H bending signals were observed in the wavelength range of 1256–1332 cm^{-1}. Signs of C-N stretching were also found [56,57].

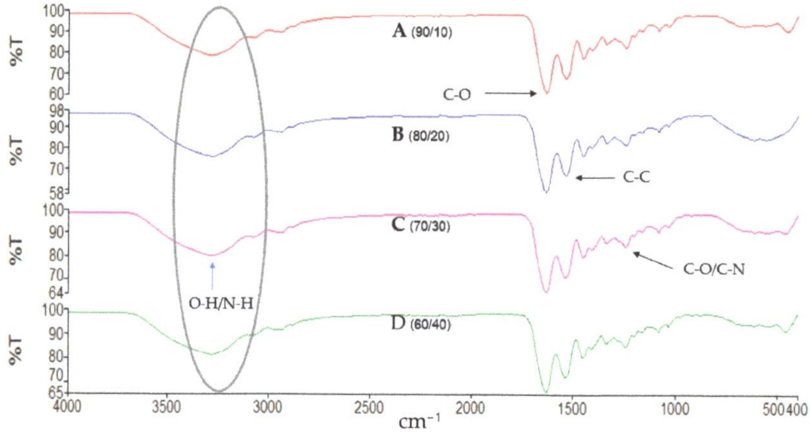

Figure 9. The FTIR spectrum of the gelatin/$Na_2Ti_3O_7$ nanocomposite scaffold.

However, no new peaks were observed, indicating that the added titanate ($Na_2Ti_3O_7$) did not react with glutaraldehyde. Titanate ($Na_2Ti_3O_7$) can act as a reinforcing agent, dispersed in the hydrogel or matrix [58]. It can enhance the properties of the matrix and the composite materials.

3.11. Effects of the Different Techniques Used for the Fabrication of a Scaffold: Freeze-Drying Technique and Salt-Leaching Technique

The scaffolds prepared using the salt-leaching technique exhibit a porous nature that depends on the size of the salt crystals used. The porous networks in these scaffolds are interconnected, and the porous morphology is homogeneous. According to the results of the present study, the porosity of the prepared samples ranged from 75% to 81%. Previous studies [59–61] have not reported any scaffolds with a porosity exceeding 85%. Moreover, the porosity of the prepared scaffold was reported to be lower than that of the sample with added salt (at 85%) [62]. This resulted in the agglomeration of salt particles and a loss of salt particles during the sedimentation process. The results obtained in the present study using the freeze-drying technique for scaffold preparation were comparable to those of the previous studies [51]. The freeze-drying technique produced scaffolds with similar shapes and hole sizes. The pores were also well connected, while the porosity and the pore size could be controlled better using this technique.

4. Conclusions

The gelatin/$Na_2Ti_3O_7$ nanocomposite scaffolds were fabricated using the freeze-drying technique, and then, they were characterized. The gelatin/$Na_2Ti_3O_7$ nanocomposite scaffolds were prepared from gelatin and titanate ($Na_2Ti_3O_7$). The mixture experimental-design technique was adopted as a statistical analysis tool. The results showed that titanate ($Na_2Ti_3O_7$) played an important role in improving the porosity, compressive test results, biodegradation, swelling test outcomes, porosity percentage, and pore-size values of the prepared nanocomposite scaffolds. A good porosity range of 67–85% was observed. In the compressive test, the prepared samples (gelatin:titanate = 80:20) showed a yield value of 30.57 kPa. The sample prepared using the mixture design technique (15.10% gelatine, 2% $Na_2Ti_3O_7$, and 82.9% DI water) resulted in the highest yield in the compressive test. Theoretically, a scaffold must be highly porous and have a three-dimensional shape. The pores should be interconnected to form networks throughout the scaffold to ensure adequate cell growth and the appropriate transport of nutrients and wastes. This is important because the transport of nutrients and wastes also controls the degradation rate and the water absorption capacity of the scaffold. Therefore, if the scaffold has a greater porous volume, cell-adhesion efficiency, and cell-growth volume are affected, which is consistent with the results of this study.

Author Contributions: Conceptualization, R.S., W.S. and K.J.; methodology, R.S. and W.S. writing—original draft preparation, R.S.; writing—review and editing, W.S. All authors have read and agreed to the published version of the manuscript.

Funding: This research was funded by the King Mongkut's Institute of Technology Ladkrabang, grant number 2564–02–01–017, and the APC was funded by the School of Engineering, King Mongkut's Institute of Technology Ladkrabang, Thailand.

Institutional Review Board Statement: Not applicable.

Informed Consent Statement: Informed consent was obtained from all subjects involved in the study.

Data Availability Statement: The data presented in this study are available in this article.

Acknowledgments: The authors thank the Industrial Engineering Department, School of Engineering, King Mongkut's Institute of Technology Ladkrabang (KMITL), Thailand.

Conflicts of Interest: The authors declare no conflict of interest.

References

1. Khan, M.U.A.; Haider, S.; Shah, S.A.; Abd Razak, S.I.; Hassan, S.A.; Kadir, M.R.A.; Haider, A. Arabinoxylan-co-AA/HAp/TiO_2 Nanocomposite Scaffold a Potential Material for Bone Tissue Engineering: An Invitro Study. *Int. J. Biol. Macromol.* **2020**, *151*, 584–594. [CrossRef] [PubMed]
2. Mao, J.S.; Zhao, L.G.; Yin, Y.J.; De Yao, K. Structure and properties of bilayer chitosan–gelatin scaffolds. *Biomaterials* **2003**, *24*, 1067–1074. [CrossRef] [PubMed]
3. Gomes, S.R.; Rodrigues, G.; Martins, G.G.; Henriques, C.M.R.; Silva, J.C. In vitro evaluation of crosslinked electrospun fish gelatin scaffolds. *Mater. Sci. Eng. C* **2013**, *33*, 1219–1227. [CrossRef] [PubMed]
4. Pezeshki-Modaress, M.; Mirzadeh, H. Gelatin–GAG electrospun nanofibrous scaffold for skin tissue engineering: Fabrication and modeling of process parameters. *Mater. Sci. Eng. C* **2015**, *48*, 704–712. [CrossRef]
5. Zeugolis, D.I.; Khew, S.T.; Yew, E.S.Y.; Ekaputra, A.K.; Tong, Y.W.; Yung, L.-Y.L.; Hutmacher, D.W.; Sheppard, C.; Raghunath, M. Electrospinning of pure collagen nano-fibers Just an expensive way to make gelatin. *Biomaterials* **2008**, *29*, 2293–2305. [CrossRef]
6. Huang, Z.-M.; Zhang, Y.Z.; Ramakrishna, S.; Lim, C.T. Electrospinning and mechanical characterization of gelatin nanofibers. *Polymer* **2004**, *45*, 5361–5368. [CrossRef]
7. Pek, Y.S.; Spector, M.; Yannas, I.V.; Gibson, L.J. Degradation of a collagen–chondroitin-6-sulfate matrix by collagenase and by chondroitinase. *Biomaterials* **2004**, *25*, 473–482. [CrossRef]
8. Genchi, G.; Nuhn, H.; Liakos, I.; Marino, A.; Marras, S.; Athanassiou, A.; Mattoli, V.; Desai, T. Titanium dioxide nanotube arrays coated with laminin enhance C2C12 skeletal myoblast adhesion and differentiation. *RSC Adv.* **2016**, *22*, 18502–18514. [CrossRef]
9. Smith, B.S.; Yoriya, S.; Johnson, T.; Popat, K.C. Dermal fibroblast and epidermal keratinocyte functionality on titania nanotube arrays. *Acta Biomater.* **2011**, *7*, 2686–2696. [CrossRef]
10. Fan, X.; Chen, K.; He, X.; Li, N.; Huang, J.; Tang, K.; Li, Y.; Wang, F. Nano-TiO_2/collagen-chitosan porous scaffold for wound repairing. *Int. J. Biol. Macromol.* **2016**, *91*, 15–22. [CrossRef]
11. Cordero-Arias, L.; Cabanas-Polo, S.; Gilabert, J.; Goudouri, O.M.; Sanchez, E.; Virtanen, S.; Boccaccini, A.R. Electrophoretic deposition of nanostructured TiO_2/alginate and TiO_2-bioactive glass/alginate composite coatings on stainless steel. *Adv. Appl. Ceram.* **2014**, *113*, 42–49. [CrossRef]
12. Kulkarni, M.; Mazare, A.; Gongadze, E.; Perutkova, S.; Kralj-Iglic, V.; Milosev, I.; Schmuki, P.; Iglic, C.; Mozetic, M. Titanium nanostructures for biomedical applications. *Nanotechnology* **2015**, *26*, 062002. [CrossRef] [PubMed]
13. Fu, S.-Y.; Lauke, B. Characterization of tensile behavior of hybrid short glass fiber/calcite particle/ABS composites. *Compos. Part A: Appl. Sci. Manuf.* **1998**, *29*, 575–583. [CrossRef]
14. Lai, S.-M.; Li, H.-C.; Liao, Y.-C. Properties and preparation of compatibilized nylon 6 Nanocomposites/ABS blends: Part II—Physical and thermal properties. *Eur. Polym. J.* **2007**, *43*, 1660–1671. [CrossRef]
15. Yousefi, M.; Salavati-Niasari, M.; Gholamian, F.; Ghanbari, D.; Aminifazl, A. Polymeric nanocomposite materials: Synthesis and thermal degradation of acrylonitrile–butadiene–styrene/tin sulfide (ABS/SnS). *Inorg. Chim. Acta* **2011**, *371*, 1–5. [CrossRef]
16. Rezwan, K.; Chen, Q.Z.; Blaker, J.J.; Boccaccini, A.R. Biodegradable and bioactive porous polymer/inorganic composite scaffolds for bone tissue engineering. *Biomaterials* **2006**, *27*, 3413–3431. [CrossRef]
17. Nam, Y.J.; Stöver, T.; Hartman, S.S.; Altschuler, R.A. Upregulation of glial cell line-derived neurotrophic factor (GDNF) in the rat cochlea following noise. *Hear. Res.* **2000**, *146*, 1–6. [CrossRef]

18. Hou, Q.; Grijpma, D.W.; Feijen, J. Porous polymeric structures for tissue engineering prepared by coagulation, compression molding, and salt leaching technique. *Biomaterials* **2003**, *24*, 1937–1947. [CrossRef]
19. Mikos, A.G.; Thorsen, A.J.; Czerwonka, L.A.; Bao, Y.; Langer, R.; Winslow, D.N.; Vacanti, J.P. Preparation and characterization of poly (L-lactic acid) foams. *Polymer* **1994**, *35*, 1068–1077. [CrossRef]
20. Lo, H.; Ponticiello, M.S.; Leong, K.W. Fabrication of Controlled Release Biodegradable Foams by Phase Separation. *Tissue Eng.* **1995**, *1*, 18–25. [CrossRef]
21. Mooney, D.J.; Baldwin, D.F.; Suh, N.P.; Vacanti, J.P.; Langer, R. A novel approach to fabricate porous sponges of poly (D, L-lactic-co-glycolic acid) without the use of organic solvents. *Biomaterials* **1996**, *17*, 1417–1422. [CrossRef]
22. Leong, K.F.; Cheah, C.M.; Chua, C.K. Solid freeform fabrication of three-dimensional scaffolds for engineering replacement tissues and organs. *Biomaterials* **2003**, *24*, 2363–2378. [CrossRef]
23. Whang, K.; Thomas, C.H.; Healy, K.E.; Nuber, G. A novel method to fabricate bioabsorbable scaffolds. *Polymer* **1995**, *36*, 837–842. [CrossRef]
24. Hoque, M.E.; Nuge, T.; Yeow, T.K.; Nordin, N.; Prasad, R.G.S.V. Gelatin Based Scaffolds for Tissue Engineering—A review. *Polym. Res. J.* **2015**, *9*, 15–32.
25. Cho, Y.S.; Kim, B.S.; You, H.K.; Cho, Y.S. A novel technique for scaffold fabrication: SLUP (salt leaching using powder). *Curr. Appl. Phys.* **2014**, *14*, 371–377. [CrossRef]
26. Bruton, A.R. Manufacturing, and Performance of Titanium Dioxide Ultra-High Molecular Weight Polyethylene nanocomposite Materials. Master's Thesis, University of Delaware, Newark, DE, USA, 2006.
27. Giraldo, L.F.; Echeverri, M.; López, B.L. Reinforcement of Polyamide 6 with Nanoparticles. *Macromol. Symp.* **2007**, *258*, 119–128. [CrossRef]
28. Zhang, Z.; Yang, J.L.; Friedrich, K. Creep resistant polymeric nanocomposites. *Polymer* **2004**, *45*, 3481–3485. [CrossRef]
29. Matsuyama, T.; Tomoda, R.; Nakajima, T.; Wake, H. Photo electrochemical sterilization of microbial cells by semiconductor powders. *FEMS Microbial. Lett.* **1985**, *29*, 211–214. [CrossRef]
30. Kiema, G.K.; Colgan, M.J.; Brett, M.J. Dye-sensitized solar cells incorporating obliquely deposited titanium oxide layers. *Sol. Energy Mater. Sol. Cells* **2005**, *85*, 321–331. [CrossRef]
31. Walsh, F.C.; Bavykin, D.V.; Torrente-Murciano, L.; Lapkin, A.A.; Cressey, B.A. Synthesis of novel composite materials via the deposition of precious metals onto protonated titanate (TiO_2) nanotubes. *Trans. Inst. Met. Finish.* **2006**, *6*, 293–299. [CrossRef]
32. Bavykin, D.V.; Parmon, V.N.; Lapkin, A.A.; Walsh, F.C. The effect of hydrothermal condition on the mesoporous structure of TiO_2 nanotube. *J. Mater. Chem.* **2004**, *14*, 3370–3377. [CrossRef]
33. Fu, Y.; Wei, Z.D.; Chen, S.G.; Li, L.; Feng, Y.C.; Wang, Y.Q.; Ma, X.L.; Liao, M.J.; Shen, P.K.; Jiang, S.P. Synthesis of Pd/TiO_2 nanotubes/Ti for oxygen reduction reaction in acidic solution. *J. Power Sources* **2009**, *189*, 982–987. [CrossRef]
34. Dechakiatkrai, C.; Chen, J.; Lynam, C.; Phanichphant, S.; Wallace, G.G. Photocatalytic degradation of methanol using titanium dioxide/single walled carbon nanotube composite. *J. Electrochem. Soc.* **2007**, *154*, A407–A411. [CrossRef]
35. Madhugiri, S.; Sun, B.; Smirniotis, P.G.; Ferraris, J.P.; Balkus, K.J., Jr. Electrospun mesoporous titanium dioxide fibers. *Microporous Mesoporous Mater.* **2004**, *69*, 77–83. [CrossRef]
36. Saujanya, C.; Radhakrishnan, S. Structure development and crystallization behavior of PP/nanoparticulate composite. *Polymer* **2001**, *42*, 6723–6731. [CrossRef]
37. Zheng, Y.; Zhengb, Y.; Ning, R. Effects of nanoparticles SiO_2 on the performance of nanocomposites. *Mater. Lett.* **2003**, *57*, 2940–2944. [CrossRef]
38. Esthappan, S.K.; Kuttappan, S.K.; Joseph, R. Effect of titanium dioxide on the thermal aging of polypropylene. *Polym. Degrad. Stab.* **2012**, *97*, 615–620. [CrossRef]
39. Akkapeddi, M.K. Glass fiber reinforced polyamide-6 nanocomposites. *Polym. Compos.* **2000**, *21*, 576–585. [CrossRef]
40. Kasuga, T.; Hiramatsu, M.; Hoson, A.; Sekino, T.; Niihara, K. Formation of titanium oxide nanotube. *Langmuir* **1998**, *14*, 3160–3163. [CrossRef]
41. Kasuga, T.; Hiramatsu, M.; Hoson, A.; Sekino, T.; Niihara, K. Titania nanotubes prepared by chemical processing. *Adv. Mater.* **1999**, *11*, 1307–1311. [CrossRef]
42. Chen, Q.; Zhou, W.; Du, G.H.; Peng, L.M. Titanate nanotubes made via a single alkali treatment. *Adv. Mater.* **2002**, *14*, 1208–1211. [CrossRef]
43. Kiatkittipong, K.; Scott, J.; Amal, R. Hydrothermally synthesized titanate nanostructures: Impact of heat treatment on particle characteristics and photocatalytic properties. *ACS Appl. Mater. Interfaces* **2011**, *3*, 3988–3996. [CrossRef] [PubMed]
44. Yuan, Z.Y.; Colomer, J.F.; Su, B.L. Titanium oxide nanoribbons. *Chem. Phys. Lett.* **2002**, *363*, 362–366. [CrossRef]
45. Kiatkittipong, K.; Ye, C.; Scott, J.; Amal, R. Understanding Hydrothermal Titanate Nanoribbon Formation. *Cryst. Growth Des.* **2010**, *10*, 3618–3625. [CrossRef]
46. Zhang, Y.X.; Li, G.H.; Jin, Y.X.; Zhang, Y.; Zhang, J.; Zhang, L.D. Hydrothermal synthesis and photoluminescence of TiO_2 nanowires. *Chem. Phys. Lett.* **2002**, *365*, 300–304. [CrossRef]
47. Yoshida, R.; Suzuki, Y.; Yoshikawa, S. Syntheses of TiO_2 (B) nanowires and TiO_2 anatase nanowires by hydrothermal and post-heat treatments. *J. Solid-State Chem.* **2005**, *178*, 2179–2185. [CrossRef]
48. Armstrong, A.R.; Armstrong, G.; Canales, J.; Bruce, P.G. TiO_2 -B nanowires. *Angew. Chem. Int. Ed.* **2004**, *43*, 2286–2288. [CrossRef]

49. Byrappa, K.; Yoshimura, M. *Handbook of Hydrothermal Technology—A Technology for Crystal Growth and Materials Processing*; William Andrew Publishing/Noyes: Norwich, NY, USA, 2001.
50. Ma, R.; Bando, Y.; Sasaki, T. Nanotubes of lepidocrocite titanates. *Chem. Phys. Lett.* **2003**, *380*, 577–582. [CrossRef]
51. Sangkatip, R.; Sriseubsai, W.; Kiatkittipong, K.; Jongwuttanaruk, K. Optimization of Salt-Leaching Parameters for Gelatin/$Na_2Ti_3O_7$ Scaffolds Using a Mixture Design Experiment. *Polymers* **2022**, *14*, 559. [CrossRef]
52. Sangkatip, R.; Sriseubsai, W.; Kiatkittipong, K. Preparation, and Characterization of the $Na_2Ti_3O_7$: ABS/$Na_2Ti_3O_7$ Composites. *Key Eng. Mater.* **2017**, *753*, 39–43. [CrossRef]
53. Myers, R.H.; Montgomery, D.C. *Response Surface Methodology Process and Product Optimization Using Designed Experiments*, 2nd ed.; John Wiley and Son, Inc.: New York, NY, USA, 2002.
54. Hajiali, H.; Karbasi, S.; Hosseinalipour, M.; Rezaie, H. Preparation of a novel biodegradable nanocomposite scaffold based on poly (3-hydroxybutyrate)/bioglass nanoparticles for bone tissue engineering. *J. Mater. Sci. Mater. Med.* **2010**, *21*, 2125–2132. [CrossRef]
55. Chen, H.; Huang, J.; Yu, J.; Liu, S.; Gu, P. Electrospun chitosan-graft-poly (ε-caprolactone)/poly (ε-caprolactone) cationic nanofibrous mats as potential scaffolds for skin tissue engineering. *Int. J. Biol. Macromol.* **2011**, *48*, 13–19. [CrossRef]
56. Mohd Amin, M.C.I.; Ahmad, N.; Halib, N.; Ahmad, I. Synthesis and characterization of thermo- and pH-responsive bacterial cellulose/acrylic acid hydrogels for drug delivery. *Carbohydr. Polym.* **2012**, *88*, 465–473. [CrossRef]
57. Rokhade, A.P.; Agnihotri, S.A.; Patil, S.A.; Mallikarjuna, N.N.; Kulkarni, P.V.; Aminabhavi, T.M. Semi-interpenetrating polymer network microspheres of gelatin and sodium carboxymethyl cellulose for controlled release of ketorolac tromethamine. *Carbohydr. Polym.* **2006**, *65*, 243–252. [CrossRef]
58. Ooi, S.Y.; Ahmad, I.; Amin, M.C.I.M. Cellulose nanocrystals extracted from rice husks as a reinforcing material in gelatin hydrogels for use in controlled drug delivery systems. *Ind. Crops Prod.* **2015**, *93*, 227–234. [CrossRef]
59. Wiwatwongwana, F.; Promma, N. Characterization of Gelatin-Carboxymethylcellulose Scaffolds. In Proceedings of the 4th International Conference on Chemical Materials and Process (ICCMP 2018), 100th Building, Faculty of Engineering, Chulalongkorn University, Bangkok, Thailand, 23–25 May 2018. (Oral presentation).
60. Jongwuttanaruk, K.; Surin, P.; Wiwatwongwana, F. Characterization of Gelatin/CMC Scaffolds by Electrospinning and Comparison with Freeze Dry Techniques. In Proceedings of the 4th International Conference on Chemical Materials and Process (ICCMP 2018), 100th Building, Faculty of Engineering, Chulalongkorn University, Bangkok, Thailand, 23–25 May 2018. (Oral presentation).
61. Ma, J.; Wang, H.; He, B.; Chen, J. A preliminary in vitro study on the fabrication and tissue engineering applications of a novel chitosan bilayer material as a scaffold of human no fetal dermal fibroblasts. *Biomaterials* **2001**, *22*, 331–336. [CrossRef]
62. Leea, S.B.; Kima, Y.H.; Chonga, M.S.; Hong, S.H.; Leea, Y.M. Study of gelatin-containing artificial skin V: Fabrication of gelatin scaffolds using a salt-leaching method. *Biomaterials* **2005**, *26*, 1961–1968. [CrossRef]

MDPI AG
Grosspeteranlage 5
4052 Basel
Switzerland
Tel.: +41 61 683 77 34

Polymers Editorial Office
E-mail: polymers@mdpi.com
www.mdpi.com/journal/polymers

www.ingramcontent.com/pod-product-compliance
Lightning Source LLC
Chambersburg PA
CBHW061846210926
51850CB00006B/983
9783725852598